Bioprocess and Analytics Development for Virus-based Advanced Therapeutics and Medicinal Products (ATMPs)

Saurabh Gautam • Abhilash I. Chiramel •
Roland Pach
Editors

Bioprocess and Analytics Development for Virus-based Advanced Therapeutics and Medicinal Products (ATMPs)

Springer

Editors
Saurabh Gautam
Boehringer Ingelheim International GmbH/ ViraTherapeutics GmbH
Innsbruck, Austria

Abhilash I. Chiramel
Boehringer Ingelheim Pharma GmbH & Co. KG
Biberach an der Riß, Germany

Roland Pach
F. Hoffmann-La Roche Ltd
Basel, Switzerland

ISBN 978-3-031-28488-5 ISBN 978-3-031-28489-2 (eBook)
https://doi.org/10.1007/978-3-031-28489-2

This Springer imprint is published by the registered company Springer Nature Switzerland AG
The registered company address is: Gewerbestrasse 11, 6330 Cham, Switzerland

This book is dedicated to all the patients around the world whose only hope is a breakthrough therapy or futuristic medicine.

Preface

It was the peak of COVID-19 pandemic in Europe in 2020–2021, when the first thoughts of compiling and editing a book about bioprocess and analytics for virus-based therapeutics, appeared in the mind of the editor. Although there are books available on topics of bioprocessing and analytics, but often with a focus on biologics or protein-based therapeutics. There was a clear lack of literature with a holistic approach from a CMC perspective when it came to virus-based therapeutics. Moreover, the book also aims to bring knowledge in a condensed form to a broader, interested audience for the benefit of the patients.

The development of innovative therapeutics for treating rare diseases brings the complexities and challenges of manufacturing. Especially, manufacturing a drug in an economical fashion readily accessible for all patients around the world. It is almost inevitable that the drugs will be un-economical (at least for the payer) due to intrinsic costs involved in R&D and manufacturing of revolutionary therapeutics, such as advanced therapeutics and medicinal products (ATMPs) including oncolytic viruses, cell and gene therapeutics, etc.

This book is a step in the direction of more collaboration, sharing knowledge, and merging the smaller knowledge bubbles into knowledge hubs, to manufacture cutting-edge therapeutics in an economical manner for patients of the wide world and not just "key markets." Therefore, the readers can expect to read about broader and principle understanding of the topics along with the assortment of options available for bioprocessing and analytics when developing virus-based therapeutics. The title of the book "Bioprocess and analytics development for virus-based advanced therapeutics and medicinal products (ATMPs)" is clear enough that the focus of the book is on virus therapeutics. However, it is important to mention that the general topics or techniques which have no difference when it comes to various therapeutics, i.e., virus or protein therapeutics, are not covered within this book. To give an example, pH determination techniques remain the same when it comes to a therapeutic solution containing viruses or proteins. Similarly, visual inspection guidelines are also the same for virus or other therapeutics.

The book starts with Chap. 1 introducing the topics and providing a historical view of the same, especially about the challenges and how the field of virus therapeutics has evolved over time. The chapter also provides a comprehensive overview of current state and shortcomings in identifying critical quality attributes (CQAs) and critical process parameters (CPP) for manufacturing virus therapeutics. Thereafter, the book is broadly subdivided into three sub-sections: upstream process development (USP), downstream process development (DSP), and analytics development strategies, all of which practically cannot be separated from each other.

Chapter 2 initiates the upstream sub-section and provides a comprehensive overview of the topics and techniques for upstream development of virus-based therapeutics. Afterwards, Chap. 3 further discusses upstream processes for viral vectors with a special focus on bioreactor technologies and processes. This is then followed by Chap. 4 on challenges and strategies in cell line development for virus-based therapeutics. It is important to emphasize that various topics in this book cannot be isolated from each other as they are a part of a single bioprocess operation. Therefore, there would be overlaps within several topics, e.g. the harvest and clarification strategies are covered within Chaps. 2 and 3 as well as Chap. 6 later.

Upstream is followed by downstream and there is a comprehensive discussion on various downstream strategies and related topics in Chap. 5. Filtration techniques are an intrinsic part of any bioprocess and are almost always employed for various downstream applications. Such downstream applications of filtration techniques include feed clarification, bioburden reduction, buffer exchange, concentration, capture, purification, etc., during the manufacturing of virus therapeutics. In this direction, Chap. 6 discusses filtration principles and techniques in detail. Chapter 7 continues the discussion on another important downstream technique being chromatography. This chapter encompasses a sublime discussion on various chromatographic techniques available for downstream processing and purification of virus therapeutics. Chapter 8 is an excellent extension of Chap. 7 on chromatography techniques but with a focus on the state-of-the-art convective flow chromatography technologies (e.g., membrane and monolith chromatography) with their pros and cons when compared to traditional chromatographic techniques.

As mentioned above, Chap. 1 starts with a good overview and introduction of CQAs and analytical techniques along with E2E (end to end) process used for characterizing virus therapeutics. In Chaps. 9 and 10, the authors discuss in more detail the various state-of-the-art analytical techniques and potency assays used for characterization and analytics of virus particles. Chapter 11 further comprehensively describes different assays for quantification of virus infectivity. Similarly, Chap. 12 focuses on another critical aspect of virus analytics and bioprocessing, which are product-related impurities, their impact on the process and techniques used to characterize them. Process analytical technologies (PAT) are an important part of any bioprocess and manufacturing operation today along with Quality by Design (QbD). Chapter 13 describes an excellent overview of both PAT and QbD with respect to virus-based therapeutics. Next set of chapters provide a spotlight on two powerful techniques which have significant potential for characterizing virus particles. Chapter 14 appraises the technique of electron microscopy used for characterizing morphology and contents of virus particles. Furthermore, Chap. 15 discusses in detail about a technique which has only been established or being used routinely for therapeutic molecules other than viruses. The technique is nuclear magnetic resonance (NMR), used for high-resolution biophysical characterization of small molecules and proteins, however with limited applications so far for virus-based therapeutics.

Last set of chapters provide a broader overview of the whole bioprocess and analytics landscape with specific examples like oncolytic measles virus (Chap. 16) and virus like particles (Chap. 17). Finally, a very important topic of viral safety, both in terms of therapeutic applications and bioprocessing/manufacturing of virus-based ATMPs, is discussed in Chap. 18.

It is also important to highlight the fact that a very important topic of formulation development and drug product (DP) processes is not within the scope of this book. Although, several principles described herein could be adopted directly for DP process development of virus-based therapeutics.

To conclude this preface, we would like to extend a heartfelt thanks and acknowledgment to the families, colleagues, and collaborators in helping this book come true. Without bearing with us and supporting this endeavor, the book would not have been possible.

Innsbruck, Austria — Saurabh Gautam
Biberach an der Riß, Germany — Abhilash I. Chiramel
Basel, Switzerland — Roland Pach

Contents

Contributors

Juliana Albrecht Pharma Technical Development Biologics Europe (PTDE), F. Hoffmann-La Roche Ltd., Basel, Switzerland

Manidipa Banerjee School of Interdisciplinary Research, Indian Institute of Technology, New Delhi, India
Kusuma School of Biological Sciences, Indian Institute of Technology, New Delhi, India

Amritanshu Bhushan Merck Life Science, Mumbai, India

Björn Boshof Institute of Bioprocess Engineering and Pharmaceutical Technology (IBPT), University of Applied Sciences Mittelhessen (THM), Gießen, Germany

Sascha Bossow Clinical Cooperation Unit Virotherapy, German Cancer Research Center (DKFZ), Heidelberg, Germany

Amelie Boulais Sartorius Stedim France FMT, Aubagne, France

Ramona Businger Development Viral Therapeutics Center, Boehringer Ingelheim GmbH & Co KG, Ochsenhausen, Germany

Emmanuelle Cameau PALL France SAS, Saint Germain en Laye, France

Paul Cashen Sartorius Stedim Biotech, Hertfordshire, UK

Abhilash I. Chiramel Virus Therapeutics Center, Boehringer Ingelheim Pharma GmbH & Co. KG, Biberach an der Riß, Germany

Marie-Laure Collignon PALL Belgium, Hoegaarden, Belgium

Louis Crowley Sartorius North America Inc, Bohemia, NY, USA

Peter Czermak Institute of Bioprocess Engineering and Pharmaceutical Technology (IBPT), University of Applied Sciences Mittelhessen (THM), Gießen, Germany
Faculty of Biology and Chemistry, Justus-Liebig University of Giessen, Giessen, Germany

Martin Dass Boehringer Ingelheim Pharma GmbH & Co KG, Biberach an der Riß, Germany

Shyamtanu Datta Pharma Technical Development Cell and Gene Therapy, F. Hoffmann-La Roche Ltd., Basel, Switzerland

Dustin Eckhardt Institute of Bioprocess Engineering and Pharmaceutical Technology (IBPT), University of Applied Sciences Mittelhessen (THM), Gießen, Germany

Saurabh Gautam Boehringer Ingelheim International GmbH, Ingelheim, Germany
ViraTherapeutics GmbH, Rum, Austria

Ingo H. Gorr Boehringer Ingelheim Pharma GmbH & Co KG, Biberach, Germany

Tanja A. Grein ProBioGen AG, Berlin, Germany

Joost van Haasteren Pharma Technical Development Cell & Gene Therapy (PTC), F. Hoffmann-La Roche Ltd., Basel, Switzerland

Orsolya Hamusics Boehringer Ingelheim Pharma GmbH & Co KG, Biberach an der Riß, Germany

Michael Hawley Material and Analytical Sciences, Boehringer Ingelheim Pharmaceuticals, Inc., Ridgefield, CT, USA

Daniel Hochdorfer Development Viral Therapeutics Center, Boehringer Ingelheim GmbH & Co KG, Ochsenhausen, Germany

Dominik Hotter Development Viral Therapeutics Center, Boehringer Ingelheim GmbH & Co KG, Ochsenhausen, Germany

Kerem Irfan Pall Europe Limited, Portsmouth, UK

Friedrich Käß Boehringer Ingelheim Pharma GmbH & Co.KG, Ingelheim, Germany

Wajihul Hasan Khan Department of Chemical Engineering, Indian Institute of Technology, New Delhi, India

Ryan Kissinger Kissinger Medical Visuals, Nelson, BC, Canada

Jan-Philip Klee Institute of Bioprocess Engineering and Pharmaceutical Technology (IBPT), University of Applied Sciences Mittelhessen (THM), Gießen, Germany

Irini Skaripa Koukelli Pharma Technical Development Biologics Europe (PTDE), F. Hoffmann-La Roche Ltd., Basel, Switzerland

Ratish Krishnan MilliporeSigma, Burlington, MA, USA

Jennifer J. Labisch Sartorius Lab Instruments GmbH & Co. KG, Göttingen, Germany

Maja Leskovec Sartorius BIA Separations d.o.o., Ajdovscina, Slovenia

Daniel Loewe Boehringer Ingelheim Therapeutics GmbH, Ochsenhausen, Germany

David Loong Merck Pte Ltd., Singapore, Singapore

Keven Lothert Department of Life Science Engineering, Institute of Bioprocess Engineering and Pharmaceutical Technology, University of Applied Sciences Mittelhessen (THM), Giessen, Germany

Alexandra H. E. Machado Pharma Technical Development Biologics Europe (PTDE), F. Hoffmann-La Roche Ltd., Basel, Switzerland

David Mainwaring Pall Europe Limited, Portsmouth, UK

Paromita Majumder Pall Europe Limited, Portsmouth, UK

Nicholas Marchand Pall Corporation, Westborough, MA, USA

Katy McLaughlin Sartorius Stedim UK Ltd., Epsom, Surrey, UK

Piergiuseppe Nestola Sartorius Stedim Switzerland AG, Tagelswangen, Switzerland

Judy King Man Ng ViraTherapeutics GmbH/Boehringer Ingelheim, Rum, Austria

Roland Pach Pharma Technical Development Analytics Biologics, F. Hoffmann-La Roche Ltd., Basel, Switzerland

Santosh Kumar Pandey Merck KGaA, Darmstadt, Germany

Andreia Pedregal PALL Belgium, Hoegaarden, Belgium

Elaine E. Peters Pharma Technical Development Cell and Gene Therapy USA, Genentech, Inc., South San Francisco, CA, USA

Aurelia Raducanu Pharma Technical Development Cell & Gene Therapy (PTC), Roche Diagnostics GmbH, Penzberg, Germany

Abhilasha K. Rani School of Interdisciplinary Research, Indian Institute of Technology, New Delhi, India

Anurag S. Rathore School of Interdisciplinary Research, Indian Institute of Technology, New Delhi, India
Department of Chemical Engineering, Indian Institute of Technology, New Delhi, India

Philippe Ringler Biozentrum, University of Basel, Basel, Switzerland

Pankaj Salvi Pall India Pvt. Ltd., Pune, Maharashtra, India

Denise Salzig Institute of Bioprocess Engineering and Pharmaceutical Technology (IBPT), University of Applied Sciences Mittelhessen (THM), Gießen, Germany
Faculty of Biology and Chemistry, Justus-Liebig University of Giessen, Giessen, Germany

Mojca Tajnik Sbaizero Sartorius BIA Separations d.o.o., Ajdovscina, Slovenia

Claire Scanlan MilliporeSigma, Burlington, MA, USA

Matthias Schad Boehringer Ingelheim Pharma GmbH & Co.KG, Ingelheim, Germany

Johannes Solzin Development Viral Therapeutics Center, Boehringer Ingelheim GmbH & Co KG, Ochsenhausen, Germany

Bart Spiesschaert Boehringer Ingelheim International GmbH, Ingelheim, Germany
ViraTherapeutics GmbH, Rum, Austria

Guy Ungerechts Clinical Cooperation Unit Virotherapy, German Cancer Research Center (DKFZ), Heidelberg, Germany
Department of Medical Oncology, National Center for Tumor Diseases (NCT) and Heidelberg University Hospital, Heidelberg, Germany
Cancer Therapeutics Program, Ottawa Hospital Research Institute, Ottawa, ON, Canada

Laurens Vergauwen Merck KGaA, Darmstadt, Germany

Robert Whitfield Pall Europe Limited, Portsmouth, UK

Michael W. Wolff Department Life Science Engineering, Institute of Bioprocess Engineering and Pharmaceutical Technology, University of Applied Sciences Mittelhessen (THM), Giessen, Germany
Bioresources, Fraunhofer Institute for Molecular Biology and Applied Ecology (IME), Giessen, Germany

Dongyue Xin Material and Analytical Sciences, Boehringer Ingelheim Pharmaceuticals, Inc., Ridgefield, CT, USA

Chapter 1
Bioprocessing and Analytical Development for Virus-Based Therapeutics

Abhilash Chiramel, Roland Pach, and Saurabh Gautam

1.1 Introduction

Besides transposons, viruses from an evolutionary perspective are ancient molecular parasites which have evolved to pass on genetic information by parasitically infecting host cells, spread through an organism in a cell type- and tissue-specific fashion. Importantly, host cells infected by viruses have evolved potent and sophisticated barriers to infection, which in turn regulate viral tropism (Nomaguchi et al. 2012). The ability of viruses to function as gene vectors amenable for genetic engineering to transfer target-specific genetic payload makes them an ideal tool for gene therapy and can serve as molecular devices (Lundstrom 2018).

It was only in the late nineteenth century when sciences realized that viruses existed and acknowledged their influence in shaping the history and evolution of life on earth (van Kammen 1999). Majority of living organisms are infected by viruses, with approximately 8% of the human genome harboring sequences derived from previous viral infections (Grandi and Tramontano 2018). Over the centuries viruses

Abhilash Chiramel and Roland Pach contributed equally with all other contributors.

A. Chiramel
Virus Therapeutics Center, Boehringer Ingelheim Pharma GmbH & Co. KG, Biberach an der Riß, Germany

R. Pach
Pharma Technical Development Analytics Biologics, F. Hoffmann-La Roche Ltd., Basel, Switzerland

S. Gautam (✉)
Boehringer Ingelheim International GmbH, Ingelheim, Germany

ViraTherapeutics GmbH, Rum, Austria
e-mail: saurabh.gautam@boehringer-ingelheim.com

S. Gautam et al. (eds.), *Bioprocess and Analytics Development for Virus-based Advanced Therapeutics and Medicinal Products (ATMPs)*,
https://doi.org/10.1007/978-3-031-28489-2_1

have caused devastating global pandemics resulting in the loss of significant parts of the human population. At the same time, over decades lifesaving vaccines have been developed based on attenuated live-replicating versions of pathogenic viruses. A first breakthrough was the successful usage of an attenuated vaccine in 1796 when the English doctor Edward Jenner used cowpox virus as vaccine against human small-pox virus directly demonstrating the value of viruses as therapeutic agents (Gurwith et al. 2020). As a result, uncovering the molecular and cell biology of virus infections has gone hand in hand with new technological developments such as reverse engineering of viruses to manipulate viral genomes to produce viral vaccines, oncolytic viruses, and gene-therapy vectors (Kelly and Russell 2007). Employing viruses for therapy belongs to a new class of modalities termed virotherapies or advanced therapy medicinal products (ATMPs) (Iglesias-Lopez et al. 2019).

From drug delivery and manufacturing perspective, another therapeutic platform is the virus-like particles (VLP), which are high-order, nanoscale structures of assembled viral proteins that lack endogenous viral genetic cargo (Nooraei et al. 2021). Particularly, the development of VLP-based vaccines has been successfully implemented for immunization against various infectious agents (Roldão et al. 2014). Analogous to viruses, VLPs can be either enveloped or non-enveloped particles (Nooraei et al. 2021). The synthesis of VLPs includes cell-based expression of the viral-derived glycoproteins which self-associate and can be produced using diverse platforms like the mammalian cell culture, baculovirus/insect cell culture system, microbial fermentation, and plants. As potential drug carriers, they can encapsulate any kind of payload like DNA/RNA, proteins, or even small molecules.

However, challenges in tissue- or cell-specific targeting of VLP limits its usage particularly for gene therapy but not as a vaccine technology wherein the uptake by antigen-presenting cells is facilitated by the particular structure. Modifications on VLP surface may enable more effective and specific targeting of VLPs as described for the viral vectors (Zdanowicz and Chroboczek 2016).

Two major reasons why developing virus-based (or VLP) therapeutics have been challenging are, first, the dearth of a mechanistic understanding of the biology of diverse viral vectors derived from a range of family of viruses including *rhabdoviridae, herpesviridae lentiviruses, adenoviridae, retroviridae,* and *parvoviridae*, and second, the lack of robustness in bioprocessing and manufacturing of virus-based products.

Another challenge in producing therapeutic viruses is the challenge in defining the precise nature of the therapeutic substance and product. Viruses can either be "non-enveloped" or "enveloped" meaning deriving a host cell membrane studded with viral glycoproteins to envelope the capsid containing the viral genome (Wang et al. 2020). Since viruses are entirely reliant on host cells to replicate and depend on an interplay between viral and host cell proteins (HCPs) to regulate infectivity and stability of virions, characterizing the nature of HCPs that are proviral is poorly defined. The lack of complete characterization of drug substance and product (therapeutic virus) adds on another challenge to manufacturing, about bridging one process of production to another, without affecting the nature of the drug

(Moleirinho et al. 2020). Therefore, different types of viral vectors have specific bioprocessing requirements and to an extent the process becomes the product (Moleirinho et al. 2020). Finally, strategies adopting continuous bioprocessing would be the cornerstone in therapeutic virus manufacturing once virus-specific bioprocessing requirements are established. This book focuses on reviewing existing strategies and evolving paradigms currently available to manufacture virus-based therapeutics, including oncolytic viruses and viral vectors for gene therapy. Importantly, the book will cover all aspects of bioprocessing including cell line development, strategies for upstream and downstream processing, scalability, purification methods, formulation, analytics, host cell, and product-related impurities.

1.2 History of Virus-Based ATMPs

The use of viruses as therapeutic agents has evolved over the past 150 years with interests in the field varying over time (Fig. 1.1) (Kelly and Russell 2007; Letarov 2020). A sequence of reports from the mid-1800s demonstrated tumor regressions in patients who contracted virus infections (Kelly and Russell 2007). Nearly 20 years after the discovery of viruses, viruses that infect bacteria termed bacteriophages were demonstrated to be effective in lysing bacterial cells. Discovery of bacteriophages not only revolutionized our understanding of molecular biology and virology but was evaluated as therapeutic options for cholera and dysentery in the 1920s–1930s (Emiliano and Shawna 2007; Letarov 2020). The popularity of phage therapy (PT) grew worldwide with several leading pharmaceutical companies in the USA, England, and France starting commercial production of these preparations (Summers 2001). Despite showing evidence of high clinical efficacy, quality control issues in bacteriophage manufacturing were the main reason leading to discrediting phage therapy by the end of the 1930s. In addition, the discovery, use, and manufacturability of antibiotics overshadowed development of PT (Zaffiri et al. 2012). The rise of antimicrobial resistance is a daunting medical concern and has reinvigorated the use of PT as a revolutionary technology and an alternative to antibiotics (Summers 2001).

An interest in employing viruses for cancer therapy emerged in the 1950s and 1970s, with the use of hepatitis B virus, West Nile virus, adenoviruses, and non-attenuated mumps virus resulting in occasional regression in tumor size and localized necrosis. However, it was only in November 2005 that the first oncolytic virus was approved by Chinese regulators for the genetically modified oncolytic adenovirus H101 (Garber 2006). Currently, Imlygic or talimogene laherparepvec (T-Vec) is the only oncolytic virus approved by both the EMA and FDA (Lauer and Beil 2022). However, many more oncolytic viruses are in various clinical trials including adenoviruses, poxviruses, herpes simplex virus-1 (HSV-1), coxsackieviruses, measles virus, vesicular stomatitis virus (VSV), Newcastle disease virus (NDV), and reovirus (Kaufman et al. 2015; Lauer and Beil 2022; Russell et al. 2022). Clinical data over the past decade have revealed that oncolytic virus therapy

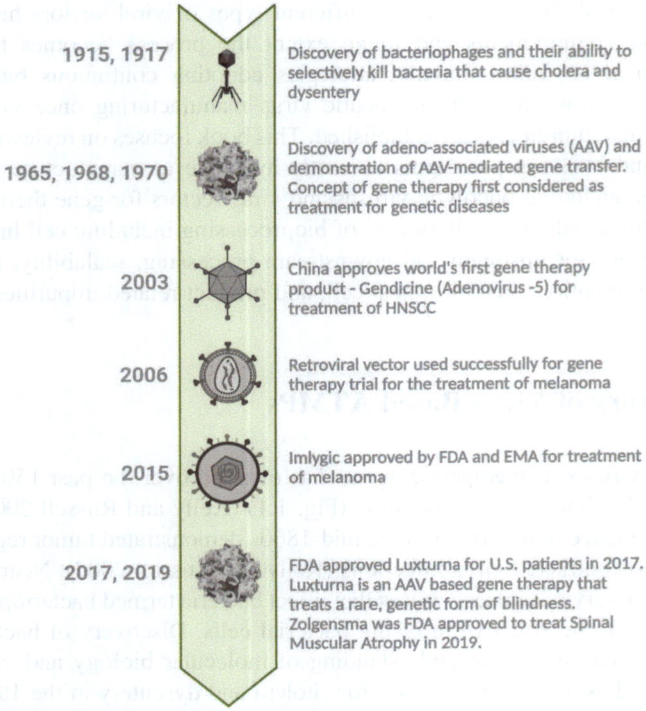

Fig. 1.1 Milestones in the development of virus-based ATMPs. A diverse range of viruses have been tested and developed to assess their therapeutic potential in immuno-oncology and gene therapy. The timeline highlights key efforts and important milestone achieved in virotherapy. Graphic created with BioRender.com

is likely benefiting from an integrated therapy concept enhancing the virus infection-mediated anti-tumor immunity. This being a key learning compared to initial assumptions that oncolysis is the predominant mode of action (MoA) to control cancer disease in humans. Nevertheless, when used in combination with immune checkpoint inhibitors and immune modulators, oncolytic viruses could be an additional pillar to integrative oncology (Bommareddy et al. 2018).

As mentioned above, the other facet to virus-based therapeutics is the fact that viruses make excellent vehicles for gene delivery (Haan et al. 2020). Viruses can be used to selectively target cell types, tissues, and organs of a diverse range of species. Furthermore, viruses can be forced to evolve and genetically manipulated to overcome barriers to infection and thus broaden tropism. This plasticity allows for genetic engineering of organisms to correct genetic deficiencies or confer resistance to disease phenotypes (Zolotukhin and Vandenberghe 2022). Several viral vectors have been evaluated and designed based on efficacy and safety for gene therapy, and of these the most popular vectors include retroviruses and AAV (Haan et al. 2020). AAV vector-based therapies have the advantage and are often used when non-integration of genomic material is planned. On the other hand, the *retroviridae* family of viruses offer two flavors of vectors to engineer T cells for chimeric antigen receptor (CAR)-T cell therapy: gamma retroviruses (GRVs) and lentiviruses (LVs), respectively. Both these vectors have been widely used in research and in clinical studies due to their ability to integrate genetic information within the host cell genome and ensure stable expression of transgenes in immune cells (Escors and Breckpot 2010; Weber et al. 2020). CAR-T cell therapy being based on a viral transduction system is by far the most effective immunotherapy to reach the market. Briefly, T cells are genetically modified by transducing cells with GRVs or LVs to ectopically express a CAR receptor, thus allowing them to selectively target cancer cells within the tumor. This form of immunotherapy has been particularly successful to treat blood cancers of CD19-positive B-cell malignancies with three commercially available products including Kymriah, Yescarta, and Brexucabtagene (Weber et al. 2020). Under the lens of gene therapy, AAV-derived vectors are fundamental tools to facilitate cell type-, tissue-, and organ-specific expression of transgenes. The approval of drugs such as Luxturna and Zolgensma has catalyzed the possibilities of gene therapies to not only treat rare diseases but also leap forward efforts in personalized medicine (Lana and Strauss 2019; Wang et al. 2020). Although cell and gene therapies are the newest pillars of modern medicine, many challenges remain to be addressed in optimizing bioprocessing and manufacturing to reduce cost and high variability in the manufacturing process (Weber et al. 2020).

1.3 Bioprocess Development for Virus-Based ATMPs

The use of viruses to treat human diseases is not a new idea (Fig. 1.1). However, it became popular in recent years with the advent of cell and gene therapies. Therefore, the lack of historical knowledge and experience is clearly reflected in challenges faced with bioprocess and analytics development, especially compared with new biological entities (NBEs) or biosimilars in recent years (Moleirinho et al. 2020; Srivastava et al. 2021).

Bioprocessing of viruses is complex and requires systematic and coordinated steps both upstream and downstream (Fig. 1.2). However, the use of inefficient technologies such as adherent cells and traditional chromatographic columns for the production of viruses does not support effective scale-up and commercial manufacturing strategies. This is amplified by the fact that viruses are significantly larger in size as compared to macromolecules such as proteins and are also much more complex. Viruses as therapeutics are very often sensitive to the local environment and stresses, such as shear stress (Grein et al. 2019; Morgan et al. 2020). Therefore, it is of utmost importance during bioprocess development and scale-up to have strategies to minimize adverse impact on the therapeutic especially during processes and techniques used in stirred tank bioreactors, chromatography, and filtration.

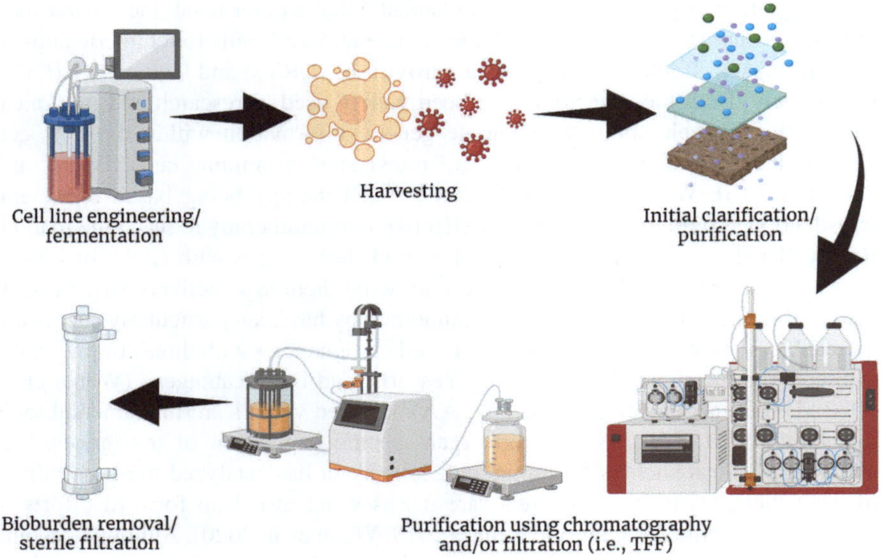

Fig. 1.2 A schematic representation of upstream and downstream workflow used for bioprocessing/manufacturing of virus particles. Graphic created with BioRender.com

The processes for manufacturing of virus therapeutics are mostly based on techniques such as cell culture, chromatography, and/or filtration (Fig. 1.2). Suspension cultures are preferred due to ease in scaling and monitoring of cell cultures; however, continued innovation does not completely rule out the adherent cell cultures being used more often. Innovation is slow when it comes to new bioprocess technologies for virus therapeutics (Fig. 1.3). However, fixed-bed bioreactor technologies such as iCellis™ or scale-X™ systems are popular strategies to drive the current efforts in bioprocess and manufacturing (Lesch et al. 2021). Moreover, adaptation of viruses to suspension cell lines, similar to the manufacturing of protein and antibody therapeutics, is a step in the right direction (Bauler et al. 2020), if not linked to changes in cell targeting properties and/or viral genome stability.

Another important aspect during bioprocess development and manufacturing of virus-based ATMPs is the formulation and drug product process (Srivastava et al. 2021). It is of utmost importance to keep in mind the requirements based on critical quality attributes (Table 1.1) of the viral vector-based active pharmaceutical ingredient (API) wherein factors such as shear stress might play an additional role as compared to other types of therapeutics (Grein et al. 2019).

Only a holistic and end-to-end view to bioprocessing starting from raw materials to the final device gives a complete picture and basis for a robust process. Keeping this holistic view in mind, bioprocesses must be developed and carried out with a very fine balance of purity, yield, and costs (Fig. 1.4). Increased complexity and number of steps in a bioprocess are generally detrimental to the yield of drug substance and often inversely proportional to the costs required for manufacturing.

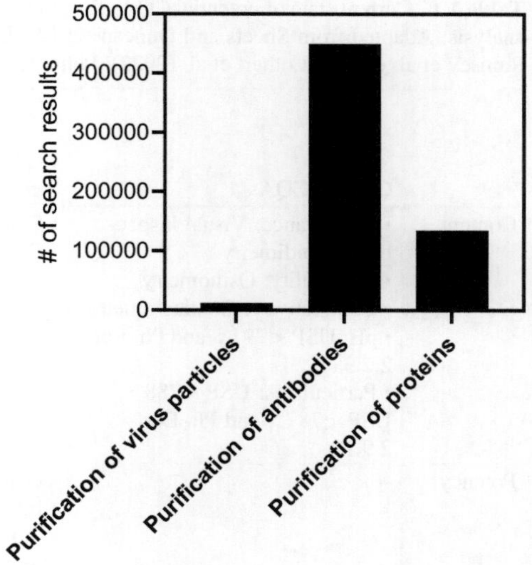

Fig. 1.3 PubMed search (https://pubmed.ncbi.nlm.nih.gov) carried out on September 2, 2022, for keywords specified on the graph reveals the slow pace at which innovation is moving with regard to bioprocessing and manufacturing of virus-based therapies

These challenges can only be overcome with more understanding of these novel therapeutics and further bioprocess innovation.

1.4 Analytics for Virus-Based ATMPs

High complexity of the virus-based ATMPs and often their multiple MoAs require an incremental and experimental approach for defining specifications and control strategy. Therefore, consequently there is no 1:1 adaption of the successful strategies and concepts in medicinal products development of biologics. This complexity in development of viral ATMPs is mirrored and driven by:

- The complex manufacturing process highlighted by its multifarious biological and technological processes.
- Intensive and sophisticated testing/control strategy in particular the multiple potency assay(s).
- The evolving new regulatory environment in the cell and gene therapy (CGT) field.

As industry, we are still in a learning phase, which is highlighted by the low numbers of existing licensed products due to limited product and process understanding, as well as the described application refusal triggered by the CMC chapters, e.g., comparability, impurities, safety, and potency (Barkholt et al. 2019). So, the demonstration of an adequate control strategy covering content, identity, safety, purity, and potency of the final ATMP product and process understanding are crucial (Ramsey et al. 2021; Wright 2021).

Table 1.1 Current state of potential CQAs for virus therapeutics and respective techniques for their analysis. Adapted from Sheets and Duncan (2014), Bracewell et al. (2021), Gimpel et al. (2021), Ramsey et al. (2021), Lothert et al. (2022), Petiot (2022), Peters and Pach (2022)

	General CQAs	CQAs specific for virus therapeutics	Relevant techniques for virus--specific CQAs
Content	• Appearance: Visual inspection, turbidimetry • Osmolality: Osmometry • Conductivity: Conductometry • pH: USP <791> and Ph. Eur. 2.2.3 • Particulates: USP <788>, USP <787> and Ph. Eur. 2.9.19	• Genomic titer • Capsid titer	• qPCR/ddPCR • ELISA
Potency	–	• Genomic titer • Infectivity • Transgene transcription/expression • Functional activity (transduction) • In vivo potency • PTMs	• qPCR/ddPCR • Infectivity assay, e.g., $TCID_{50}$, plaque assay • ELISA • FACS • Transgene reporter/functionality assay • Mass spectrometry • In vivo potency assay • WB • HPLC
Identity	–	• Capsid identity • Payload identity	• ELISA • qPCR/ddPCR • PCR/restriction digestion • Sequencing (Sanger/NGS) • Mass Spectrometry • SDS-PAGE/CE • WB • NMR
Process-related impurities	• Residual nuclease/benzonase: ELISA • Leachables/extractables: HPLC, mass spectrometry, GC, FID, TCD, SEC • Host cell protein: ELISA • Host cell DNA: qPCR/ddPCR • Residual chromatographic ligands: ELISA, SPR	–	–

(continued)

Table 1.1 (continued)

	General CQAs	CQAs specific for virus therapeutics	Relevant techniques for virus--specific CQAs
	• Other residual process elements, e.g., BSA: ELISA, SPR, SDS-PAGE		
Product-related impurities	• Host EVs: Cryo-EM, FACS, HPLC, AUC	• Empty capsid • Deformed or aggregated virus particles • Truncated or mutated viral proteins • Residual helper pDNA or virus products • Non-infectious virus particles or VLPs/EVs • Non-wild-type virus variants, e.g., mutated, PTMs	• Electron microscopy • AUC • Mass spectrometry/peptide mapping • SDS-PAGE/CE • HPLC • Mass photometry • qPCR/ddPCR • NGS • DLS • SEC/IEX-MALS • Infectivity assay, e.g., $TCID_{50}$, plaque assay • NMR
Safety	• Bioburden: USP <61> and Ph. Eur 2.6.12 • Sterility: USP <71> and Ph. Eur. 2.6.1 • Mycoplasma: Agar-Cultivable Mycoplasma Assay, qPCR. • Appearance: Visual inspection, turbidimetry • Endotoxin: LAL • Viral and microbiological safety, e.g., adventitious agents: EP 2.6.16, 21CFR, qPCR, RT-qPCR, NGS	• Virus therapeutic safety • Replication competent virus in case of non-replicating vectors (e.g., presence of rep or cap sequences)	• Cell culture-based assays • In vivo safety testing • Southern blotting • qPCR

Process and product analytics are key to achieve maximum productivity and best quality of the final therapeutic virus drug product. Analytical strategies to cover three areas of engineering have to be in the focus: first, process engineering starting from raw material until fill and finish (Robert et al. 2017; Selvaraj et al. 2021); second, cell line engineering to cover development of producer cell lines (Farson et al. 2001; Gao et al. 2002; Kovesdi and Hedley 2010; Martin et al. 2013; Tan et al. 2021); and third, design of the vector and therapeutic cargo (Naso et al. 2017). Therefore, a sound

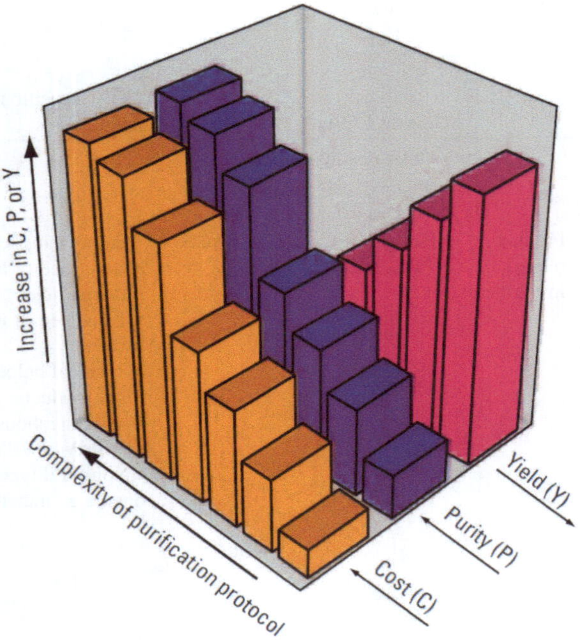

Fig. 1.4 Correlation among complexity, cost, purity, and yield of a bioprocess. Reprinted with permission from Mondal et al. (2006). Copyright 2006 by American Chemical Society

understanding and identification of the critical quality attributes (CQAs) of the product and associated structure–function relationships to corresponding critical process parameters is crucial. Figure 1.5 provides an overview of how process and product analytics can reveal essential and non-essential CQAs by mapping key attributes along two parallel streams. Some attributes are overlapping, others are assigned to the single stream and are influenced by the other.

In summary, the control strategy as per ICH Q10 guidance recommends a planned set of controls, derived from current product and process understanding that assures process performance and product quality. The key factors influencing the determination of pre-CQAs and therefore the control strategy include vector type, production platform, process and product knowledge, analytics, product variants, contaminants, routes of administration, and dose.

A potential process map and assigned attributes of a plasmid DNA-based AAV manufacturing platform is depicted below (Fig. 1.6) in the context of an end-to-end (E2E) perspective, starting from raw material until fill and finish step (Peters and Pach 2022).

Quality attributes related to the activity of the vector's payload/cargo can be only demonstrated indirectly by testing functionality of the transgene by in vitro cell-based assays or in vivo models. Overall, from the host cell perspective, two major classes of quality attributes can be distinguished for gene therapy products:

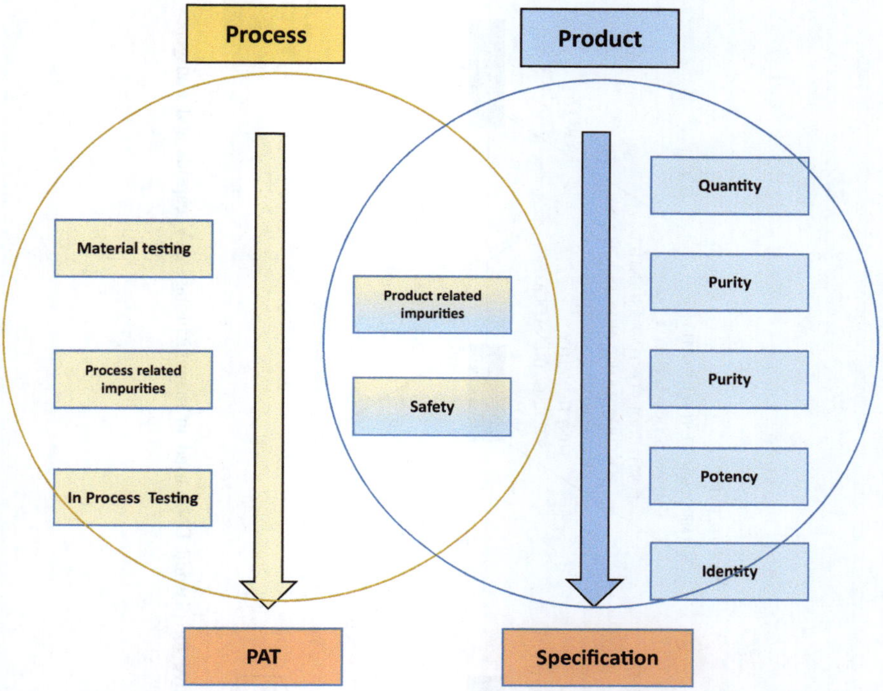

Fig. 1.5 Functional relation between process and product and some key attributes. The schematic highlights the cross-talk between both streams. Abbreviations: *PAT* process analytical technologies, *IPT* in-process testing

- Extracellular attributes: all quality attributes that relate to targeting and tropism of the vector, including infectivity or transducibility, stability, capsid identity, quantity, payload identity and activity, and immunogenicity
- Intracellular attributes: attributes related to the viral vector and encoded payload, intracellular processing of the payload in the target cell cover aspects such as translocation, transcription, expression, post-translational modifications (PTM), sorting, and potency.

As an example of successful optimization and development of viral vectors to assess the extracellular quality attributes are the significant reduction of immunogenicity and improved tropism of AAV-based drugs for second-generation vectors (Domenger and Grimm 2019; Lotfinia et al. 2019). Current efforts focus on improving the intracellular attributes of next-generation AAV vectors.

The molecular engineering of these transgenic expression cassettes and its flanking regions like promotors/enhancers and inverted terminal repeat (ITR) sequences affect a number of factors related to these viral vectors. The factors include capsid assembly, genome encapsulation (packaging), transduction efficiency, and cell type and tissue specificity expression and finally allow the use of

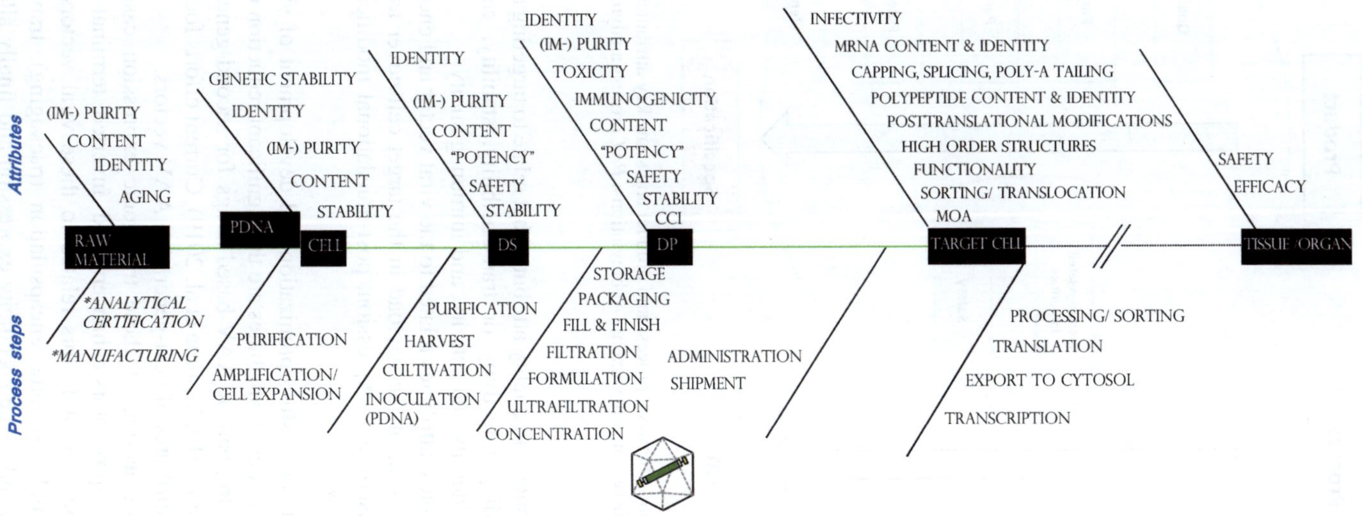

Fig. 1.6 Mapping of process steps versus attributes from an E2E perspective for plasmid DNA used in manufacturing AAV capsids and transgenes. Abbreviations: *DS* drug substance, *DP* drug product, *pDNA* plasmid DNA

lower and safer vector doses (Domenger and Grimm 2019). Exemplary highlighted here in the case of AAV is the product-related impurity like the wrongly packed or empty viral particles (Wright 2021). Characterization of the rep genes (Mietzsch et al. 2021) and ITR flanking regions (Yan et al. 2005; Earley et al. 2020) allows for regulation of rep expression and ITR engineering to increase the packaging efficiency and better transduction as well. These modifications may augment the future development of AAV vectors and reduce burden on process and analytics and eventually improve purification and yield. These efforts may allow for a leaner analytical strategy related to improved ratios for empty versus full capsids. Briefly, on the molecular engineering side, more efforts in the design of the DNA sequence of the transgenic payload and its flanking regions enable particularly a better control of the product-related impurities and related product variants.

According to the FDA and EMA's recently revised guideline for cell and gene therapy, changes were made for viral vector analytics to include analytical procedures specific to impurities, replication, titer, and infectivity (U.S. FDA 2022). In general, the regulators and authorities require detailed information and regulatory documentation inclusive of the above-mentioned parameters, aspects of which are expected to increase as the field evolves and develops. Given the diversity that exists in the virus-therapeutics landscape and technologies employed in manufacturing them, higher expectations for specificity, sensitivity, and accuracy exceed the limits of available analytics. To this end, analytical technologies to characterize viral vectors for gene therapy and oncolytic viruses continue to evolve in order to assign CQAs (Table 1.1), highlighting identity, potency, purity, and safety of these products (Gurwith et al. 2020; Moaven et al. 2021; Tustian and Bak 2021). The heterogeneity observed in virus-derived medicinal products presents a unique challenge to ensure safety and efficacy, particularly important for cell and gene therapies.

When developing analytics for virus-derived medicinal products, it is important to consider the nature of the platform, i.e., replication incompetent or competent. If the vector is replication competent, analytics employed to determine virus titer is entirely different in contrast to replication incompetent vectors (Keiser et al. 2021; Hochdorfer et al. 2022). Of relevance, cell-based assays developed for determining virus titers are also indicative of viral infectivity, mode of action, and the therapeutic dose for efficacy (Wang et al. 2020). By nature, viruses are morphologically diverse entities (Jardetzky et al. 2017), thus understanding and analyzing biophysical and biochemical properties of viral vectors are paramount in defining the nature of these therapeutic products. Depending on the virus, morphology is a direct function of coding capacity (Jardetzky et al. 2017). Therefore, using suitable analytical technologies to determine changes in virus morphology, aggregation, and biochemical composition, when subjected to different stages of bioprocessing and manufacturing, is crucial in ensuring the quality of the product (Hajba and Guttman 2020).

Another important aspect to consider in the manufacturing of next generation of therapeutic viruses is a holistic approach of process analytical technology (PAT) (Pais et al. 2019). PAT and Quality by Design (QbD) framework has driven efforts in high-quality analytics that can generate nearer to "real-time" data to advise process development decisions and achieve higher yields and viral vector safety profiles. The lag time between sampling and results from available assays can vary from days to months, which is not conducive for effective design of experiment (DOE). Furthermore, the limited resolution inherent to specific analytical assays may fail to reveal nuances in bioprocessing to result in stepwise process optimization (Pais et al. 2019). A key goal for viral vector manufacturing is to incorporating novel and accurate analytical tools. Importantly, these methods require validation before they are implemented in clinical manufacturing by regulatory agencies. Key to this will be the usage of parallel orthogonal methods such as incorporation of artificial intelligence (AI) and machine learning to ensure integrity and robustness of data. As emphasized (Table 1.1), a myriad of analytical technologies is currently available to manufacture therapeutic viruses. This book attempts to cover all aspects of analytics that are important from the perspective of the viral vectors and the bioprocesses required for their production.

References

Barkholt L, Voltz-Girolt C, Raine J, Salmonson T, Schüssler-Lenz M (2019) European regulatory experience with advanced therapy medicinal products. Nat Rev Drug Discov 18(1):8–9. https://doi.org/10.1038/nrd.2018.200

Bauler M, Roberts JK, Wu C-C, Fan B, Ferrara F, Yip BH, Diao S, Kim Y-I, Moore J, Zhou S, Wielgosz MM, Ryu B, Throm RE (2020) Production of lentiviral vectors using suspension cells grown in serum-free media. Mol Ther Methods Clin Dev 17:58–68. https://doi.org/10.1016/j.omtm.2019.11.011

Bommareddy PK, Shettigar M, Kaufman HL (2018) Integrating oncolytic viruses in combination cancer immunotherapy. Nat Rev Immunol 18(8):498–513. https://doi.org/10.1038/s41577-018-0014-6

Bracewell DG, Smith V, Delahaye M, Smales CM (2021) Analytics of host cell proteins (HCPs): lessons from biopharmaceutical mAb analysis for Gene therapy products. Curr Opin Biotech 71: 98–104. https://doi.org/10.1016/j.copbio.2021.06.026

Domenger C, Grimm D (2019) Next-generation AAV vectors—do not judge a virus (only) by its cover. Hum Mol Genet 28(R1):R3–R14. https://doi.org/10.1093/hmg/ddz148

Earley LF, Conatser LM, Lue VM, Dobbins AL, Li C, Hirsch ML, Samulski RJ (2020) Adeno-associated virus serotype-specific inverted terminal repeat sequence role in vector transgene expression. Hum Gene Ther 31(3–4):151–162. https://doi.org/10.1089/hum.2019.274

Escors D, Breckpot K (2010) Lentiviral vectors in gene therapy: their current status and future potential. Arch Immunol Ther Ex 58(2):107–119. https://doi.org/10.1007/s00005-010-0063-4

Farson D, Witt R, McGuinness R, Dull T, Kelly M, Song J, Radeke R, Bukovsky A, Consiglio A, Naldini L (2001) A new-generation stable inducible packaging cell line for lentiviral vectors. Hum Gene Ther 12(8):981–997. https://doi.org/10.1089/104303401750195935

Fruciano DE, Bourne S (2007) Phage as an antimicrobial agent: D'herelle's heretical theories and their role in the decline of phage prophylaxis in the West. Can J Infect Dis Medical Microbiol 18(1):19–26. https://doi.org/10.1155/2007/976850

Gao G, Lu F, Sanmiguel JC, Tran PT, Abbas Z, Lynd KS, Marsh J, Spinner NB, Wilson JM (2002) Rep/Cap gene amplification and high-yield production of AAV in an A549 cell line expressing Rep/Cap. Mol Ther 5(5):644–649. https://doi.org/10.1006/mthe.2001.0591

Garber K (2006) China approves world's first oncolytic virus therapy for cancer treatment. J Natl Cancer Inst 98(5):298–300. https://doi.org/10.1093/jnci/djj111

Gimpel AL, Katsikis G, Sha S, Maloney AJ, Hong MS, Nguyen TNT, Wolfrum J, Springs SL, Sinskey AJ, Manalis SR, Barone PW, Braatz RD (2021) Analytical methods for process and product characterization of recombinant adeno-associated virus-based gene therapies. Mol Ther Methods Clin Dev 20:740–754. https://doi.org/10.1016/j.omtm.2021.02.010

Grandi N, Tramontano E (2018) Human endogenous retroviruses are ancient acquired elements still shaping innate immune responses. Front Immunol 9:2039. https://doi.org/10.3389/fimmu.2018.02039

Grein TA, Loewe D, Dieken H, Weidner T, Salzig D, Czermak P (2019) Aeration and shear stress are critical process parameters for the production of oncolytic measles virus. Front Bioeng Biotechnol 7:78. https://doi.org/10.3389/fbioe.2019.00078

Gurwith M, Condit RC, Excler J-L, Robertson JS, Kim D, Fast PE, Drew S, Wood D, Klug B, Whelan M, Moore TM, Khuri-Bulos N, Smith ER, Chen RT, Kochhar S, V3SWG F the BCVVVSWG (2020) Brighton Collaboration Viral Vector Vaccines Safety Working Group (V3SWG) standardized template for collection of key information for benefit-risk assessment of live-attenuated viral vaccines. Vaccine 38(49):7702–7707. https://doi.org/10.1016/j.vaccine.2020.09.042

Haan PD, Diemen FRV, Toscano MG (2020) Viral gene delivery vectors: the next generation medicines for immune-related diseases. Hum Vacc Immunother 17(1):1–8. https://doi.org/10.1080/21645515.2020.1757989

Hajba L, Guttman A (2020) Recent advances in the analysis full/empty capsid ratio and genome integrity of adeno-associated virus (AAV) gene delivery vectors. Curr Mol Med 20(10):806–813. https://doi.org/10.2174/1566524020999200730181042

Hochdorfer D, Businger R, Hotter D, Seifried C, Solzin J (2022) Automated, label-free TCID50 assay to determine the infectious titer of virus-based therapeutics. J Virol Methods 299:114318. https://doi.org/10.1016/j.jviromet.2021.114318

Iglesias-Lopez C, Agustí A, Obach M, Vallano A (2019) Regulatory framework for advanced therapy medicinal products in Europe and United States. Front Pharmacol 10:921. https://doi.org/10.3389/fphar.2019.00921

Jardetzky T, Kuhn R, Lamb R (2017) Editorial overview: virus structure and functions. Curr Opin Virol 24:ix. https://doi.org/10.1016/j.coviro.2017.07.006

Kaufman HL, Kohlhapp FJ, Zloza A (2015) Oncolytic viruses: a new class of immunotherapy drugs. Nat Rev Drug Discov 14(9):642–662. https://doi.org/10.1038/nrd4663

Keiser PT, Anantpadma M, Staples H, Carrion R, Davey RA (2021) Automation of infectious focus assay for determination of filovirus titers and direct comparison to plaque and TCID50 assays. Microorg 9(1):156. https://doi.org/10.3390/microorganisms9010156

Kelly E, Russell SJ (2007) History of oncolytic viruses: genesis to genetic engineering. Mol Ther 15(4):651–659. https://doi.org/10.1038/sj.mt.6300108

Kovesdi I, Hedley SJ (2010) Adenoviral producer cells. Viruses 2(8):1681–1703. https://doi.org/10.3390/v2081681

Lana MG, Strauss BE (2019) Chimeric antigen receptor T cells, development and production. Methods Mol Biology 2086:61–67. https://doi.org/10.1007/978-1-0716-0146-4_4

Lauer UM, Beil J (2022) Oncolytic viruses: challenges and considerations in an evolving clinical landscape. Future Oncol 18(24):2713–2732. https://doi.org/10.2217/fon-2022-0440

Lesch HP, Valonen P, Karhinen M (2021) Evaluation of the single-use fixed-bed bioreactors in scalable virus production. Biotechnol J 16(1):2000020. https://doi.org/10.1002/biot.202000020

Letarov AV (2020) History of early bacteriophage research and emergence of key concepts in virology. Biochem Mosc 85(9):1093–1112. https://doi.org/10.1134/s0006297920090096

Lotfinia M, Abdollahpour-Alitappeh M, Hatami B, Zali MR, Karimipoor M (2019) Adeno-associated virus as a gene therapy vector: strategies to neutralize the neutralizing antibodies. Clin Exp Med 19(3):289–298. https://doi.org/10.1007/s10238-019-00557-8

Lothert K, Eilts F, Wolff MW (2022) Quantification methods for viruses and virus-like particles applied in biopharmaceutical production processes. Expert Rev Vaccines 21:1029–1044. https://doi.org/10.1080/14760584.2022.2072302

Lundstrom K (2018) Viral vectors in gene therapy. Diseases 6(2):42. https://doi.org/10.3390/diseases6020042

Martin J, Frederick A, Luo Y, Jackson R, Joubert M, Sol B, Poulin F, Pastor E, Armentano D, Wadsworth S, Vincent K (2013) Generation and characterization of adeno-associated virus producer cell lines for research and preclinical vector production. Hum Gene Ther Method 24(4):253–269. https://doi.org/10.1089/hgtb.2013.046

Mietzsch M, Eddington C, Jose A, Hsi J, Chipman P, Henley T, Choudhry M, McKenna R, Agbandje-McKenna M (2021) Improved genome packaging efficiency of adeno-associated virus vectors using rep hybrids. J Virol 95(19):e00773–e00721. https://doi.org/10.1128/jvi.00773-21

Moaven O, Mangieri CW, Stauffer JA, Anastasiadis PZ, Borad MJ (2021) Strategies to develop potent oncolytic viruses and enhance their therapeutic efficacy. JCO Precis Oncol 5(5):733–743. https://doi.org/10.1200/po.21.00003

Moleirinho MG, Silva RJS, Alves PM, Carrondo MJT, Peixoto C (2020) Current challenges in biotherapeutic particles manufacturing. Expert Opin Biol Ther 20(5):451–465. https://doi.org/10.1080/14712598.2020.1693541

Mondal K, Gupta MN, Roy I (2006) Affinity-based strategies for protein purification. Anal Chem 78(11):3499–3504. https://doi.org/10.1021/ac0694066

Morgan BA, Manser M, Jeyanathan M, Xing Z, Cranston ED, Thompson MR (2020) Effect of shear stresses on adenovirus activity and aggregation during atomization to produce thermally stable vaccines by spray drying. ACS Biomater Sci Eng 6(7):4304–4313. https://doi.org/10.1021/acsbiomaterials.0c00317

Naso MF, Tomkowicz B, Perry WL, Strohl WR (2017) Adeno-associated virus (AAV) as a vector for gene therapy. Biodrugs 31(4):317–334. https://doi.org/10.1007/s40259-017-0234-5

Nomaguchi M, Fujita M, Miyazaki Y, Adachi A (2012) Viral tropism. Front Microbiol 3:281. https://doi.org/10.3389/fmicb.2012.00281

Nooraei S, Bahrulolum H, Hoseini ZS, Katalani C, Hajizade A, Easton AJ, Ahmadian G (2021) Virus-like particles: preparation, immunogenicity and their roles as nanovaccines and drug nanocarriers. J Nanobiotechnol 19(1):59. https://doi.org/10.1186/s12951-021-00806-7

Pais DAM, Portela RMC, Carrondo MJT, Isidro IA, Alves PM (2019) Enabling PAT in insect cell bioprocesses: in situ monitoring of recombinant adeno-associated virus production by fluorescence spectroscopy. Biotechnol Bioeng 116(11):2803–2814. https://doi.org/10.1002/bit.27117

Peters E, Pach R (2022) E2E perspective of adequate (pre-) cQAS & cognate control strategy: challenges and chances. In: 4th Gene and Cell Therapy; Quality Development, Manufacturing & Commercialization, Berlin

Petiot E (2022) Bioprocessing of viral vaccines. CRC Press, pp 201–224. https://doi.org/10.1201/9781003229797-8

Ramsey JP, Khatwani SL, Lin M, Boregowda R, Surosky R, Ramelmeier RA (2021) Overview of analytics needed to support a robust gene therapy manufacturing process. Curr Opin Biomed Eng 20:100339. https://doi.org/10.1016/j.cobme.2021.100339

Robert M, Chahal PS, Audy A, Kamen A, Gilbert R, Gaillet B (2017) Manufacturing of recombinant adeno-associated viruses using mammalian expression platforms. Biotechnol J 12(3):1600193. https://doi.org/10.1002/biot.201600193

Roldão A, Mellado MCM, Castilho LR, Carrondo MJ, Alves PM (2014) Virus-like particles in vaccine development. Expert Rev Vaccines 9(10):1149–1176. https://doi.org/10.1586/erv.10.115

Russell SJ, Bell JC, Engeland CE, McFadden G (2022) Advances in oncolytic virotherapy. Commun Med 2(1):33. https://doi.org/10.1038/s43856-022-00098-4

Selvaraj N, Wang C-K, Bowser B, Broadt T, Shaban S, Burns J, Saptharishi N, Pechan P, Golebiowski D, Alimardanov A, Yang N, Mitra G, Vepachedu R (2021) Detailed protocol for the novel and scalable viral vector upstream process for AAV gene therapy manufacturing. Hum Gene Ther 32(15–16):850–861. https://doi.org/10.1089/hum.2020.054

Sheets RL, Duncan PA (2014) Role of analytics in viral safety. In: Vaccine analysis: strategies, principles and control. Springer, pp 397–443. https://doi.org/10.1007/978-3-662-45024-6_10

Srivastava A, Mallela KMG, Deorkar N, Brophy G (2021) Manufacturing challenges and rational formulation development for AAV viral vectors. J Pharm Sci 110(7):2609–2624. https://doi.org/10.1016/j.xphs.2021.03.024

Summers WC (2001) BACTERIOPHAGE THERAPY. Annu Rev Microbiol 55(1):437–451. https://doi.org/10.1146/annurev.micro.55.1.437

Tan E, Chin CSH, Lim ZFS, Ng SK (2021) HEK293 cell line as a platform to produce recombinant proteins and viral vectors. Front Bioeng Biotechnol 9:796991. https://doi.org/10.3389/fbioe.2021.796991

Tustian AD, Bak H (2021) Assessment of quality attributes for adeno-associated viral vectors. Biotechnol Bioeng 118(11):4186–4203. https://doi.org/10.1002/bit.27905

U.S.FDA (2022) Cellular & gene therapy guidances | FDA. https://www.fda.gov/vaccines-blood-biologics/biologics-guidances/cellular-gene-therapy-guidances. Accessed 26 Sep 2022

van Kammen A (1999) Beijerinck's contribution to the virus concept–an introduction. Arch Virol Suppl 15:1–8. https://doi.org/10.1007/978-3-7091-6425-9_1

Wang D, Wang K, Cai Y (2020) An overview of development in gene therapeutics in China. Gene Ther 27(7–8):338–348. https://doi.org/10.1038/s41434-020-0163-7

Weber EW, Maus MV, Mackall CL (2020) The emerging landscape of immune cell therapies. Cell 181(1):46–62. https://doi.org/10.1016/j.cell.2020.03.001

Wright JF (2021) Quality control testing, characterization and critical quality attributes of adeno-associated virus vectors used for human gene therapy. Biotechnol J 16(1):2000022. https://doi.org/10.1002/biot.202000022

Yan Z, Zak R, Zhang Y, Engelhardt JF (2005) Inverted terminal repeat sequences are important for intermolecular recombination and circularization of adeno-associated virus genomes. J Virol 79(1):364–379. https://doi.org/10.1128/jvi.79.1.364-379.2005

Zaffiri L, Gardner J, Toledo-Pereyra LH (2012) History of antibiotics. From salvarsan to cephalosporins. J Invest Surg 25(2):67–77. https://doi.org/10.3109/08941939.2012.664099

Zdanowicz M, Chroboczek J (2016) Virus-like particles as drug delivery vectors. Acta Biochim Pol 63(3):469–473. https://doi.org/10.18388/abp.2016_1275

Zolotukhin S, Vandenberghe LH (2022) AAV capsid design: a Goldilocks challenge. Trends Mol Med 28(3):183–193. https://doi.org/10.1016/j.molmed.2022.01.003

Chapter 2
Viral Vector Upstream Processing and Clarification

Emmanuelle Cameau, Marie-Laure Collignon, Andreia Pedregal, Paromita Majumder, Pankaj Salvi, Kerem Irfan, Nicholas Marchand, David Mainwaring, and Robert Whitfield

2.1 Introduction

The requirement to produce a sufficient number of doses to supply the market has led the industry to scale up processes. The regulatory environment has been evolving at the same time, developing requirements to safely bring products to patients. The main industry challenge is the translation of the processes conducted in research and development (R & D), within the microliters range to hundreds of liters required at large scale. This brings new challenges and opportunities for innovation.

There are a variety of methods for viral vector production used in gene therapy applications and depending on which are used they have their own inherent challenges (discussed under Sect. 2.3 "Cell Lines and Plasmids").

In Sect. 2.2 "Choosing the Right Gene Therapy Vector," we detail packaging capacity of several target viral vectors used in the market and several other critical characteristics that help manufacturers choose the right vector based on the target indication to be treated.

E. Cameau (✉)
PALL France SAS, Saint Germain en Laye, France
e-mail: emmanuelle_cameau@europe.pall.com

M.-L. Collignon · A. Pedregal
PALL Belgium, Hoegaarden, Belgium

P. Majumder · K. Irfan · D. Mainwaring · R. Whitfield
Pall Europe Limited, Portsmouth, UK

P. Salvi
Pall India Pvt. Ltd., Pune, Maharashtra, India

N. Marchand
Pall Corporation, Westborough, MA, USA

S. Gautam et al. (eds.), *Bioprocess and Analytics Development for Virus-based Advanced Therapeutics and Medicinal Products (ATMPs)*,
https://doi.org/10.1007/978-3-031-28489-2_2

There are three main transfection types:

- Biological: use of virus-mediated transfection, e.g., baculovirus. It is very efficient, but the DNA packaging is limited, and it requires the use of *Spodoptera frugiperda* (Sf9) insect cells.
- Chemical: use of cationic lipid, cationic polymer, calcium phosphate, or liposome. There is no viral particle involved, and a number of commercial products are available in Good Manufacturing Practice (GMP) grade. It may cause cell toxicity and variability on transient transfection processes reducing production consistency. The most used in GMP production is polyethylenimine (PEI).
- Physical: requires direct injection, electroporation, laser, or sonoporation. There is no vector or chemical substance used; however, it may require trained personnel to operate the equipment and can be laborious; some examples are microneedle, gene gun, and phototransfection (Chong et al. 2021).

At manufacturing scale, chemical transfection reagents seem to be one of the preferred choices because of their simplicity and availability of GMP graded products. However, production variability and high cost due to GMP graded material for transient transfection reagents and plasmid are the main challenges and these costs add up onto the final value of the therapy restricting the use. Among the options, PEI is widely used already for the last three decades.

Transfection at large scale in bioreactors can be challenging. Process optimization is needed to ensure high efficiency. Typical transient transfection processes for adherent or suspension mode (function of the target vector to be produced and its characteristics) are described in Sect. 2.5 "Viral Vector Upstream Processing."

Time to market can be crucial for some gene therapy indications, and while research on optimization of new methods for viral vector production is ongoing, manufacturers typically launch clinical trials using the market golden standard transient transfection. In Sect. 2.4 "Cell Culture Systems for Adherent and Suspension Cells," we review different methods of production for different viral vectors, including cell line type and their characteristics.

2.2 Choosing the Right Gene Therapy Vector

There are several different gene therapy vectors currently in use. Table 2.1 summarizes viruses used in gene and cell therapy, the cost, tropism, packaging size, and transgene expression time.

The size of the vector insert, and whether integration or long-lasting effect is needed, dictates which viral vector is used.

The host immune response depends on the vector choice; e.g., adeno-associated virus (AAV) triggers a mild immune response compared to the higher immunogenicity of adenovirus serotype 5 (Ad5).

Another factor to consider is the cargo or gene of interest (GoI) inside the virus. Some constructs may have a specific tissue promoter activated only in the cells

Table 2.1 Common viral gene therapy vectors used in clinical trials and features

	Adenovirus (Ad)	Adeno-associated virus (AAV)	Herpesvirus (HSV)	Retrovirus (RV)	Letivirus (LV)	Vaccinia[a]	Human respiratory syncytial virus (RSV)[a]
Family	Adenoviridae	Parvoviridae	Hepesviridea	Retroviridae	Retroviridae	Poxviridae	Pneumoviridae
Genome	dsDNA	ssDNA	dsDNA	ssRNA	ssRNA	dsDNA	ssRNA (−ve)
Coat	Naked	Naked	Enveloped	Enveloped	Enveloped	Enveloped	Enveloped
Genome	38–39 kb	5 kb	120–200 kb	3–9 kb	3–9 kb	130–280 kb	15 kb
Packaging capacity	7.5 kb	4.5 kb	>30 kb	9 kb	8 kb	25 kb	Approx 5 kb
Tropism	Dividing and non-dividing cells	Dividing and non-dividing cells	Dividing and non-dividing cells	Dividing cells	Dividing and non-dividing cells	Dividing and non-dividing cells	Dividing and non-dividing cells
Genome interaction	Non-integrating	Non-integrating or stable	Non-integrating	Integrating	Integrating	Non-integrating	Non-integrating
Transgene expression	Transient	Long-lasting	Long-lasting	Long-lasting	Long-lasting	Transient	Long-lasting
Immune response	High	Very low	High	Moderate	Low	High	High

[a]Not reached clinical stage yet

containing certain transcriptional factors which can activate protein translation of the cargo or GoI generating tissue or cell specificity expression. The vector tropism for the targeted cells will facilitate its transduction and cargo delivery into the cells.

The increasing knowledge on genetic and molecular biology of the herpes simplex virus (HSV) has enabled the development of potential new gene therapy vectors. Different aspects of HSV biology render this virus attractive: broad host range—infects dividing and nondividing cells, delivery and expression of genes into central nervous system (CNS) cells, transportability along axons, and establishment of latent infections as an episome with prolonged gene expression (Manservigi et al. 2010).

2.3 Cell Lines and Plasmids

One of the biggest challenges facing the gene therapy industry is the need to develop production processes that allow for faster transfer into clinical development phases as well as significantly reducing the overall costs to develop a therapeutic product (Carrondo and Cruz 2010).

The most commonly used vector platforms for in vivo gene therapy applications, including adeno-associated viruses, adenoviruses, and lentiviruses, come with significant production issues, in terms of culture format (adherent vs suspension), product yield and quality, and process scalability, for gene therapy manufacturers to overcome. One way the industry is looking to tackle these issues is through the development of new and improved upstream platforms (e.g., cell lines and plasmids) (Merten et al. 2014).

One of the key considerations in terms of viral vector production is the amount of vector to be produced, which in turn dictates the cell line platform and cell culture systems that are fit for purpose. Some of the main upstream production platforms and processes for viral vectors include mammalian cells infected, transfected packaging, or producer grown in either suspension or adherence.

As summarized in Table 2.2, the production of viral vectors using transient transfection platforms remains expensive, mainly due to issues with scalability and the need for expensive GMP-grade plasmids. While switching over to other production platforms, such as packaging and producer cell lines, may lower the overall cost of goods sold (COGS), the generation of these cell lines takes significantly longer.

For transient transfection-based production processes, utilizing either adherent or suspension HEK293 cells, the transfection is performed by the insertion of multiple plasmids via the use of a transfection agent, such as PEI. By comparison, in infection-based AAV production methods, either adenovirus or baculoviruses are added to the host cell line at a specific multiplicity of infection (MOI) or concentration per cell. Finally, for the generation of stable producer cells, such as suspension HEK293 cell lines, all the genetic elements that are needed to generate full virus particles are stably incorporated into the host cell genome. These different types of processes are described in detail in Sect. 2.5 “Viral Vector Upstream Processing.”

Table 2.2 Overview of the different cell line platforms used for production of most commonly manufactured viral vectors (Gálvez et al. 2012; Tang et al. 2015; Naso et al. 2017; Chen et al. 2020; Lesch et al. 2021; Cashen and Manser 2021)

	AAV: Transient Production	LV: Transient production	AAV: Baculovirus infection of insect cells	AAV: Co-infection of mammalian cells	AAV & LV: Producer cell lines	Ad: Infection of mammalian cells
Host cell line	HEK293 Adherent/ suspension	HEK293 Adherent/ suspension	Sf9 Suspension	BHK Suspension	HEK293 Adherent/ suspension	HEK293 Adherent/ suspension
Production method	Multiple plasmid transfection—transient	Multiple plasmid transfection—transient	Multiple baculovirus infection method	Herpes virus or adenovirus co-infection	Stable integration of genes	Infection
Plasmids/ infection	2 or 3 plasmids: GoI/ITR REP/CAP Helper	4 or 5 plasmids: GoI ENV REV GAG-POL	REP/CAP integrated, GoI/ITR—BEV Helper—BEV	REP/CAP—rHSV1, GoI/ITR—rHSV2, Helper gene—rHSV	Stable integration of: AAV—GoI, REP/CAP, Helper LV—GoI, ENV, REV, GAG-POL	Ad5
Productivity range	10^3–10^5 vg/cell	10^6–10^7 TU/mL	10^2–10^5 vg/cell	10^4–10^5 vg/cell	AAV: 10^4–10^5 vg/cell LV: 10^7 10^8 TU/mL	10^{10}–10^{11} VP/mL
Pros	Quick and easy to make AAV at small scales	Quick and easy to make LV at small scales	Insect cell line safety, efficient large-scale production	No need for stable cell line, efficient large-scale production	Same plasmids can be used, efficient large-scale production	No need for stable cell line, efficient large-scale production
Cons	Triple transfection—low scalability, need for GMP quality plasmids	Four-plasmid transfection—low scalability, need for GMP quality plasmids	Low stability of BEV	Need for helper virus, HSV sensitivity	New stable cell line produced for each project	Higher immune response
Cost	Relatively high COGS	Relatively high COGS	Lower COGS	Lower COGS	Lower COGS	Lower COGS

2.3.1 Adherent vs Suspension Cell Lines

From a process perspective, there are key differences between anchorage-dependent adherent cells and cells that grow in shaking suspension conditions. For routine sub-culture of adherent cell lines, the medium must be removed from the cells and the cell layer washed and then treated enzymatically to detach the cells from the surface. After dilution/inactivation of the enzymes the cells are then plated on a new surface in fresh media. This process is both laborious and time-consuming, whereas suspension cells are simply diluted into fresh culture media in a new vessel.

Typically, adherent cell lines require the inclusion of serum in the culture media for growth and to maintain cell attachment to the vessel surface. Despite the widespread use of serum-containing media, there is a push toward the use of serum-free media, especially in suspension culture where cell attachment is not required. Sourcing large volumes of quality serum is costly and the serum itself is a potential source of adventitious agents. In addition, the removal of serum is advantageous from a downstream processing perspective, as serum proteins are required to be removed from the crude viral preparation.

In terms of future cell line platforms, there is expected to be a shift toward utilizing suspension cell lines that grow in serum-free media for viral vector production. In addition to the biosafety aspects of serum-free culture, suspension cell lines can be more easily scaled up using rocking and stirred-tank bioreactor platforms, reducing manufacturing timelines and costs (Merten 2004).

2.3.2 Transient vs Stable Production

In terms of viral vector manufacturing, several different production strategies are currently being employed. The most used strategy for AAV production is based on the helper-virus-free transient transfection method which uses three plasmids. The GoI is on one plasmid flanked by inverted terminal repeat (ITR) elements. A second plasmid contains the rep/cap genes for the AAV. A third plasmid contains the Ad helper genes. These are all transiently transfected into a host cell line such as HEK293. While this transient multiple-plasmid transfection method is fairly simple in terms of plasmid construction and transfection process, there are significant downsides in terms of process scalability and production costs (Wang et al. 2003; Lock et al. 2010).

To overcome the challenges faced by transient transfection production methods, the development of more "stable" production platforms is becoming more common in the gene therapy (GT) industry. For stable AAV and LV packaging and producer cell lines, some or all of the genetic elements for viral vector production are stably inserted into the HEK293 genome. In general, packaging cell lines have all the required elements to assemble a full viral particle stably integrated (such as the rep-cap and helper genes), apart from the GoI, which is added transiently. By comparison, producer cell lines also have the GoI that is integrated into the host cell genome, thus removing the need for further transfection events (Bulcha et al. 2021).

2.3.3 New Technology Platforms: Example Cell Line Systems

As the GT industry looks more toward using stable production cell lines, some interesting new technology platforms have emerged that have been generated using different cell line engineering strategies. These systems remove the requirement for high-quality plasmid DNA and transfection reagents during the production process and eliminate the potential batch to batch variability that can be introduced by transient transfection. The CEVEC ELEVECTA® Technology Platform, for the large-scale production of AAV vectors, utilizes an inducible promoter system with all the required genetic components stably integrated into the cell (Escandell et al. 2022). Another platform that is underpinned by the use of novel inducible promoters is the AskBio AAV platform. As well as novel promoters, the AskBio platform uses a high-yield suspension cell line to increase AAV productivity (Suarez 2021). Looking at the production of LV vectors, Oxford Biomedica has developed its LentiVector® platform, using various cell and vector engineering strategies. As part of the platform, the LentiStable™ inducible stable packaging and producer cell lines were designed to enable cost-effective and scalable manufacturing processes (Stewart et al. 2011).

In summary, transient transfection (based on multiple-plasmid transfection) is currently the primary viral vector production method that is used in the gene therapy industry. This method is expected to remain dominant for the next 3–5 years particularly for indications that have a smaller patient population and require a lower dosage. Transient transfection also allows therapy developers to get to market quickly. When looking to the future, the industry is expected to shift toward packaging and producer cell lines to enable increased scalability and lower overall production costs, particularly for indications with larger patient populations and higher dosing strategies (Comisel et al. 2021).

2.4 Cell Culture Systems for Adherent and Suspension Cells

There are two options for the choice of cell culture systems for viral vector production, adherent or suspension. Most cell lines used to support viral vectors production are inherently adherent (e.g., HEK293, Vero), and it can take time to adapt existing cell lines to efficiently grow in suspension and reach comparable specific viral vector productivities as in adherent mode. The use of commercially available suspension cell lines can remove this bottleneck but is associated with additional licensing/costs.

Moreover, there are still challenges associated with transient transfection of cells at large scale using the current gold standard method for viral vector production (Merten et al. 2014), and process development is usually required to optimize its efficiency and practicality at larger scales. Thus, it is important to understand the advantages and disadvantages of the use of adherent and suspension cultures for a particular viral vector product and cell line, and the required process development needed to make the manufacturing process viable. It is important to start with the understanding of the targeted indications' market and required speed to market, so patients have access to the therapy. Choices around the right trade-off between speed to market and time to develop the process can be considered together with a potential switch to a suspension process. Today several viral vector manufacturing processes use transient transfection of adherent cells, as this provides a fast track to market. For indications that require a treatment with high yield of viral vector and have a longer acceptable time for introduction in the market, cell line adaptation to suspension or development of a stable producer cell line can have benefits in the longer term. Manufacturers can choose to enter the market using a transient transfection adherent cell process while concomitantly developing, characterizing, and validating a stable producer suspension cell line that could be used in a second-generation process.

When it comes to viral vectors produced using adherent cell culture, there are currently three main single-use technologies to consider: flatware [roller bottles (RB), multi-trays (MT)], fixed-bed bioreactors (Pall's iCELLis® bioreactor system, Univercells Scale-X™, Corning Ascent™), and microcarriers in suspension bioreactors [rocking platform, stirred-tank reactor (STR)] (Lesch et al. 2021). The major advantages of bioreactors over flatware are their automation capabilities leading to process control and their capabilities associated with scale-up. For adherent cells, the surface area on which cells grow limits further expansion when the culture reaches a confluent state. Thus, manufacturing scale would rely on either multiplying the number of culture units (scale-out) or using successively larger units (scale-up). The scale-out approach leads to an increase in the number of manual operations and associated contamination risks, as well as increased labor costs.

Moreover, footprint requirements are not to be disregarded with an important cost associated with cleanroom and equipment, i.e., incubators and biosafety cabinets. Thus, scale-up using bioreactors came to address important challenges linked to gene therapy products production scale. Viral vector manufacturers have been using flatware for process development and small-scale needs. Multi-trays are often used and can provide up to 2.5 m^2 surface area per unit, while roller bottles provide up to 0.17 m^2. These last ones can be applied with some degree of automation to address challenges in gas exchange. More recently, HYPERstacks® (Corning) with reduced footprint and membranes for gas exchange were developed, providing up to 1.8 m^2 per unit. However, all these systems normally require a scale-out strategy for manufacturing scale.

The use of bioreactors can streamline viral vector processes, increasing production scale, allowing for fewer operations, lower contamination risks. and reduced facility space, and culminating normally in lower costs. The cost benefits of the iCELLis® fixed-bed bioreactor system over multi-trays flatware have been shown in a cost modeling for an AAV study that estimates a 40–50% cost of goods reduction for 1000 L scale-out MT processes (Cameau et al. 2019). Bioreactors also allow process monitoring and control which can lead not only to process robustness but to improved productivities and easier traceability which can comply with regulatory burdens. Furthermore, the perfusion capabilities of fixed-bed bioreactors can be suitable for shear and thermosensitive viral vectors (such as lentivirus), an additional benefit over STRs where a biomass retention device is required. The iCELLis® fixed-bed bioreactor system has gained some attention in the market of gene therapy as it is used in the manufacturing of Zolgensma®, an AAV viral vector gene product used to target spinal muscular atrophy. Central to the iCELLis® bioreactor technology is the use of a compact fixed bed, filled with proprietary macrocarriers. This matrix is made of polyethylene terephthalate (PET) macrocarriers and provides a surface area for cell adhesion and cell growth up to 500 m^2. This is the surface equivalent of 3000 RB (0.17 m^2 each) or 780 ten-layer MT. The range of scales, from small to industrial scale, of the iCELLis® fixed-bed bioreactor technology can allow for fast viral vector development to industrial scale production and easy tech transfer of existing flatware processes. The largest industrial scale, 500 m^2, is comparable with a 500–1000 L suspension stirred-tank reactor (STR). The same hardware and consumable can be used with different fixed-bed height allowing operation at 66–500 m^2. Thus, the iCELLis® fixed-bed bioreactor can be a suitable technology for manufacturers looking into flexible production scales for viral vectors.

Moving from fixed-bed bioreactors into rocking or STR bioreactors with microcarriers for anchorage of adherent cell cultures is a potential option for viral vector manufacturing, as it combines the advantages of adherent and suspension cultures, enabling the suspension culture of cells in adherent state. However, microcarrier culture has many challenges like finding the optimal stirring and aeration conditions to provide cells with oxygen, while keeping shear stress in acceptable range, or the implementation of media exchange strategies while keeping cell microcarriers in the bioreactor (Gregoriades et al. 2000). Process development is time-consuming, which explains microcarriers' very low implementation in GT commercial scale compared to vaccine manufacturing. Time is probably better used to adapt cells directly into suspension. Moreover, transfection of cells still needs to be better understood and characterized. These reasons contributed to this suspension-like culture mode not being the mode of choice when it comes to viral vector manufacturing.

In recent years, there has been a drive to move from adherent to suspension viral vector production, due to biosafety aspects, i.e., removal of serum and the well-defined scalability of suspension processes. There are two main categories to produce viral vectors in suspension bioreactors: rocker/shaking-type and STRs. Rocking and shaking-type bioreactors involve a moving platform to generate mixing and mass transfer, these systems have no direct gas sparging, and gas transfer occurs via the headspace instead. The lack of gas sparging can be seen as an advantage if damage from bubble bursting is a concern; however, rocking/shaker-type bioreactors have limited scale and scalability, with a maximum volume of 200 L (Merten et al.

2014). Single-use STRs offer greater flexibility with scale, with standard volumes up to 2000 L and even as high as 6000 L, and scale-up through greater control of mixing and mass transfer through agitation and gas sparging (ABEC 2019). STRs are well characterized due to them being extensively used for decades and are considered the gold standard for commercial production of mAbs and recombinant proteins.

The advantage of suspension bioreactors over adherent includes the increased scale, simpler seed train, and the ability to easily measure cell density. Compared to adherent bioreactors, suspension processes tend to operate at smaller volumes for the same total cell density. Due to the nature of adherent processes, and the need for a short time to market, little development is often put into media types and volumes used, and there is often a need for serum containing processes to aid with cell attachment. This results in unoptimized media volumes and less concentrated media types versus suspension and increased risk of adventitious agents and purification burden. For example, for a typical suspension process, transfection up to 2×10^6 cells/mL in a 2000 L STR of chemically defined media compared to a 2000 m^2 process that produces the same total number of cells would use between 3200 and 4400 L of serum-containing Dulbecco's Modified Eagle Medium (DMEM) (media to surface area ratio of 0.16 to 0.22 mL/cm^2). In addition to typically larger, unoptimized volumes, a larger footprint is required, either for the incubators and class II biological cabinets or, in the case of a fixed-bed bioreactor, the bioreactor and the media tanks. The seed train for suspension bioreactors is simpler, as it does not require any cell detachment and allows us to culture cells to a higher cell density in the same volume compared to adherent cultures. For adherent bioreactors, sampling to measure the cell density is more difficult, requiring either an additional vessel being set up to sacrifice for cell counting or the use of methods which are not representative or can lead to errors, such as macrocarrier sampling or inference through online and at-line variables such as glucose concentration or capacitance, whereas a suspension bioreactor can be directly sampled for a cell count.

The drawback of suspension bioreactors over adherent, specifically for viral vectors, includes the difficulty of performing media exchanges while retaining cells. If media exchanges or removal of temperature- and shear-sensitive products is required for suspension processes, e.g., lentivirus where the removal from the bioreactor environment can be beneficial (Yoganathan et al. 2020), then, additional large bore ports, additional perfusion devices and pumps, and more advanced process monitoring and control are required, which can complicate the process (Bielser et al. 2018). For adherent processes, media exchanges and perfusion to remove product from the bioreactor can easily be achieved as the cells are held in place on the growth surface. The need for media exchange in suspension can be overcome through media selection/design and concentrated nutrient or medium feeds. Currently for viral vector processes, complex media additions are not required, and a base media alone is sufficient. However, for some processes such as lentivirus production perfusion may be required.

Unlike adherent systems, STRs require direct gas sparging for gas transfer; however, the gas rates required for viral vector production do not approach levels that may be considered a concern (Chaudhary et al. 2020). Despite these low gas rates, foam production can be a potential issue. The effect of additives such as antifoam on transfection needs to be assessed before use, and therefore, the control of foaming may need to be achieved via selecting the right combination of agitation and gassing strategy. A potential drawback to suspension bioreactors is that often early, small-scale viral vector development work is carried out in adherent mode. However, the switch to suspension cells or starting the development work in suspension cells can be relatively straightforward. Another potential challenge with suspension bioreactors is that at the larger volumes the addition of transfection complexes can become problematic. This is because the timescale for transfection complex mixing and addition does not change with scale, requiring large flow rates to add large volumes over the same time as for the small and large scale. For example, at the 2000 L scale, 200 L for complexes may be needed to be added to the bioreactor within 5–15 min (Pall Corporation and Polyplus 2020). However, this challenge also exists with adherent when considering a similar total cell density; it is only that this total cell density is more easily achieved in suspension. At the larger volumes, e.g., 2000 L, gravity addition is not practical, due to the need to lift large volumes, such as 200 L, above the top of the bioreactor height and the need for multiple or large tubing bore connections, and therefore, pumping is required. Adjustments may be required to complexation times when pumping into the bioreactor to account for the transfer time, i.e., a reduction in the hold-step to account for the additional time to pump (Pezzoli et al. 2017), and avoiding the step down of the internal diameter (ID) of the tubing to avoid creating turbulence, which may over mix the complexes and therefore cause the complexes to aggregate and become too large to be taken up by cells efficiently (González-Domínguez et al. 2019).

Adherent bioreactors might have benefits over suspension bioreactors regarding the harvest step. With adherent bioreactors, the burden on clarification can be reduced as the cell debris is usually retained in the fixed bed which acts as a filter. The benefits are particularly noticeable when the product is extracellular (as for lentivirus) and the cell culture does not require a chemical lysis step, as the cells are retained on the growth surface (see Sect. 2.5.3 "Viral Vector Harvest").

Finally, process innovation is actively focusing not only on bioreactor optimization, but on cell culture media formulation (by the implementation of chemically defined media and elimination of animal serum media, source of adventitious contamination), cell lines that can be grown at higher densities, optimized set of plasmids, and transfection reagents that can support viral vector production.

2.5 Viral Vector Upstream Processing

As previously described, the production of GT vectors can be accomplished in a variety of approaches such as development of stable vector producer cell lines, transient transfection of human cells, or infection-based systems. Consequently, their production process approach varies dependent on the nature of the platform, i.e., constitutive, induced, or transient vector production, adherent or suspension cell culture, use of serum-free or serum-containing media, and the scale limitation. As these viral vectors are derived from viruses that naturally infect human cells or other mammalian cells, their production requires culturing of mammalian cells or, for some systems, insect cells in cell culture bioreactor systems.

This section describes the process map of various vector production methods (AAV, LV, Ad, etc.) focusing first on the adherent platform and then on the suspension platform. Standard general recommendations related to the technology are shared, while each GT process has specific critical process parameters impacting critical process attributes.

2.5.1 Viral Vector Production in Adherent Platform

Multiple fixed-bed bioreactors are now available on the market for commercial production of viral vector as already mentioned earlier (see Sect. 2.4 "Cell Culture Systems for Adherent and Suspension Cells"). Moreover, Lesch et al. have reviewed all the options available on the market and summarize yields and titers for the most common GT processes using these technologies (Lesch et al. 2021). While fixed-bed bioreactor models differ on design, in all of them, adherent mammalian cells attach to non-woven microfibers. These microfibers are packed into several centimeters high fixed beds with optimal geometry and porosity. Cell culture media circulates through the fixed bed. Critical parameters are all expressed in relation to the fixed-bed surface:

- The cell concentration expressed in cells/cm^2 at the time of inoculation and time of transfection, infection, or induction
- The media volume to surface ratio expressed in mL/cm^2
- The media flow rate through the fixed bed to homogeneously distribute cells in the fixed bed after inoculation and provide them with metabolites (O_2, glucose, etc.) while keeping low level of shear stress.

The process in a fixed-bed bioreactor can be split into five steps (Fig. 2.1):

- Cell inoculation and attachment phase
- Cell growth until reaching the optimal cell concentration (cells/cm^2)
- Transient transfection, infection, or induction
- Viral vector production
- Harvest

The total process duration will vary depending on the cell seeding density and the optimal cell concentration for starting the production phase. In general, the duration is 7 days for transient transfection and for infection. It may be longer for the induction of producer cell lines. Powers et al. reported a process using a stable packaging HEK 293T cell line to produce GPRTG-EF1α-hγ-OPT LV that lasts

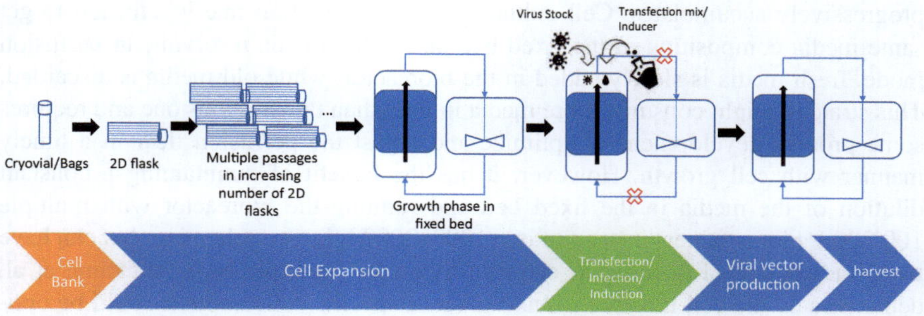

Fig. 2.1 Typical upstream adherent viral vector production process flow

14 days, with 5 days of cell growth and 9 days of production by inducing cells with doxycycline media-free (Powers et al. 2020).

It has been reported that compared to 2D flatware systems, the 3D structure of the fixed bed allows cell inoculation in the range 5×10^3 to 1×10^4 cells/cm^2 without experiencing a long lag phase (Lesch et al. 2015; Valkama et al. 2018). Despite this low cell concentration, a minimum of 10^{10} cells are still required to inoculate one commercial fixed-bed bioreactor (~500 m^2). Mammalian cells, like HEK293, are first expanded in MTs, inoculated at a range typically 1–3 × 10^4 cells/cm^2 and harvested 3–4 days later in a range of 1×10^5 to 3×10^5 cells/cm^2. The seed train lasts approximately 2–4 weeks.

Aiming toward a uniform cell distribution in a fixed-bed bioreactor is key to increase process performance. One of the most important factors to achieve this is to have a good cell inoculation step. Key points of attention at this step are:

- Harvesting cells from MT while they are in exponential phase, minimizing time from harvest to inoculation or cell transfer to maintain high viability, and paying attention to the harvest process to get single cell suspension and avoid cell aggregation.
- The selection of the appropriate cell culture media circulation speed (linear speed) through the fixed bed is also a key parameter to push cells through it and get uniform distribution.
- After becoming trapped in the fixed bed, cells require a few hours to effectively attach and spread on the non-woven fibers.

The growth phase typically lasts between 3 and 5 days to reach the optimal cell density, depending on the starting cell density and the optimal cell density for transfection, infection, or induction. Multiple strategies can be applied to maintain cells in a good environment: operating in batch mode or perfusion. In the first case, a large reservoir containing a total volume corresponding to 0.15–0.2 mL/cm^2 is connected to the fixed-bed bioreactor, and this volume is recirculated multiple times per day. Nutrients are progressively depleted while by-products are accumulating. For instance, glucose concentration decreases over time, while lactate progressively accumulates. Cell culture media recirculation rate is adjusted to get same media composition in the fixed bed and in the media reservoir. In perfusion mode, fresh media is slowly added in the bioreactor, while old media is discarded. This strategy might consume more media in total than the previous one and requires some process development to optimize and adjust the perfusion flow in a timely manner with cell growth. However, it has the benefit of maintaining a constant dilution of the media in the fixed bed and running the bioreactor with multiple 100/200 L biocontainers instead of one 1000–2000 L large-scale mixer but can have the disadvantage of needing to connect/disconnect multiple bags. Valkama et al. demonstrated the cell metabolism and glucose-specific consumption could be optimized by fine-tuning the level of glucose in the fixed-bed bioreactor via perfusion; however, the impact on productivity was minor (Valkama et al. 2018).

The next step is transient transfection, infection, or induction depending on the GT process.

The transfection step consists of adding multiple plasmids (usually three or four) to the bioreactor. To enter the cells, they are commonly complexed with PEI as this chemical approach is easily scalable in commercial fixed-bed technology, though some process-specific challenges exist (see Sect. 2.4 "Cell Culture Systems for Adherent and Suspension Cells"). Transient transfection with PEI is currently the gold standard method to produce AAV and LV vectors in HEK293 cells (Leinonen et al. 2019; Nass et al. 2019; Lesch et al. 2021). Formation of PEI-DNA complexes is usually optimized in static conditions. However, while cells are attached to a fixed support in the fixed-bed technology, the environment is dynamic with media constantly circulating through it. Therefore, optimal processes in 2D flatware might not give the equivalent viral vector titer in fixed-bed reactors. In fixed-bed bioreactors, cells can create 3D aggregates; cells in the center of this aggregate may not be transfected. Better transfection efficiency seems to be obtained at a lower cell concentration (1–2 × 10^5 cells/cm^2) compared to 2D flatware when moving to iCELLis® fixed-bed bioreactor (Valkama et al. 2018). The quantity of DNA per cell can be revisited, most of the time with a lower quantity (Powers et al. 2016) to the benefit of the overall process cost.

The PEI/DNA ratio, plasmid ratio, PEI and DNA concentration, media composition to form the complex, complex formation time, method to add the complex in the bioreactor, and internal media recirculation in the fixed bed are critical parameters to optimize for fixed-bed bioreactors. Lab-scale fixed-bed bioreactors allow this process optimization to be carried out with more acceptable costs. The application note published by Pall and Polyplus is a reference document covering this topic (Pall Corporation and Polyplus 2020). The transfection mix is added in the bioreactor by gravity or using a peristaltic pump. For some PEI reagents, time is a crucial parameter as it impacts the complex size and capabilities to enter the cells. This constraint must be managed while scaling up the process in large-scale fixed-bed bioreactors. Indeed, 10–20 L or more of transfection mix volume must be transferred into the bioreactor. During the transfection step, the media recirculation with the external mixer is usually stopped to avoid some of the transfection mix escaping from the fixed-bed bioreactor and being wasted. The media volume to surface area ratio is thus lower for a few hours, resulting in a quick drop of the glucose concentration and accumulation of some by-product. Stirring conditions may be adjusted at this time, to achieve a balance between promoting contact between cells and the transfection mix while not aggregating the transfection particles resulting in the formation of larger particles, which cannot enter the cells anymore. Once the step is over, the recirculation is restarted.

The infection step consists of adding a virus with a specific MOI in the bioreactor. Runs in bench-scale bioreactors allow for the optimization of the process and cover similar considerations as transfection regarding optimal cell concentration, cell growth phase, and internal media recirculation in the fixed bed. The media recirculation with the external mixer is also stopped during this step. However, compared to transfection while tens of liters are usually added into the bioreactor, the virus stock is concentrated, requiring the addition of only few liters. As for transfection processes, this is easily performed by gravity transfer or using a peristaltic pump and is used for adenovirus production in HEK293 (Lesch et al. 2015) and HSV production in Vero cells (Knop and Harrell 2007). As outlined previously, some producer cell lines are induced to produce the viral vector by adding or removing a component in the culture media or switching to another production media. Powers et al. illustrated the induction of a LV production process by replacing the DMEM media with media without doxycycline (Powers et al. 2020). While media exchange is easily performed in bench-scale fixed-bed bioreactors, another strategy needs to be considered at clinical/commercial scale. This is discussed with a broader scope in the next paragraph. Some producer cell lines do not need an induction step. This is illustrated by Wang et al. for a γ retroviral vector produced with PG13 or HEK 293 VEC packaging cells (Wang et al. 2015). The decision to initiate the production phase is thus based on cost analysis and some regulatory questions, as discussed in the next paragraph.

A few hours after transfection, infection, or induction step, cells start to produce the viral vector. Many growth media still contain animal origin components like FBS to boost cell growth, while regulatory affairs request proof of absence of any animal origin components in the final product. An advantage of fixed-bed technologies is the possibility to easily perform a full media exchange. This can be done before or after the transfection, infection, or induction step. When performed before, it has the advantage to bring back glucose level to a high value and dilute any by-products before the external reservoir is disconnected. This puts cells in optimal conditions and can be beneficial for the transfection/infection cell efficiency for some processes. When performed after, it has the benefit of eliminating toxic components or residual DNA from the transfection step, improving conditions for the production step, or easing the downstream processing (DSP). Draining and refilling the bioreactor with fresh media is time-consuming and puts cells in non-optimal conditions (pH, DO, glucose, etc.). A fast perfusion is a better approach and consists of flushing the bioreactor vessel with fresh media until the FBS concentration in the fixed bed or any other component of interest reaches zero or an acceptable low value.

For most viral vector processes, cells reach maximum viral vector production 2–4 days after transfection or infection and prolonging the production step over that time might not be cost-efficient. Optimal harvest time is another parameter that should be defined during process optimization at bench scale. The media strategy during the production phase depends on the stability of the vector in the bioreactor conditions (shear stress, temperature) and on the nature of the viral vector, either secreted by the cells or accumulating in them, requiring lysing the cells at the harvest. The media volume to surface area ratio is increased back to a range of 0.1–0.2 mL/cm^2 during the production phase by implementing an external media recirculation with a large-scale reservoir or by perfusion. Advanced discussion regarding the best media strategy and recommendations for harvest is available in Sect. 2.5.3 "Viral Vector Harvest" of this chapter.

2.5.2 Viral Vector Production in Suspension Platform

Though there are appreciable differences in mammalian cell lines and viral vector systems for gene therapy applications (as described in Sect. 2.3 "Cell Lines and Plasmids"), the basic steps of suspension upstream manufacturing essentially remain the same requiring some process optimization around the way the vector is being produced, i.e., transient transfection, infection, and induction/constitutive.

As mentioned previously in Sect. 2.4 "Cell Culture Systems for Adherent and Suspension Cells," dynamic systems such as shake flasks, rocking motion, and stirred-tank bioreactors can be used to cultivate suspension adapted cells. Small-scale cultures are typically performed in shake flasks usually for initial process development studies or used for the amplification of cell seed for initiation of large-scale suspension cultures. Typically, the seed train for a suspension manufacturing process would include two main steps, the initial cell expansion in shake flasks and subsequent expansion in rocking motion or stirred-tank bioreactors. Passive control of the temperature, pH, and dissolved oxygen (DO) is achieved by keeping shake flasks on a shaking platform in an incubator at 37 °C and in equilibrium with a gas mix of 5–10% CO_2 in air.

The final production can be achieved in single-use rocking motion bioreactors available as large as 200 L or stirred-tank bioreactors available in stainless steel or single use at 2000 L and more (Merten et al. 2014). For transfection-based systems, the final production volume for suspension cultures needs careful consideration as the efficient transfection of large amounts of cells is considered a bottleneck in clinical manufacturing that is susceptible to variation. Moreover, scaling up to larger volumes of transient transfection suspension cultures is limited due to considerable costs related to the preparation of GMP or high-quality grade plasmids as the number of cells to be transfected is higher. For infection-based systems, the MOI and the time of infection would be critical parameters affecting the vector productivity. For inducer-based systems, time of induction would be critical to avoid early cell growth inhibition due to cytotoxic proteins.

A generic process flow for suspension viral vector production is illustrated in Fig. 2.2.

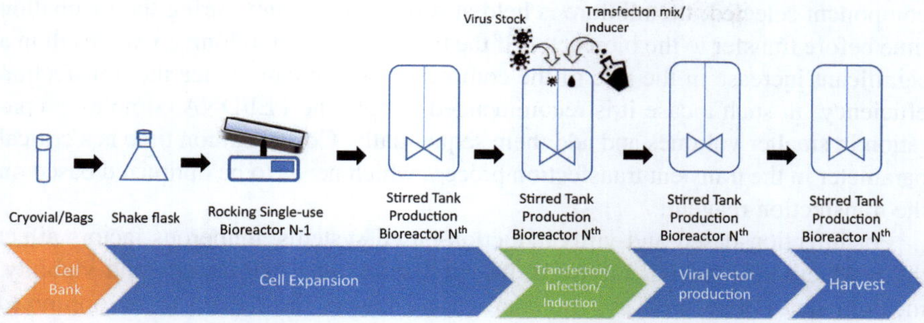

Fig. 2.2 Typical upstream suspension viral vector production process flow

The final upstream production stage (N^{th}) for viral vectors can be split into four main steps:

- Cell inoculation
- Cell expansion/growth
- Transient transfection/infection/induction
- Viral vector production
- Lysis/harvest

The growth phase begins with the inoculation of cells. Alternately, the last seed train step can be performed in the production bioreactor at a lower starting volume, and subsequently, once the cells have reached the desired concentration, fresh media is added to dilute the culture to the final volume. The growth phase is usually run as a batch process. A fed-batch strategy can be an alternative to maintain the viability for production, but this requires some process optimization. Perfusion strategies are also an alternative, but are more complex methods for viral vector production, especially in case of transient transfected based retroviral vectors and stable producer cell line (Merten 2004). Perfusion can enable media exchange post-transfection and avoid vector inactivation due to an extended residence time.

Transient transfection is usually done when the cells are actively dividing, and the cell concentration is around 2×10^6 cells/mL (Durocher et al. 2007). Large-scale production until now has only been economically feasible when using either calcium phosphate ($CaPO_4$) precipitation or PEI-mediated transfection. Since calcium phosphate has two major disadvantages, the requirement of serum or albumin to reduce $CaPO_4$ cytotoxicity and the need to perform a complete medium exchange prior to transfection (Pham et al. 2006), PEI becomes a more suitable transfection reagent. The optimal volume of the transfection mix should be between 5% and 10% of the bioreactor volume to balance PEI and DNA solubility and large volume handling. To prepare this transfection mix, the first step is to dilute in two separate containers, containing half of the total volume, the plasmid DNA and PEI component, in a transfection compatible media. The complexation can be done by the transfer of the media containing PEI into the DNA by gravity or pumping. Depending on the PEI component selected, the mixture is held at room temperature during the incubation time before transfer to the bioreactor. If the transfer time is too long it may result in a significant increase in the size of the complexes, which can reduce the transfection efficiency; in such a case it is recommended to split the PEI/DNA complex preparation in smaller volumes and add them sequentially. Complexation time is a critical parameter in the transient transfection process which needs to be optimized based on the transfection reagent.

For infection-based and virus induction-based systems, numerous factors affect the productivities such as the MOI, the viral stock titer, cell density and viability, medium renewal as well as temperature, and hence, these need to be optimized during process development. In contrast to transient transfection, viral stocks can be added directly to the production bioreactor without the same time and volume-based constraints, based on the targeted MOI. The productivity of non-perfusion adenovirus processes may be limited by the "cell density effect." A decrease in cell-specific productivity has been reported to drop at infection cell densities exceeding 2×10^6 cells/mL; hence, optimizing the right infection cell densities is critical for a set media and feed regime (Kamen and Henry 2004). The production of adenoviral vectors and retroviral vectors has been reviewed respectively by Nadeau et al. and Merten (Nadeau and Kamen 2003; Merten 2004).

In the case of inducible stable producer cell lines, such as CAP cells of the ELEVECTA® (CEVEC) AAV production platform, the production of viral vectors is enabled under the control of a doxycycline-inducible promoter and abolishes the need of transient transfection or helper virus infection (Mainwaring et al. 2021).

The production phase starts after the transfection, infection, or induction step but would not be distinctive in the case of constitutive stable producers. The viral vector production typically peaks within 48 h post-transfection and declines from 72 h. In some transient transfection processes, endonucleases like Benzonase™ or DENARASE®(c-LEcta) may be added 24 h after transfection to digest the remaining plasmid and host cell DNA in suspension. In addition, operating conditions like pH, DO, and stirring setpoint can be varied between the cell culture growth phase and the production phase to promote viral vector production.

An approach to increase yields in suspension cultures is the use of the invertebrate Sf9 insect cells using the baculovirus (BV) system. An example of clinical manufacturing of AAV is the AAV-1-based drug Glybera® (uniQure). The method originally configured by Urabe (Urabe et al. 2002) involves suspension cultivation of Sf9 cells and infection with three recombinant baculoviruses (a Rep-baculovirus, a VP-baculovirus, and an AAV ITR vector genome baculovirus). Simplified two-BV system by combining the rep and cap expression cassettes and single BV infection systems with stable cell lines having integrated copies of AAV rep and cap have also been reported (Aslanidi et al. 2009; Smith et al. 2009). The advantages of this system are a lower cost and complexity of producing vector compared to transiently transfected mammalian cells. Production usually involves generating the infectious virus stocks of respective baculovirus vectors and generating a large volume of producer cells to be infected for AAV production. The use of recombinant baculovirus seed stocks presents manufacturing challenges, particularly with scale-up, i.e., time-consuming titration of recombinant baculovirus stocks by plaque assay, separation of baculovirus from the AAV during the DSP, and loss of infectivity upon prolonged storage in serum-free medium.

2.5.3 *Viral Vector Harvest*

Harvest is a crucial step of the upstream process, and its optimization should ensure the recovery of viral vector particles produced is maximized to be further processed during the downstream purification.

The harvest process is, in part, determined by the cell line attributes, the cell culture method (suspension or adherent), and the viral vector characteristics. Some enveloped viral vectors, as lentivirus, are rather unstable, and require a production mode strategy to tackle this. Others are non-lytic, as AAV, and require the addition of a specific chemical lysis buffer to disrupt the cell membrane and release the vector. This adds additional steps to the upstream harvest step and challenges to the subsequent downstream purification steps.

2.5.3.1 Viral Vector Characteristics and Harvest

The different harvest strategies available to achieve optimal harvest are primarily influenced by the type of viral vector. Non-lytic viral vectors which remain inside the producer cell (intracellular) require cell membrane disruption prior to clarification. Others are lytic and can be harvested directly from the culture supernatant (extracellular), which eliminates complexity to both upstream and downstream processes (lower host cell-related impurities). In other cases, viral vectors might be lytic and prone to non-specific binding to the cell or cellular debris. These viral vectors require additional harvest strategies to release them prior to clarification. Table 2.3 summarizes the main characteristics of typical viral vectors and their production.

2.5.3.2 Harvest Window

Understanding the optimal time to harvest the viral vector is key to process optimization to maximize product yields. Harvesting too early may result in decreased productivity and harvesting too late may result in low cell viability and associated protease release into the supernatant that may impact vector yield and/or quality. Understanding viral production kinetics is thus crucial.

When the optimal harvest window is identified, controlled cell lysis can be achieved (for intracellular vectors) followed by clarification and further downstream purification. This should help tackle yield losses associated with vector degradation.

Table 2.3 Characteristics of common viral vectors (Pall Corporation 2022)

Virus family	Common subtypes	Enveloped	Type	Typical distribution	Typical cell line	Cultivation mode	Lysis recommended
Ad	Ad5	No	Non-Lytic	Intra	HEK293	Adherent/suspension	Yes
AAV	AAV2 AAV5 AAV8 AAV9	No	Non-Lytic	Intra/extra	HEK293	Adherent/suspension	Yes Yes No No
LV (retrovirus)	n/a	Yes	Lytic	Extra	HEK293 (ex-vivo)	Adherent/suspension	No
Other	HSV-1 HSV-2 Vaccinia	Yes	Budding	Intra/extra/sticky	Vero	Adherent	Maybe

2.5.3.3 Non-lytic Viral Vector Harvest

Non-lytic viral vectors can either be extracellular, like some excreted AAV serotypes (AAV8, AAV9) or intracellular like most AAV serotypes used today in the market (AAV2, AAV5) (Vandenberghe et al. 2010). When extracellular, as mentioned before, the vector can be harvested directly from the supernatant. It might also be that the vector is partially extracellular and partially intracellular. In these cases, estimating the proportion of each fraction and assessing the benefits of the yield increase after mediated cell lysis is necessary as this extra step comes with extra costs (additional time in GMP room, labor, and DSP burdens to remove the lysis reagents).

The harvest of intracellular non-lytic viral vectors presents extra challenges as it requires cell membrane-controlled lysis to release the vector. The chosen lysis method should not only allow for effective cell membrane disruption but maintain the viral vector quality (functionality), be scalable, and minimize impact on downstream processing.

From a biochemical point of view there are several cell lysis strategies possible; nevertheless, only a few are viable at manufacturing scale.

The cell lysis strategies most suitable to manufacturing scale are a pH shift, osmotic shock, or a detergent cell membrane digestion. Different strategies can also be conjugated. Process development using a Design-of-Experiment (DOE)-based approach is recommended to identify the best method and associated parameters (duration, concentrations, mixing).

Some viral vectors can be prone to aggregation especially in the presence of host cell nucleic acids because of the mediated cell lysis. This phenomenon can impact vector yield; thus, addition of certain salts in the harvest bulk is commonly used to help mitigate this phenomenon (NaCl or $MgSO_4$) as well as the use of a nuclease. Table 2.4 here below provides a summary of the different lysis methods available.

2.5.3.4 Lytic Viral Vectors' Harvest

When lytic viral vectors bud out of cells, they "coat" themselves with glycoproteins present in the membrane of the producer cell line. This "envelope," present in viral vectors such as lentiviruses, makes them particularly sensitive to physical parameters such as temperature or shear. Thus, strategies to minimize their degradation are required to maximize their functional titer. One of the most common strategies is the implementation of a perfusion mode during the production phase of the vector. This will allow continuous harvest of the viral vector from the bioreactor, minimizing retention times of the vector in the bioreactor and associated exposure to high temperatures and shear caused by the impeller/bubble damage. The vector can be cooled down immediately as it is harvested in the harvest tank at 4 °C.

Table 2.4 Different cell lysis strategies (Brown and Audet 2008; Pall Corporation 2022)

Lysis method	Typical range	Advantages (+)/disadvantages (−)	Comments
pH shift or alkaline lysis	pH 11.5–12.5	+: Suitable for all kinds of cells −: Slow process (6–12 h) −: High pH can damage vector integrity	Mostly used for isolating plasmid DNA Not widely used for viral vectors
High pressure homogenization	n/a	+: Avoids addition of chemicals that need to be removed by downstream process −: Only works on suspension or detached cells. Cannot be used on fixed-bed bioreactors	Not very common for viral vectors
Hyperosmotic shock	$(NH_4)_2SO_4$ 0.5–1.5 M	+: Works well −: Can be damaging to salt-sensitive vectors +: Easy to implement in suspension and adherent processes	
Hypoosmotic shock	~No salt (ex: WFI)	+: Works very well +: Low cost −: Hard to be implemented in suspension	Some salt might need to be added after cell lysis to prevent vector aggregation
Detergents	Tween 20 (0.1–1%)	+: Very efficient −: Needs to be removed during downstream processing	This method is commonly used in combination with osmotic shock or pH shift
Enzyme lysis (nuclease)	0.5–2 h	+: Reduces RNA/DNA quantity increasing capacity on clarification filters and/or capture chromatography −: Removal of nuclease through the downstream process must be proven, validated, and tested with the proper analytical methods	Usually performed in addition to the selected lysis method Bioreactor pH must be adjusted to ~8.0–9.0 and $MgCl_2$ added (1–2 mM) to promote nuclease efficiency

If perfusion is used, the media in flow rate should be optimized at small scale during process development. This should take into consideration the harvest window and the volume to be processed in the downstream.

For lentivirus transfection processes using the iCELLis® bioreactor, it is common to perfuse media over the last 48–72 h post-transfection, at a reduced flow rate, to minimize the overall volume to be processed in the downstream. Perfusion has been proven to increase significantly LV yields (Manceur et al. 2017), probably due to the advantages highlighted above, and possibly preventing reinfection of the HEK293 cells.

For suspension cell cultures, the harvest of lytic sensitive vectors can also be done by perfusion using alternating flow tangential technology or hollow fibers, even if most processes today use batch-mode cultures or partial harvests. Partial harvest methods consist of removing a fraction of the cell culture supernatant and adding new fresh media to the culture, at a specific time after the transfection (or induction step), prior to full culture harvest. As for the perfusion, the partial harvest is then stored at cold temperatures or immediately processed for downstream depending on viral vector half-life data analysis at the storage temperature. This results in an increased volume to be treated in the downstream compared to the batch approach so assessment of costs versus benefits should be considered.

Whatever approach is chosen, a nuclease step is commonly put in place to digest host cell nucleic acids and potential remaining pDNA used for transfection. This helps mitigate burdens on the DSP.

2.6 Clarification of Viral Vectors

In any cell-based bioprocess clarifying the product from the cells and large cellular debris will be a critical step bridging production and purification. The first basic requirement of the step is to remove insoluble particles down to a level that will prevent fouling of subsequent unit operations (often including chromatography) as well as to remove any bioburden introduced during processing. The second basic requirement of the clarification step is to achieve a high yield of the product and to maintain product quality. Beyond these basic requirements that will apply to all processes, there are several others that are more process specific such as facility footprint, process economics, and process-related impurity removal.

Many of the harvest technologies used in recombinant protein bioprocessing (Liu et al. 2010) have translated well to gene therapy processing. However, there are some notable differences for clarifying gene therapy products which will be highlighted below. The first key factor in determining the best technology for harvest is the bioreactor mode of operation. In perfusion cell culture, the product can be continuously clarified from the bioreactor as it is produced. This mode can have yield and product quality benefits, particularly for products that show poor stability at the elevated temperatures in a bioreactor or are toxic to the host cells at higher concentrations. To continuously harvest from a bioreactor the most commonly used technology is a hollow fiber filter connected to a pump that allows for either alternating tangential flow filtration (ATF) or tangential flow filtration (TFF) (Wang et al. 2017; Nikolay et al. 2020). These filters have pore sizes in between the product size and the cell size allowing the cells to be retained in the bioreactor, while the product passes through the filter and is collected. Perfusion harvest is ideal for stable producer cell lines which can continuously express and secrete the viral vector of interest. However, there are examples of short-duration perfusion harvests using transiently transfected cultures. Most gene therapy processes today are produced in batch or fed-batch cell culture, for which there are several harvest technologies available. Centrifuges separate product from cells based on a difference in density and are widely used in protein processing, particularly at large scales.

Ultracentrifuges have proven effective at development scales and can even separate out empty viral capsids, a particularly challenging impurity (Qu et al. 2015). However, neither have been widely adopted in gene therapy processing likely due to a combination of factors including limited scale-down options for process development (centrifuges), limited scale-up options for manufacturing (ultracentrifuges), limited single-use options, and some concern for shear exposure impacting product quality. There are some examples of new technologies being applied in the literature such as macroporous hollow fibers (Williams et al. 2020) and acoustic separators (Gränicher et al. 2020), but these have also not yet been widely adopted. The dominant technology employed in gene therapy clarification today is normal flow filtration using some combination of depth and membrane filtration. Typically, cell culture fluid is first run through one or more depth filters to remove the cells and large insoluble impurities and then through a membrane filter to remove some of the finer particles and any bioburden. These filters are single use and have a long history in bioprocessing with well-defined scalability and robustness. They are available in a wide range of chemistries, pore sizes, and pore structures to be tailored to each set of process requirements. It is worth noting that there has also been some work looking at precipitation and flocculation (typically in combination with a depth and membrane filter process) to improve footprint, cost, and host cell protein and DNA reduction.

Developing an optimal clarification process often starts with an evaluation of the feedstock to be clarified and any process restrictions from manufacturing. Starting with the feedstock evaluation, there are several characteristics that can help guide process development including the cell density and viability, whether they are grown in adherent or suspension culture, and whether the cells need to be lysed to release the product. All of these will impact the load and size distribution of the particles that need to be cleared. A lysed suspension culture could give relatively high particle loads with wide size distributions requiring multiple grades of depth filtration, whereas a product secreted from an adherent culture at high viability may be clean enough to go directly onto a membrane filter. Manufacturing constraints could include a footprint limitation or requirement to fit with existing capital equipment. Ability to clean in place can impact technology selection; however, given the relatively small batch sizes many manufacturers are opting for more single-use consumables to reduce operator time and validation time and enable multi-product manufacturing facilities.

With a technology selected, process inputs can be developed to optimize process performance. Key performance indicators that are typically tracked include pressure, filtrate turbidity, yield, and product quality. Product-related impurities such as aggregates or fragments are often unaffected by normal flow filtration processes, but a benefit of a depth filtration process is that they often show some adsorption of process-related impurities such as host cell proteins (HCP) and DNA (Khanal et al. 2018; Nejatishahidein et al. 2020). Filtrate turbidity is often tracked as an indicator of the pool clarity and risk of fouling a subsequent purification step.

Key process inputs for normal flow filtration processes can include filter selection, loading (volume per unit area), flux (flow rate per unit area), and scale. Both depth and membrane filters are available in a wide range of chemistries, pore sizes, and pore gradients. Many processes will require some form of prefiltration to protect the membrane filter. The most common options are depth filters made of cellulose, inorganic filter aids, and resins. They are available in a wide range of pore sizes, are cost-effective, and show high capacities for particle removal. Furthermore, they have some adsorptive capacity to bind host cell impurities. There are many examples of adeno-associated virus (AAV) processes that successfully employ these depth filters with relatively high capacities and yields (Raghavan et al. 2019; Hejmowski et al. 2021). However, there is also some evidence that the diatomaceous earth (DE; a commonly used filter aid) can bind lentiviral vectors (Labisch et al. 2021). In some cases, this can be avoided with specially designed filters such as Pall's Seitz V100P which has no DE and is low charge. In other cases, these filters can be replaced by synthetic options such as polypropylene or glass fiber. Synthetic filters such as these tend to be more inert and can provide higher product yield but also show relatively low volumetric capacity and have limited device options for large-scale manufacturing. Membrane filtration is typically done with a polymeric membrane filter rated down to 0.2 μm, although in some cases a larger vector may benefit from a more open pore size (Taylor et al. 2021). In processes that include long intermediate pool hold times a sterilizing-grade filter may be required to eliminate the risk of bioburden growth. However, many gene therapy processes are targeting fast processing times to improve product quality, and closed processing to reduce bioburden and endotoxin introduction, and are therefore able to harvest using bioburden reduction filters without a full sterile claim. Flux and loading are fairly straightforward to optimize for membrane filters as they are set based on the membrane differential pressure. These parameters are more challenging for depth filters as they can have two separate modes of failure: pressure rise and turbidity breakthrough. Furthermore, flux rate can impact the level of impurity adsorption in cellulose-based depth filters. Once a process is ready to transfer to manufacturing, it is standard to assess scalability and apply an appropriate safety factor based on the results (Lutz 2009; Krupp et al. 2017).

2.7 Conclusion and Perspectives

As outlined in this chapter, there are many ways to manufacture viral vectors for gene therapy applications. One could use a suspension or an adherent cell line, associated respectively with the appropriate bioreactor technology and process. Depending on the dosage, the target population, and the mode of action of the viral vector therapy, the scale needed will differ. However, whatever the vector, the GOI, the cell system used, the production platform (adherent or suspension), and the bioreactor, the key challenge remains the upstream productivity increase required to make all these therapies more accessible to the wider public. Many companies are already looking into this, with improved vectors, improved cell lines, and improved technologies, such as bioreactors, cell culture media, and transfection reagents. How future viral vector manufacturing will look when all of these efforts come to fruition remains to be seen.

References

ABEC (2019) ABEC advances single-use bioreactor volumes to 6,000 liters. ABEC. https://www.abec.com/2019/09/10/abec-advances-single-use-bioreactor-volumes-to-6000-liters/. Accessed 23 Aug 2022

Aslanidi G, Lamb K, Zolotukhin S (2009) An inducible system for highly efficient production of recombinant adeno-associated virus (rAAV) vectors in insect Sf9 cells. Proc Natl Acad Sci U S A 106:5059–5064. https://doi.org/10.1073/pnas.0810614106

Bielser J-M, Wolf M, Souquet J, Broly H, Morbidelli M (2018) Perfusion mammalian cell culture for recombinant protein manufacturing – a critical review. Biotechnol Adv 36:1328–1340. https://doi.org/10.1016/j.biotechadv.2018.04.011

Brown RB, Audet J (2008) Current techniques for single-cell lysis. J R Soc Interface 5:S131–S138. https://doi.org/10.1098/rsif.2008.0009.focus

Bulcha JT, Wang Y, Ma H, Tai PWL, Gao G (2021) Viral vector platforms within the gene therapy landscape. Signal Transduct Target Ther 6:53. https://doi.org/10.1038/s41392-021-00487-6

Cameau E, Pedregal A, Glover C (2019) Cost modelling comparison of adherent multi-trays with suspension and fixed-bed bioreactors for the manufacturing of gene therapy products. Cell Gene Ther Insights 5:1663–1674. https://doi.org/10.18609/cgti.2019.175

Carrondo M, Cruz P (2010) Hot topic: GMP production of gene and cell therapy therapeutics. Curr Gene Ther 10:413–413

Cashen P, Manser B (2021) Quality by design (QbD) for adeno-associated virus (AAV) a framework for a QbD assessment for AAV products within the chemistry manufacturing and controls (CMC). Documentation

Chaudhary G, Luo R, George M, Tescione L, Khetan A, Lin H (2020) Understanding the effect of high gas entrance velocity on Chinese hamster ovary (CHO) cell culture performance and its implications on bioreactor scale-up and sparger design. Biotechnol Bioeng 117:1684–1695. https://doi.org/10.1002/bit.27314

Chen YH, Pallant C, Sampson CJ, Boiti A, Johnson S, Brazauskas P, Hardwicke P, Marongiu M, Marinova VM, Carmo M, Sweeney NP, Richard A, Shillings A, Archibald P, Puschmann E, Mouzon B, Grose D, Mendez-Tavio M, Chen MX, Warr SRC, Senussi T, Carter PS, Baker S, Jung C, Brugman MH, Howe SJ, Vink CA (2020) Rapid lentiviral vector producer cell line generation using a single DNA construct. Mol Ther Methods Clin Dev 19:47–57. https://doi.org/10.1016/j.omtm.2020.08.011

Chong ZX, Yeap SK, Ho WY (2021) Transfection types, methods and strategies: a technical review. PeerJ 9:e11165. https://doi.org/10.7717/peerj.11165

Comisel R-M, Kara B, Fiesser FH, Farid SS (2021) Gene therapy process change evaluation framework: transient transfection and stable producer cell line comparison. Biochem Eng J 176:108202. https://doi.org/10.1016/j.bej.2021.108202

Durocher Y, Pham PL, St-Laurent G, Jacob D, Cass B, Chahal P, Lau CJ, Nalbantoglu J, Kamen A (2007) Scalable serum-free production of recombinant adeno-associated virus type 2 by transfection of 293 suspension cells. J Virol Methods 144:32–40. https://doi.org/10.1016/j.jviromet.2007.03.014

Escandell JM, Pais DA, Carvalho SB, Vincent K, Gomes-Alves P, Alves PM (2022) Leveraging rAAV bioprocess understanding and next generation bioanalytics development. Curr Opin Biotechnol 74:271–277. https://doi.org/10.1016/j.copbio.2021.12.009

Gálvez J, Lecina M, Solà C, Cairó JJ, Gòdia F (2012) Optimization of HEK-293S cell cultures for the production of adenoviral vectors in bioreactors using on-line OUR measurements. J Biotechnol 157:214–222. https://doi.org/10.1016/j.jbiotec.2011.11.007

González-Domínguez I, Grimaldi N, Cervera L, Ventosa N, Gòdia F (2019) Impact of physicochemical properties of DNA/PEI complexes on transient transfection of mammalian cells. New Biotechnol 49:88–97. https://doi.org/10.1016/j.nbt.2018.09.005

Gränicher G, Coronel J, Trampler F, Jordan I, Genzel Y, Reichl U (2020) Performance of an acoustic settler versus a hollow fiber–based ATF technology for influenza virus production in perfusion. Appl Microbiol Biotechnol 104:4877–4888. https://doi.org/10.1007/s00253-020-10596-x

Gregoriades N, Clay J, Ma N, Koelling K, Chalmers JJ (2000) Cell damage of microcarrier cultures as a function of local energy dissipation created by a rapid extensional flow. Biotechnol Bioeng 69:171–182. https://doi.org/10.1002/(sici)1097-0290(20000720)69:2<171::aid-bit6>3.0.co;2-c

Hejmowski A, Olson M, MacIntyre A, Rose A, Boenning K, Huato J, Schofield M, Kavara A, Marchand N, Collins M (2021) Enrichment of full rAAV capsids in a scalable, reproducible viral vector manufacturing platform. Cell Gene Ther Insights:1563–1579. https://doi.org/10.18609/cgti.2021.211

Kamen A, Henry O (2004) Development and optimization of an adenovirus production process. J Gene Med 6(Suppl 1):S184–S192. https://doi.org/10.1002/jgm.503

Khanal O, Singh N, Traylor SJ, Xu X, Ghose S, Li ZJ, Lenhoff AM (2018) Contributions of depth filter components to protein adsorption in bioprocessing. Biotechnol Bioeng 115:1938–1948. https://doi.org/10.1002/bit.26707

Knop DR, Harrell H (2007) Bioreactor production of recombinant herpes simplex virus vectors. Biotechnol Prog 23:715–721. https://doi.org/10.1021/bp060373p

Krupp AU, Please CP, Kumar A, Griffiths IM (2017) Scaling-up of multi-capsule depth filtration systems by modeling flow and pressure distribution. Sep Purif Technol 172:350–356. https://doi.org/10.1016/j.seppur.2016.07.028

Labisch JJ, Bollmann F, Wolff MW, Pflanz K (2021) A new simplified clarification approach for lentiviral vectors using diatomaceous earth improves throughput and safe handling. J Biotechnol 326:11–20. https://doi.org/10.1016/j.jbiotec.2020.12.004

Leinonen HM, Lipponen EM, Valkama AJ, Hynynen H, Oruetxebarria I, Turkki V, Olsson V, Kurkipuro J, Samaranayake H, Määttä A-M, Parker NR, Ylä-Herttuala S, Lesch HP (2019) Preclinical proof-of-concept, analytical development, and commercial scale production of lentiviral vector in adherent cells. Mol Ther Methods Clin Dev 15:63–71. https://doi.org/10.1016/j.omtm.2019.08.006

Lesch HP, Heikkilä KM, Lipponen EM, Valonen P, Müller A, Räsänen E, Tuunanen T, Hassinen MM, Parker N, Karhinen M, Shaw R, Ylä-Herttuala S (2015) Process development of adenoviral vector production in fixed bed bioreactor: from bench to commercial scale. Hum Gene Ther 26:560–571. https://doi.org/10.1089/hum.2015.081

Lesch HP, Valonen P, Karhinen M (2021) Evaluation of the single-use fixed-bed bioreactors in scalable virus production. Biotechnol J 16:e2000020. https://doi.org/10.1002/biot.202000020

Liu HF, Ma J, Winter C, Bayer R (2010) Recovery and purification process development for monoclonal antibody production. MAbs 2:480–499. doi:https://doi.org/10.4161/mabs.2.5.12645

Lock M, Alvira M, Vandenberghe LH, Samanta A, Toelen J, Debyser Z, Wilson JM (2010) Rapid, simple, and versatile manufacturing of recombinant adeno-associated viral vectors at scale. Hum Gene Ther 21:1259–1271. https://doi.org/10.1089/hum.2010.055

Lutz H (2009) Rationally defined safety factors for filter sizing. J Membr Sci 341:268–278. https://doi.org/10.1016/j.memsci.2009.06.015

Mainwaring D, Joshi A, Coronel J, Niehus C (2021) Transfer and scale-up from 10 L BioBLU® to Allegro™ STR 50 and STR 200 bioreactors. Cell Gene Ther Insights 7:1347–1362. https://doi.org/10.18609/cgti.2021.178

Manceur AP, Kim H, Misic V, Andreev N, Dorion-Thibaudeau J, Lanthier S, Bernier A, Tremblay S, Gélinas A-M, Broussau S, Gilbert R, Ansorge S (2017) Scalable lentiviral vector production using stable HEK293SF producer cell lines. Hum Gene Ther Methods 28:330–339. https://doi.org/10.1089/hgtb.2017.086

Manservigi R, Argnani R, Marconi P (2010) HSV recombinant vectors for gene therapy. Open Virol J 4:123–156. https://doi.org/10.2174/1874357901004030123. PMID: 20835362; PMCID: PMC2936037

Merten O-W (2004) State-of-the-art of the production of retroviral vectors. J Gene Med 6(Suppl 1): S105–S124. https://doi.org/10.1002/jgm.499

Merten O-W, Schweizer M, Chahal P, Kamen A (2014) Manufacturing of viral vectors for gene therapy: Part I. Upstream processing. Pharm Bioprocess 2:183–203. https://doi.org/10.4155/pbp.14.16

Nadeau I, Kamen A (2003) Production of adenovirus vector for gene therapy. Biotechnol Adv 20: 475–489. https://doi.org/10.1016/s0734-9750(02)00030-7

Naso MF, Tomkowicz B, Perry WL, Strohl WR (2017) Adeno-associated virus (AAV) as a vector for gene therapy. BioDrugs 31:317–334. https://doi.org/10.1007/s40259-017-0234-5

Nass S, Nambiar B, Mattingly M, Woodcock D, O'Riordan C (2019) Evaluation of AAV vector production from the iCELLis fixed bed bioreactor vessel. BioInsights. https://insights.bio/immuno-oncology-insights/journal/article/22/evaluation-of-aav-vector-production-from-the-icellis-fixed-bed-bioreactor-vessel. Accessed 9 Sep 2022

Nejatishahidein N, Borujeni EE, Roush DJ, Zydney AL (2020) Effectiveness of host cell protein removal using depth filtration with a filter containing diatomaceous earth. Biotechnol Prog. https://doi.org/10.1002/btpr.3028

Nikolay A, de Grooth J, Genzel Y, Wood JA, Reichl U (2020) Virus harvesting in perfusion culture: choosing the right type of hollow fiber membrane. Biotechnol Bioeng. https://doi.org/10.1002/bit.27470

Pall Corporation (2022) USD3843-Harvesting Viral Vectors – a summary of process choices and solutions

Pall Corporation, Polyplus (2020) Guide for DNA transfection in iCELLis® bioreactor systems for gene therapy vector manufacturing

Pezzoli D, Giupponi E, Mantovani D, Candiani G (2017) Size matters for in vitro gene delivery: investigating the relationships among complexation protocol, transfection medium, size and sedimentation. https://www.nature.com/articles/srep44134. Accessed 23 Aug 2022

Pham PL, Kamen A, Durocher Y (2006) Large-scale transfection of mammalian cells for the fast production of recombinant protein. Mol Biotechnol 34:225–237. https://doi.org/10.1385/MB:34:2:225

Powers AD, Piras BA, Clark RK, Lockey TD, Meagher MM (2016) Development and optimization of AAV hFIX particles by transient transfection in an iCELLis(®) fixed-bed bioreactor. Hum Gene Ther Methods 27:112–121. https://doi.org/10.1089/hgtb.2016.021

Powers AD, Drury JE, Hoehamer CF, Lockey TD, Meagher MM (2020) Lentiviral vector production from a stable packaging cell line using a packed bed bioreactor. Mol Ther Methods Clin Dev 19:1–13. https://doi.org/10.1016/j.omtm.2020.08.010

Qu W, Wang M, Wu Y, Xu R (2015) Scalable downstream strategies for purification of recombinant adeno-associated virus vectors in light of the properties. Curr Pharm Biotechnol 16:684–695. https://doi.org/10.2174/1389201016666150505122228

Raghavan B, Collins M, Walls S, Lambropoulos A, Bergheim-Pietza S (2019) Development of gene therapy purification processes. Cell Gene Ther Insights 5:1311–1322. https://doi.org/10.18609/cgti.2019.137

Smith RH, Levy JR, Kotin RM (2009) A simplified baculovirus-AAV expression vector system coupled with one-step affinity purification yields high-titer rAAV stocks from insect cells. Mol Ther 17:1888–1896. https://doi.org/10.1038/mt.2009.128

Stewart HJ, Fong-Wong L, Strickland I, Chipchase D, Kelleher M, Stevenson L, Thoree V, McCarthy J, Ralph GS, Mitrophanous KA, Radcliffe PA (2011) A stable producer cell line for the manufacture of a lentiviral vector for gene therapy of Parkinson's disease. Hum Gene Ther 22:357–369. https://doi.org/10.1089/hum.2010.142

Suarez L (2021) Innovation in rAAV gene therapy is a need: AAV capsids. BioInsights. https://www.insights.bio/cell-and-gene-therapy-insights/journal/article/2300/Innovation-in-rAAV-gene-therapy-is-a-need-AAV-capsids. Accessed 15 Sep 2022

Tang Y, Garson K, Li L, Vanderhyden BC (2015) Optimization of lentiviral vector production using polyethylenimine-mediated transfection. Oncol Lett 9:55–62. https://doi.org/10.3892/ol.2014.2684

Taylor N, Ma W, Kristopeit A, Wang SC, Zydney AL (2021) Evaluation of a sterile filtration process for viral vaccines using a model nanoparticle suspension. Biotechnol Bioeng 118:106–115. https://doi.org/10.1002/bit.27554

Urabe M, Ding C, Kotin RM (2002) Insect cells as a factory to produce adeno-associated virus type 2 vectors. Hum Gene Ther 13:1935–1943. https://doi.org/10.1089/10430340260355347

Valkama AJ, Leinonen HM, Lipponen EM, Turkki V, Malinen J, Heikura T, Ylä-Herttuala S, Lesch HP (2018) Optimization of lentiviral vector production for scale-up in fixed-bed bioreactor. Gene Ther 25:39–46. https://doi.org/10.1038/gt.2017.91

Vandenberghe LH, Xiao R, Lock M, Lin J, Korn M, Wilson JM (2010) Efficient serotype-dependent release of functional vector into the culture medium during adeno-associated virus manufacturing. Hum Gene Ther 21:1251–1257. https://doi.org/10.1089/hum.2010.107

Wang Z, Ma H-I, Li J, Sun L, Zhang J, Xiao X (2003) Rapid and highly efficient transduction by double-stranded adeno-associated virus vectors in vitro and in vivo. Gene Ther 10:2105–2111. https://doi.org/10.1038/sj.gt.3302133

Wang X, Olszewska M, Qu J, Wasielewska T, Bartido S, Hermetet G, Sadelain M, Rivière I (2015) Large-scale clinical-grade retroviral vector production in a fixed-bed bioreactor. J Immunother 38:127–135. https://doi.org/10.1097/CJI.0000000000000072

Wang S, Godfrey S, Ravikrishnan J, Lin H, Vogel J, Coffman J (2017) Shear contributions to cell culture performance and product recovery in ATF and TFF perfusion systems. J Biotechnol 246:52–60. https://doi.org/10.1016/j.jbiotec.2017.01.020

Williams T, Goodyear O, Davies L, Knevelman C, Bransby M, Miskin J, Mitrophanous K (2020) Lentiviral vector manufacturing process enhancement utilizing TFDF™ technology. Cell Gene Ther Insights 6:455–467. https://doi.org/10.18609/cgti.2020.053

Yoganathan P, Lee AC, Hyatt N, Bell J, Becheau O, Lundeen T, Gupta S, Laskowski A (2020) Perfusion enables increased lentivirus production using the iCELLis® bioreactor system. Cytotherapy 22:S150. https://doi.org/10.1016/j.jcyt.2020.03.308

Chapter 3
Upstream Processing of Viral Therapeutics: From Host Cell Expansion to Virus Production

Orsolya Hamusics and Daniel Loewe

3.1 Introduction

Viruses are powerful weapons and emerging therapeutics either as vectors for gene therapy or as oncolytic viruses for the treatment of uncurable cancers. While the concept of viral vectors was first introduced in the early 1980s (Mackett et al. 1982; Panicali and Paoletti 1982), first observations of oncolytic virus activity were made in the end of the nineteenth century. For influenza, the oncolytic characteristic was discovered as the severity of cancer temporarily improved after infection with influenza (Dock 1904). Further, in the beginning of the twentieth century, treatment with live attenuated rabies virus was correlated with the regression of malignant melanoma (Pack 1950). Additionally, spontaneous remission of Burkitt's lymphoma and the regression of Hodgkin's disease were reported (Bluming and Ziegler 1971; Taqi et al. 1981). Finally, the progress of science, the better understanding of biological mechanisms, and the systematic use of these biologics have evolved the area of virus-based therapies. Some examples of oncolytic viruses are measles, vesicular stomatitis, herpes simplex, and adenovirus. The most used representatives in gene therapy are adeno-associated viruses (AAV) and lentiviruses.

While for vaccines typically low titers in a single dose are needed for the stimulation of the immune system, much higher virus titers are necessary for gene and oncolytic therapy. From an industrial perspective, this shifts the attention to the

Orsolya Hamusics and Daniel Loewe contributed equally.

O. Hamusics
Boehringer Ingelheim Pharma GmbH & Co. KG, Biberach an der Riß, Germany

D. Loewe (✉)
Boehringer Ingelheim Therapeutics GmbH, Ochsenhausen, Germany
e-mail: daniel.loewe@boehringer-ingelheim.com

S. Gautam et al. (eds.), *Bioprocess and Analytics Development for Virus-based Advanced Therapeutics and Medicinal Products (ATMPs)*,
https://doi.org/10.1007/978-3-031-28489-2_3

manufacturing process, which should cover this demand for the clinical trials and market supply in an efficient manner. Most flu vaccines are still produced in chicken egg-based processes, and as recently as in 2012, the first cell-based influenza vaccine was approved by the US Food and Drug Administration (FDA) (Nuwarda et al. 2021). Using roller bottles or cell stacks for cell-based virus production is labor-intensive and only scalable to a certain level. Additionally, these production systems are poorly controllable.

Therefore, the industrial focus is on developing an efficient manufacturing process using controllable and truly scalable intensified processes. Within this chapter, a first view is set on the different cell lines used for virus production, which have a significant influence on the process design. Further, state-of-the-art production systems and production modes with respect to the achievable titers are presented. Development approaches and tools promoted by regulatory authorities are also described.

3.2 Cell Lines and Seed Train

Virus vaccine production processes are based on different cell lines. This comprises adherent cell lines [MRC-5, WI-38, HEK293, BHK-21, Vero, MDCK, and CEF cells (Genzel 2015)] and suspension cells [HEK293SF-3F6 (Elahi et al. 2019), PBG. PK2.1 (Gränicher et al. 2019), A549 (Moreira et al. 2020), Sf9, HeLa S3, SF-BMAdR (derived from A549 cells) (Elahi et al. 2019), and suspension-adapted Vero cells (Kiesslich et al. 2021)]. Viral vectors for gene and oncolytic therapies are predominantly produced in adherent Vero (Yang et al. 2019a; Offersgaard et al. 2021) and HEK293T (Yang et al. 2019a) cells.

In general, the use of continuous cell lines (CCL) has still a short history. In the 1950s only primary cells were approved to produce biological products for human use. CCLs were first approved in the late 1970s (Petricciani 1995). With this approval, a new era has begun for the virus production. The Vero cell line is used in commercial vaccine production processes for over 40 years (Hegde 2015). It has a long-term and well-known safety profile. The HEK293 cell line and its derived sub-cell lines are highly efficient in transfection and infection-based processes. Since Vero cells are from non-human origin (African green monkey), there is a risk that the virus product encounters undesired post-translational modifications. This risk is reduced by using cell lines of human origin, such as the HEK293 cell line (Tan et al. 2021).

Today, only three virus-based products are approved by the authorities and available for commercial supply. Luxturna (Mullard 2018) and Zolgensma (Mullard 2020) are AAV-based therapeutics for gene therapy, which are produced in HEK293 cells (Tan et al. 2021). Presently, only one oncolytic virus, a herpes simplex virus, is available on the market (Imlygic). This virus is produced in Vero cells according to the manufacturer, Amgen Europe B.V. (EMA 2016).

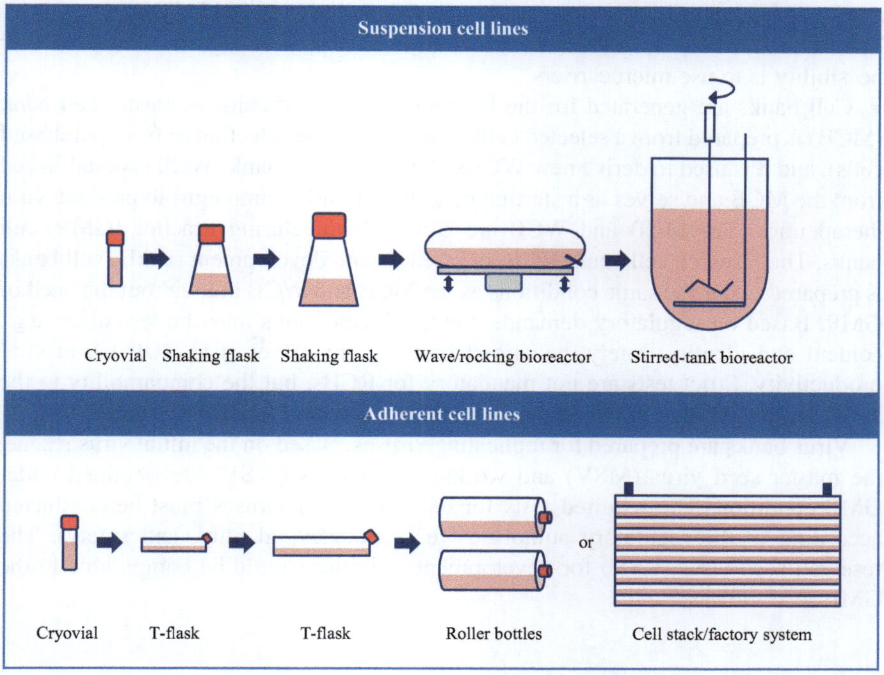

Fig. 3.1 Possible seed trains for suspension cell lines (e.g., HEK293F) and adherent cell lines (e.g., Vero, HEK293T)

Table 3.1 Comparison of suspension or adherent cell lines for viral vector production

	Adherent cells	Suspension cells
Seed train	Complex and labor-intensive	Easy
Scalability	Difficult	Easy
Growth	Limited by surface area and nutrient, inhibited by toxic waste products	Limited by nutrient, inhibited by toxic waste products
Equipment	T-flask, roller bottle, cell factory, wave or stirred-tank bioreactor with microcarrier	Shaking flask, wave bioreactor, stirred-tank bioreactor

Vero cells are adherent in nature, and thus, they need a surface to grow (adherent cell line). Nevertheless, several studies have demonstrated the feasibility to adapt adherent Vero cells to suspension growth (suspension cell line) (Paillet et al. 2009; Shen et al. 2019; Lee et al. 2020). The HEK293 cell line is commercially available both as adherent (e.g., HEK293T) and suspension cells (e.g., HEK293F).

The seed train considerably differs for adherent and suspension cells (Fig. 3.1 and Table 3.1). Subculturing adherent cell lines is more complex than handling suspension cells due to the cell attachment and detachment steps. This requires more time and labor in comparison to suspension cells (Madeline et al. 2015; Shen et al. 2019). T-flasks, roller bottles, and cell stacks can be used to provide a growth area. Another possibility is to use microcarriers.

Cell banks are generated for the long term storage of cells. A master cell bank (MCB) is prepared from a selected cell clone (after clone selection or from purchased cells), and it is used to derive new WCBs. A working cell bank (WCB) is established from the MCB and serves as a starting material for GMP campaigns to produce viral therapeutics. The MCB and WCB are Good Manufacturing Practice (GMP) cell banks. The research cell bank (RCB, or research and development (R&D) cell bank) is prepared under the same conditions as the MCB and WCB but without the need of GMP. Based on regulatory demands, the GMP cell banks must be tested for, e.g., content and identity, safety, genetic stability, purity (Byer et al. 2015), and viral productivity. Strict tests are not mandatory for RCBs, but the comparability to the MCB and WCB regarding cell growth and productivity should be assured.

Virus banks are prepared for replicating viruses. Based on the initial virus rescue, the master seed virus (MSV) and working seed virus (WSV) are obtained under GMP condition. The required tests for the GMP seed viruses must be conducted according to the regulatory authorities (e.g., potency, identity, and safety). The research seed virus (RSV) for development activities should be comparable to the GMP seed viruses.

3.3 Modes of Viral Vector Production

Viral vectors are divided into two major categories: replication-competent and replication-deficient viruses. Contrary to the replication-deficient viruses, replication-competent viruses are able to multiplicate in cells. Replication-deficient viruses can be produced in three different ways, by (1) transient transfection, (2) stable transfection/stable producer cell line, and (3) using a helper virus (Table 3.2).

The primarily applied method, due to the ease of adaption, is the transient transfection. This method is scalable, but it can be challenging to provide the number of plasmids for large-scale production. Additionally, transfection reagents are toxic to cells. With stable cell lines, the scale-up is easier. Nevertheless, the genetic instability of the cell line is of great concern because it might negatively influence the product quality and thus, presents risk for the patient. Using a helper virus has the major disadvantage of introducing a contaminant into the system, which must be removed from the viral product, together with other impurities (e.g., host cell DNA and proteins).

Cell line optimization for replication-competent viruses is gaining more and more interest. But until now, less work has been done in designing production cell lines for virus propagation, which is a normal process for protein production with CHO cells (Genzel 2015). The cellular pathways that are involved in the antiviral mechanism of the host cell vary with the respective virus-cell system. These cellular pathways must

Table 3.2 Comparison of different production strategies for replication-deficient viruses. Adapted from Andari and Grimm (2021)

	Pros	Cons
Transient transfection	– Easy to adapt for new viral vectors	– Difficult to scale up since high amount of DNA and transfection reagent are needed – Toxicity
Stable producer cell line	– Scale-up easily possible	– Genetic instability might be an issue – Generation of a new stable cell line needs a lot of know-how
Helper virus	– Scalable	– Contamination with helper virus – Instability of helper virus

be identified first to knock-out the appropriate gene(s). For antibodies, the scientific progress is much ahead, so that timelines for cell line generation are reduced as much as possible.

3.4 Generic Virus Production Process

Typically, virus production processes are divided into two phases: (1) cell growth and (2) virus production. Cell propagation is needed to provide sufficient cell number for the time of infection (TOI) or transfection. Only healthy cells can lead to a satisfying virus production process. Therefore, cells need an optimal nutrient supply during the growth phase (Tapia et al. 2016).

The TOI or time of transfection are important parameters defining the end of the cell propagation and the start of the virus production. This timepoint is usually chosen during the exponential growth phase. In adherent cultures, the coverage of the growth surface (confluency) is often used as indicator for the optimal TOI. For example, an influenza virus process yielded higher when cells were infected below 60% confluency (Bock et al. 2011), with a higher ratio of dividing to non-dividing cells (Bock et al. 2009). In contrast, the recombinant HSV-1 yield was higher when cells were infected at confluences above 80% (Ozuer et al. 2002a). Several viruses spread by cell-to-cell contact [e.g., measles virus and LCMV (Mothes et al. 2010; Lapošová et al. 2013)], which is facilitated at a higher cell confluence.

Another important parameter for infection-based processes is the optimal multiplicity of infection (MOI), defined by the amount of virus added to the viable cells (virus per cell). Normally, a higher MOI does not lead to a higher virus titer in the process but can accelerate the virus production (Elahi et al. 2019; Kiesslich et al. 2021). The MOI should be high enough to counterbalance the virus degradation rate in the medium until the virus infects the cell (Tapia et al. 2016). Too high MOIs can trigger the formation of defective interfering virus particles (DIPs) by the error-prone virus replication. This can negatively influence the application by inducing an antiviral immune response in the host cell (Yang et al. 2019b; Huo et al. 2020).

Furthermore, DIPs suppress the replication of infectious virus particles (Yang et al. 2019b).

For transfection-based production processes, the transfection reagent with the optimal number of plasmids and its ratio to the cells as well as the medium need to be considered (Dekevic et al. 2022; Guan et al. 2022). While transient transfection is often performed in adherent cell cultures, switching to suspension cells is preferred to achieve an easy scale-up (Tan et al. 2021).

The time of harvest (TOH) should be chosen at or shortly before the peak virus concentration to avoid virus degradation. Depending on the type of virus process, cell-associated viruses might need to be liberated for instance by hypotonic buffers, by deionized water lyses (Laposova et al. 2017), or by adding ionic additives to the cultures (Kuroda et al. 2020). While the detergent Triton X-100 was commonly used for cell lysis and virus release (especially for AAV), Triton-X is on the REACH (Registration, Evaluation, Authorisation and Restriction of Chemicals) authorization list of the European Union (EU) since January 4, 2021. Therefore, Polysorbate 20 might become a promising alternative (Moleirinho et al. 2018).

3.5 Bioreactor Systems

Mainly three types of bioreactors are used for virus production: stirred-tank bioreactor (STR), fixed-bed bioreactor (FBR), and wave bioreactor (Table 3.3). While the FBR is designed for adherent cell-based processes, the STR and wave bioreactor can be used for either adherent or suspension cells. Disposable bioreactors are usually used to produce viral vectors with capacities up to 2000 L (STR). The advantages of disposable bioreactors are the reduced validation costs and contamination risk. Furthermore, the capital investment is reduced, and the downtimes between processes is shortened.

3.5.1 Stirred-Tank Bioreactor

Stirred-tank bioreactors are commonly used in laboratory and industrial settings. Based on that, a lot of effort has been made to understand the hydrodynamics and scalability of STRs. This makes the STR the best characterized production system. Process parameters [e.g., pH, dissolved oxygen (DO)] can be accurately measured and controlled in the homogenous bioreactor environment.

For an easy scale-up from lab to commercial scale, the geometrical similarity of the bioreactors—the height-to-diameter ratio and the stirrer diameter to bioreactor-diameter—is of high importance. Deviations can lead to a different behavior with regard to the oxygen transfer, mixing, and bubble dispersion. The impeller, agitation, and aeration also impact the process. Therefore, the volumetric power input (P/V), the impeller tip speed (u_{tip}), the oxygen mass transfer coefficient

Table 3.3 Common bioreactors for viral vector production

Bioreactor	Remark
Stirred-tank bioreactor (STR)	
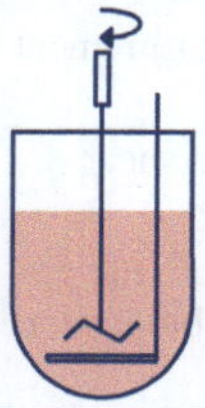	– For the cultivation of suspension and adherent cells (with microcarriers) – Examples for available systems: – Xcellerex XDR, Cytiva (up to 2000 L) – Biostat STR, Sartorius (up to 2000 L) – Mobius, Merck (up to 2000 L) – CSR, ABEC [up to 6000 L (Stanton 2019)]
Fixed-bed bioreactor (FBR)	
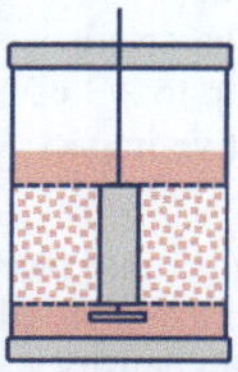	– For the cultivation of adherent cell lines – Examples for available systems: – BioBLU 5p, Eppendorf (up to 18 m^2 growth surface) – iCellis, PALL (up to 500 m^2 growth surface, 70 L) – Scale-X, Univercells Technologies (up to 600 m^2 growth surface) – TideCell, ESCO Healthcare (up to 1560 m^2 growth surface, 100 L)
Wave bioreactor	
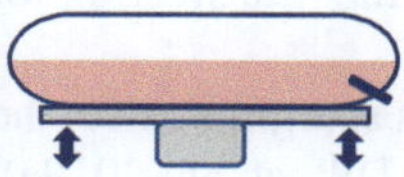	– For the cultivation of suspension and adherent cells (with microcarriers) – Examples for available systems: – Biostat RM, Sartorius (up to 200 L) – HyPerforma Rocker, Thermo Fisher Scientific (up to 200 L) – WAVE Rocker, Cytiva (up to 500 L)

(k_La), and the oxygen transfer rate (OTR) are common scale-up parameters (Table 3.4).

Establishing scale-down models are not only useful for process development, optimization and characterization but also for scale-up. (Gallo-Ramírez et al. 2015). Large-scale bioreactors are mainly limited in the operational space, which defines the design space for small-scale bioreactors.

The most applied impeller types in cell culture processes are pitch-blade and marine impellers. The pitch-blade impeller provides efficient mixing by creating radial and axial flows, and it is suitable for shear-sensitive microcarrier and suspension cultures. The marine impeller produces an axial flow and, thus, has a lower mixing efficiency. It is considered to be more gentle than the pitch-blade impeller.

If the shear sensitivity of the cell line hampers the scalability, shear protectants can be used to improve the cell culture robustness (Tharmalingam et al. 2008). Cell aggregation is additionally an important factor for viral vector production because it might hinder the accessibility to virus infection. Cell aggregation can be reduced with commercial anti-clumping agents or surfactants such as Poloxamer 188. The latter one also protects the cells from shear forces in the bioreactor. Especially for

Table 3.4 Common scale-up parameters for stirred-tank bioreactors

Parameter	Formula	Remark
Volumetric power input	$\frac{P}{V_L} = 2 \cdot \pi \cdot n \cdot \frac{M}{V_L} = N_p \cdot d^5 \cdot n^3 \cdot \frac{\rho}{V_L}$	- Defines the effective power input by the stirrer into the system - Calculated based on torque measurement or impeller-specific data
Impeller tip speed	$u_{tip} = \pi \cdot d \cdot n$	- Impacts cells by the introduced shear stress
Oxygen transfer rate (OTR)	$\text{OTR} = k_L a \cdot (c_L^* - c_L)$	- Defines the oxygen transfer from the gas phase to the liquid phase in the bioreactor - Depends on the mass transfer coefficient ($k_L a$)
Kolmogorov eddy size	$\eta = \left[\frac{\nu^3}{\frac{N_p \cdot d^5 \cdot n^3}{V_L}}\right]^{1/4}$	- Defines the swirling of a fluid, which depends on the stirrer characteristics - Eddy size up to cell size or up to 2/3 of the microcarrier size has negative impact on the cell growth

c_L oxygen concentration in the bulk liquid; c_L^* oxygen concentration at the liquid/gas interface; d impeller diameter; $k_L a$ oxygen mass transfer coefficient; M momentum; n impeller speed; N_p impeller power number; P power; V_L liquid volume; ν kinematic viscosity; ρ broth density

virus production, the interference of these supplements on the transfection or infection must be evaluated in advance.

Foaming is introduced into the bioreactor by agitation and gas sparging. Also, the release of proteins by lysed cells supports foam formation (Dill et al. 2019a). Foaming might block the exhausting air filter, which can lead to contamination and pressure increase in the bioreactor (Routledge et al. 2014). Foaming becomes an important issue in high cell density cultures where high amounts of proteins are secreted to the medium. Additionally, a high gassing rate is needed to maintain the dissolved oxygen level in the bioreactor (Routledge 2012). Foaming can be reduced by adding surfactants either before foaming as a prevention or after foaming with the intent to collapse the foam (Abdolahi et al. 2005). Surfactants are usually silicone-based. Antifoam agents, mostly with chemical nature, might interact with the membrane of the host cell. This interaction can influence the virus–host interaction. In the study of Dill et al., a concentration-dependent negative effect of a simethicone-based antifoam agent was shown on the virus production (Dill et al. 2019a).

3.5.2 *Microcarrier Systems*

The microcarrier technology allows the cultivation of adherent cells in bioreactors that are typically used for suspension cell cultures (e.g., STR but also wave bioreactor). Wezel pioneered the growth of anchorage-dependent cell lines in an STR by using microcarriers to produce polio and rabies virus vaccines (Wezel 1967).

Presently, the largest microcarrier cultivation is scaled to 6000 L for influenza vaccine production (Barrett et al. 2014).

Various manufacturers supply microcarriers, either microporous or solid. Cytodex 1 (GE Healthcare) is the most widely used microcarrier for viral vector production. Cytodex 1 microcarriers are solid cross-linked dextran beads, presenting positive charges throughout the matrix. These microcarriers have a median size of ~170 μm and a specific surface area of 3000 $cm^2\ g^{-1}$ (Kiesslich and Kamen 2020). In most processes, Cytodex 1 was used in a concentration of 3 $g\ L^{-1}$ (=9000 $cm^2\ L^{-1}$) for the production of Peste de Petits ruminants vaccine (Silva et al. 2008), rabies virus with Vero cells (Trabelsi et al. 2006; Rourou et al. 2007, 2014) and measles virus with MRC-5 cells (Rourou et al. 2014), or Vero cells (Weiss et al. 2015; Grein et al. 2018, 2019). 6 $g\ L^{-1}$ microcarrier was applied for rabies virus production in Vero cells (Trabelsi et al. 2006). Other microcarriers were also used for virus production, such as SoloHill microcarriers (PALL), CellBIND (Corning), Synthemax (Corning), and Fibracell (Eppendorf). A detailed overview of commercially available microcarriers was given by Rafiq et al. (2016). These microcarriers were screened, for example, to produce the oncolytic H-1 protoparvovirus with NB-324 K cells (Wohlfarth et al. 2021). The production of different viruses (e.g., Sindbis virus in CHO cells) with microcarrier technology showed mostly an increased performance compared to roller bottle cultures (Giard et al. 1977).

The scale-up of microcarrier processes is challenging because a higher impeller speed is needed to suspend the microcarriers in larger volumes. This leads to higher shear forces that counteract the cell growth (Sousa et al. 2015). Aeration is important to supply the cells with enough oxygen, but aeration with sparger was shown to hamper the cell growth (Rourou et al. 2007) and virus production (Grein et al. 2019). On the other side, low oxygen partial pressure can limit the virus productivity (Lim et al. 1999). The shear stress is sufficiently large to damage cells if the eddy size (Table 3.4) is similar to the cell size (suspension cells) or microcarrier size (adherent cells). For process optimization, microcarrier concentration, ratio of seeding cells to microcarriers, stirrer speed, and media components are important factors to consider (Silva et al. 2008, Leber et al. 2017, Kiesslich and Kamen 2020).

Another drawback of microcarrier cultivations is the difficult handling at a higher scale to provide large amounts of cells for inoculation and from an economic perspective microcarriers are expensive. For higher scales, the challenges with cell detachment must be sufficiently solved and/or the bead-to-bead transfer must highly be optimized (Marcelino et al. 2006, Gallo-Ramírez et al. 2015, Chen et al. 2021). For Vero and HEK293T cells, the microcarrier process was successfully transferred from spinner flask to wave bioreactors within the seed train and then to 200 L single-use bioreactors (Xcellerex disposable reactor from Cytiva) (Yang et al. 2019a). Process transfers from static cultures to stirred microcarrier cultures can lead to increased volumetric productivity as shown by, e.g., Peste des Petits ruminants virus (Silva et al. 2008).

3.5.3 Wave Bioreactor

Wave bioreactors are single-use semi-controllable bioreactors with sizes up to 500 L. Wave bioreactors are called after the continuously generated waves in the bag caused by the rocker plate. These generated waves agitate the medium within the bag that enables the homogenous distribution of nutrients and extracellular bioproducts as well as gas dispersion. The rocking angle and speed can be adjusted. Oxygen supply is ensured by head-space aeration. Wave bioreactors are easier to operate than an STR, and the fluid shear is reduced due to the absence of gas bubbles. Furthermore, the DO and pH are controllable in these systems.

Wave bioreactors are often used in the seed train due to the ease of use and scalability, where, e.g., shaking flasks are limited. While wave bioreactors are mostly applied with suspension cell lines, adherent cells can also be cultivated in wave bioreactors with microcarriers. Wave bioreactors were used for the production of influenza virus with MDCK cells, grown on Cytodex 1 (Genzel et al. 2006). The Aujeszky's disease virus was successfully produced in wave bioreactor with suspension BHK-21 cells (Slivac et al. 2006). Grieger et al. reported an AAV production process with transfected HEK293 cells in a wave bioreactor (Grieger et al. 2016). Wave bioreactors were also applied to produce AAV with suspension BHK-21 cells by using a recombinant herpes simplex virus as a helper virus and by exchanging the medium 2 h post-infection. Additionally, several medium exchange strategies (e.g., centrifugation, TFF, or integrated filtration) were tested (Thomas et al. 2009). Another example is the transfection-based retroviral vector process with adherent HEK293T cells that were grown on FibraCell microcarriers in a wave bioreactor (Loo et al. 2012).

3.5.4 Fixed-Bed Bioreactor

Anchorage-dependent cells can also be cultivated in fixed-bed bioreactors (FBR), which provides compact cell growth surface. Presently, different formats of FBRs are available: BioBLU 5p (Eppendorf), iCellis (PALL), Scale-X (Univercells Technologies), and TideXCell (ESCO Healthcare). The commercially available formats were discussed in a recently published article by Lesch et al. The iCellis and Scale-X bioreactors use magnetic stirrers (Lesch et al. 2021). The TideXCell and the smaller-scale CelCradle bioreactors utilize the tide motion technology wherein the cells are alternately exposed to medium and air through the upward and downward motion of the medium (Offersgaard et al. 2021). The CelCradle was successfully applied to produce hepatitis C virus by using adherent Huh7.5 cells (Offersgaard et al. 2022). Additionally, the hepatitis delta virus-like particle production in BHK-21 cells was shown (Chen et al. 2005).

Fixed-bed bioreactors are mostly preferred over microcarrier systems because of the minimal contamination risk, higher homogeneity of batches, and reduced complexity. One of the biggest commercially available fixed-bed bioreactors is the Scale-X from Univercells Technologies. Here, 600 m^2 cell growth surface is provided with a total bioreactor volume of 70 L (approximately 8.57 m^2L^{-1}). In comparison, a stirred-tank bioreactor with 3 g L^{-1} Cytodex 1 (specific growth surface of 3000 cm^2 g^{-1}) can provide 63 m^2 (approximately 0.9 m^2 L^{-1}) surface area in a 70 L format.

The use of the Scale-X FBR for a VSV-based Ebola vaccine production (rVSV-ZEBOV) resulted in a similar virus titer, but a higher cell-specific productivity and a lower ratio of genomic to infectious titer compared to the microcarrier process format (Kiesslich et al. 2020). Leinonen et al. compared the Scale-X hydro FBR with the iCellis Nano and proposed that the FBRs are interchangeable. This was confirmed for transfection-based lentiviral and adenoviral vector production processes with HEK293T cells using perfusion (Leinonen et al. 2020; Martínez-Molina et al. 2020).

Further, the iCellis FBR was successfully applied several times to produce lentiviral vectors, either by transfection (Valkama et al. 2018; Leinonen et al. 2019) or by using a stable packaging cell line (Powers et al. 2020). Lentiviral vector production with transfection was more beneficial with perfusion than with media recirculation (Leinonen et al. 2019). The production of retroviral vectors in iCellis was also compared with cell factories. Production in FBR supported a higher titer per harvest, which was summed for the entire bioreactor run (Wang et al. 2015).

3.5.5 Further Bioreactor Systems

Besides the predominantly used production systems—either fixed-bed or stirred-tank bioreactors—there are more bioreactor formats available on the market, but up to now rarely discussed in the literature.

For adherent cells, the vertical wheel bioreactor (PBS Biotech) with microcarriers can be an alternative to the STR, supplying reactor volumes from 0.1 to 500 L. With lower shear stress impact, a sufficient mixing can be achieved with the rotating wheel. This process format is truly scalable according to the manufacturer. Adenovirus production in a vertical wheel bioreactor led to a slightly higher virus titer and quality than in stirred-tank bioreactor (Sousa et al. 2015).

While no scientific publication is available so far, a new variety of bioreactor, called HyPerforma DynaDrive (Thermo Fisher Scientific), is available on the market. The single-use bioreactor is unique with its cuboid shape [like the Allegro single-use bioreactor supplied by PALL (Schirmer et al. 2019)], which increases the mixing efficacy according to the manufacturer. The promoted benefit of this system is the high turndown ratio (ratio of maximum to minimum operating volume) of at least 10:1 [and up to 20:1, compared to the usual 5:1 ratio (Müller et al. 2022)], which is ensured by the impeller design. The impeller mimics a helix structure with impeller elements within the helix, inducing lower shear stress (Thermo Fisher

Scientific 2022a). Single-use systems from 50 to 5000 L are available where a power input of up to 80 W m^{-3} is feasible (Thermo Fisher Scientific 2022b).

3.6 Bioreactor Operation Modes

Various bioreactor operation modes are available to produce viruses from batch to continuous cultivation mode (Fig. 3.2).

The batch cultivation is the simplest cultivation technique where the initial filling with medium provides metabolites for the entire run (Fig. 3.2). Batch cultivation is often preferred for virus production due to its simplicity and robustness. Cultivation systems such as roller bottles or cell factories are naturally construed for batch processes. One disadvantage of batch cultivations is that large production volumes might be required to supply the clinical studies. Therefore, it is a common strategy to intensify the production by increasing the cell density of the cultures by using, e.g., fed-batch, repeated-batch, and perfusion cultivation techniques. A general challenge with higher-cell density cultivations is the so-called cell density effect, which refers to the decrease in the cell-specific virus yield in cultures with increased cell

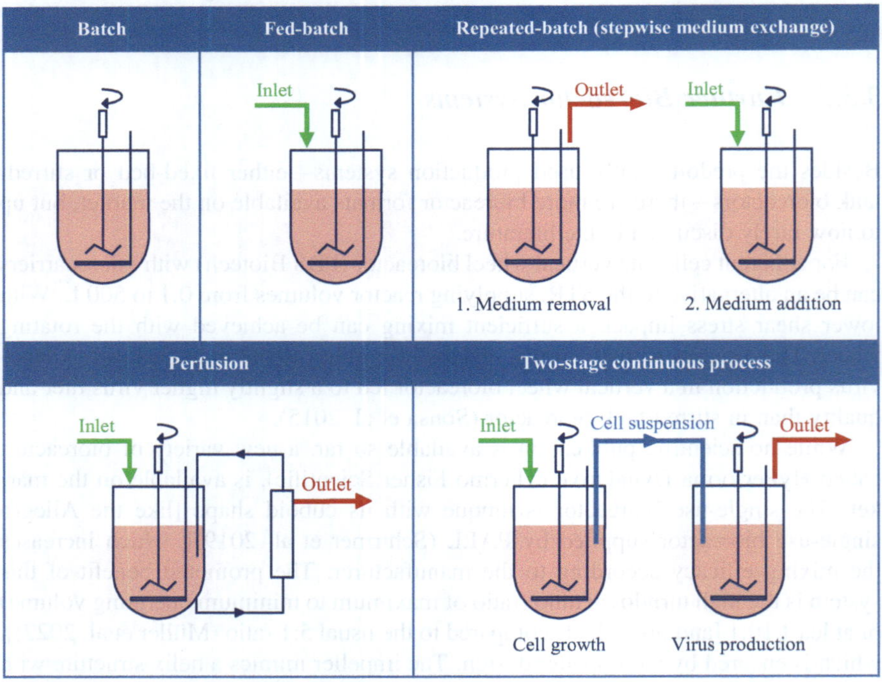

Fig. 3.2 Bioreactor operation modes for virus production

concentration (Thomassen et al. 2014; Dill et al. 2019b). Besides nutrient limitation (Maranga et al. 2003), inhibitory by-products can also hamper the virus productivity (Ozuer et al. 2002b; Kamen and Henry 2004). Vero cells tend to form multilayers once the growth surface is covered, which has also been associated with decreased cell-specific virus yield (Thomassen et al. 2014).

Cultivation in fed-batch mode prolongs the cell growth phase due to the concentrated feed medium, which is added during the cultivation, either stepwise or continuously (Fig. 3.2). The higher cell density can improve the virus yield. Developing a fed-batch process requires caution since the accumulation of inhibitory by-products is enhanced and might impair the virus yield.

One possibility to overcome this limitation is the repeated-batch mode. In repeated-batch (or semi-perfusion) cultivation, the medium is stepwise exchanged by first removing the spent medium and then adding fresh medium (Fig. 3.2). This technique can be easily implemented in adherent cultures without a cell retention device. In microcarrier cultures, the medium can be quickly exchanged after the beads settled (Thomassen et al. 2014). Cell damage through the medium exchange can be minimized by slowly feeding the new medium into the bioreactor, for example based on a cell-specific feed rate. This technique was referred to as repeated fed-batch cultivation by the authors (Bock et al. 2011).

Perfusion is an advanced bioreactor operation mode. In perfusion, the cell culture medium is continuously exchanged to fresh medium (Fig. 3.2). This technique not only provides nutrients for the cell growth and virus production, but also removes inhibitory waste products from the cultures. In the meantime, suspension cells are retained in the bioreactor using cell retention systems. Most often alternating tangential flow (ATF) (Genzel et al. 2014; Gränicher et al. 2019, 2021; Coronel et al. 2020), tangential flow filtration (TFF) (Gálvez et al. 2012), acoustic settler (Coronel et al. 2020; Gränicher et al. 2021), and inclined settler (Coronel et al. 2020; Alvim et al. 2021; Gränicher et al. 2021) are used for cell retention.

The major advantage of the perfusion is that it supports cell growth to high cell densities while maintaining high viability. For example, while cultivating HEK-293S cells in batch mode enabled infection at 1×10^6 cells mL^{-1}, cultivation with perfusion increased the cell density to 7.8×10^6 cells mL^{-1} for the infection with an adenoviral vector (Gálvez et al. 2012). 1×10^9 cells mL^{-1} is the highest theoretical cell density, which can be achieved based on calculations from packed tissue geometries (Ozturk 1996). Most often the aeration capacity limits the achievable cell concentration (Gálvez et al. 2012). At high cell densities, difficulties with stirring might also occur (Bielser et al. 2018). With perfusion, a smaller bioreactor volume is sufficient to achieve a comparable yield to batch and fed-batch processes. For this smaller-scale, single-use equipment is available, which simplifies the production process (Croughan et al. 2015; Bielser et al. 2018).

The perfusion rate is either specified as the exchanged medium volume per working volume per day (VVD) (Yuk et al. 2004) or specifically as cell-specific perfusion rate (CSPR) (Schwarz et al. 2020). Besides fixed perfusion rates (Schwarz et al. 2020), a dynamic perfusion operation mode is also possible and can be based on the cell growth rate, oxygen uptake rate (OUR), and nutrient concentration

(Gálvez et al. 2012; Genzel et al. 2014; Gränicher et al. 2019). Using dynamic perfusion, the cell-specific yield of the adenoviral vector production improved by a factor of four compared to the batch production without medium exchange (Gálvez et al. 2012). One drawback of the perfusion cultivation mode is the high medium demand, which is challenging in terms of storage and handling in large-scale production. Therefore, efforts have been made to concentrate or enrich the perfusion medium (Lin et al. 2017; Ozturk 1996) to reduce the medium consumption.

Perfusion can be used to continuously harvest viruses in the permeate (Petiot et al. 2011; Manceur et al. 2017; Coronel et al. 2020). For this purpose, the hollow fiber filter should be carefully chosen to enable the viruses to enter the permeate while cells are retained in the bioreactor. A critical parameter for the membrane selection is not only the membrane pore size but also the membrane material and related structural and physicochemical properties that influence filter fouling and thus, the effective membrane pore size (Nikolay et al. 2020). Continuous virus harvest is especially helpful for highly unstable viruses that degrade under the physicochemical conditions of the bioreactor. Besides, proteases that are released into the supernatant after cell lysis can further enhance virus degradation (Petiot et al. 2011). The harvested virus can be stored in cooled harvest tanks [e.g., at 2–8 °C (Petiot et al. 2011)] or directly purified in the downstream process. Generally, two strategies are applied to minimize the virus loss at infection through the perfusion. One strategy is to increase the MOI to ensure virus penetration into cells (Gálvez et al. 2012). Alternatively, the perfusion is stopped for a short time to allow the virus to infect the cells (Gálvez et al. 2012; Genzel et al. 2014).

Perfusion processes can also be used for only cell expansion, either as a pre-stage (N−1) to the N-stage virus production process or for the generation of high-density cell banks. Both approaches shorten the time of the seed train expansion. Additionally, the N−1 perfusion enables higher seeding cell densities for the N-stage bioreactor, which reduces the process time (Tao et al. 2011, 2012; Stepper et al. 2020; Rittershaus et al. 2022).

Perfusion was developed in the late 1980s to improve the productivity of the cell culture processes (Croughan et al. 2015). At that time, perfusion was mainly applied to produce labile products (Croughan et al. 2015). However, the rapidly changing and increasing market requires manufacturing flexibility—in terms of changing demand, new products, and manufacturing capacity—and reduced capital investments (Bielser et al. 2018). Thus, future manufacturing processes are envisioned as intensified and integrated continuous processes according to the biomanufacturing technology roadmap of the BioPhorum Operations Group (BioPhorum 2017). Perfusion is the fundament to reach this milestone.

Continuous production can increase the volumetric productivity and operational flexibility and decrease the manufacturing costs (Bielser et al. 2018; Gutiérrez-Granados et al. 2018). For influenza virus and Modified Vaccinia Ankara (MVA) virus, a two-segmented process with two bioreactors was applied with continuous medium supply (Frensing et al. 2013; Tapia et al. 2017; Gränicher et al. 2021). Here, one bioreactor was devoted for cell growth and another one for virus propagation. The bioreactors were connected by a continuous flow to provide cells for the virus-containing bioreactor (Fig. 3.2). For the influenza virus, this process format led to a periodical accumulation of DIPs (Frensing et al. 2013). These DIPs are reasoned by the "von-Magnus effect" and negatively impact the product quality (Magnus 1951). Decreasing the DIP content of the viral seed could not hinder the DIP generation. Using another cell line to decrease the *de novo* DIP generation rate only delayed the DIP accumulation (Frensing et al. 2013). For the MVA virus production, no DIP accumulation was observed. The results indicated that the time yield and space-time yield of the two-stage system can be at least as efficient as a batch process when the production is maintained for at least two batch cycles (Tapia et al. 2017). One drawback of the two-stage system is the low robustness and the high process complexity. It is also questionable whether the footprint can be reduced with this system considering that two bioreactors are needed for the continuous setup.

Tubular bioreactors (or plug-flow bioreactors) also enable continuous processing (Tapia et al. 2019), where the process progress is related to the location in the bioreactor. The use of a tubular bioreactor for continuous influenza production led to two times higher space-time yield compared to the batch production. Additionally, the DIP-induced oscillations in the yield could be avoided (Tapia et al. 2019). While the process format is in the early stages of development, it could be applied for the production of other viral vectors.

3.7 Process Development Concepts

Quality by Design (QbD) became an important part of bioprocess development and is supported by key regulatory authorities (Rathore 2014). In the traditional development approach, the process is first defined, and if the product quality expectations are met, the process is locked. With the QbD approach, the quality target product profile (QTPP) and the deriving critical quality attributes (CQA) are first defined that influence the product efficacy or safety. During process development, the influence of critical process parameters (CPP) and critical material attributes (CMA) is evaluated on the CQA. Then, the design space of CPPs and CMAs is established to consistently achieve the targeted product quality (Fig. 3.3) (Rathore 2014; Manser and Glenz 2021).

Various tools are available to follow the QbD approach. For example, risk assessment can be performed to identify CPPs. By performing Design of Experiments (DoE), the impact of selected process parameters as well as their interactions is statistically evaluated (Rathore 2014). The QbD approach is fully exploited by defining the design space with response surface models. The QbD principle with the help of DoE was successfully applied to pseudorabies virus production (Nie et al. 2020) and DNA vaccine production (Hocharoen et al. 2020). Besides this, the DoE approach itself is a very powerful tool to accelerate process development. DoE was used for creating a high-yielding recombinant AAV production platform (Zhao et al. 2020) as well as optimized transfection protocols for virus-like particles (VLP) (Lavado-García et al. 2020; Boix-Besora et al. 2022).

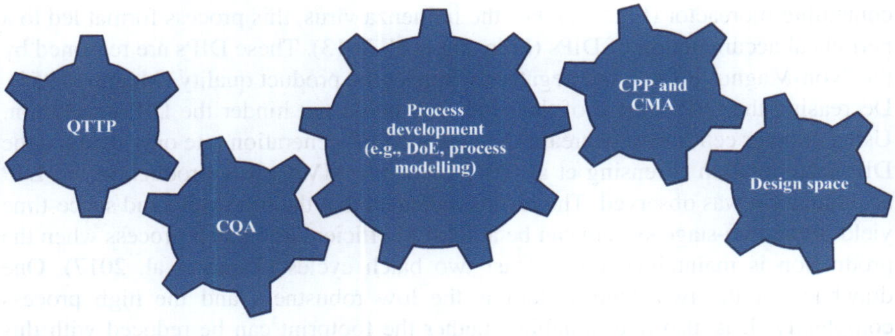

Fig. 3.3 The Quality by Design (QbD) concept. First the Quality target product profile (QTPP) and the deriving critical quality attributes (CQA) are defined that are relevant for product efficacy and safety. During process development the design space of critical process parameters (CPP) and critical material attributes (CMA) is defined to assure that the product quality is consistently achieved. Adapted from Manser and Glenz (2021)

Mechanistic models are also compatible with the QbD concept. Mechanistic models are based on physical, chemical, and biological principles (Tsopanoglou and Val 2021) and can be considered as the summary of process knowledge (Gernaey et al. 2010). Generating a mechanistic model is time intensive. However, once it is established and validated it can be used to mathematically describe the relationship between CPPs and CQAs and to define the design space (Gernaey et al. 2010). Deviations within the design space of critical attributes do not need to be reported for the regulatory authorities (Gernaey et al. 2010; Fernandes et al. 2012). Several models have been published for virus production that could aid the QbD approach (Ursache et al. 2015; Jiang et al. 2019; Nguyen et al. 2021). QbD modeling was applied to support the production of an RNA vaccine (Berg et al. 2021).

Multivariate data analysis (MVDA) can be used to examine variables and interactions despite of missing data, experimental error, or noise (Kirdar et al. 2007; Thomassen et al. 2010). MVDA was applied to extract knowledge from the historical data of inactivated polio vaccine manufacturing (Thomassen et al. 2010). It also helped to compare different production batches and evaluate process scalability (Kirdar et al. 2007).

Process analytical technology (PAT) provides the basis for the QbD approach. PAT is a mechanism to design, analyze, and control the manufacturing process by measuring product quality attributes, raw material attributes, or process parameters (Rathore and Winkle 2009). Different analytical tools are available to monitor attributes such as viable cell density and medium components (e.g., Raman spectroscopy, near-infrared spectroscopy, flow cytometry) as summarized by Rathore (2014). For monitoring the virus production, dielectric spectroscopy was successfully used in suspension and adherent cell cultures for several enveloped and non-enveloped viruses (Emma and Kamen 2013; Petiot et al. 2017; Grein et al. 2018). Contrary to the traditional fixed controlling strategy, controlling with PAT is dynamic and enables process adjustment to variabilities in the input parameters.

Consequently, the product quality remains consistent (Rathore 2014). While PAT methods are often only used for process monitoring, there are also examples for dynamic controlling. For instance, monitoring the cell growth with a dielectric spectroscopy sensor allowed to determine the TOI accurately and led to a consistent rAAV productivity (Pais et al. 2020). A dielectric spectroscopy sensor was also used to control the medium exchange rate in perfusion cultures (Hein et al. 2021).

3.8 Outlook

Virus vector-based therapies are an emerging and promising class of modalities. For a therapeutic effect, high virus doses are needed, which poses challenges for the manufacturing process. Furthermore, the rapidly changing and increasing market calls for manufacturing flexibility with reduced manufacturing and investment costs. Consequently, the focus of process development has shifted from the traditional batch processes toward intensified virus production processes, underlined by the growing number of published studies. Additionally, first continuous production approaches have been reported for several viruses. This should provide a good basis for the industry to move toward intensified and continuous manufacturing processes.

Product quality and process understanding have also become important aspects of upstream process development. There are major advances in tools (such as DoE, process modeling, MVDA) to investigate the impact of process variables on the product quality. PAT has also gained high interest for online process monitoring even though it has rarely been implemented for real-time decision-making. With time, however, the industry will measure up to the complexity of PAT-based control.

Overall, these ongoing activities pave the way for establishing highly efficient processes with consistent product quality to improve outcomes of clinical trials and solve unmet medical needs in the future.

References

Abdolahi F, Moosavia MA, Vatani A (2005) The mechanism of action of antifoams. J Appl Sci 5: 1122–1129. https://doi.org/10.3923/jas.2005.1122.1129

Alvim RGF, Lima TM, Silva JL et al (2021) Process intensification for the production of yellow fever virus-like particles as potential recombinant vaccine antigen. Biotechnol Bioeng 118: 3581–3592. https://doi.org/10.1002/bit.27864

Andari JE, Grimm D (2021) Production, processing, and characterization of synthetic AAV gene therapy vectors. Biotechnol J 16:2000025. https://doi.org/10.1002/biot.202000025

Barrett PN, Mundt W, Kistner O, Howard MK (2014) Vero cell platform in vaccine production: moving towards cell culture-based viral vaccines. Expert Rev Vaccines 8:607–618. https://doi.org/10.1586/erv.09.19

Berg D, Kis Z, Behmer CF et al (2021) Quality by design modelling to support rapid RNA vaccine production against emerging infectious diseases. NPJ Vaccines 6:65. https://doi.org/10.1038/s41541-021-00322-7

Bielser J-M, Wolf M, Souquet J et al (2018) Perfusion mammalian cell culture for recombinant protein manufacturing – a critical review. Biotechnol Adv 36:1328–1340. https://doi.org/10.1016/j.biotechadv.2018.04.011

BioPhorum (2017) Biomanufacturing technology roadmap. https://www.biophorum.com/download/overview/. Accessed 8 Aug 2022

Bluming Avrum Z, Ziegler John L (1971) Regression of Burkitt's lymphoma in association with measles infection. Lancet 298:105–106. https://doi.org/10.1016/s0140-6736(71)92086-1

Bock A, Sann H, Schulze-Horsel J et al (2009) Growth behavior of number distributed adherent MDCK cells for optimization in microcarrier cultures. Biotechnol Progr 25:1717–1731. https://doi.org/10.1002/btpr.262

Bock A, Schulze-Horsel J, Schwarzer J et al (2011) High-density microcarrier cell cultures for influenza virus production. Biotechnol Progr 27:241–250. https://doi.org/10.1002/btpr.539

Boix-Besora A, Lorenzo E, Lavado-García J et al (2022) Optimization, production, purification and characterization of HIV-1 GAG-based virus-like particles functionalized with SARS-CoV-2. Nato Adv Sci Inst Se 10:250. https://doi.org/10.3390/vaccines10020250

Byer H, Wang W, Mogilyanskiy L (2015) Mastering cell bank production. BioPharm Int 28:20–24

Chen Y-H, Wu J-C, Wang K-C et al (2005) Baculovirus-mediated production of HDV-like particles in BHK cells using a novel oscillating bioreactor. J Biotechnol 118:135–147. https://doi.org/10.1016/j.jbiotec.2005.02.018

Chen S, Sato Y, Tada Y et al (2021) Facile bead-to-bead cell-transfer method for serial subculture and large-scale expansion of human mesenchymal stem cells in bioreactors. Stem Cell Transl Med 10:1329–1342. https://doi.org/10.1002/sctm.20-0501

Coronel J, Gränicher G, Sandig V et al (2020) Application of an inclined settler for cell culture-based influenza A virus production in perfusion mode. Front Bioeng Biotechnol 8:672. https://doi.org/10.3389/fbioe.2020.00672

Croughan MS, Konstantinov KB, Cooney C (2015) The future of industrial bioprocessing: batch or continuous? Biotechnol Bioeng 112:648–651. https://doi.org/10.1002/bit.25529

Dekevic G, Tasto L, Czermak P, Salzig D (2022) Statistical experimental designs to optimize the transient transfection of HEK 293T cells and determine a transfer criterion from adherent cells to larger-scale cell suspension cultures. J Biotechnol 346:23–34. https://doi.org/10.1016/j.jbiotec.2022.01.004

Dill V, Ehret J, Zimmer A et al (2019a) Cell density effects in different cell culture media and their impact on the propagation of foot-and-mouth disease virus. Viruses 11:511. https://doi.org/10.3390/v11060511

Dill V, Zimmer A, Beer M, Eschbaumer M (2019b) Investigation of cell culture conditions for optimal foot-and-mouth disease virus production. BMC Biotechnol 19:33. https://doi.org/10.1186/s12896-019-0527-5

Dock G (1904) The influence of complicating diseases upon leukaemia.*. Am J Med Sci. https://doi.org/10.1097/00000441-190404000-00001

Elahi SM, Shen CF, Gilbert R (2019) Optimization of production of vesicular stomatitis virus (VSV) in suspension serum-free culture medium at high cell density. J Biotechnol 289:144–149. https://doi.org/10.1016/j.jbiotec.2018.11.023

EMA (2016) Imlygic, INN-talimogene laherparepvec. In: Imlygic: EPAR – Product Information. https://www.ema.europa.eu/en/documents/product-information/imlygic-epar-product-information_en.pdf. Accessed 9 Jun 2022

Emma P, Kamen A (2013) Real-time monitoring of influenza virus production kinetics in HEK293 cell cultures. Biotechnol Progr 29:275–284. https://doi.org/10.1002/btpr.1601

Fernandes RL, Bodla VK, Carlquist M et al (2012) Measurement, monitoring, modelling and control of bioprocesses. Adv Biochem Eng Biotechnol 132:137–166. https://doi.org/10.1007/10_2012_166

Frensing T, Heldt FS, Pflugmacher A et al (2013) Continuous influenza virus production in cell culture shows a periodic accumulation of defective interfering particles. PLoS One 8:e72288. https://doi.org/10.1371/journal.pone.0072288
Gallo-Ramírez LE, Nikolay A, Genzel Y, Reichl U (2015) Bioreactor concepts for cell culture-based viral vaccine production. Expert Rev Vaccines 14:1181–1195. https://doi.org/10.1586/14760584.2015.1067144
Gálvez J, Lecina M, Solà C et al (2012) Optimization of HEK-293S cell cultures for the production of adenoviral vectors in bioreactors using on-line OUR measurements. J Biotechnol 157:214–222. https://doi.org/10.1016/j.jbiotec.2011.11.007
Genzel Y (2015) Designing cell lines for viral vaccine production: where do we stand? Biotechnol J 10:728–740. https://doi.org/10.1002/biot.201400388
Genzel Y, Olmer RM, Schäfer B, Reichl U (2006) Wave microcarrier cultivation of MDCK cells for influenza virus production in serum containing and serum-free media. Vaccine 24:6074–6087. https://doi.org/10.1016/j.vaccine.2006.05.023
Genzel Y, Vogel T, Buck J et al (2014) High cell density cultivations by alternating tangential flow (ATF) perfusion for influenza A virus production using suspension cells. Vaccine 32:2770–2781. https://doi.org/10.1016/j.vaccine.2014.02.016
Gernaey KV, Lantz AE, Tufvesson P et al (2010) Application of mechanistic models to fermentation and biocatalysis for next-generation processes. Trends Biotechnol 28:346–354. https://doi.org/10.1016/j.tibtech.2010.03.006
Giard DJ, Thilly WG, Wang DI, Levine DW (1977) Virus production with a newly developed microcarrier system. Appl Environ Microb 34:668–672. https://doi.org/10.1128/aem.34.6.668-672.1977
Gränicher G, Coronel J, Pralow A et al (2019) Efficient influenza A virus production in high cell density using the novel porcine suspension cell line PBG.PK2.1. Vaccine 37:7019–7028. https://doi.org/10.1016/j.vaccine.2019.04.030
Gränicher G, Tapia F, Behrendt I et al (2021) Production of Modified Vaccinia Ankara virus by intensified cell cultures: a comparison of platform technologies for viral vector production. Biotechnol J 16:2000024. https://doi.org/10.1002/biot.202000024
Grein TA, Loewe D, Dieken H et al (2018) High titer oncolytic measles virus production process by integration of dielectric spectroscopy as online monitoring system. Biotechnol Bioeng 115: 1186–1194. https://doi.org/10.1002/bit.26538
Grein TA, Loewe D, Dieken H et al (2019) Aeration and shear stress are critical process parameters for the production of oncolytic measles virus. Front Bioeng Biotechnol 7:78. https://doi.org/10.3389/fbioe.2019.00078
Grieger JC, Soltys SM, Samulski RJ (2016) Production of recombinant adeno-associated virus vectors using suspension HEK293 cells and continuous harvest of vector from the culture media for GMP FIX and FLT1 clinical vector. Mol Ther 24:287–297. https://doi.org/10.1038/mt.2015.187
Guan J-S, Chen K, Si Y et al (2022) Process improvement of adeno-associated virus production. Front Chem Eng 4. https://doi.org/10.3389/fceng.2022.830421
Gutiérrez-Granados S, Gòdia F, Cervera L (2018) Continuous manufacturing of viral particles. Curr Opin Chem Eng 22:107–114. https://doi.org/10.1016/j.coche.2018.09.009
Hegde NR (2015) Cell culture-based influenza vaccines: a necessary and indispensable investment for the future. Hum Vacc Immunother 11:1223–1234. https://doi.org/10.1080/21645515.2015.1016666
Hein MD, Arora P, Marichal-Gallardo P et al (2021) Cell culture-based production and in vivo characterization of purely clonal defective interfering influenza virus particles. BMC Biol 19:91. https://doi.org/10.1186/s12915-021-01020-5
Hocharoen L, Noppiboon S, Kitsubun P (2020) Process characterization by definitive screening design approach on DNA vaccine production. Front Bioeng Biotechnol 8:574809. https://doi.org/10.3389/fbioe.2020.574809

Huo C, Tian J, Cheng J et al (2020) Safety, immunogenicity, and effectiveness of defective viral particles arising in mast cells against influenza in mice. Front Immunol 11:585254. https://doi.org/10.3389/fimmu.2020.585254

Jiang Y, Welle JE van der, Rubingh O, et al (2019) Kinetic model for adherent Vero cell growth and poliovirus production in batch bioreactors. Process Biochem 81:156–164. doi:https://doi.org/10.1016/j.procbio.2019.03.010

Kamen A, Henry O (2004) Development and optimization of an adenovirus production process. J Gene Med 6:S184–S192. https://doi.org/10.1002/jgm.503

Kiesslich S, Kamen AA (2020) Vero cell upstream bioprocess development for the production of viral vectors and vaccines. Biotechnol Adv 44:107608. https://doi.org/10.1016/j.biotechadv.2020.107608

Kiesslich S, Losa JPV-C, Gélinas J-F, Kamen AA (2020) Serum-free production of rVSV-ZEBOV in Vero cells: microcarrier bioreactor versus scale-X™ hydro fixed-bed. J Biotechnol 310:32–39. https://doi.org/10.1016/j.jbiotec.2020.01.015

Kiesslich S, Kim GN, Shen CF et al (2021) Bioreactor production of rVSV-based vectors in Vero cell suspension cultures. Biotechnol Bioeng 118:2649–2659. https://doi.org/10.1002/bit.27785

Kirdar AO, Conner JS, Baclaski J, Rathore AS (2007) Application of multivariate analysis toward biotech processes: case study of a cell-culture unit operation. Biotechnol Progr 23:61–67. https://doi.org/10.1021/bp060377u

Kuroda S, Miyagawa Y, Sato Y et al (2020) Protocol optimization for the production of the non-cytotoxic JΔNI5 HSV vector deficient in expression of immediately early genes. Mol Ther Methods Clin Dev 17:612–621. https://doi.org/10.1016/j.omtm.2020.03.014

Lapošová K, Pastoreková S, Tomášková J (2013) Lymphocytic choriomeningitis virus: invisible but not innocent. Acta Virol 57:160–170. https://doi.org/10.4149/av_2013_02_160

Laposova K, Oveckova I, Tomaskova J (2017) A simple method for isolation of cell-associated viral particles from cell culture. J Virol Methods 249:194–196. https://doi.org/10.1016/j.jviromet.2017.09.014

Lavado-García J, Cervera L, Gòdia F (2020) An alternative perfusion approach for the intensification of virus-like particle production in HEK293 cultures. Front Bioeng Biotechnol 8:617. https://doi.org/10.3389/fbioe.2020.00617

Leber J, Barekzai J, Blumenstock M et al (2017) Microcarrier choice and bead-to-bead transfer for human mesenchymal stem cells in serum-containing and chemically defined media. Process Biochem 59:255–265. https://doi.org/10.1016/j.procbio.2017.03.017

Lee D-K, Park J, Seo D-W (2020) Suspension culture of Vero cells for the production of adenovirus type 5. Clin Exp Vaccine Res 9:48–55. https://doi.org/10.7774/cevr.2020.9.1.48

Leinonen HM, Lipponen EM, Valkama AJ et al (2019) Preclinical proof-of-concept, analytical development, and commercial scale production of lentiviral vector in adherent cells. Mol Ther Methods Clin Dev 15:63–71. https://doi.org/10.1016/j.omtm.2019.08.006

Leinonen HM, Lepola S, Lipponen EM et al (2020) Benchmarking of scale-X bioreactor system in lentiviral and adenoviral vector production. Hum Gene Ther 31:376–384. https://doi.org/10.1089/hum.2019.247

Lesch HP, Valonen P, Karhinen M (2021) Evaluation of the single-use fixed-bed bioreactors in scalable virus production. Biotechnol J 16:2000020. https://doi.org/10.1002/biot.202000020

Lim HS, Chang KH, Kim JH (1999) Effect of oxygen partial pressure on production of animal virus (VSV). Cytotechnology 31:265–270. https://doi.org/10.1023/a:1008060502532

Lin H, Leighty RW, Godfrey S, Wang SB (2017) Principles and approach to developing mammalian cell culture media for high cell density perfusion process leveraging established fed-batch media. Biotechnol Progr 33:891–901. https://doi.org/10.1002/btpr.2472

Loo JCM, Swaney WP, Grassman E et al (2012) Critical variables affecting clinical-grade production of the self-inactivating gamma-retroviral vector for the treatment of X-linked severe combined immunodeficiency. Gene Ther 19:872–876. https://doi.org/10.1038/gt.2012.37

Mackett M, Smith GL, Moss B (1982) Vaccinia virus: a selectable eukaryotic cloning and expression vector. Proc Natl Acad Sci 79:7415–7419. https://doi.org/10.1073/pnas.79.23.7415

Madeline B, Ribaud S, Xenopoulos A et al (2015) Culturing a Duck ES-derived cell line in single-use bioreactors. BioProcess Int 3:26–33

Magnus P (1951) Propagation of the PR8 strain of influenza A virus in chick embryos. Acta Pathologica Microbiol Scand 28:278–293. https://doi.org/10.1111/j.1699-0463.1951.tb03693.x

Manceur AP, Kim H, Misic V et al (2017) Scalable lentiviral vector production using stable HEK293SF producer cell lines. Hum Gene Ther Method 28:330–339. https://doi.org/10.1089/hgtb.2017.086

Manser B, Glenz M (2021) Technical regulatory topic: quality by design in continuous bioprocesses. https://www.pall.com/content/dam/pall/biopharm/regulatory/sme/ustr-3491-quality-by-design-in-continuous-bioprocesses.pdf. Accessed 9 Jun 2022

Maranga L, Brazão TF, Carrondo MJT (2003) Virus-like particle production at low multiplicities of infection with the baculovirus insect cell system. Biotechnol Bioeng 84:245–253. https://doi.org/10.1002/bit.10773

Marcelino I, Sousa MFQ, Veríssimo C et al (2006) Process development for the mass production of Ehrlichia ruminantium. Vaccine 24:1716–1725. https://doi.org/10.1016/j.vaccine.2005.08.109

Martínez-Molina E, Chocarro-Wrona C, Martínez-Moreno D et al (2020) Large-scale production of lentiviral vectors: current perspectives and challenges. Pharm 12:1051. https://doi.org/10.3390/pharmaceutics12111051

Moleirinho MG, Rosa S, Carrondo MJT et al (2018) Clinical-grade oncolytic adenovirus purification using polysorbate 20 as an alternative for cell lysis. Curr Gene Ther 18:366–374. https://doi.org/10.2174/1566523218666181109141257

Moreira AS, Silva AC, Sousa MFQ et al (2020) Establishing suspension cell cultures for improved manufacturing of oncolytic adenovirus. Biotechnol J 15:1900411. https://doi.org/10.1002/biot.201900411

Mothes W, Sherer NM, Jin J, Zhong P (2010) Virus cell-to-cell transmission. J Virol 84:8360–8368. https://doi.org/10.1128/jvi.00443-10

Mullard A (2018) 2017 FDA drug approvals. Nat Rev Drug Discov 17:81–85. https://doi.org/10.1038/nrd.2018.4

Mullard A (2020) 2019 FDA drug approvals. Nat Rev Drug Discov 19:79–84. https://doi.org/10.1038/d41573-020-00001-7

Müller J, Teale M, Steiner S et al (2022) Cell culture engineering and technology, in appreciation to Professor Mohamed Al-Rubeai. Cell Eng:401–429. https://doi.org/10.1007/978-3-030-79871-0_13

Nguyen TNT, Sha S, Hong MS et al (2021) Mechanistic model for production of recombinant adeno-associated virus via triple transfection of HEK293 cells. Mol Ther Methods Clin Dev 21:642–655. https://doi.org/10.1016/j.omtm.2021.04.006

Nie J, Sun Y, Han F et al (2020) Rapid process development of serum-free pseudorabies virus production with the Quality by Design approach. Cytotechnology 72:283–293. https://doi.org/10.1007/s10616-020-00377-9

Nikolay A, Grooth J, Genzel Y et al (2020) Virus harvesting in perfusion culture: choosing the right type of hollow fiber membrane. Biotechnol Bioeng 117:3040–3052. https://doi.org/10.1002/bit.27470

Nuwarda RF, Alharbi AA, Kayser V (2021) An overview of influenza viruses and vaccines. Nato Adv Sci Inst Se 9:1032. https://doi.org/10.3390/vaccines9091032

Offersgaard A, Hernandez CRD, Pihl AF et al (2021) SARS-CoV-2 production in a scalable high cell density bioreactor. Nato Adv Sci Inst Se 9:706. https://doi.org/10.3390/vaccines9070706

Offersgaard A, Hernandez CRD, Pihl AF et al (2022) High-titer hepatitis C virus production in a scalable single-use high cell density bioreactor. Nato Adv Sci Inst Se 10:249. https://doi.org/10.3390/vaccines10020249

Ozturk SS (1996) Engineering challenges in high density cell culture systems. Cytotechnology 22:3–16. https://doi.org/10.1007/bf00353919

Ozuer A, Wechuck JB, Goins WF et al (2002a) Effect of genetic background and culture conditions on the production of herpesvirus-based gene therapy vectors. Biotechnol Bioeng 77:685–692. https://doi.org/10.1002/bit.10162

Ozuer A, Wechuck JB, Russell B et al (2002b) Evaluation of infection parameters in the production of replication-defective HSV-1 viral vectors. Biotechnol Progr 18:476–482. https://doi.org/10.1021/bp010176k

Pack GT (1950) Note on the experimental use of rabies vaccine for melanomatosis. M Archiv Dermatol Syphilol 62:694–695. https://doi.org/10.1001/archderm.1950.01530180083015

Paillet C, Forno G, Kratje R, Etcheverrigaray M (2009) Suspension-Vero cell cultures as a platform for viral vaccine production. Vaccine 27:6464–6467. https://doi.org/10.1016/j.vaccine.2009.06.020

Pais DAM, Brown C, Neuman A et al (2020) Dielectric spectroscopy to improve the production of rAAV used in gene therapy. Process 8:1456. https://doi.org/10.3390/pr8111456

Panicali D, Paoletti E (1982) Construction of poxviruses as cloning vectors: insertion of the thymidine kinase gene from herpes simplex virus into the DNA of infectious vaccinia virus. Proc Natl Acad Sci 79:4927–4931. https://doi.org/10.1073/pnas.79.16.4927

Petiot E, Jacob D, Lanthier S et al (2011) Metabolic and kinetic analyses of influenza production in perfusion HEK293 cell culture. BMC Biotechnol 11:84. https://doi.org/10.1186/1472-6750-11-84

Petiot E, Ansorge S, Rosa-Calatrava M, Kamen A (2017) Critical phases of viral production processes monitored by capacitance. J Biotechnol 242:19–29. https://doi.org/10.1016/j.jbiotec.2016.11.010

Petricciani JC (1995) The acceptability of continuous cell lines: a personal and historical perspective. Cytotechnology 18:9–13. https://doi.org/10.1007/bf00744314

Powers AD, Drury JE, Hoehamer CF et al (2020) Lentiviral vector production from a stable packaging cell line using a packed bed bioreactor. Mol Ther Methods Clin Dev 19:1–13. https://doi.org/10.1016/j.omtm.2020.08.010

Rafiq QA, Coopman K, Nienow AW, Hewitt CJ (2016) Systematic microcarrier screening and agitated culture conditions improves human mesenchymal stem cell yield in bioreactors. Biotechnol J 11:473–486. https://doi.org/10.1002/biot.201400862

Rathore AS (2014) QbD/PAT for bioprocessing: moving from theory to implementation. Curr Opin Chem Eng 6:1–8. https://doi.org/10.1016/j.coche.2014.05.006

Rathore AS, Winkle H (2009) Quality by Design for biopharmaceuticals. Nat Biotechnol 27:26–34. https://doi.org/10.1038/nbt0109-26

Rittershaus ESC, Rehmann MS, Xu J et al (2022) N-1 perfusion platform development using a capacitance probe for biomanufacturing. Bioengineering 9:128. https://doi.org/10.3390/bioengineering9040128

Rourou S, van der Ark A, van der Velden T, Kallel H (2007) A microcarrier cell culture process for propagating rabies virus in Vero cells grown in a stirred bioreactor under fully animal component free conditions. Vaccine 25:3879–3889. https://doi.org/10.1016/j.vaccine.2007.01.086

Rourou S, Ayed YB, Trabelsi K et al (2014) An animal component free medium that promotes the growth of various animal cell lines for the production of viral vaccines. Vaccine 32:2767–2769. https://doi.org/10.1016/j.vaccine.2014.02.040

Routledge SJ (2012) Beyond de-foaming: the effects of antifoams on bioprocess productivity. Comput Struct Biotechnol J 3:1–7. https://doi.org/10.5936/csbj.201210014

Routledge SJ, Poyner DR, Bill RM (2014) Antifoams: the overlooked additive? Biopharm Bioprocess 2:103–106. https://doi.org/10.2217/pbp.14.5

Schirmer C, Müller J, Steffen N et al (2019) Animal cell biotechnology, methods and protocols. Methods Mol Biol 2095:169–186. https://doi.org/10.1007/978-1-0716-0191-4_10

Schwarz H, Zhang Y, Zhan C et al (2020) Small-scale bioreactor supports high density HEK293 cell perfusion culture for the production of recombinant Erythropoietin. J Biotechnol 309:44–52. https://doi.org/10.1016/j.jbiotec.2019.12.017

Shen CF, Guilbault C, Li X et al (2019) Development of suspension adapted Vero cell culture process technology for production of viral vaccines. Vaccine 37:6996–7002. https://doi.org/10.1016/j.vaccine.2019.07.003

Silva AC, Delgado I, Sousa MFQ et al (2008) Scalable culture systems using different cell lines for the production of Peste des Petits ruminants vaccine. Vaccine 26:3305–3311. https://doi.org/10.1016/j.vaccine.2008.03.077

Slivac I, Srček VG, RadoŠević K et al (2006) Aujeszky's disease virus production in disposable bioreactor. J Biosci 31:363–368. https://doi.org/10.1007/bf02704109

Sousa MFQ, Silva MM, Giroux D et al (2015) Production of oncolytic adenovirus and human mesenchymal stem cells in a single-use, Vertical-Wheel bioreactor system: impact of bioreactor design on performance of microcarrier-based cell culture processes. Biotechnol Progr 31:1600–1612. https://doi.org/10.1002/btpr.2158

Stanton D (2019) ABEC launches 6,000 L single-use bioreactor. https://bioprocessintl.com/bioprocess-insider/upstream-downstream-processing/abec-breaks-plastic-ceiling-again-with-6000-l-single-use-bioreactor/. Accessed 26 Sep 2022

Stepper L, Filser FA, Fischer S et al (2020) Pre-stage perfusion and ultra-high seeding cell density in CHO fed-batch culture: a case study for process intensification guided by systems biotechnology. Bioproc Biosyst Eng 43:1431–1443. https://doi.org/10.1007/s00449-020-02337-1

Tan E, Chin CSH, Lim ZFS, Ng SK (2021) HEK293 cell line as a platform to produce recombinant proteins and viral vectors. Front Bioeng Biotechnol 9:796991. https://doi.org/10.3389/fbioe.2021.796991

Tao Y, Shih J, Sinacore M et al (2011) Development and implementation of a perfusion-based high cell density cell banking process. Biotechnol Progr 27:824–829. https://doi.org/10.1002/btpr.599

Tao Y, Yusuf-Makagiansar H, Shih J et al (2012) Novel cholesterol feeding strategy enables a high-density cultivation of cholesterol-dependent NS0 cells in linear low-density polyethylene-based disposable bioreactors. Biotechnol Lett 34:1453–1458. https://doi.org/10.1007/s10529-012-0915-6

Tapia F, Vázquez-Ramírez D, Genzel Y, Reichl U (2016) Bioreactors for high cell density and continuous multi-stage cultivations: options for process intensification in cell culture-based viral vaccine production. Appl Microbiol Biot 100:2121–2132. https://doi.org/10.1007/s00253-015-7267-9

Tapia F, Jordan I, Genzel Y, Reichl U (2017) Efficient and stable production of Modified Vaccinia Ankara virus in two-stage semi-continuous and in continuous stirred tank cultivation systems. PLoS One 12:e0182553. https://doi.org/10.1371/journal.pone.0182553

Tapia F, Wohlfarth D, Sandig V et al (2019) Continuous influenza virus production in a tubular bioreactor system provides stable titers and avoids the "von Magnus effect". PLoS One 14:e0224317. https://doi.org/10.1371/journal.pone.0224317

Taqi AM, Abdurrahman MB, Yakubu AM, Fleming AF (1981) Regression of Hodgkin's disease after measles. Lancet 317:1112. https://doi.org/10.1016/s0140-6736(81)92286-8

Tharmalingam T, Ghebeh H, Wuerz T, Butler M (2008) Pluronic enhances the robustness and reduces the cell attachment of mammalian cells. Mol Biotechnol 39:167–177. https://doi.org/10.1007/s12033-008-9045-8

Thermo Fisher Scientific (2022a) Application Note: Scale-up evaluation of the HyPerforma DynaDrive S.U.B.s. https://assets.thermofisher.com/TFS-Assets/BPD/Application-Notes/evaluation-of-hyperforma-dynadrive-application-note.pdf. Accessed 23 Jul 2022

Thermo Fisher Scientific (2022b) Application Note: Streamlining seed train scale-up through utilization of the turndown ratio of the HyPerforma DynaDrive S.U.B. https://assets.thermofisher.com/TFS-Assets/BPD/Application-Notes/dyna_drive_single_use_bioreactor_turn_down_ratio_application_note.pdf. Accessed 23 Jul 2022

Thomas DL, Wang L, Niamke J et al (2009) Scalable recombinant adeno-associated virus production using recombinant herpes simplex virus type 1 coinfection of suspension-adapted mammalian cells. Hum Gene Ther 20:861–870. https://doi.org/10.1089/hum.2009.004

Thomassen YE, van Sprang ENM, van der Pol LA, Bakker WAM (2010) Multivariate data analysis on historical IPV production data for better process understanding and future improvements. Biotechnol Bioeng 107:96–104. https://doi.org/10.1002/bit.22788

Thomassen YE, Rubingh O, Wijffels RH et al (2014) Improved poliovirus d-antigen yields by application of different Vero cell cultivation methods. Vaccine 32:2782–2788. https://doi.org/10.1016/j.vaccine.2014.02.022

Trabelsi K, Rourou S, Loukil H et al (2006) Optimization of virus yield as a strategy to improve rabies vaccine production by Vero cells in a bioreactor. J Biotechnol 121:261–271. https://doi.org/10.1016/j.jbiotec.2005.07.018

Tsopanoglou A, del Val IJ (2021) Moving towards an era of hybrid modelling: advantages and challenges of coupling mechanistic and data-driven models for upstream pharmaceutical bioprocesses. Curr Opin Chem Eng 32:100691. https://doi.org/10.1016/j.coche.2021.100691

Ursache RV, Thomassen YE, van Eikenhorst G et al (2015) Mathematical model of adherent Vero cell growth and poliovirus production in animal component free medium. Bioproc Biosyst Eng 38:543–555. https://doi.org/10.1007/s00449-014-1294-2

Valkama AJ, Leinonen HM, Lipponen EM et al (2018) Optimization of lentiviral vector production for scale-up in fixed-bed bioreactor. Gene Ther 25:39–46. https://doi.org/10.1038/gt.2017.91

Wang X, Olszewska M, Qu J et al (2015) Large-scale clinical-grade retroviral vector production in a fixed-bed bioreactor. J Immunother Hagerstown Md 38:127–135. https://doi.org/10.1097/cji.0000000000000072

Weiss K, Gerstenberger J, Salzig D et al (2015) Oncolytic measles viruses produced at different scales under serum-free conditions. Eng Life Sci 15:425–436. https://doi.org/10.1002/elsc.201400165

Wezel AL (1967) Growth of cell-strains and primary cells on micro-carriers in homogeneous culture. Nature 216:64–65. https://doi.org/10.1038/216064a0

Wohlfarth D, Frehtman V, Müller M et al (2021) Upstream process optimization and micro- and macrocarrier screening for large-scale production of the oncolytic H-1 protoparvovirus. Appl Microbiol Biot 105:9113–9124. https://doi.org/10.1007/s00253-021-11642-y

Yang J, Guertin P, Jia G et al (2019a) Large-scale microcarrier culture of HEK293T cells and Vero cells in single-use bioreactors – PMC. AMB Express 70. https://doi.org/10.1186/s13568-019-0794-5

Yang Y, Lyu T, Zhou R et al (2019b) The antiviral and antitumor effects of defective interfering particles/genomes and their mechanisms. Front Microbiol 10:1852. https://doi.org/10.3389/fmicb.2019.01852

Yuk IHY, Olsen MM, Geyer S, Forestell SP (2004) Perfusion cultures of human tumor cells: a scalable production platform for oncolytic adenoviral vectors. Biotechnol Bioeng 86:637–642. https://doi.org/10.1002/bit.20158

Zhao H, Lee K-J, Daris M et al (2020) Creation of a high-yield AAV vector production platform in suspension cells using a Design-of-Experiment approach. Mol Ther Methods Clin Dev 18:312–320. https://doi.org/10.1016/j.omtm.2020.06.004

Chapter 4
An Overview of the Cell Biology and Genetic Barriers to Virus Infections with Implications in Cell Line Development for Virotherapy

Abhilash I. Chiramel, Ryan Kissinger, and Ingo H. Gorr

4.1 Introduction

Utilizing viruses as therapeutics belongs to a new class of treatments, termed *virotherapy* (Letarov 2020; Moaven et al. 2021; Haan et al. 2020) (Fig. 4.1). Viruses can be genetically manipulated to function as oncolytic agents for immuno-oncology and vectors for cell and gene therapy (Moaven et al. 2021; Haan et al. 2020; Chung et al. 2020). Although virus-based therapeutics has the potential to be a new pillar of modern medicine, many challenges need to be addressed from optimizing bioprocessing to reducing cost and variability in manufacturing (Perry and Rayat 2021; Pais et al. 2019).

For virotherapy to be an accessible and feasible modality, maximizing production in a cost-effective manner is of utmost importance. Virotherapy exploits both live replicating and non-replicating viral vectors for oncolysis, immunomodulation, and gene delivery. A sound understanding of virology, cell biology, and immune responses to virus infection is crucial to develop therapeutic viruses. Cell line selection and development is the first step in biopharmaceutical manufacturing. Production of viral vectors—replication incompetent or competent—requires cell line selection in combination with cell line engineering strategies to maximize the virus producing capability of cells. For viruses, major barriers to infection are imposed by (1) availability of appropriate entry receptor(s), (2) cell's innate antiviral immune response, and (3) cellular metabolic and catabolic pathways. Hence, three

A. I. Chiramel (✉) · I. H. Gorr
Boehringer Ingelheim Pharma GmbH & Co KG, Biberach, Germany
e-mail: abhilash.chiramel@boehringer-ingelheim.com

R. Kissinger
Kissinger Medical Visuals, Nelson, BC, Canada

S. Gautam et al. (eds.), *Bioprocess and Analytics Development for Virus-based Advanced Therapeutics and Medicinal Products (ATMPs)*,
https://doi.org/10.1007/978-3-031-28489-2_4

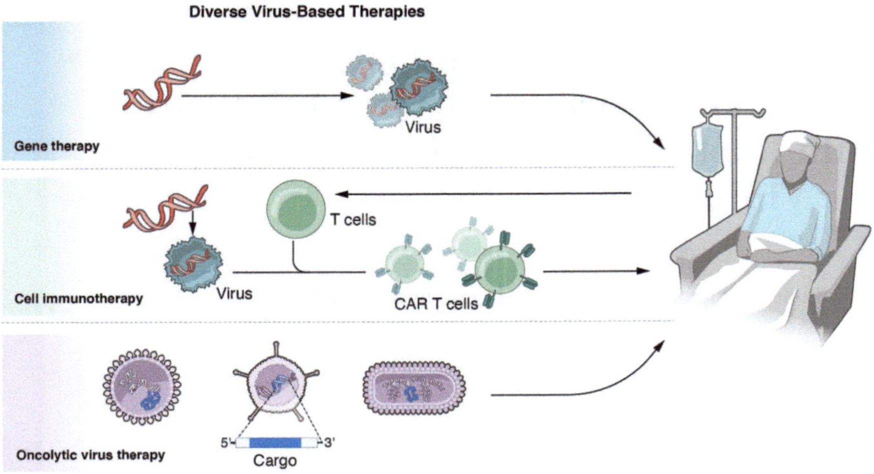

Fig. 4.1 Diverse viruses for different virotherapies. Adeno-associated viruses (AAV), adenoviruses, retroviruses for CAR-T cell therapy, and oncolytic viruses for cancer therapy

main requirements must be met to successfully initiate infection of cells: a minimal number of infectious virus particles, access of particles to susceptible and permissive cells, and dampened or inactive cellular antiviral defenses. Virus spread beyond the site of infection depends on the initial infection dose, availability of viral receptors on cells, relative rates of induction of antiviral signaling, and release of infectious particles. Importantly, most viruses are restricted to specific cell types. Thus, selecting cell line and engineering cells to be permissive to infection is crucial to optimize therapeutic virus production. Viruses are entirely dependent on host cells and incorporate host cell proteins within the virion or as a part of the lipid-rich viral envelope. Profiling biochemical modifications and host cell proteins (HCPs) which regulate virus infectivity is key in defining active pharmaceutical ingredients (APIs) essential for making therapeutic viruses. Lastly, it is important to consider virus morphology and whether the therapeutic virus is non-enveloped or enveloped. This chapter summarizes the biology of viral vectors currently employed in immuno-oncology and gene and cell therapy. Moreover, we focus on emphasizing barriers to virus infection and strategies employed to engineer innate immune effectors to dampen antiviral pathways and enhance cellular permissiveness to virus infection.

4.2 Barriers to Virus Infection

Virus infections are often compared to and in terminology associated with warfare. Virus–host interactions entail opposing forces, each with evolving strategies to defeat the other. A key concept in virus–host interactions is the term *tropism*: the cellular and anatomical parameters which allow cell types to support all stages of the viral life cycle. Tropism is mainly dictated by three parameters: *susceptibility*—cells must have receptors that viruses can bind to and promote entry or lack potent basally expressed genetic restriction factors; *permissiveness*—host cell gene products and pathways that facilitate competition of the infectious cycle, the availability of which may be cell type specific; and *intrinsic and innate immune defenses*—evolutionarily conserved genes specialized to recognizing and eliminate virus infections. Considering all these opposing forces to infection, viruses must continuously breach host barriers to attach and enter cells, establish genome replication, assemble infectious virions, and be released from cells. In principle, for every host defense, there will exist a viral offense. This section will highlight key barriers to infection including the cell biology and intrinsic and innate immunity as major modes that can restrict all stages of virus life cycle.

4.2.1 *The Life Cycle of Viruses:* Entry, Replication, Gene Expression, Assembly, and Release

Virus entry into cells As a rule of thumb, viruses can only infect cells to which they bind. Cell surface binding occurs via *attachment factors* and *virus entry receptors*. *Attachment factors* facilitate concentration of viral particles on the cell's surface, thus enhancing entry and infection (Fig. 4.2). Importantly, interactions with attachment factors are not highly specific. Many viruses engage multiple *attachment factors* and *entry receptors*. They interact with them in parallel or tandem or employ

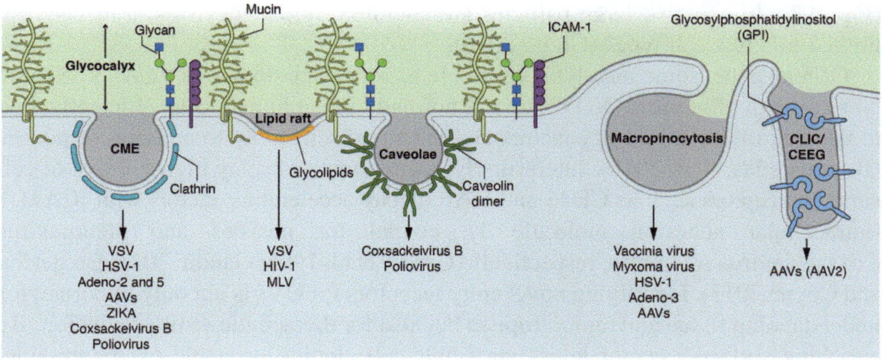

Fig. 4.2 Routes of virus entry. Therapeutic viruses are derived from a diverse family of viruses and have different mechanisms to enter cells, including clathrin-mediated endocytosis (CME), lipid drafts, caveolae-dependent entry, micropinocytosis, and GPI-anchored protein-enriched endosomal compartment (GEEC) (McCann and Castellino 2022; Nonnenmacher and Weber 2011; Greber and Flatt 2019; Gerlier et al. 1995; Dimitrov 2000). Abbreviations: *ICAM-1* intracellular adhesion molecule 1, *CLIC/GEEG* clathrin-independent carriers/GPI-anchored protein-enriched endosomal compartment, *VSV* vesicular stomatitis virus, *HSV-1* herpes simplex virus-1, *adeno* adenovirus, *AAV* adeno-associated virus, *HIV-1* human immunodeficiency virus-1, *MLV* murine leukemia virus

different receptors for entry into different cell types. When multiple receptors are essential for entry, the one engaged first is termed *receptor* and the subsequent ones *co-receptors* (Mercer et al. 2010). In majority of the cases, identity, distribution, and biochemical activity of these cell surface components determines which cell types, tissues, and organisms a virus can infect. For most viruses, the first cellular barrier to overcome is the glycocalyx, a layer of glycoconjugates that covers the external surface of cells (Mercer et al. 2010). The glycocalyx is composed of glycoproteins, glycolipids, and proteoglycans (Fig. 4.2). The composition and thickness of the glycocalyx is variable and is cell type dependent. Through interactions with oligosaccharides, many viruses use the glycocalyx for initial attachment. The next barrier to overcome is the plasma membrane (PM) which is highly sensitive to stimuli and viruses take advantage of this during entry (Harrison 2015). Importantly, cells may be susceptible to virus infection if specific viral receptor(s) are present and functional on the cell surface or within subsequent endocytic compartments (Pillay and Carette 2017). On breaching the PM, viruses enter the lumen of endosomes or the endoplasmic reticulum (ER) by receiving cues in the form of exposure to low pH, proteolytic cleavage and priming of viral proteins, and/or association with host cell proteins (McCann and Castellino 2022). Only primed or activated viruses penetrate the vacuolar membrane, thus depositing the viral genome, capsid, or the entire viral particle into the cytosol with access to the nucleus (Fig. 4.2). Although cell surface binding and receptor usage are paramount to virus entry, the route or endocytic mechanism usurped plays a crucial role in allowing cells to become susceptible to virus infections (Ghigo 2010). Given the diversity in viruses and the range of cells they infect, endocytosis can occur via several different mechanisms (Fig. 4.2: pinocytosis, i.e., endocytic uptake of fluids, solutes, and small particles, clathrin-mediated endocytosis, macropinocytosis, and caveolar/raft-mediated mechanisms (Fig. 4.2). In contrast, large particles are engulfed by phagocytosis, a mechanism limited to a few cell types (Ghigo 2010).

Of relevance, some oncolytic viruses [e.g., vesicular stomatitis virus (VSV) and Newcastle disease virus (NDV)] do not normally infect human cells efficiently, due to the lack of essential entry factors and increased sensitivity to antiviral responses (Bommareddy et al. 2018). Interestingly, many tumors display higher levels of cell surface receptors such as CD46 and DAF (decay accelerating factor) with ICAM-1 (intracellular adhesion molecule 1), crucial for measles and picornavirus Coxsackievirus A21 entry, respectively (Gerlier et al. 1995; Gaudin 2021; Bergelson and Coyne 2013). Identifying novel entry receptors for OVs is not only important for understanding tissue and tumor tropism but also for therapeutic virus production. By enriching surface receptor levels via single-cell cloning or stable overexpression, producer cells can be rendered more susceptible early on in infection. Moreover, oncolytic viral variants capable of binding entry receptors with high affinity can be engineered or adapted to improve targeting and potency of OVs. Viral glycoprotein-dependent receptor-mediated virus entry is not the only mode viruses use to infect cells. A body of work has emerged in the past decade, demonstrating that viruses also infect cells as "units of virions" enclosed in host cell-derived extracellular vesicles (EVs) and exosomes (Kerviel et al. 2021). The membranes of these vesicles

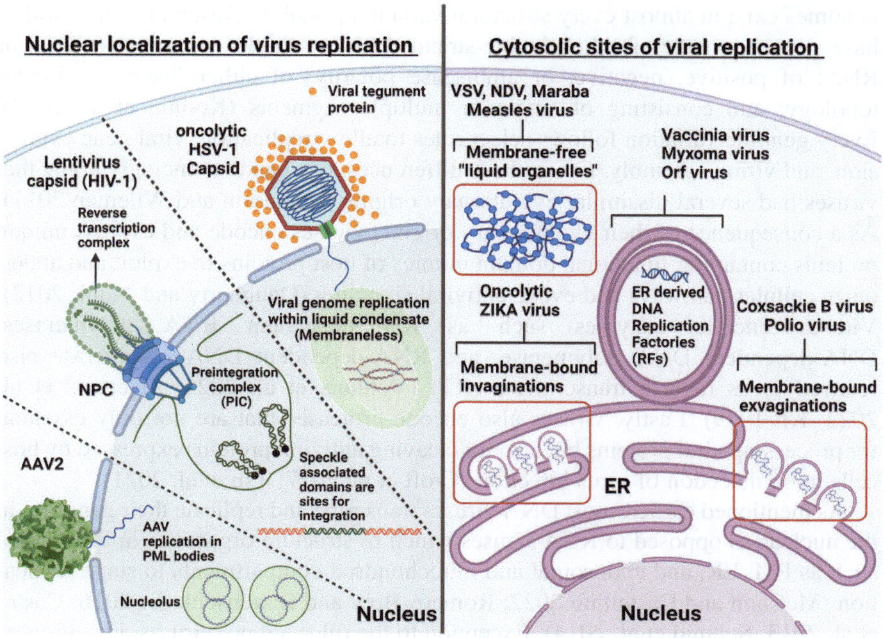

Fig. 4.3 Sites of viral genome replication. Life cycle of viruses that replicate in the nucleus and cytosol. AAV, lentiviruses, and oncolytic herpesviruses replicate their genomes within the nucleus (nucleolus, speckle associated domains, or membraneless liquid condensates). Many RNA viruses replicate in the cytoplasm using ER-derived membranes and membrane-free liquid organelles

mimic apoptotic bodies as they expose phosphatidylserine (PS) on their surface which bind dedicated PS receptors to enhance virus uptake (Croft et al. 2017; Moller-Tank and Maury 2014). In summary, appreciating the different routes viruses employ to enter cells is important to consider when screening virus producer cells and what determines their susceptibility to infection.

Intracellular Sites of Viral Genome Replication and Gene Expression After successfully penetrating cells and delivery of viral genomes into the cytosol, viruses replicate in different locations within the cell (Moss 2013; Romero-Brey and Bartenschlager 2016). To accomplish expression, replication, and spread of viral genes, both DNA and RNA viruses have evolved a range of genetic strategies and replicative cycles to exploit the biology of host cells. Despite their restricted coding capacity, viruses encode all necessary proteins to reprogram host cells and transform them into viral factories (Netherton and Wileman 2011). Furthermore, most RNA viruses replicate viral RNA in distinct subcellular compartments, whereas the majority of DNA viruses exploit the nucleus to amplify their genomes (Netherton and Wileman 2011) (Fig. 4.3). Of relevance, the most striking aspect of viruses at a molecular level is their genome diversity and replication approaches. As opposed to cellular genomes, which comprise consistently of double-stranded DNA, viral genomes exist in almost every structural variation possible. Different virus families have genomes made of either double-stranded (ds) or single-stranded (ss) DNA or RNA; of positive, negative, or ambisense polarity; of either linear or circular topology; and consisting of single or multiple segments (Koonin et al. 2022). Every genome variation follows select rules to allow replication, viral gene expression, and virion assembly. Given these differences, one can convincingly argue that viruses had several dissimilar evolutionary origins (Netherton and Wileman 2011). As a consequence of their evolutionary origins, viruses encode and express unique proteins containing functional domain mimics of host proteins to exploit and antagonize cellular pathways and evade antiviral signaling (Daugherty and Malik 2012). Viruses encode enzymes such as RNA-dependent RNA polymerases, DNA-dependent DNA polymerase, and RNA-dependent DNA polymerase also referred to as reverse transcriptase (RT) (Velthuis et al. 2021; Maertens et al. 2022; Kit 1979). Lastly, viruses also encode proteases that are not only essential for processing viral proteins but also for cleaving antiviral proteins expressed by host cells upon detection of virus infection (Croft et al. 2017; Tsu et al. 2021).

As mentioned earlier, most DNA viruses transcribe and replicate their genomes in the nucleus as opposed to RNA viruses which re-structure organelles in the cytosol such as PM, ER, and endosomal and mitochondrial compartments to stage replication (McCann and Castellino 2022; Romero-Brey and Bartenschlager 2016; Castro et al. 2013; Schmid et al. 2014). Exception to the rules are poxviruses, iridoviruses, and African swine fever virus, which all replicate their DNA genomes partly or completely within the cytosol (Schmid et al. 2014; Wolff and Bárcena 2021). Further exceptions to this generalization are RNA viruses like retroviruses, which integrate DNA copies of their genomes into host chromosomes, and orthomyxoviruses (influenza viruses), bornaviruses, and plant-infecting rhabdoviruses, which replicate their linear negative-sense RNA genomes in the nucleus (Lucic et al. 2021). Distinct subcellular sites of replication offer viruses diverse opportunities and challenges to co-opt cellular components and pathways for the synthesis and trafficking of viral proteins, genomes, virion assembly, and release of infectious progeny.

Localization of Virus Assembly and Release of Infectious Virions Viruses assemble to adopt different morphologies (icosahedral, spherical, or helical) (Luque and Castón 2020). Virus assembly is the second to last stage of the life cycle where newly formed viral capsids and assembly proteins package genomes to initiate maturation of nascent virions to infectious viruses. Non-enveloped and enveloped viruses assemble using different cellular strategies (Casjens and King 1975). Assembly of each virion at a well-defined point within the cell occurs on membrane-derived platforms (Casjens and King 1975). On the other hand, to enable productive infection, cells need to have the necessary cellular factors (e.g. proteases) and secretory pathways to assemble and release infectious virus. This section will focus on the cell biology of how widely used viral vectors such as adenoviruses and AAV (non-enveloped) and HSV-1 and lentiviruses (enveloped) assemble infectious virions. A detailed review on virus assembly and release is beyond the scope of this chapter and can be found elsewhere (Casjens and King 1975). As alluded earlier, viruses encode a limited number of gene products and thus depend on host cells not only for biosynthesis of viral genomes and proteins that comprise the virus particle but also for intracellular sorting mechanisms that viruses rely on to deliver viral components to sites of virion assembly. Since these are the same sorting mechanisms that host cells employ to demarcate subcellular organelles (e.g., ER, mitochondria, nucleus, Golgi, and PM), viral proteins via "viral mimicry" contain targeting signals like components of those organelles (Cook and Cristea 2019; Amara and Mercer 2015).

For adenoviruses (AdV), replication of viral DNA in the nucleus coupled with the production of a large number of adenovirus structural polypeptides sets the stage for virus assembly. Being non-enveloped icosahedral in morphology, AdVs have a diameter between 65 and 95 nm and comprise multiple proteins, termed hexon, penton, IIIa, VI, VII, VIII, and IX (de Pablo and Martín 2022). The major capsid protein (hexon) dominates the virion surface, followed by viral proteins VII, VI, IX, VIII, IIIa, and penton. AdV capsids package genomes ~ 26–45 kbps in length with positively charged virus proteins. Importantly, proteins involved in genome replication, packaging, and virus maturation (adenovirus maturation protein [AVP]) are bound to the genome within the capsid shell. The role of cellular factors essential for AdV assembly is poorly understood and requires further investigation to define proviral virus–host interphases for AdVs. Release of infectious AdV is coordinated by several processes (Lucic et al. 2021). One such example is the cleavage of intermediate filament protein cytokeratin K18 by viral proteases L3-23K (Chen et al. 1993). Cytokeratin K18 cleavage perturbs the intermediate filament system, thus compromising the mechanical integrity of cells and making infected cells more susceptible to lysis and virus release. Another viral mechanism that enables virus release involves the viral protein E3, referred to as adenovirus death protein which kills cells upon accumulating within cells during late stage of infection (Georgi and Greber 2020).

In the case of AAV, capsids assemble from combinations of three overlapping virus proteins (VPs): VP1, VP2, and VP3. Importantly, AAV capsid proteins are not glycosylated but are variably phosphorylated. Regions within VP1 contain a calcium-dependent phospholipase A2 (PLA_2) enzymatic activity. Furthermore, the N-terminus of VP1 contains residues that mimic nuclear localization motifs required of cell entry and capsid assembly. VP2 has a nuclear export signal (NES) to enable exit of assembled capsid from the nucleus. On translation in the cytosol, capsid proteins are transported to the nucleus, where they assemble. Capsid proteins can self-assemble into capsids that are structurally similar to regular capsids and can package viral DNA when expressed with a replicating genome. Capsids are infectious, only when they contain the largest of the minor capsid proteins VP1 and VP2. This is likely due to VP1's PLA_2 activity and nuclear localization and VP2 nuclear export functions crucial for regulating virus infectivity. Incorporation of DNA occurs by insertion of viral single-stranded DNA (ssDNA) into preassembled capsids. After genome packaging and assembly, capsids may be retained within the nucleus and transported to the cytosol or out of cells dependent on the phosphorylation state of capsid proteins. Recent studies have discovered the requirement of two additional viral proteins important for assembly, which result from frameshifted open reading frames (ORFs) within the *cap* gene, namely, assembly-activating protein (AAP) and membrane-associated protein (MAAP) (Sonntag et al. 2010; Earley et al. 2017; Elmore et al. 2021). Importantly, these findings reveal that AAP supports AAV assembly in a serotype-specific fashion and MAAP regulates virus assembly across serotypes. Depending on the serotype, AAV can be secreted more efficiently by cells; however very little is known about cellular pathways that control AAV secretion (Nóbrega et al. 2020). Virus assembly and release is more complicated for enveloped viruses, when considering viral vectors and oncolytic viruses derived from HIV-1 (retroviruses) and HSV-1 (herpesviruses), respectively. For retroviruses, assembly of the particle is primarily driven by the *gag* precursor protein (Krebs et al. 2021). Expression of *gag* alone is sufficient for virion formation and mediates assembly and release of an immature "bald" particle—lacking infectivity and the "ḥair" provided by the envelope protein, *env*. Lentiviral vectors are usually pseudotyped with glycoproteins of gp160 or vesicular stomatitis virus G (VSV-G) protein to modify tropism. Typically, there are 1200–1800 *gag* proteins per virion; though this number may vary from virus to virus (Krebs et al. 2021). Given the central role in virus particle assembly, *gag* proteins have been subjected to extensive mutational analyses to define subdomains required for various steps of the process. Notably, short stretches of *gag* are sufficient to assemble virions. Three domains, a membrane-binding (M) domain, an interaction domain, and a late-assembly (L) domain, are critical for assembly. Importantly, these assembly domains need not be a part of any cleavage products of the precursor, since unprocessed *gag* can promote assembly of virions (Krebs et al. 2021). The viral RNA genome is incorporated into virions via interactions of the nucleocapsid (NC) domain of *gag* with specific packaging signals, termed Psi sequences near the 5′ end of the genome (Hanson et al. 2022). Maturation of infectious viruses requires the proteolytic cleavage *gag* mediated by viral protease PR (Pornillos and Ganser-Pornillos 2019). *Gag* proteins rearrange to form the characteristic virion core, comprised of capsid proteins (CA) surrounding the dimeric viral RNA condensed with the NC protein (Krebs et al. 2021). Retrovirus maturation is a complex process and occurs at the cell surface. The initially formed particles have a characteristic immature morphology: round particles, but on release from cells, the morphology of these immature particles changes to exhibit a more condensed structure, containing a central core separated from the surrounding envelope (Krebs et al. 2021). Depending on the retrovirus, the capsid core can be spherical or conical within the envelope and is characteristic of the virus genera: round for *alpha-* and *gammaretroviruses*, cylindrical for the *betaretroviruses* and conical for *lentiviruses* (Pornillos and Ganser-Pornillos 2019). Finally, retrovirus assembly is not only regulated by viral proteins but requires intricate interactions of viral proteins with the ESCRT (endosomal sorting complexes required for transport) machinery to assembly infectious virions. A detailed overview of these interactions is described elsewhere (Votteler and Sundquist 2013).

Assembly of large, enveloped DNA viruses as in the case of HSV-1 offers yet another stream of highly complex assembly and release events. Briefly, HSV virions consists of major elements: (1) an electron-opaque core encompassing the viral DNA, (2) an icosahedral capsid surrounding the core, (3) an unstructured proteinaceous layer, termed the *tegument* that surrounds the capsid, and (4) an outer lipid bilayer studded with spikes on the surface (Ahmad and Wilson 2020). The first step of nascent HSV particles is capsid assembly. The HSV capsid is around 125 nm in diameter and comprises mainly of viral proteins VP26, VP23, VP19C, VP5, and pUL6 (important for genome packaging) (Ahmad and Wilson 2020). There are additional viral proteins that structurally support the main capsid proteins and play a crucial role in ensuring capsid integrity. Capsids assemble as procapsids around a scaffold complex. The viral DNA genomes are packaged into preformed procapsids and have packaging signals (termed pac sequences) (Crump 2018). This complex interplay between viral proteins and DNA results in large structural changes leading to rearrangements of spherical procapsids into stable icosahedral capsids containing the viral genome (referred to as C-capsids or nucleocapsids). Only capsids that are packaged with viral DNA can exit the nucleus by escaping the boundaries of the nuclear envelope (Ahmad and Wilson 2020). For herpes viruses, the nuclear envelope is a formidable barrier and is composed of two phospholipid bilayers—the inner nuclear membrane facing the nucleoplasm and the outer membrane facing the cytosol (Lucic et al. 2021). Transport of nucleocapsids into the cytosol is accomplished by budding of capsids at the inner nuclear membrane to forming primary enveloped particles followed by fusion to the other nuclear membrane to release capsid into the cytosol. Disruption of the nuclear lamina to allow nucleocapsids to transit the nuclear membrane is facilitated through phosphorylation of lamins by viral and cellular kinases such as protein kinase C (PKC) isoforms, demonstrating the requirement of a virus-host interphase for functional virus assembly (Ahmad and Wilson 2020). Following nuclear exit, viral capsids are coated with a proteinaceous layer called the tegument, which connects viral genome containing capsids to the viral envelope, which in HSV contains 24 viral proteins (Owen et al. 2015). The structure of the viral tegument is undefined. The last stage of HSV assembly and maturation is secondary envelopment, a process by which nucleocapsids along with tegument proteins are layered with viral envelope proteins. Secondary envelopment occurs at the trans-Golgi network (TGN) after which infectious virions leave cells using either the endomembrane system or secretory pathway (Ahmad and Wilson 2020; Crump 2018; Owen et al. 2015).

In summary, while design of viral vectors is key in engineering therapeutic viruses, an understanding of virology and cell biology is crucial in selecting cell lines and engineering host dependency or antiviral factors to optimize therapeutic virus production. Lastly, producing higher titer viruses requires an understanding of what regulates virus infectivity within cells and on the virus particle itself to assess vector quality. Thus, an extensive understanding of virus–host cell interactions is important for several reasons, including defining critical quality attributes (CQAs) that regulate infectivity and immunogenicity of specific viral platforms.

4.2.2 Early Host Responses to Virus Infections: Cell-Autonomous (Intrinsic) and Innate Immunity

Cell-autonomous (intrinsic) and innate immune responses are integral to antiviral defense since they are rapidly activated to function within minutes to hours after infection. The instant response to infection consists of two processes: sensing or detecting invading viruses and sounding the alarm to "switch on" an antiviral program. What delineates intrinsic from innate immunity is not clear and under debate, but intrinsic defenses are constitutively active and "ready to go" and do not require transcriptional and translational regulation of genes. In contrast, elements of innate immunity are synthesized in cells in response to infection or are immune cells such as professional antigen-presenting cells (e.g., macrophages and dendritic cells that aid in educating T and B cells). Intrinsic immunity can directly restrict viral entry, replication, translation, assembly, and release, thus rendering a cell non-permissive to a specific class or species of viruses. At large, intrinsic immunity is conferred by restriction factors preexisting in certain cell types and has evolved to recognize specific viral components. Innate immunity encodes genes that recognize patterns and are referred to as patter recognition receptors (PRRs) that inhibit virus infection indirectly by inducing "interfering molecules," termed interferons (IFNs) which upregulated the expression of interferon-stimulated genes (ISGs) (Fig. 4.4).

A very important concept in understanding virus–host interphases is the evolutionary principle of the Red Queen hypothesis (RQH) proposed in 1973, which states that in order to survive, species must repetitively adapt, evolve, and proliferate in environments against ever-evolving opposing species (Morran et al. 2011; Strotz

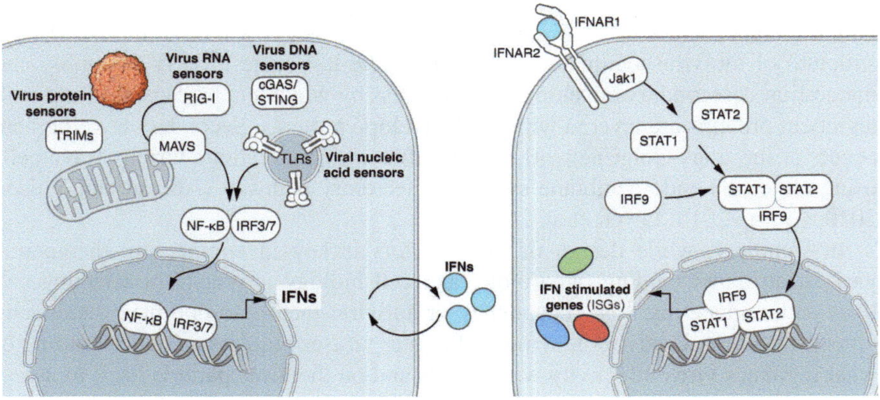

Fig. 4.4 Antiviral immunity to virus infection. DNA and RNA viral genomes and proteins can be sensed by dedicated and evolutionary conserved innate immune sensors to trigger the production of interferons (IFNs) which bind interferon α and β receptor subunits 1 and 2 (IFNAR1 and 2) on the surface of cells to upregulate interferon-stimulated genes (ISGs). The consequence of these events leads to the activation of an antiviral state in infection and uninfected cells, priming the environment to prevent the spread of viral infection

et al. 2018). Positively selected residues in virus entry receptors and host-defense genes highlight molecular recognition interphases between host and viral proteins, which have adapted and counter-adapted through a series of Red Queen conflicts (Daugherty and Malik 2012). In this section, we will highlight mechanisms by which cells respond to virus infections and how viruses antagonize these responses to establish infection. Understanding virus–host interactions under the lens of evolution is extremely valuable not only while designing viral vectors of gene delivery but also what genetic considerations should be accounted for in cell line selection and engineering to employ strategies to produce high titer and infectious viruses.

Cell-Autonomous (Intrinsic) Responses to Virus Infections Cellular intrinsic defenses are diverse, including self-killing (apoptosis), self-eating (autophagy), restriction factors, and RNA interference. Apart from being a normal biological process used to eliminate particular cells during biological development, apoptosis is a powerful intrinsic response to infection, abolishing a viral factory to synthesize hundreds of new infectious particles. Virus particles engage cell receptors, and the production of infectious virions requires many host metabolic pathways. Once infected, host cells stop dividing and become quiescent, thus unable to provide enzymes and other raw materials needed to support virus production. As a result, some viral proteins such as adenoviral E1A will trigger cells to switch from a resting state to enter the cell cycle. In response, cell cycle check-point proteins may induce apoptosis to counteract viral manipulation. Many viral genomes encode strategies that block cell-initiated apoptosis. Adenovirus strains lacking the gene encoding for functional E1B protein greatly enhanced cytopathic effects. The influence of E1B to prevent apoptosis which is thought to be p53-dependent could be complemented by ectopic expression of cellular Bcl-2 which is functionally similar to E1B of adenoviruses (Chiou et al. 1994). Another example of inhibiting apoptosis by viruses is the case of herpesviruses where a virally expressed non-coding RNA binds and inhibits the mitochondrial protein complex that triggers apoptosis (Zhao et al. 2010). Viral apoptotic mimicry is a common strategy the multiple viruses employ to enable entry, replication, and immune evasion (Amara and Mercer 2015).

Intrinsic responses to infection do not always result in cell death (Yan and Chen 2012). Cells degrade and turnover cytosolic contents by engulfing them in demarcated compartments that fuse with degradative lysosomes. The process is highly evolutionarily conserved and is termed autophagy (in Greek, "to eat oneself"). Autophagy and endo-lysosomal pathways may funnel into the lysosome to target incoming viral particles for degradation. This mode of autophagy is referred to as xenophagy (selective degradation of viruses, bacteria, and parasites) (Chiramel et al. 2013). Autophagy is triggered by stressors, such as nutrient starvation or viral infection, including stress-induced alterations in translation (Deretic 2021). Additionally, autophagy can block virus replication by transferring viral nucleic acids to endo-lysosomal compartments to stimulate innate immunity. Depending on the virus, autophagy is known to be either pro- or antiviral (Chiramel et al. 2013; Deretic 2021). Viruses have devised elegant strategies to block autophagic degradation by preventing fusion of autophagosomes with lysosomes and using autophagosomal membranes as sites of viral genome replication (Chiramel et al. 2013).

As a part of the "ready-to-go" intrinsic response, cells across diverse species have evolved specialized genes to restrict every single step of the virus life cycle (Yan and Chen 2012; Liu et al. 2018). These genes or factors are termed restriction factors (RFs). RFs can target viral genomes as in the case of zinc-finger antiviral protein (ZAP) (Ficarelli et al. 2021). A critical feature of the immune response is to distinguish self from non-self and elicit responses only to the latter. Viral components, using viral mimicry, can adopt characteristics of host cells to avoid immune detection. One characteristic mimicked by certain viruses is the relative scarcity of CpG dinucleotides or islands (Ficarelli et al. 2021). Traditionally, host-encoded methyltransferases methylate C residues of CpG pairs and deamination converts methylated C's into T's, thereby purging the genome of CpG pairs. Some DNA viruses (alphaherpesviruses) and RNA viruses (HIV-1) have CpG islands in genomes and are subjected to ZAP-mediated degradation. Moreover, HSV-1 encodes UL4, a viral antagonist of ZAP (Ficarelli et al. 2021). Similar to ZAP, another cytidine deaminase called apolipoprotein B mRNA-editing enzyme, catalytic polypeptide-like 3 (APOBEC3) mediated the deamination of cytosine/cytidine to uridine. APOBEC3 can target both DNA and RNA viral genomes (Sadeghpour et al. 2021). APOBEC3 resides in the cytosol and can bind mature viral RNAs, allowing them to be packages into budding viral particles. Once newly infectious viruses infect cells, APOBEC3 protein can both hinder reverse transcription and deaminate the single-stranded DNA intermediate product. The genomes of viruses that are targeted by this mechanism encode counterstrategies to evade the antiviral activity of APOBEC3. Vif proteins from primate lentiviruses bind to APOBEC3 and recruit a ubiquitin ligase that degrades the RF via the proteasome (Sadeghpour et al. 2021). Apart from directly targeting viral genomes, RFs inhibit trafficking of viral components (e.g., myxoma virus resistance [MX]), block release of virus particles (e.g., tetherin), and selectively block virus infection in a virus-specific fashion (tripartite motif [TRIM] proteins) (Yan and Chen 2012). Briefly, MX proteins which belong to dynamin superfamily of guanosine triphosphates (GTPases) are localized in perinuclear regions in the cytosol and prevent the trafficking of incoming retroviruses and herpesviruses (Haller et al. 2007). Tetherin, or bone marrow stromal antigen 2 (BST2) is a membrane-associated protein constitutively expressed in some human cells and prevents the budding of enveloped viruses from the cell surface, thereby exerting broad antiviral activity (Neil 2013). Finally, TRIM proteins are selectively antiviral depending on the family of viruses (van Gent et al. 2018). TRIMs display a diverse range of subcellular localization and antiviral activity profiles. Being E3 ubiquitin ligases, TRIM proteins engage proteasomal or autophagosomal degradation to restrict viruses at different steps of their life cycle. There are other pathways that serve as intrinsic responses to virus infection but are beyond the scope of this chapter but are well described elsewhere (Yan and Chen 2012).

Continuum between Intrinsic and Innate Immunity As pointed out earlier, intrinsic immune effectors respond rapidly, not requiring new gene expression. As will be described in the next section, cells respond to virus infection by secreting IFNs which trigger antiviral responses in bystander cells preparing them to engage newly formed invading viruses (Liu et al. 2018; Schoggins 2019). These ready to respond cells are activated in advance and many of the effectors of intrinsic immunity (e.g., MX and TRIM proteins) can also be upregulated by the presence of IFNs to limit the spread of virus infections (Fig. 4.4).

The Innate Immune Response: Sensing Viruses to Co-ordinate the Antiviral State Innate immunity is followed or acts in concert with intrinsic immunity. Usually, the innate immune response is thought to be triggered a few hours after infection and culminates with the activation of the *antiviral state* (Fig. 4.4). Triggering of the innate immune system also recruits dendritic cells and macrophages, ready to engage invading pathogens. The first step in this response is pathogen sensing, carried out by pathogen sensors which reside in subcellular compartments or traverse the cytoplasmic space to detect viral components (nucleic acids and proteins). Innate microbial sensing mechanism is based on pattern recognition receptors (PRRs) (Iwasaki and Medzhitov 2015). Importantly, all cell types that can be infected by any given class of viruses express corresponding PRRs (Iwasaki and Medzhitov 2015). PRRs detect conserved microbial structures shared by entire classes of microbes. The innate immune sensors detect specific structural features of viral RNA and DNA that are distinct from the host nucleic acid (Iwasaki and Medzhitov 2015). These include long double-stranded RNA, RNAs containing 5-triphosphate, and unmethylated CpG motifs in viral DNA genomes. Recognition of these structural features is insufficient to reliably distinguish host from viral nucleic acids. Thus, additional aspects help to determine their source. As examples, localization of viral but not host nucleic acids in the endo-lysosomes is unusual and more so the paucity of host cell DNA in the cytosol is a signal to trigger antiviral responses. Innate immune sensors of viral RNA and DNA are present in these locations, where they can be triggered to activate the antiviral responses. In general, PRRs can be broadly classified into toll-like receptors (TLRs) and cytosolic sensors. Sensing of extracellular pathogens is mediated by TLRs expressed on the plasma membrane or with endosomal compartments (e.g., TLR4, 7, 8, and 9) (Iwasaki and Medzhitov 2015). Detection of cytosolic viral RNA is achieved through retinoic acid-inducible gene 1 (RIG-I)-like receptors (RLRs), whereas sensing of cytosolic viral DNA is accomplished by AIM2-like receptors (ALRs) and cyclic-GMP-AMP synthase (cGAS)/stimulator of interferon genes (STING) pathway (Iwasaki and Medzhitov 2015).

In addition to pattern recognition of viral components, signatures of virus infection can be detected indirectly, through alteration in normal cellular processes, such as inhibition of host protein synthesis, altered ion channel activity, or ER stress. Detection of viral pathogens by PRRs results in the production and secretion of interferons (IFNs). Most PPR stimulated pathways converge on two critical cellular transcription activators, interferon regulatory factors (IRFs), and nuclear factor kappa-light-chain-enhancer of activated B cells (NF-kB) (Liu et al. 2018; Schoggins 2019). IRFs and NF-kB then translocate into the nucleus and initiate transcription of IFN genes. Locally released IFN activates a global antiviral state and is orchestrated by a series of signaling cascades to upregulate ISGs, which directly inhibit all stages of virus life cycle. Cells that bind and respond to IFN fail to support virus infection and are programmed to adopt an *antiviral state* (Liu et al. 2018). ISGs serve multiple functions. As such, PRR, IRFs, and several components of the Janus kinase-signal transducer and activation of transcription (JAK-STAT) pathway are expressed basally but are also ISGs and reinforce IFN response (Schoggins 2019). Importantly, transcription of many genes is IFN inducible, but expression profiles and concentration of gene products may vary dependent on the cell type (Schoggins 2019). The exact number of ISGs that are encoded by the human genome is unknown. Several products of ISGs possess broad-spectrum antiviral activities, but precise mechanisms of antiviral function are understood only for a few. IFN secretion can also induce cell death of infected and bystander cells (Schoggins 2019). Thus, evaluating the interferon competence or immunocompetence of virus producer cells is extremely crucial to cell line selection to produce therapeutic viruses.

4.3 Cell Line Selection and Engineering Strategies to Produce Therapeutic Viruses

When developing therapeutic viruses such as live-replication competent oncolytic viruses, the cell surface expression and abundance of attachment and entry factors (receptor or co-receptors) are critical access parameters to define cell types susceptible to virus infection. On selection of a suitable cell type, pools can be subjected to single-cell cloning and profiled for optimal receptor expression as a first step to generate cell clones to serve as high virus producer cells. Alternatively, partially susceptible cells can be engineered to ectopically express entry factors to promote infection.

Currently in industry HEK293 cell lines are utilized to produce recombinant virus-like particles and viral vector cell and gene therapy. A body of work over the years has generated molecular characterization of HEK293 variants, which has allowed re-engineering of cells to become more transfectable and permissive for production of viral vectors. Hand in hand, there have been significant advances in the bioprocessing conditions, including adaptation of cells to suspension cultures and growing cells in serum-free media. These newly developed features enable the potential for large-scale manufacture of virus-based therapies at commercially conducive costs. This section will briefly describe the origin of HEK293 cells and the available variants. Furthermore, we will review engineering approaches adopted for HEK239s so far and suggest possible other targets that can be engineered to augment the production of viral vectors for cell and gene therapy.

HEK293 Origins and Variants HEK293s were initially generated in 1973, by immortalization of cells through the introduction of adenoviral 5 genes E1A and E1B into the genome of HEK293 cells (Formas-Oliveira et al. 2020). The introduction of E1A and E1B into the HEK293 genome transformed this cell line to favor enhanced cell growth to increase protein production. Currently, there are at least six variants of HEK293s used in laboratories (Formas-Oliveira et al. 2020). These variants have been engineered or selected to generate clones with specific functional properties connected to augment the quality or yield of proteins and further adapted to culture environments for large-scale manufacturing. In order to generate HEK293 clones with specialized properties for production of therapeutic viruses, an in-depth understanding of the fundamental biology of variants needs examination to further engineer cells to produce cell lines specific to certain viral vectors.

HEK293s for Production of Virus-like Particles (VLPs) and Viral Vectors Given the ease of transfection of HEK293 cells by ectopic expression of plasmids encoding viral proteins, HEK293s have been used for VLP production for vaccine development for HIV-1 and ZIKA virus. HEK293s are utilized for the production of viral vectors for multiple cell and gene therapies, thus providing cell platforms for high titer yield of recombinant lentiviral, adenoviral, and AAV vectors. Recombinant AAV is readily produced in HEK293s because of the expression for E1A and E1B essential for particle assembly. Furthermore, it might be worthwhile generating clones of HEK293 cells with high expression profiles of E1A and E1B to further boost the production of AAV and examine the ratios of empty versus full capsid from such clones.

Engineering HEK293 Cells to Produce Lentiviruses Retroviruses such as gammaretroviruses and lentiviruses (γ-RV and LV) are the most commonly used vectors in gene therapy. Although HEK293 cells are the go-to for virus production, low particle yield is a significant challenge in realizing their therapeutic applications. A recent study underlined the role of endoplasmic reticulum (ER) processing and anti-apoptotic mechanisms that directly influence vector productivity. Stable overexpression of ER processing genes dramatically increased γ-RV volumetric productivity by 97% and LV productivity by up to 53% (Formas-Oliveira et al. 2020). These findings demonstrate that ER processing as well anti-apoptotic genes can be expressed in combination to further augment retroviral vector production.

Engineering Intrinsic and Innate Immune Efforts to Disarm Antiviral Defense for Virus Production As described extensively early on in the chapter, host cells deploy a diverse range of mechanisms to protect cells from virus infections. Before cell line selection or subcloning efficient virus producing cells, it is important to investigate the immunocompetence of producer cells. This examination will provide an insight into the basal antiviral status of cells. A range of methods can be employed to investigate which arms of innate immunity are functional in producer cells. Treatment of cells with viral DNA or RNA mimic allows us to monitor secretion of IFNs in the supernatants of cultures to determine whether producer cells are capable of secreting antiviral molecules. Moreover, pretreatment of producer with

IFNs prior to transfection of plasmids encoding viral proteins of adenoviruses, AAV, or lentiviruses can determine whether a preexisting *antiviral state* negatively impacts the production of recombinant viruses. In the case that producer cells are able to trigger an antiviral response, engineering cell to knockout key antiviral factors using CRISRP-Cas9 or zinc-finger nucleases may dramatically increase virus production.

4.4 Summary

This chapter is aimed at covering important aspects regarding the biology of different viral vector currently used or engineered as therapeutic viruses. Furthermore, we expanded the multiple mechanisms by which a diverse range of viruses enter, replicate, assemble, and release from cells. This biology is central to the development of the next generation of viral vectors and more importantly the host cell pathways that are essential to producing infectious virions. Next, we explored the major cellular barriers that exist to prevent infection and production of viruses from cells. The roles of intrinsic and innate immunity to limit the spread of viruses are of direct relevance when considering methods by which one can engineer cells to increase not only virus production but also the quality of produced viruses. Another advantage to disable antiviral pathways in cells may contribute to limiting the occurrence of viral variants, especially in the case of live-replicating oncolytic viruses. The era of virotherapy is in its embryonic stage and therefore an understanding of virology and cell biology is crucial to produce efficacious and affordable virus-based therapies.

References

Ahmad I, Wilson DW (2020) HSV-1 cytoplasmic envelopment and egress. Int J Mol Sci 21:5969

Amara A, Mercer J (2015) Viral apoptotic mimicry. Nat Rev Microbiol 13:461–469

Bergelson JM, Coyne CB (2013) Viral entry into host cells. Adv Exp Med Biol 790:24–41

Bommareddy PK, Shettigar M, Kaufman HL (2018) Integrating oncolytic viruses in combination cancer immunotherapy. Nat Rev Immunol 18:498–513

Casjens S, King J (1975) Virus assembly. Annu Rev Biochem 44:555–611

Castro IF, Volonté L, Risco C (2013) Biogenesis and architecture of virus factories. Cell Microbiol 15:24–34

Chen PH, Ornelles DA, Shenk T (1993) The adenovirus L3 23-kilodalton proteinase cleaves the amino-terminal head domain from cytokeratin 18 and disrupts the cytokeratin network of HeLa cells. J Virol 67:3507–3514

Chiou SK, Tseng CC, Rao L, White E (1994) Functional complementation of the adenovirus E1B 19-kilodalton protein with Bcl-2 in the inhibition of apoptosis in infected cells. J Virol 68:6553–6566

Chiramel AI, Brady NR, Bartenschlager R (2013) Divergent roles of autophagy in virus infection. Cells 2:83–104

Chung YH, Cai H, Steinmetz NF (2020) Viral nanoparticles for drug delivery, imaging, immunotherapy, and theranostic applications. Adv Drug Deliver Rev 156:214–235

Cook KC, Cristea IM (2019) Location is everything: protein translocations as a viral infection strategy. Curr Opin Chem Biol 48:34–43

Croft SN, Walker EJ, Ghildyal R (2017) Picornaviruses and apoptosis: subversion of cell death. Mbio 8:e01009–e01017

Crump C (2018) Human herpesviruses. Adv Exp Med Biol 1045:23–44

Daugherty MD, Malik HS (2012) Rules of engagement: molecular insights from host-virus arms races. Annu Rev Genet 46:677–700

de Pablo PJ, Martín CS (2022) Seeing and touching adenovirus: complementary approaches for understanding assembly and disassembly of a complex virion. Curr Opin Virol 52:112–122

Deretic V (2021) Autophagy in inflammation, infection, and immunometabolism. Immunity 54: 437–453

Dimitrov DS (2000) Cell biology of virus entry. Cell 101:697–702

Earley LF et al (2017) Adeno-associated virus (AAV) assembly-activating protein is not an essential requirement for capsid assembly of AAV serotypes 4, 5, and 11. J Virol 91

Elmore ZC et al (2021) The membrane associated accessory protein is an adeno-associated viral egress factor. Nat Commun 12:6239

Ficarelli M, Neil SJD, Swanson CM (2021) Targeted restriction of viral gene expression and replication by the ZAP antiviral system. Ann Rev Virol 8:1–19

Formas-Oliveira AS, Basílio JS, Rodrigues AF, Coroadinha AS (2020) Overexpression of ER protein processing and apoptosis regulator genes in human embryonic kidney 293 cells improves gene therapy vectors production. Biotechnol J 15:1900562

Gaudin Y (2021) Penetration of non-enveloped viruses. Nat Microbiol 6:1343–1344

Georgi F, Greber UF (2020) The Adenovirus Death Protein – a small membrane protein controls cell lysis and disease. FEBS Lett 594:1861–1878

Gerlier D, Varior-Krishnan G, Devaux P (1995) CD46-mediated measles virus entry: a first key to host-range specificity. Trends Microbiol 3:338–345

Ghigo E (2010) A dilemma for viruses and giant viruses: which endocytic pathway to use to enter cells? Intervirology 53:274–283

Greber UF, Flatt JW (2019) Adenovirus entry: from infection to immunity. Ann Rev Virol 6:1–21

Haan PD, Diemen FRV, Toscano MG (2020) Viral gene delivery vectors: the next generation medicines for immune-related diseases. Hum Vacc Immunother 17:1–8

Haller O, Staeheli P, Kochs G (2007) Interferon-induced Mx proteins in antiviral host defense. Biochimie 89:812–818

Hanson HM, Willkomm NA, Yang H, Mansky LM (2022) Human retrovirus genomic RNA packaging. Viruses 14:1094

Harrison SC (2015) Viral membrane fusion. Virology 479:498–507

Iwasaki A, Medzhitov R (2015) Control of adaptive immunity by the innate immune system. Nat Immunol 16:343–353

Kerviel A, Zhang M, Altan-Bonnet N (2021) A new infectious unit: extracellular vesicles carrying virus populations. Annu Rev Cell Dev Bi 37:1–27

Kit S (1979) Viral-associated and induced enzymes. Pharmacol Therapeut 4:501–585

Koonin EV, Dolja VV, Krupovic M (2022) The logic of virus evolution. Cell Host Microbe 30: 917–929

Krebs A-S, Mendonça LM, Zhang P (2021) Structural analysis of retrovirus assembly and maturation. Viruses 14:54

Letarov AV (2020) History of early bacteriophage research and emergence of key concepts in virology. Biochem Mosc 85:1093–1112

Liu BC, Sarhan J, Poltorak A (2018) Host-intrinsic interferon status in infection and immunity. Trends Mol Med 24:658–668

Lucic B, de Castro IJ, Lusic M (2021) Viruses in the nucleus. CSH Perspect Biol 13:a039446

Luque D, Castón JR (2020) Cryo-electron microscopy for the study of virus assembly. Nat Chem Biol 16:231–239

Maertens GN, Engelman AN, Cherepanov P (2022) Structure and function of retroviral integrase. Nat Rev Microbiol 20:20–34
McCann N, Castellino FJ (2022) Cell entry and unusual replication of SARS-CoV-2. Curr Drug Targets 23:1539–1554
Mercer J, Schelhaas M, Helenius A (2010) Virus entry by endocytosis. Annu Rev Biochem 79:803–833
Moaven O, Mangieri CW, Stauffer JA, Anastasiadis PZ, Borad MJ (2021) Strategies to develop potent oncolytic viruses and enhance their therapeutic efficacy. JCO Precis Oncol 5:733–743
Moller-Tank S, Maury W (2014) Phosphatidylserine receptors: enhancers of enveloped virus entry and infection. Virology 468:565–580
Morran LT, Schmidt OG, Gelarden IA, Parrish RC II, Lively CM (2011) Running with the Red Queen: host-parasite coevolution selects for biparental sex. Science 333:216–218
Moss B (2013) Poxvirus DNA replication. CSH Perspect Biol 5:a010199
Neil SJD (2013) Intrinsic immunity. Curr Top Microbiol 371:67–104
Netherton CL, Wileman T (2011) Virus factories, double membrane vesicles and viroplasm generated in animal cells. Curr Opin Virol 1:381–387
Nóbrega C, Mendonça L, Matos CA (2020) A handbook of gene and cell therapy. Springer, pp 39–90. https://doi.org/10.1007/978-3-030-41333-0_3
Nonnenmacher M, Weber T (2011) Adeno-associated virus 2 infection requires endocytosis through the CLIC/GEEC pathway. Cell Host Microbe 10:563–576
Owen DJ, Crump CM, Graham SC (2015) Tegument assembly and secondary envelopment of alphaherpesviruses. Viruses 7:5084–5114
Pais DAM, Portela RMC, Carrondo MJT, Isidro IA, Alves PM (2019) Enabling PAT in insect cell bioprocesses: in situ monitoring of recombinant adeno-associated virus production by fluorescence spectroscopy. Biotechnol Bioeng 116:2803–2814
Perry C, Rayat ACME (2021) Lentiviral vector bioprocessing. Viruses 13:268
Pillay S, Carette JE (2017) Host determinants of adeno-associated viral vector entry. Curr Opin Virol 24:124–131
Pornillos O, Ganser-Pornillos BK (2019) Maturation of retroviruses. Curr Opin Virol 36:47–55
Romero-Brey I, Bartenschlager R (2016) Endoplasmic reticulum: the favorite intracellular niche for viral replication and assembly. Viruses 8:160
Sadeghpour S, Khodaee S, Rahnama M, Rahimi H, Ebrahimi D (2021) Human APOBEC3 variations and viral infection. Viruses 13:1366
Schmid M, Speiseder T, Dobner T, Gonzalez RA (2014) DNA virus replication compartments. J Virol 88:1404–1420
Schoggins JW (2019) Interferon-stimulated genes: what do they all do? Ann Rev Virol 6:1–18
Sonntag F, Schmidt K, Kleinschmidt JA (2010) A viral assembly factor promotes AAV2 capsid formation in the nucleolus. Proc Natl Acad Sci 107:10220–10225
Strotz LC et al (2018) Getting somewhere with the Red Queen: chasing a biologically modern definition of the hypothesis. Biol Lett 14:20170734
Tsu BV et al (2021) Running with scissors: evolutionary conflicts between viral proteases and the host immune system. Front Immunol 12:769543
van Gent M, Sparrer KMJ, Gack MU (2018) TRIM proteins and their roles in antiviral host defenses. Ann Rev Virol 5:385–405
Velthuis AJW, Grimes JM, Fodor E (2021) Structural insights into RNA polymerases of negative-sense RNA viruses. Nat Rev Microbiol 19:303–318
Votteler J, Sundquist WI (2013) Virus budding and the ESCRT pathway. Cell Host Microbe 14: 232–241
Wolff G, Bárcena M (2021) Multiscale electron microscopy for the study of viral replication organelles. Viruses 13:197
Yan N, Chen ZJ (2012) Intrinsic antiviral immunity. Nat Immunol 13:214–222
Zhao J et al (2010) Cytomegalovirus β2.7 RNA transcript protects endothelial cells against apoptosis during ischemia/reperfusion injury. J Hear Lung Transplant 29:342–345

Chapter 5
Overview of Current Downstream Processing for Modern Viral Vectors

Paul Cashen and Katy McLaughlin

5.1 Introduction

Two of the most common viral vectors for gene therapy applications are lentiviruses (LVs) and adeno-associated viruses (AAVs). Their major differences are summarized in Table 5.1.

An important characteristic is that LVs are enveloped, whereas AAVs are non-enveloped. As a result, AAVs are more stable with respect to processing technologies and less sensitive to shear and perturbations in conditions such as buffer pH, conductivity, and temperature (Segura et al. 2011). The presence of an envelope in LVs reduces their stability window, requiring a fast processing time from upstream to final formulation and affecting the technologies that can be used to purify the product (Carmo et al. 2009; McCarron et al. 2016). Additionally, the size of LVs makes 0.2 μm filtration steps more difficult, leading to product loss during sterile filtration.

5.1.1 Viral Vector Purification Challenges

Historically, viruses have been contaminants in the production of other products, such as monoclonal antibodies and recombinant proteins. Therefore, many established technologies have been adapted to perform efficient virus removal,

P. Cashen (✉)
Sartorius Stedim Biotech, Hertfordshire, UK
e-mail: paul_cashen@pall.com

K. McLaughlin
Sartorius Stedim UK Ltd., Longmead Business Centre, Surrey, UK

S. Gautam et al. (eds.), *Bioprocess and Analytics Development for Virus-based Advanced Therapeutics and Medicinal Products (ATMPs)*,
https://doi.org/10.1007/978-3-031-28489-2_5

Table 5.1 Overview of typical product characteristics for lentiviruses and adeno-associated viruses

	Lentivirus	Adeno-associated virus
Envelope	Yes	No
Stability	Low	High
Virus size (nm)	~100	~20
Insert size (kb)	~7.5	~4.7

ensuring the high purity of the target product. This becomes a challenge when a typically unwanted impurity (the virus) is the desired product requiring separation from other product-related and process-related impurities.

Another issue is the relative lack of technology capable of purifying these products at high purity compared to conventional products. Monoclonal antibodies and recombinant protein production approaches are increasingly focused on process intensification (strategies to make existing processes more economical, productive, and cost-effective). In contrast, the current challenge with the downstream processing of viral vectors is identifying what technologies exist on the market that could work, as well as developing new technologies to increase the recovery and purity of the vector across the entire downstream process. For reference, an unoptimized monoclonal antibody downstream process could relatively comfortably achieve a product yield of ~95% without further optimization; a fully optimized downstream process for a lentiviral vector is likely to achieve a ~30% product yield on average (Elizalde and Ramírez 2021).

These issues are further compounded when product purity is considered. This is also a challenge in viral products due to three main factors: the current lack of available technology, the relatively small stability window of viral vectors, and the idea that viruses are traditionally seen as impurities in downstream processing, making it difficult to separate them from other common impurities, such as host cell proteins (HCPs) and host cell DNA (hcDNA). Another issue to consider is the general reliability of analytical assays for viral vectors, which are seen as slow, low throughput, and highly variable (van der Loo and Wright 2016; Do Minh and Kamen 2021).

Finally, the production of varying levels of empty capsids—that lack the viral genome and the genes of interest—in the final product is also an important consideration for AAV products. Separating empty capsids from full capsids is necessary to avoid unwanted immune responses in patients. However, this poses an additional purification challenge: empty and full capsids are highly similar and require high-resolution methods for their separation. As AAV particles differ in structure and dynamics from serotype to serotype (Venkatakrishnan et al. 2013), standardization of empty/full separation methods is difficult.

5.2 Purification Processes at a Glance

Due to the relatively low number of downstream steps used in a viral vector production process (typically four for LVs and six for AAVs), each purification step must be highly efficient (Hebben 2018; Moreira et al. 2021). A typical recombinant protein or monoclonal antibody process can consist of ten or more purification steps, with many orthogonal approaches (e.g., size exclusion, hydrophobic interaction, ion exchange, and low pH holds). Therefore, greater focus can be given to steps that specifically target impurity removal (i.e., HCP removal or separation of isomers). Often, these impurity removal steps do not come at the cost of overall product recovery, as monoclonal antibody processes tend to achieve high purity and process yields. Such recoveries are simply not possible for viral vector processes, where each purification step is a delicate balance between product purity and product recovery.

Both AAV and LV production processes consist of the same core technologies: clarification by filtration, anion-exchange chromatography (AEX), tangential flow filtration (TFF), and sterile filtration. The main difference between AAV (Fig. 5.1) and LV (Fig. 5.2) processes is the inclusion of affinity chromatography for AAV purification and the potential inclusion of a virus filtration step. While not commonly used in many current AAV processes, virus filters with a retention rating of 50 nm could be implemented in future downstream processes to address viral clearance (Hermens and Smith 2018), given that at 20 nm in size, AAV particles can freely pass through a wider pore virus filter. DNA digestion (using endonuclease) is typically implemented at the point of harvest during upstream processing (Yang et al. 2012). Lentiviral processes may have an additional endonuclease step later during downstream steps (Merten et al. 2016).

Affinity technology for LVs is still not readily available. Additionally, the presence of the envelope makes lentiviral vectors more sensitive to shear stress, generally resulting in the preferred use of hollow fibers for TFF as opposed to more traditional cassettes (Moleirinho et al. 2020). However, good results have also been demonstrated using cassettes (Valkama et al. 2020), so the choice between hollow fibers and cassettes can be evaluated on a per-product basis.

While the industry has settled on a baseline process for both vectors, some divergence exists as neither approach is currently considered optimal. Multiple process paths exist for the downstream processing of both vector types. For instance, AAV processes may necessitate the use of a hydrophobic interaction (HIC) membrane in situations where cell lysis has been initiated using salt. For LVs, mixed-mode chromatography has been considered as a way to prevent the high salt conditions the virus is exposed to during the anion-exchange chromatography (AEX) step. Both of these will be discussed in greater detail later.

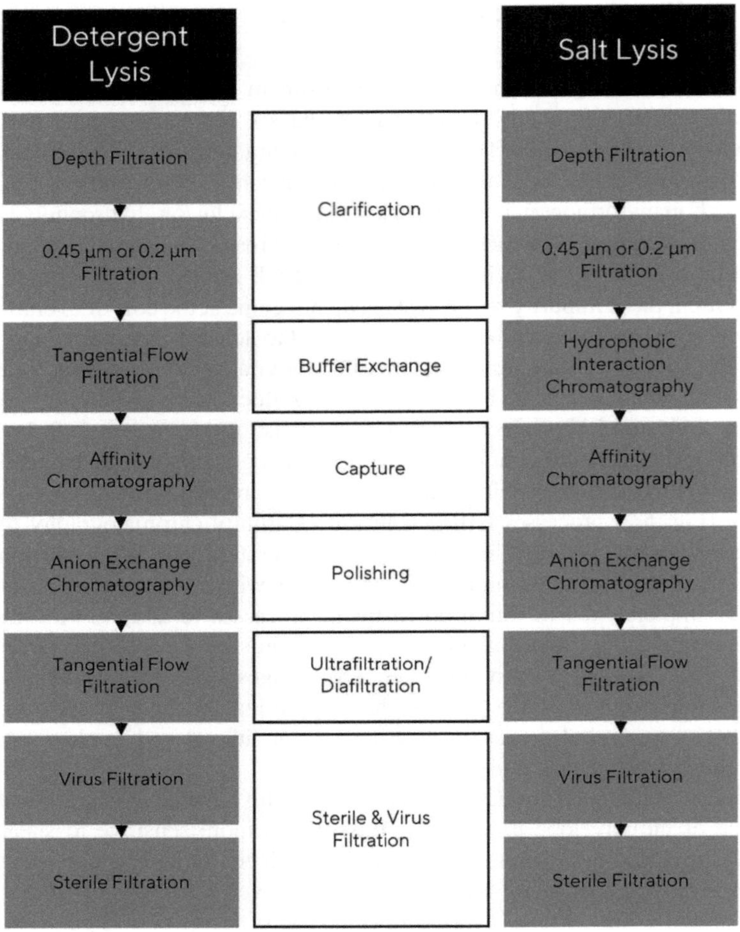

Fig. 5.1 Common process maps for downstream AAV purification with detergent lysis and how a downstream process using salt as the lysis reagent may look. Detergent lysis is the more common of the two methods, while salt lysis is rarely used

5.2.1 Challenges for Each Vector Type

Modern upstream filtration solutions are challenging for LV feedstock. Typical filtration solutions for modern bioprocesses are cellulose-based or involve the use of dematiaceous earth (DE), both of which can readily bind an enveloped vector (Khanal et al. 2018). The size of the LV makes a terminal clarification filtration step at 0.2 μm challenging due to potential losses across the filter, opening up issues with bioburden or sterility control. Filters used in lentiviral processes tend to be composed of polypropylene (PP), polyethersulfone (PES), or a form of synthetic material,

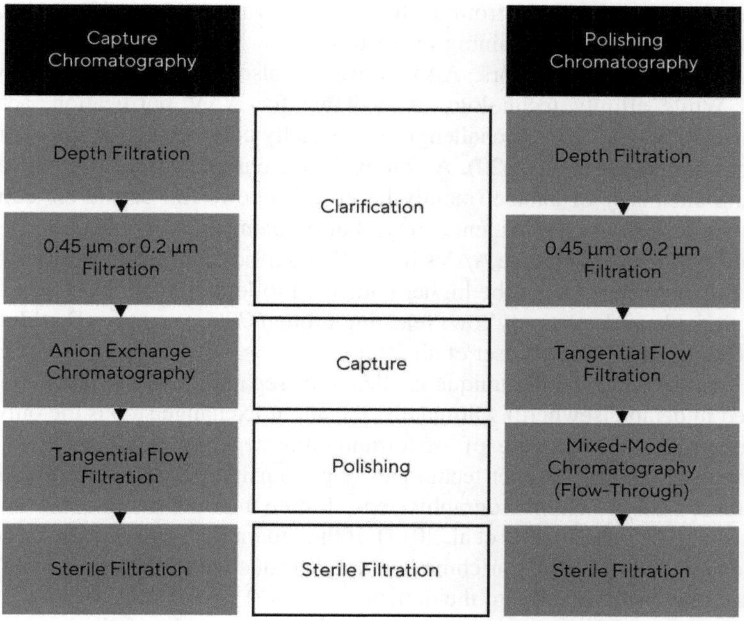

Fig. 5.2 Common process map for downstream LV purification using AEX chromatography and how a downstream process using mixed-mode chromatography as a polishing step may look. The process on the left using chromatography in the capture step is the most common

which reduces viral retention across the filter, but at the expense of impurity clearance (i.e., more impurities pass through the filter) (Perry and Rayat 2021). This creates a trade-off in which product purity is sacrificed for maximized recovery, shifting the burden of impurity clearance further downstream. LV processes typically only contain two purification steps after clarification; additional impurities can significantly influence their efficiency.

Lentiviral vector purification processes generally lack an affinity step, as adequate technologies do not yet exist. Work has been done in the past using heparin affinity capture (Segura et al. 2005, 2007). However, this requires resin-based chromatography, which is not always suitable for large viral vectors (as the pore size of the resin is too small for large vectors to access the ligands). Additionally, the adverse effects associated with heparin can create safety and image concerns (European Commission 2011).

Currently, the only way to adequately separate LVs from impurities is through size and charge. While there have been some instances where cation exchange (CEX) has been used (Merten et al. 2014), buffer adjustment is required to facilitate the binding of the LV to CEX media, as LVs have a negative charge at neutral pH. Due to the small stability window of LVs, infectivity can become compromised by the shift to low pH. Ion exchange, therefore, has been limited to mainly just anion exchange. Without both types of ion exchange in a process, it is difficult to fully remove all HCP and hcDNA from the feed stream, which means relying on the final TFF step to remove any remaining impurities.

Similar to lentiviral vectors, AAV processes also have chromatography challenges. While affinity technology is available for AAV purification, achieving consistent step yields can be challenging, especially considering the many different serotypes (Mietzsch et al. 2020). Anion-exchange chromatography is typically the polishing chemistry of choice (mainly because it allows full capsid enrichment to further increase purity and patient safety), but problems persist. Like LVs, it can be incredibly difficult to separate AAVs from all impurities, and while yields across the anion-exchange step tend to be higher compared to lentiviral vectors, they are still highly variable and relatively low, reaching around 70% recovery (Davidoff et al. 2004; Okada et al. 2009; Rieser et al. 2021).

AAV processes have the unique challenge of separating empty and full capsids (covered in detail elsewhere). Ultimately, the anion-exchange step is the only large-scale operation unit capable of performing this separation (Green et al. 2002; Kaludov et al. 2002), as other techniques (e.g., density gradient centrifugation) do not scale as easily as chromatography steps, despite their use being demonstrated in large-scale processes (Fiedler et al. 2017). If the separation is inadequate or does not provide enough full capsid enrichment, empty capsids will be passed down through the rest of the process and into the patient.

Concentrating LV and AAV bulk drug substances can be challenging, as typically, the product has to be concentrated up to 10,000-fold to achieve titers that can be administered in a reasonable volume. Combined with the difficulty in impurity removal, concentrating the vector product sufficiently without inducing aggregation or precipitation of the product is a delicate balance.

5.2.2 *Challenges Depending on Cell Culture Mode*

While similar problems persist downstream for the purification of both suspension and adherent cell cultures, several factors must be considered during process development for each mode. Suspension-based upstream processes generally produce greater cell densities and, as a result, more impurities in the form of HCPs and hcDNA. Another point to consider is the potential for cell viability to decrease during viral vector production, which also leads to a higher impurity load in the culture. These increased impurity profiles usually require a more robust clarification setup, achieved either with a larger surface area or with multiple clarification filters, usually up to a maximum of three.

For adherent processes where the viral vector is secreted (LVs and some AAV serotypes), and therefore no lysis is needed, a large portion of cellular impurities are either immobilized within the upstream process platform or, in the case of fixed-bed bioreactors, retained within the bioreactor. Consequently, the feedstream's impurity profile and turbidity may be reduced so that a single filter train is sufficient to clarify it to a low enough turbidity for processing in the subsequent step (Raghavan et al.

2019). In the case of suspension processes (or any processes where cells are lysed), if the impurities are not sufficiently dealt with during clarification, a bottleneck is created during the next unit operation. This potentially limits the binding capacity of chromatography sorbents, as the virus begins to compete with other closely related proteins for binding sites, or it can affect the throughput of a TFF membrane, as more impurities can contribute to faster fouling of the device.

5.3 Lysis Strategy and Endonuclease Additions

While enveloped viral vectors (such as LVs) exit the host cell through budding, naked viruses (such as many serotypes of AAV) are often stuck inside the cell. Chemical or enzymatic lysis is needed to disintegrate the cell membrane and release the viral particles into the feedstream. Following cell lysis, endonuclease is usually added to digest nucleic acids, reducing the viscosity and turbidity of the feedstream and aiding future purification steps.

5.3.1 Cell Lysis

Lysis is only relevant to a set of AAV serotypes that are not secreted from the cell and therefore have no way of making it into the feedstream. Due to the presence of fractured cells, processes requiring lysis will result in feedstreams with a lower particle size distribution than processes where the product is secreted from intact cells. As a consequence, depending on the cellular concentration, clarification filters with a lower retention rating may be needed to remove host cell impurities from a lysed upstream process. For processes that do not involve lysis, assuming cell viability is high, most intact cells can be removed with a filter with a larger retention rating.

The two most commonly used lysis buffers are either detergent-based or salt-based. Detergent (e.g., Triton) is typically preferred as it is cheap, is easy to remove, and does not pose any adverse effects on the product or process (Ma et al. 2020). However, above certain quantities, detergents, such as Triton X-100, present an EHS consideration for disposal and patient safety (Hunter et al. 2022). In fact, as of 2017, the Triton X-100 family was added in an amendment to the Registration, Evaluation, Authorisation and Restriction of Chemicals (REACH) list (The European Commission 2017). An alternative to detergent, salt (typically ammonium sulfate, in large quantities, i.e., ~1.5 M), can be used as a lysis buffer (McNally et al. 2020).

In either case, the lysis buffer must be removed further downstream through dedicated removal steps or as a consequence of additional purification. In the case of ammonium sulfate, removal is typically performed through a dedicated HIC step prior to capture chromatography. While the presence of high levels of ammonium sulfate does not pose the same risks as detergent, the requirement for an additional step to remove the salt and the additional EHS considerations needed for its disposal (National Center for Biotechnology Information 2022) are two factors to consider when choosing the appropriate lysis buffer.

5.3.2 *Nuclease Addition*

Impurities present within the feedstream can reduce the efficiency of various downstream purification steps. Endonucleases are enzymes added to reduce the size and quantity of any nucleic acids present in the upstream process. Limiting the amount of nucleic acid in the final product is a key requirement when designing a downstream process for viral vectors (U.S. Food and Drug Administration (FDA) 2020). Currently, the limit for nucleic acid is set at 10 ng/dose and 100 base pairs (bp) (Tanaka et al. 2020). This is difficult to achieve in viral-based therapeutics due to the relatively low amount of virus in the final formulation, a problem that is further exacerbated in high therapeutic dose products.

There are various potential addition points for nuclease within a viral vector process, all with different benefits and drawbacks. While addition at the point of harvest (within the bioreactor) is the most common (Buclez et al. 2016; De Rooij et al. 2019), it is also the most expensive, given that the endonuclease needs to be added at a specific concentration, and the process volume is generally at its largest in upstream. Nuclease additions can also be employed post-clarification filtration and during TFF, post-concentration, and prior to diafiltration (Ausubel et al. 2012; Soldi et al. 2020). In general, adding endonuclease at a specific point in the downstream process only shifts the bottleneck (of impurity removal) by placing a greater burden and emphasis on removal to the other purification steps.

While the harvest step is often seen as the most optimal point for nuclease addition, there is a potential issue of cleaving a portion of nucleic acid strands to smaller sizes than the retention ratings specified in the clarification step, resulting in more impurities being able to flow freely through the clarification step and into the subsequent downstream steps. This is exacerbated further in situations where a PES or synthetic filter is used, which offers lower impurity removal than, for example, a cellulose-based filter. Another point to consider is that harvest conditions (37 °C) are typically viewed as optimal for nuclease activity, meaning no adjustment time is needed at the point of addition.

As the addition of nuclease during the harvest step helps to reduce the overall turbidity of the feedstream, post-harvest nuclease addition (such as post-clarification or during TFF) means that the clarification step will be challenged with a feedstream of higher turbidity. A common solution to this issue is to either increase the overall surface area of the filter train or add a second or third filter to act as a pre-filter. Typically, this is viewed as a less optimal approach due to the unnecessary complications created during the clarification step, particularly in the case of post-clarification addition, where the volume of material is unchanged and, therefore, no cost savings are made on the amount of nuclease required.

Moreover, as it is not feasible to heat a large volume of product mid-process at manufacturing scale, the nuclease is often required to incubate for longer periods, usually overnight, increasing processing time (and running the risk of vector degradation given the relative instability of lentiviral vectors). However, adding nuclease after filtration may result in a product of higher purity, given that the majority of longer nucleic acids will hopefully have been retained by the clarification filters, as opposed to being cleaved to such a small size that they may freely pass through. This decision must be made on a case-by-case basis, depending on the user's specific needs.

The final nuclease addition point that could be considered is during the TFF step. Usually, this is considered as a secondary addition point and is typically employed in processes where the amount of nucleic acid removed from the first addition is not sufficient for the final product. Added post-concentration, the nuclease addition at this point is at a relatively low process volume, usually 10- to 100-fold less than the volume at harvest, requiring much less nuclease than the first addition point. The nuclease would be incubated in recirculation at room temperature before being removed into the appropriate processing buffer by diafiltration. Like the post-clarification addition, the nuclease added here will likely have reduced activity compared to at harvest due to the difference in temperature (room temperature vs. cell culture temperature). However, given that this is typically an additional step, acting on a much smaller amount of nuclease (likely at the end of the process, where the product is at its most pure), there is less need for strong endonuclease action.

Salt-active nuclease (SAN) is an endonuclease with optimum activity at high concentrations of salt at high pH. Its stability in high salt means it can be potentially included at the point of AEX elution for lentiviral vector processes. If intending to include a second nuclease addition point, then incubation of the product with nuclease at the elution step allows for a quicker process by eliminating the need for two sets of diafiltration during the TFF step (one diafiltration to exchange the product into a nuclease-suitable buffer and then the second diafiltration to remove the nuclease). As lentiviral vectors are relatively unstable at high salt concentrations, the use of SAN and incubation duration must be determined on a per-product basis. SAN has been shown to have optimal activity above 150 mM NaCl (Farley and Mitrophanous 2019), while lentivirus infectivity begins to drop above approximately 0.8 M NaCl, suggesting there is a window for optimization of salt concentration.

5.4 Clarification by Filtration

This chapter will only consider clarification of harvest material by filtration and not by other means, such as centrifugation. This is due to its greater prevalence in the industrial manufacture of viral vectors, as well as its relative ease of scalability compared to other techniques.

The aim of the clarification step is to remove as many cellular contaminants as possible while allowing for transmission of the product (virus) across the filter medium. The filtered material must also be of a low enough turbidity so that it does not negatively impact the subsequent process steps. The most significant factors to consider when selecting media to achieve this are the material of the filter(s) and their specific retention rating. In addition, sufficient turbidity reduction may need to be achieved with multiple filters. For any given process, two filters are usually preferred, although a third may be added to help increase filter performance. The material of construction of the filter is highly dependent on the type of vector (typically, whether there is the presence of an envelope) and will be discussed further below.

Beyond material choices, it is important to choose a filter train that provides retention ratings suitable for both the product and the process. While the type of vector is the most critical consideration for the choice of material for the filter, the retention ratings needed are typically defined by the upstream process as a whole. For example, host cell type, specific process and product-related impurities, and the presence or absence of a lysis step are all important considerations. Generally, the first coarser filter in the filter train is tasked with removing a lot of cellular debris, while the second finer filter should help clear the smaller impurities that may otherwise pass through the first filter. Tables 5.2 and 5.3 provide an overview of the various clarification filtration solutions available for LVs and AAVs.

Careful consideration should be given to ensure the retention ratings of all filters in a particular train are complementary and do not cause unnecessary burden to the other filters in the train. For example, using a filter with too low of a retention rating could cause the filter to block and/or foul, resulting in product loss and decreased product purity as impurities begin to break through the filter into the feedstream and interfere in the purification process, by, for example, competing with binding sites on chromatography media. Conversely, if the retention rating of the filter is too high, it cannot perform adequately, as large portions of the feedstream that should otherwise be excluded are now free to pass into the subsequent downstream processing steps.

A major factor in achieving a robust and reproducible clarification step is ensuring that the feed material produced upstream is as controlled and consistent as possible. As clarification filtration is primarily dictated by size (of the filter channels and the particles in the feedstream), it stands to reason that the more consistent a feedstream—with respect to the size of the particles within it—the more consistent the filtration step. For this reason, it is crucial to understand the particle size distribution of the feedstream. Particle size distribution affects what filter retention rating is suitable for a specific feedstream (van Reis and Zydney 2007; Chan et al. 2021). A feedstream with a low particle size distribution will likely require filters with smaller retention ratings and vice versa. Cell shearing and cell death release intracellular proteins and fractured cells into the feedstream, lowering the distribution. Therefore, cell shearing and death must be limited to ensure a consistent particle size distribution from batch to batch. Should this occur unexpectedly, the distribution could fall below the retention rating of the filter, leading to the problems mentioned above.

Table 5.2 Filtration solutions for the clarification of lentiviruses

Supplier	Filter	Material	Retention Rating (μm)
Merck Millipore	Millistak+ D0HC	DE + cellulose	0.5–8
	Millistak+ CE35	Cellulose fibers	2.5–4
	Clarisolve 60HX	Polypropylene	7.5–60
	Milligard SMP8	Polyether sulfone (PES)	1.2/0.8
	Polygard CN06	Polypropylene	0.6
	Polysep II CGW6	Borosilicate glass microfiber/mixed esters of cellulose	1.0/0.5
Pall	Suporlife	PES	0.45
	Fluorodyne II DBL	PVDF	0.45
	Ultipor N66	Nylon 66	0.45
	PreFlow UB	Resin-bonded GF	0.45
	Supor EAV	PES	0.2
	Profile II/Star	PP	0.5
			1
			3
			5
	HDC II	PP	10
Sartorius	Sartopure PP3	PP	1.2
			3
			5
			8
	Sartopure GF+	Glass fiber	0.65
			1.2
	Sartoguard PES	PES	1.2/0.2

Any lysis procedures will further complicate this because lysis is an uncontrolled process; all cells will rupture differently, creating various subsets of particles, all with different sizes. As a consequence, any filtration train designed to handle lysed material must demonstrate robust performance with feedstreams that vary in composition from batch to batch.

Since lentiviral vectors are enveloped, it is typically not advisable to use cellulose-based filtration devices or devices that contain diatomaceous earth for the filtration of LVs (Wang 2011; Venkiteshwaran et al. 2015). This is because the envelope confers a negative charge, meaning the vector is readily retained by filters constructed from these materials. A recent study showed that while the presence of diatomaceous earth can have a negative effect on the filtration of lentiviral vectors at

Table 5.3 Filtration solutions for the clarification of adeno-associated viruses

Supplier	Filter	Material	Retention rating (μm)
Merck Millipore	Clarisolve 20MS	Polypropylene	0.5–20
	Millistak D0SP	Polyacrylic fiber and silica	0.5–8
	Millistak D0HC	DE + cellulose	
	Millistak C0SP	Polyacrylic fiber and silica	0.5–2
	Millistak C0HC	DE + cellulose	
Pall	Seitz Bio 10	Cellulose, resin	0.2–0.4
	Seitz V100P	Cellulose, perlite, resin	2–4
	Seitz HP PDH11	Cellulose, diatomaceous earth, perlite, resin	2–15
	Seitz HP PDK11		2–20
	Seitz HP PDP8		6–30
Sartorius	Sartoclear DL90	Cellulose	2–15
	Sartoclear DL75	Cellulose	2–10
	Sartoclear DL60	Cellulose	0.8–8
	Sartoclear DL20	Cellulose	0.4–8

specific concentrations, adjusting the amount so that it is more suitable for enveloped viruses allows for increased impurity reduction and throughput and enables the transmission of the vector (Labisch et al. 2021).

Generally, non-cellulose filters such as polypropylene-based, polyethersulfone-based, or synthetic filters are recommended for viral vector purification. This is because these materials support the full transmission of the vector while also excluding many impurities via size exclusion (Raghavan et al. 2019). However, it must be noted that by removing these filter additives/aids, the secondary mechanism of particle retention (charge) is also removed. Therefore, although high vector recovery can be expected from these filters, this is counterbalanced typically by lower throughputs (as filters block quicker), requiring more filtration area to clarify a given bioreactor volume, or by a reduction in impurity clearance, as smaller particles that would usually bind within the filter are now able to pass through freely. For lentiviral processes where uncharged filters must be used, the impurities that do not bind to the filter must be removed in subsequent downstream processing steps. As most clarification filtration trains will include a bioburden-reduction filtration at the terminal point, it is crucial to consider the impact of non-charged clarification filters

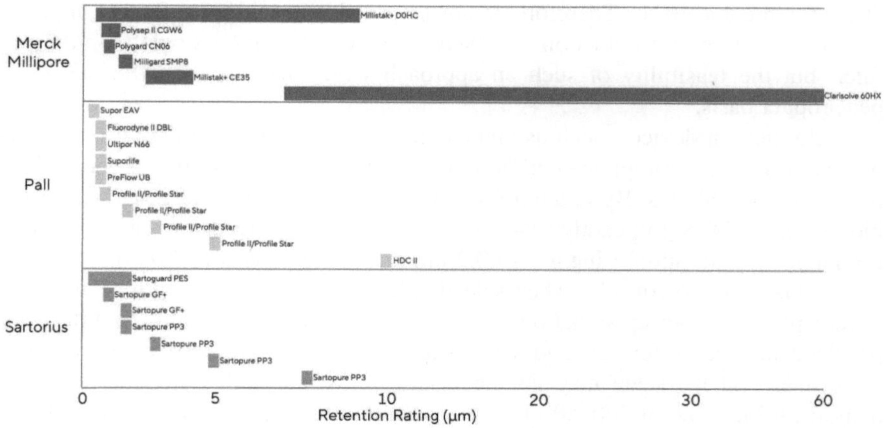

Fig. 5.3 Filtration solutions for LVs, with respect to their retention rating ranges

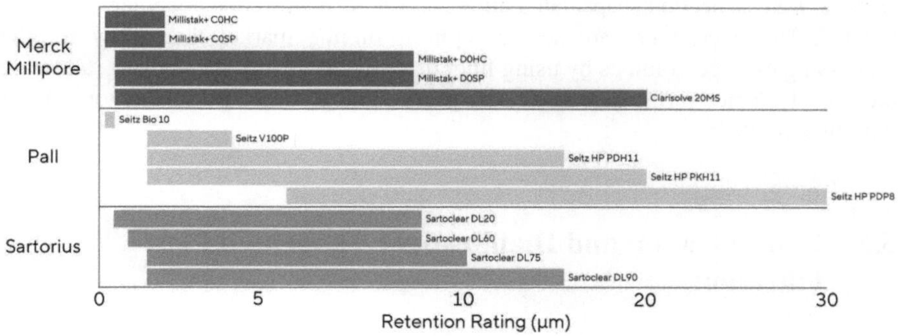

Fig. 5.4 Filtration solutions for AAVs, with respect to their retention rating ranges

on the terminal filter. As with the primary filter, it may be required to increase the surface area of the terminal secondary filter to prevent fouling (Fig. 5.3).

Clarification filtration of AAVs can be considered significantly more straightforward than for lentiviral or other enveloped vectors. As AAVs are non-enveloped, they do not bind as readily to common filter matrices, and so it is typically seen as more economical to use cellulose-based filters for these processes. The key variable is whether or not a cell lysis step is implemented, as it changes the particle size distribution within the feedstream. Often, this will affect the primary filter in the filtration train; a process with a lysis step will require a primary filter with a lower retention rating to exclude most of the lysed cell fragments (Fig. 5.4).

Regarding the choice of clarification filtration for adherent and suspension processes, generally, an adherent process is considered to be "cleaner" than a suspension process (i.e., lower in turbidity), owing to lower cell numbers and, in some cases, retention of cells either within a fixed bed or cell factory (Lennaertz et al.

2013; Baghirzade 2021). Therefore, it may be possible to achieve sufficient impurity removal and turbidity reduction in adherent processes using a single 0.45/0.2 μm filter, but the feasibility of such an approach would have to be evaluated on a per-product basis.

Cell retention devices, such as tangential flow depth filtration (TFDF), are now being considered for implementation into non-lysed viral vector processes to reduce filtration requirements. By retaining much of the cellular material generated in the bioreactor, TFDF significantly reduces the turbidity of the feedstock and can allow for single-step filtration using a 0.45/0.2 μm filter (Williams et al. 2020).

The last thing to consider when selecting a suitable filtration step for the downstream process is an appreciation of the final manufacturing scale. As filters are provided in specific formats and sizes (surface areas), a solution that shows great performance at lab scale may not necessarily have the ability to scale to a full manufacturing scale of 200–500 L without combining multiple filters in parallel or performing the filtration in multiple “sub-batches,” both of which are time-consuming and labor-intensive for the operator. This is true for capsulated filters, which have defined upper limits on their surface areas. Beyond this, if a greater surface area is needed, capsulated filters must be manifolded together to run in parallel. However, there are several options on the market that allow for the processing of large volumes by using filter discs or sheets inside a housing, allowing the use of greater surface areas and the processing of the entire bioreactor harvest bulk in a single run.

5.5 Concentration and Diafiltration: Tangential Flow Filtration

Tangential flow filtration (TFF) is usually implemented in one or two specific places within viral vector purification processes. The primary objectives are concentrating and exchanging the product into a buffer suitable for further processing or a stable buffer for sterile filtration and final formulation. In either case, the product is concentrated 10- to 100-fold, and during the process of concentration and diafiltration, low-molecular-weight impurities can be further removed from the feedstream. Table 5.4 provides an overview of common TFF solutions available for gene therapy applications.

For AAV processes, TFF is usually implemented before capture chromatography and after capsid enrichment, prior to sterile filtration (Qu et al. 2015; Dobrowsky et al. 2021). The step is performed mainly using standard TFF cassettes made from either PES or cellulose and with a molecular weight cutoff (MWCO) of 100 kDa.

If buffer exchange is not needed before capture chromatography, then it may be more effective to use single-pass TFF or inline concentration to reduce the time taken to perform the TFF step (Madsen et al. 2022). In these scenarios, a reduced process time is traded for the lack of diafiltration and potentially a lower concentration factor, as single-pass devices can be limited by how much they can concentrate a

Table 5.4 TFF solutions for gene therapy applications

Supplier	Product	Format	Material	Molecular weight cutoff (kDa)
Cytiva	ReadyToProcess	Hollow fiber	PES	100, 300, 500, 750
Merck Millipore	Pellicon Ultracel	Flat sheet cassette	Cellulose	100, 300
	Pellicon Biomax	Flat sheet cassette	PES	100, 300, 500
Pall	T-Series Omega	Flat sheet cassette	PES	100, 300
Repligen	TangenX SIUS ProStream	Flat sheet cassette	mPES	100, 300
	TangenX SIUS HyStream	Flat sheet cassette	mPES	
	Spectrum Hollow Fiber Filters	Hollow fiber	mPES	100, 300, 500, 750
Sartorius	Hydrosart Ultrafilter	Flat sheet cassette	Cellulose	100, 300
	PES Ultrafilter	Flat sheet cassette	PES	
	Hollow Fibre TFF	Hollow fiber	mPES	100, 300, 500, 750

therapeutic product due to the lack of a recirculation loop, as is seen in conventional TFF (Moleirinho et al. 2020). The elimination of the recirculation loop means that it may be more difficult to hit a target concentration. Nevertheless, there is merit in evaluating both options for pre-chromatography loading.

As with clarification and chromatography, the relative fragility of LVs can make TFF a challenging process step. Firstly, due to the size of LV particles, TFF devices with bigger MWCOs (300 kDa, 500 kDa, or even 750 kDa) can be used, as a larger cutoff can help alleviate any issues due to fouling and enable faster processing times (Schweizer and Merten 2010). Secondly, as LVs are extremely shear sensitive (due to the presence of the envelope), hollow fibers are primarily used over cassettes (Negrete et al. 2014), owing to their lower shear. TFF of LVs using cassettes has been demonstrated (Valkama et al. 2020). Smaller MWCO 100 kDa cassettes may also be preferred to larger MWCO cassettes (e.g., 300 kDa), where quick fouling is observed, likely due to partial plugging of the cassette by LVs. Nevertheless, the combination of low shear, plus their availability at higher cutoffs, makes hollow fibers the primary TFF device of choice for lentiviral purification.

TFF for lentiviral vectors is usually implemented in one of two places within the purification process and is typically determined by what other purification steps are present (Moreira et al. 2021). The most common point of use for TFF is after anion-exchange chromatography before sterile filtration. At this point in the process, the LVs would be in a high salt elution buffer that can rapidly accelerate their degradation. An important aspect of TFF here is to diafilter the vector out of the elution buffer and into a stable buffer as soon as possible. The concentration of LVs may lead to aggregation (Hollis et al. 2009), limiting the amount by which the vector can

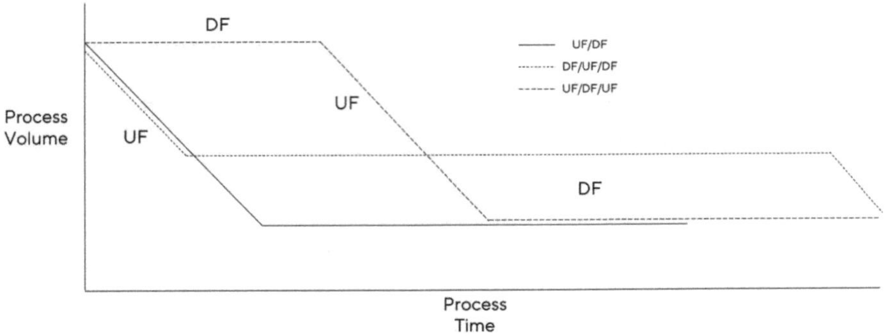

Fig. 5.5 Overview of different TFF scenarios. Diafiltration is quicker at a lower volume; therefore, traditional UF/DF is the fastest way to process the product. While slower, other options, such as DF/UF/DF or UF/DF/UF, may be needed to help preserve vector integrity

be concentrated. In these cases, at the expense of processing time, it may be worth performing an upfront diafiltration, followed by concentration and further diafiltration (DF/UF/DF), or partial concentration, followed by diafiltration and then a final concentration (UF/DF/UF). This approach has been demonstrated in prior publications (Cooper et al. 2011). By exchanging the lentiviral vector into a more stable buffer, it is possible to concentrate it further before aggregation becomes an issue (Fig. 5.5).

For AAV vectors, TFF can also be placed further upstream in the process, following clarification filtration. In these instances, TFF is usually applied where product concentration is needed to reduce loading times, typically onto a chromatography resin instead of a membrane or monolith. Generally, the most important consideration here is to reduce the volume of the feedstream to account for the longer loading times of chromatography resins. Buffer exchange is usually performed to ensure optimal loading of the vector.

For both LV and AAV vectors, if the TFF step is used as an additional nuclease addition point, the nuclease must be removed by an additional diafiltration step. This is particularly important if the TFF step directly precedes the final sterile filtration.

5.6 Chromatography: Capture Step

For the majority of AAV processes, the capture step is performed using affinity chromatography (Nass et al. 2018). There are various affinity resins on the market that match each serotype; several non-specific affinity resins that bind each serotype with similar success are also available (Toueille et al. 2018). The advantage of non-specific affinity resins is that they allow for a platform process to be built that can cover the spectrum of AAV serotypes without the need to substitute specific resins and perform subsequent process optimizations. The idea of platform processes is an attractive option in modern bioprocessing as it helps to reduce costs, development time, and on-site storage of purification consumables.

Resin-based affinity chromatography is widely implemented for AAV purification. However, it is also slow. In situations where a faster processing time is needed, monolith chromatography using cation-exchange chemistry can be considered (Rieser et al. 2021).

Due to the lack of affinity technology for lentiviral vector purification, the capture of lentiviral particles is typically performed using anion-exchange chromatography (AEX) (Segura et al. 2011). While not a capture step in the traditional sense (AEX will bind vector product and a wide range of contaminants, such as HCP and hcDNA), the AEX step is often the first purification step after clarification filtration, with the objective being to remove as many contaminants as possible. Given the size of LVs and their instability, membrane or monolith chromatography is preferred over resin-based chromatography due to the speed at which material can be processed and their open pore structure, maximizing the surface area available for large LVs to bind.

Heparin chromatography has shown some promise in the past few years as a pseudo-affinity step (Segura et al. 2007, 2010; Schweizer and Merten 2010), but the problems with using resin for such a large molecule persist. Using AEX membranes or monoliths for lentiviral capture enables the product to move from the bioreactor to the final formulation as quickly as possible, reducing the rate at which the LVs degrade. Unlike the affinity step for AAV purification, recoveries using AEX for LVs are often poor, varying from 20% to 60%. The high salt needed to elute the vector from the membrane or monolith also damages the virus, reducing the infectious titer (Olgun et al. 2019; Valkama et al. 2020; Moreira et al. 2021). Dilution of the high salt eluate must be performed as soon as possible to help maintain a high infectious titer. Typically, this is performed by eluting the LVs into a bag/bottle already pre-filled with dilution buffer to reduce the salt in the feedstream from ~1.0–1.2 M to ~0.2–0.4 M.

Separation of viral products from impurities can also be quite challenging, as HCP and hcDNA have a range of charges, some of which overlap perfectly with the vector product. This means that a subset of impurities will always be carried over into subsequent purification steps. Given that lentiviral processes typically do not have many purification steps, this can create a burden on the next downstream step to remove large amounts of impurities.

In addition to the problems this step causes to the vector, the high salt combined with the low level of the product makes quantitative assaying difficult (Logan et al. 2004), as it is challenging to reduce the level of salt within the sample enough to eliminate background interference while simultaneously maintaining the vector above the limit of quantification (LOQ). The difficulty in assaying around the AEX step, combined with very low step yields, makes the LV capture step its biggest current bottleneck.

5.7 Chromatography: Intermediate Purification, Polishing, and Empty/Full Capsid Separation

In the case of AAVs, polishing chromatography is usually synonymous with empty/full capsid separation or capsid enrichment. Not all capsids produced in the upstream bioreactor contain a fully functional vector. Empty capsids (viruses that do not contain the therapeutic payload) are also present within the feedstream. As they are structurally similar to full capsids, it is challenging to remove empty capsids from the product. However, there is a need to separate empty capsids from full capsids to ensure the feedstream has as much therapeutic drug as possible. Patients are dosed based on the amount of full capsid within the drug. Empty capsids are considered process contaminants, and their administration should be avoided as their effects on the patient are still not fully understood, primarily due to the lack of an appropriate animal model (Ronzitti et al. 2020). However, it is currently thought that empty capsids contribute significantly to increased immune responses, which are directed to the capsid antigen (Wright 2014; Pei et al. 2018). The issue of having empty capsids in the drug product becomes more significant as the amount of the final dose administered to patients increases. Currently, the technique seeing the most success is AEX chromatography, which, through careful manipulation of the elution gradient, allows for independent elution of empty capsids and full functional capsids (Urabe et al. 2006; Okada et al. 2009).

Capsid enrichment via empty/full separation is still an emerging area of gene therapy, largely due to the field still being in its relative infancy (so less data is available on techniques such as empty/full separation), but also because capsid separation varies from AAV serotype to serotype, which further complicates developments in the field (Hejmowski et al. 2022).

As mentioned previously, the narrow stability window of LVs does not allow for many different options when designing a purification process. In general, one round of chromatography is used before progressing to either TFF or sterile filtration, limiting options for a dedicated polishing step. CEX can be used for LV purification, but recent literature on its implementation is lacking. Certainly, AEX is the preferred ion-exchange technique used for LV. Where AEX is not used, and capture is performed using TFF, multi-modal chromatography using CaptoCore from Cytiva (in flow-through mode) has seen promising results for impurity removal in lentiviral vector processes (James et al. 2016; Do Minh and Kamen 2021). The lack of a high salt added to the product also helps prevent unnecessary vector degradation, which supports high infectious vector yield across the process.

5.8 Virus Filtration and Sterile Filtration

Incorporating virological safety in the development of biological products is a key aspect of ensuring safe therapeutics suitable for human administration. Viral safety is achieved using a "three-pronged approach" that includes raw material testing, virus testing of cell culture harvests, and the inclusion of several virus clearance steps in the purification process (Dinowitz 2002). The risk for viral contamination increases as higher doses of gene therapy products are required. Therefore, the necessity for viral clearance steps increases in accordance with the dose administered.

The concept of virus filtration in gene therapy processes is a relatively new area of development and consideration. In more traditional therapeutics, such as monoclonal antibodies or recombinant proteins, viral clearance is an integral part of the downstream process, ensuring that any adventitious viruses, should they enter the feedstream, can be cleared within the process to maintain the safety and efficacy of the therapeutic product (Ajayi et al. 2022). However, this becomes more difficult to manage when the therapeutic product is a virus itself, where most common viral removal applications will also act upon the product. One of the more recent developments in this field is related to virus filtration of the vector product using a virus filter with a 50 nm pore size (Long et al. 2020). While this does not contribute to a comprehensive removal of all potential viral contaminants, it does allow for some degree of control (and removal) of adventitious viruses without negatively impacting the viral product.

Beyond filtration, viral clearance has been demonstrated using chromatographic methods (Winkler et al. 2021), which could become a strong alternative to virus filtration. A previous study showed that clearance of more than 14 $\log_{10}$ of herpes simplex virus (HSV) could be achieved in an AAV process through a lysis step and two chromatography steps (Ye et al. 2014). More recently, Thermo Fisher Scientific has demonstrated log reduction values (LRV) of common adventitious viruses between three and five logs using their POROS CaptureSelect AAVX resin (ThermoFisher Scientific 2021).

Due to their relatively large size compared to AAVs, LVs cannot be filtered using a 50 nm virus filter, as the pore size is too narrow to allow full passage of the vector, leading to significant product loss. For these processes, a greater focus on closed systems (where possible) and possible upstream filtration of media should be considered. Sterile filtration of AAVs is relatively straightforward. Typically performed using a PES filter, very little product loss is observed during this step (McAlister and Lambropoulos 2022); any product loss at this point is usually a strong indicator that the AAV vectors have aggregated during a preceding purification step.

As the size of LVs (~100 nm) is close to the pore size of 0.2 μm sterile filters, it can be challenging to filter the vector product without significant losses over the filter. This is a significant obstacle in the sterile filtration of lentiviral vectors, where product losses of 50% or more can be expected (Merten et al. 2011; Valkama et al. 2020), especially if the product begins to aggregate after concentration in the TFF step. Usually, this product loss is factored into the process when accounting for the final vector titer.

5.9 Quality by Design (QbD), Identification of Critical Quality Attributes (CQAs) and Critical Process Parameters (CPPs)

While still a relatively unknown concept for gene therapy applications, several guides are beginning to emerge in the industry (Tanaka et al. 2020; Cashen and Manser 2021; Greg et al. 2021) to help users apply the principles of quality by design (QbD) in their process. In recent years, the FDA has rejected several high-profile therapies, with the general reasoning being a lack of data in the chemistry, manufacturing, and controls (CMC) documentation (Viswanathan and Puich 2021). This is also compounded by the fact that, currently, there is a lack of established regulatory frameworks for developing gene therapy products (Drago et al. 2021). Regulatory bodies are increasingly endorsing the implementation of QbD principles. The ICH guidance documents Q8–11 describe principles and tools for quality risk management and pharmaceutical development relevant to quality by design (QbD).

QbD principles are based on the idea that quality should be built into each process step rather than relying on testing at the end of the process (i.e., quality-by-testing). The product's critical quality attributes (CQAs) are defined at each stage, derived from the quality target product profile (QTPP) (of the therapeutic product), and the manufacturing process is developed with suitable critical process parameters (CPPs) to achieve those attributes (Fig. 5.6). Appropriate monitoring and control strategies are also defined to ensure quality criteria are achieved throughout the process and how CQAs are impacted at each step. Using this information to implement QbD principles helps control potential sources of variability and minimizes the risk of deviation (Tanaka et al. 2020; Tustian and Bak 2021).

QbD requires a detailed understanding of the product and process. However, since viral vector-based gene therapy is part of a relatively new field, manufacturers do not have the same information at their disposal as they might with more established protein-based modalities. Additionally, a platform process does not yet exist for viral vector production, and standard methods are not in abundance.

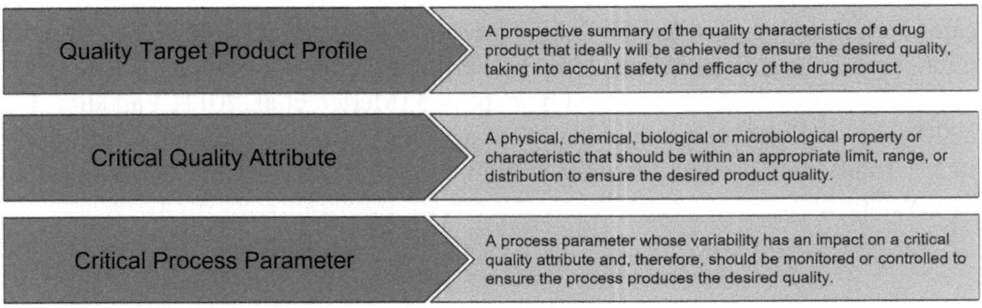

Fig. 5.6 Definitions of QTPP, CQA, and CPP, as described in ICH Q8

Variations in upstream process steps (different serotypes and production protocols) can strongly impact the profile of impurities entering the downstream process.

5.9.1 The Quality Target Product Profile

The first step in implementing QbD is to define the QTPP of the therapeutic product, which should be done during the early stages of development (Fig. 5.7). The QTPP is a summary of the attributes of a drug that should be achieved to ensure the desired quality. QTPPs can include but are not limited to dosage strength, sterility, purity, and stability (Yu et al. 2014).

Defining a robust QTPP is considered essential for a successful, safe, and efficacious drug product (Zurdo et al. 2015). The QTPP ultimately determines what CQAs within the process are relevant (and need to be controlled), and therefore, any missing information at this point can have severe knock-on effects during QbD implementation. When defining the QTPP, it is important to involve as many technical experts as possible (e.g., regulatory, development, and supply chain) from different areas of the product development life cycle so that no key information is missed. Generally, the QTPP should be revisited during the product's life cycle and updated if necessary (ICH: International Conference of Harmonization 2009).

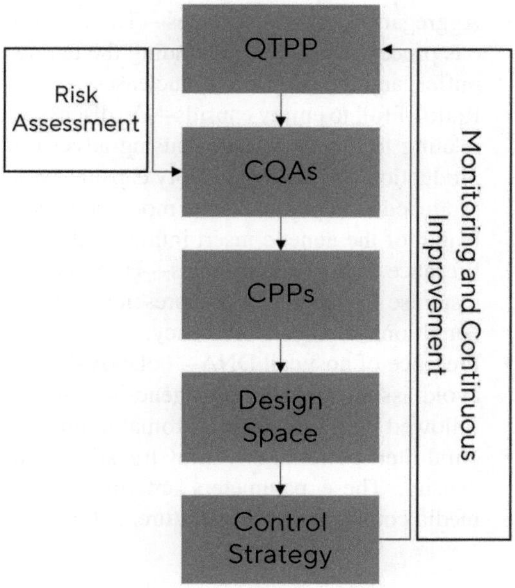

Fig. 5.7 Stages of QbD implementation. Critical quality attributes needed to meet the QTPP are determined following a risk assessment. As part of the control strategy, information is fed back to ensure the QTPP is correctly defined as process knowledge improves

5.9.2 CQAs for Viral Vector Processes

Many CQAs are product and process dependent and will differ from product to product. However, there are some standard CQAs that must be taken into account for biopharmaceutical products, which are defined in ICH Q6b:

- Identity—Should be highly specific for the drug substance and based on unique aspects of its molecular structure and/or other specific properties.
- Potency—Quantitative measure of biological activity based on the product's attributes linked to the relevant biological properties. A relevant, validated potency assay should be part of the specifications for a biotechnological or biological drug substance and/or drug product.
- Quantity—Usually measured by mass. Critical for biological products and determined using an appropriate (usually physiochemical) assay.
- Purity—Absolute purity can be difficult to determine and is usually estimated by a combination of assays, with a focus on separating the target product from product-related substances and impurities (see Fig. 5.7 for examples of process- and product-related impurities).

There are also some attributes important to many viral processes (Cashen and Manser 2021):

- Presence of non-infectious particles—Some produced viral particles will be defective and unable to perform as gene therapy vehicles. Their formation is primarily determined by pH and temperature of the cell culture, meaning defining the correct process conditions is essential. Therefore, the limited formation of non-infectious particles is a valuable mark of quality.
- Aggregation of viral particles—The extent of aggregation is influenced by various process conditions, including the temperature, pH, ionic strength of the lysis buffer, and the addition of nucleases.
- Ratio of full to empty capsids—As discussed previously, there are potential issues relating to empty capsids causing adverse immune responses in patients. AAV production can generate empty capsids, which often represent the majority of the produced viral particles. The mode of transfection, cell culture conditions, and the length of the genetic insert influence the viral packaging efficiency.
- Presence of host cell proteins—HCPs can be immunogenic and elicit an adverse response in patients. Their presence is strongly impacted by cell viability, lysis conditions, and lysis efficiency.
- Presence of host cell DNA—hcDNA should be minimized in the final product to avoid issues such as oncogenesis. Nuclease activity at the point of harvest, followed by purification (chromatography and TFF), affects this attribute.
- Viral titer—Titer is driven by transfection or infection efficiency and viable cell density. These parameters are impacted by cultivation conditions, including media composition, temperature, pH, and shear (Fig. 5.8).

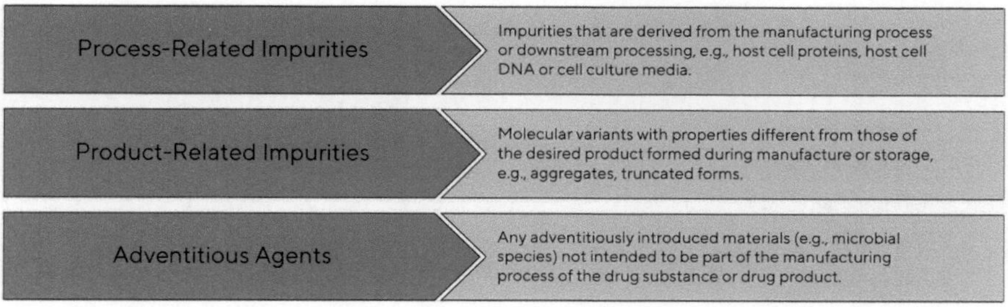

Fig. 5.8 Product-related impurities, process-related impurities, and adventitious agents, as described in ICH 6b

In addition, common adventitious agents, such as endotoxin, bioburden, mycoplasma or spiroplasma, and adventitious viruses, should be considered as CQAs for any process. Lastly, typical process-related impurities, such as detergents, leachables, and nuclease (where appropriate), should be considered. Figure 5.8 provides definitions on product-related impurities, process-related impurities and adventitious agents, as defined by ICH 6b guidelines.

Once the CQAs that ensure the safety and efficacy of the gene therapy product have been identified, manufacturers must build an understanding of how the conditions and materials at each process step influence CQAs. Building a detailed framework for each unit operation is essential to promote quality across each step and minimize the risk of process failure and product loss.

5.9.3 CQAs and CPPs across Downstream Process Steps

As discussed, an important feature of a QbD framework is understanding the relationship between process and product. Therefore, each unit operation must be considered in isolation to develop a robust production process. Every identified CQA must be addressed somewhere within the process. This involves reducing the CQA to an acceptable level or eliminating it entirely. Many CQAs will be addressed in multiple purification steps due to how they are removed or controlled, for example, by charge or size. In general, regulating specific CQAs across multiple steps ensures greater control of these attributes and helps mitigate the effect of process variance (Fig. 5.9).

CPPs are often specific to the process, but many parameters can overlap during a purification process, such as time and pressure limits. The CPPs are usually designed based on specific quality attributes of the product. For example, LVs have a short half-life and low resistance to shear, so step duration and processing pressure will always be considered CPPs across all filtration steps. In general, process duration and pressures will have more of an effect on LVs across the downstream process compared to AAVs due to the difference in their characteristics (Fig. 5.10).

	Process Steps					
Critical Quality Attributes	Clarification Filtration	Tangential Flow Filtration	Affinity Chromatography	Anion-Exchange Chromatography	Virus Filtration	Sterile Filtration
Functional viral titer	√	√	√	√	√	√
Noninfectious virus				√		
Aggregated virus	√	√	√	√	√	√
Empty capsid				√		
Residual host cell DNA	√	√	√	√		
Residual host cell protein	√	√	√	√		
Endotoxin	√				√	
Bioburden	√				√	√
Adventitious agents					√	
Adventitious viruses			√	√	√	

Fig. 5.9 Overview of CQAs per process step for LV and AAV. Note: Affinity chromatography and virus filtration are not considered part of the LV assessment

	Process Steps					
Critical Process Parameters	Clarification Filtration	Tangential Flow Filtration	Affinity Chromatography	Anion-Exchange Chromatography	Virus Filtration	Sterile Filtration
Pressure	√				√	√
Flux	√				√	√
Throughput					√	√
Duration	√[1]	√[1]		√[1]	√	√
Flow rate			√	√		
Load density			√	√		
Buffer pH		√	√	√		
Buffer conductivity		√	√	√		
Elution volume				√		
Volumetric concentration factor		√				
Transmembrane pressure		√				
Permeate flux		√				
Diafiltration volume		√				

Fig. 5.10 Overview of CPPs per process step for LV and AAV. Note: Affinity chromatography and virus filtration are not considered part of the LV assessment. [1]Duration is considered a critical process parameter for lentivirus only

5.9.3.1 Clarification

Pressure and flux are considered CPPs for the clarification step, as their control is important for a consistent product filtrate. Over-pressure or processing at a high flux may create issues such as filter fouling or complete blockage, allowing more impurities than usual to enter the feedstream.

While not a CPP itself, the choice of filter media is a crucial consideration when designing the clarification filtration step. Positively charged filters will ensure a degree of endotoxin removal and help increase hcDNA and HCP removal. However, as discussed, impurity removal must be balanced against vector recovery.

5.9.3.2 Affinity Chromatography

Affinity chromatography is important for removing HCPs, hcDNA, and protein impurities during AAV purification while also demonstrating good viral clearance capabilities. CPPs for a chromatography process include flow rate and loading capacity. Further, critical material attributes include the pH and conductivity of the processing buffers. Finding the optimal range for these factors strongly influences the binding specificity and impurity or viral clearance.

5.9.3.3 Ion-Exchange Chromatography

An ion-exchange chromatography step is used for LV capture and AAV polishing. IEX further reduces the amount of process and product-related impurities and can separate empty and full AAV capsids. Most CQAs are influenced by the conditions selected in this step, including pH, conductivity, and load density. As the design space is so broad, polishing chromatography represents a crucial point at which the target's CQAs and the influence of the CPPs should be fully understood. As LVs do not have an initial affinity step, greater emphasis is placed on the control of CQAs in this step compared to AAVs.

5.9.3.4 Other Chromatography Chemistries

In specific circumstances, other chromatography chemistries may be used, such as multi-modal or mixed-mode chromatography (MMC) or hydrophobic interaction chromatography (HIC). As these techniques deviate from the standard processes, their inclusion is usually to deal with a specific process challenge or issue. As such, CQAs and CPPs related to these process steps should be evaluated on a case-by-case basis.

5.9.3.5 Ultrafiltration and Diafiltration (UFDF)

As UFDF can be implemented in multiple areas of the purification process, feedstream variance from process to process is likely to be significant. In general, the volumetric concentration factor of the product is considered vital in preventing aggregate formation, while the choice of molecular weight cutoff of the cassette or hollow fiber will play a significant role in determining what level of impurity clearance can be achieved. Any impurities larger than the cutoff of the cassette or fiber will be retained and concentrated alongside the vector product. Nuclease may be added during TFF to help increase DNA removal, but the removal of the nuclease should also be performed (through diafiltration) within the process step. Buffer conditions, such as pH and conductivity, can be considered CPPs if they affect the aggregation or stabilization of the vector product.

5.9.3.6 Viral Filtration and Sterile Filtration

Virus filtration using a 50 nm virus filter can only be considered for AAV production. Virus filters with lower retention ratings cannot be considered for either viral process. Virus filtration is primarily used to remove adventitious agents and, while not the primary aim of the step, may also help remove aggregated viral particles. Where viral filtration of the product is not possible, viral filtration of cell culture media could be performed.

Bioburden elimination should be considered the key CQA of the sterile filtration step. As with virus filtration, sterile filtration can also help remove aggregated viral particles. However, the presence of aggregation at these late stages can lead to significant product loss and may indicate a problem further upstream, such as over-concentration of the vector.

In both filtration steps, processing pressure, flux, filtration time, and filter throughput are all CPPs that must be considered. In some cases, a final sterile filtration step is not possible due to substantial product loss. In these cases, fully closed systems are required for the downstream operations, which will strongly impact material and system selection.

5.9.4 Monitoring, Control, and Testing

The effectiveness of QbD relies on a robust monitoring and control strategy supplemented with quality control testing. Analytical tools and software continue to be developed to meet the speed and sensitivity requirements of modern bioprocessing. Additionally, orthogonal methods should be applied in parallel to meet the requirements of regulatory agencies (Food and Drug Administration 2015). Closed processing should be used as much as possible to ensure an effective bioburden control strategy and mitigate viral contamination.

A robust control strategy will include process monitoring to ensure conditions do not deviate from established CPPs, in-process testing to ensure CQAs are within an acceptable range, and lot-release testing to test the product at the end of the production process (Kelner and Bhalgat 2006; Raj and Vishal Gupta 2016; Wasalathanthri et al. 2020).

For all viral vectors, each design space is different and often not entirely understood. Each product and process must be considered on an individual basis. As our collective understanding develops and analytical tools improve, manufacturers will have an enhanced understanding of their process, allowing them to fully apply QbD principles and take full advantage of their benefits.

5.10 Current Challenges and Hurdles

In an ideal world, technologies for the downstream processing of viral vectors would reach the level of optimization achieved in monoclonal antibodies. This includes more platform-like solutions (i.e., affinity-based purification protocols for lentiviral vectors), improved overall efficiency through tailored processing tools, and limited variability between processes. The sources of process variance include differences in transfection, endonuclease, and lysis efficiency. Steps to limit these inconsistencies, as well as improvements in standardization and overall productivity, will promote increased robustness in viral vector purification processes.

Moving away from adherent to suspension cell culture brings its own set of problems, such as higher cell densities, viral titers, and impurities, all of which require improved downstream solutions. As upstream processing continues to push the boundaries of the viral product yields that can be achieved, the onus is on downstream technology to improve alongside it. Otherwise, downstream process steps could become a bottleneck impairing future progress. Product yields are increasing in LV and AAV downstream processes, but there is a desire to accomplish improved recoveries, from the typical 20–60% to something more along the lines of monoclonal antibody processes, upward of 90%.

In some instances, the technology does not yet exist or has not been sufficiently developed for large-scale applications. Fibro technology from Cytiva has demonstrated promising preliminary results (Ruscic et al. 2019; Adamson-Small et al. 2020) and alleviates many downstream processing bottlenecks at the capture step when using conventional packed-bed chromatography, such as low flow rates and diffusion limitations. This technology is yet to be implemented at commercial scale; however, supporting data and the range of devices in development (from 400 μL to 2.4 L) suggest that its implementation could be imminent. The CIMmultus™ PrimaS monolith from Sartorius has been evaluated as an alternative to empty/full capsid separation of AAV particles (Rieser et al. 2021). While monoliths display similar issues with scalability as membranes do (essentially, having a fixed volume per capsule, which does not give as much flexibility in scale-up as resins), the use of monoliths for viral vector purification seems to be increasing as their capabilities become more documented (Lock et al. 2012; Wang et al. 2019; Joshi et al. 2021).

Nereus LentiHERO by Astrea Bioseparations is a spin column device designed for the purification of LVs, achieving significant increases in vector yield vs. traditional purification techniques without compromising infectivity or viral structure (Scanlon et al. 2022). Currently, the product is only available in a spin column format, significantly reducing its ability to be used in manufacturing processes. However, work is being done on devices that can be implemented in large-scale processes.

Isolere Bio's patented IsoTag technology (Luginbuhl and Chilkoti 2019) also shows potential as an alternative to chromatography-based affinity (Isolere Bio 2022). Combining affinity and size-exclusion technology into a single step, IsoTag may help de-bottleneck the early part of downstream processing for viral vectors and many other advanced therapies. Despite being in the early stages of development, IsoTag has already been picked up by leaders in the industry (OxfordBiomedica 2021) seeking to evaluate the technology for their own platforms.

Looking at the current state of downstream LV and AAV processing, affinity technology for LVs and more scalable solutions for empty/full capsid separation for AAVs remain at the top of the agenda. Meaningful improvements across these two areas would empower manufacturers with tools to overcome the unique challenges both viral vectors face in their large-scale production.

Like upstream and downstream process development, viral vector analytics are continually improving. Driven by the increase in the number of approved viral vector therapies, there is a greater regulatory emphasis and expectation for more sensitive, specific, accurate, faster, and more robust assays. Current analytical assays, while somewhat suitable, tend to be slow, low throughput, and highly variable. Many standard assays used for the analysis of viral vector products require long preparative times, making them unsuitable for in-process assays due to their long turnaround times. With the understanding that current viral vector processes have a high degree of variability from batch to batch, the need for process analytical technology (PAT) and assays that can be implemented within the PAT framework will be a key driver of innovation in the development of novel tools.

As QbD begins to find its way into the development of new gene therapies, and as its practice is now more widespread across many different drug development pipelines, it is of the utmost importance that equal attention is given to how PAT can assist in the development of new technologies by supporting the QbD framework.

5.11 Conclusion

The use of viral vectors in gene therapy relies on their efficient production and purification. The processes currently in place are adequate but have not accomplished the efficiency levels achieved for monoclonal antibody processes. There remains significant potential for process optimization and intensification in downstream production to increase capacity and reduce production costs.

Despite relatively suboptimal processes, there remains a significant need for QbD to ensure more products make it to market, making viral vector-based therapies a more viable and accessible solution and increasing focus and development energy on next-generation techniques.

References

Adamson-Small L, Lundgren M, Guterstam P (2020) Scalable and efficient AAV production process with new fibro chromatography technology. Cell Gene Ther Insights 6(8):1249–1262. https://doi.org/10.18609/cgti.2020.138

Ajayi OO et al (2022) An updated analysis of viral clearance unit operations for biotechnology manufacturing. Curr Res Biotechnol 4:190–202. https://doi.org/10.1016/j.crbiot.2022.03.002

Ausubel LJ et al (2012) Production of CGMP-grade lentiviral vectors. Bioprocess Int 10(2):32–43. https://pubmed.ncbi.nlm.nih.gov/22707919/. Accessed 12 Oct 2022

Baghirzade R (2021) Adherent versus suspension based platforms: what is the near future of viral vector manufacturing? Cell Gene Ther Insights 7(11):1365–1371. https://doi.org/10.18609/cgti.2021.180

Buclez PO et al (2016) Rapid, scalable, and low-cost purification of recombinant adeno-associated virus produced by baculovirus expression vector system. Mol Ther Methods Clin Dev 3:16035. https://doi.org/10.1038/mtm.2016.35

Carmo M et al (2009) Stabilization of gammaretroviral and lentiviral vectors: from production to gene transfer. J Gene Med 11(8):670–678. https://doi.org/10.1002/jgm.1353

Cashen P, Manser B (2021) Quality by design (QbD) for adeno-associated virus (AAV). Bioporcess Online. https://www.bioprocessonline.com/doc/quality-by-design-qbd-for-adeno-associated-virus-aav-0001. Accessed 13 Oct 2022

Chan K et al (2021) Effect of particle size distribution on filtration performance of cell culture medium. Biotechnol Prog 37(3):e3130. https://doi.org/10.1002/btpr.3130

Cooper AR et al (2011) Highly efficient large-scale lentiviral vector concentration by tandem tangential flow filtration. J Virol Methods 177:1–9. https://doi.org/10.1016/j.jviromet.2011.06.019

Davidoff AM et al (2004) Purification of recombinant adeno-associated virus type 8 vectors by ion exchange chromatography generates clinical grade vector stock. J Virol Methods 121(2):209–215. https://doi.org/10.1016/j.jviromet.2004.07.001

De Rooij J et al (2019) Upstream and downstream solutions for AAV manufacturing. Cell Gene Ther Insights 5:1017–1029. www.insights.bio. Accessed 12 Oct 2022

Dinowitz M (2002) Viral safety and clearance evaluation--implications for process change and comparability. Dev Biol 109:121–126. https://pubmed.ncbi.nlm.nih.gov/12434921/. Accessed 1 Nov 2022

Do Minh A, Kamen AA (2021) Critical assessment of purification and analytical technologies for enveloped viral vector and vaccine processing and their current limitations in resolving co-expressed extracellular vesicles. Vaccines 9(8):823. https://doi.org/10.3390/vaccines9080823

Dobrowsky T et al (2021) AAV manufacturing for clinical use: insights on current challenges from the upstream process perspective. Curr Opin Biomed Eng 20:100353. https://doi.org/10.1016/j.cobme.2021.100353

Drago D et al (2021) Global regulatory progress in delivering on the promise of gene therapies for unmet medical needs. Mol Ther Methods Clin Dev 21:524–529. https://doi.org/10.1016/j.omtm.2021.04.001

Elizalde N, Ramírez JC (2021) Lentiviral vectors: key challenges and new developments. Cell Gene Ther Insights 7(6):667–677. https://doi.org/10.18609/cgti.2021.002

European Commission (2011) Note for guidance on minimising the risk of transmitting animal spongiform encephalopathy agents via human and veterinary medicinal products (EMA/410/01 rev.3). Official Journal of the European Union, C, pp 1–18. https://www.ema.europa.eu/en/minimising-risk-transmitting-animal-spongiform-encephalopathy-agents-human-veterinary-medicinal. Accessed 12 Oct 2022

Farley D, Mitrophanous K (2019) Viral vector production system. https://patents.google.com/patent/WO2019175600A1/en. Accessed 1 Nov 2022

Fiedler C et al (2017) Adeno-associated virus purification methods. https://patents.google.com/patent/US11000561B2/en. Accessed 1 Nov 2022

Food and Drug Administration (2015) Analytical procedures and methods validation for drugs and biologics, guidance for industry. https://www.fda.gov/regulatory-information/search-fda-guidance-documents/analytical-procedures-and-methods-validation-drugs-and-biologics. Accessed 14 Oct 2022

Green AP et al (2002) A new scalable method for the purification of recombinant adenovirus vectors. Hum Gene Ther 13(16):1921–1934. https://doi.org/10.1089/10430340260355338

Greg M et al (2021) Project A-Gene: a case study-based approach to integrating QbD principles in gene therapy CMC programs

Hebben M (2018) Downstream bioprocessing of AAV vectors: industrial challenges & regulatory requirements. Cell Gene Ther Insights 4(2):131–146. https://doi.org/10.18609/cgti.2018.016

Hejmowski AL et al (2022) Novel anion exchange membrane chromatography method for the separation of empty and full adeno-associated virus. Biotechnol J 17(2):e2100219. https://doi.org/10.1002/biot.202100219

Hermens WTJMC, Smith J (2018) Removal of contaminating viruses from AAV preparations. Patent EP2990477A1. European Patent Office. https://patents.google.com/patent/EP2990477A1/en. Accessed 12 Oct 2022

Hollis RP et al (2009) 712. Simplified large-scale production of lentiviral vectors. Mol Ther 17: S272. https://doi.org/10.1016/s1525-0016(16)39070-0

Hunter AK et al (2022) Identification of compendial nonionic detergents for the replacement of Triton X-100 in bioprocessing. Biotechnol Prog 38(2):e3235. https://doi.org/10.1002/btpr.3235

ICH: International Conference of Harmonization (2009) ICH Q8 (R2) Pharmaceutical development | European Medicines Agency. https://www.ema.europa.eu/en/ich-q8-r2-pharmaceutical-development. Accessed 14 Oct 2022

Isolere Bio (2022) Increased productivity using IsoTag AAV at various harvest scales. https://isolere.squarespace.com/articles/increased-productivity-of-aav9-affinity-capture-using-isotag-aav-at-various-harvest-scales. Accessed 14 Oct 2022

James KT et al (2016) Novel high-throughput approach for purification of infectious virions. Sci Rep 6(1):1–11. https://doi.org/10.1038/srep36826

Joshi PRH et al (2021) Development of a scalable and robust AEX method for enriched rAAV preparations in genome-containing VCs of serotypes 5, 6, 8, and 9. Mol Ther Methods Clin Dev 21:341–356. https://doi.org/10.1016/j.omtm.2021.03.016

Kaludov N, Handelman B, Chiorini JA (2002) Scalable purification of adeno-associated virus type 2, 4, or 5 using ion-exchange chromatography. Hum Gene Ther 13(10):1235–1243. https://doi.org/10.1089/104303402320139014

Kelner DN, Bhalgat MK (2006) Analytical strategy for biopharmaceutical development. In: Shukla AA, Etzel MR, Gadam S (eds) Process scale bioseparations for the biopharmaceutical industry. CRC Press, pp 395–418. https://doi.org/10.1201/9781420016024.ch13

Khanal O et al (2018) Contributions of depth filter components to protein adsorption in bioprocessing. Biotechnol Bioeng 115(8):1938–1948. https://doi.org/10.1002/bit.26707

Labisch JJ et al (2021) A new simplified clarification approach for lentiviral vectors using diatomaceous earth improves throughput and safe handling. J Biotechnol 326:11–20. https://doi.org/10.1016/j.jbiotec.2020.12.004

Lennaertz A et al (2013) Viral vector production in the integrity® iCELLis® single-use fixed-bed bioreactor, from bench-scale to industrial scale. BMC Proc 7(S6):P59. https://doi.org/10.1186/1753-6561-7-s6-p59

Lock M, Alvira MR, Wilson JM (2012) Analysis of particle content of recombinant adeno-associated virus serotype 8 vectors by ion-exchange chromatography. Hum Gene Ther Methods 23(1):56–64. https://doi.org/10.1089/hgtb.2011.217

Logan AC et al (2004) Factors influencing the titer and infectivity of lentiviral vectors. Hum Gene Ther 15(10):976–988. https://doi.org/10.1089/hum.2004.15.976

Long Y et al (2020) Optimization validation and implementation of retroviral retentive filters for production of adeno-associated virus (AAV). https://www.biopharma-reporter.com/Library/Optimization-Validation-and-Implementation-of-Retroviral-Retentive-Filters-for-Production-of-Adeno-Associated-Virus-AAV. Accessed 13 Oct 2022

Luginbuhl KM, Chilkoti A (2019) Genetically encoded polypeptide for affinity capture and purification of biologics. Canada

Ma J et al (2020) The effect of residual Triton X-100 on structural stability and infection activity of adenovirus particles. Mol Ther Methods Clin Dev 19:35–46. https://doi.org/10.1016/j.omtm.2020.08.013

Madsen E et al (2022) Single pass tangential flow filtration: critical operational variables, fouling, and main current applications. Sep Purif Technol 291:120949. https://doi.org/10.1016/j.seppur.2022.120949

McAlister M, Lambropoulos A (2022) Consideration of quality by design principles in the use of sterilizing grade filters for clarification and final sterile filtration in adeno-associated virus applications. Cell Gene Ther Insights 8(6):895–915. https://doi.org/10.18609/CGTI.2022.134

McCarron A et al (2016) Challenges of up-scaling lentivirus production and processing. J Biotechnol 240:23–30. https://doi.org/10.1016/j.jbiotec.2016.10.016

McNally DJ et al (2020) Development and optimization of a hydrophobic interaction chromatography-based method of AAV harvest, capture, and recovery. Mol Ther Methods Clin Dev 19:275–284. https://doi.org/10.1016/j.omtm.2020.09.015

Merten OW et al (2011) Large-scale manufacture and characterization of a lentiviral vector produced for clinical ex vivo gene therapy application. Hum Gene Ther 22(3):343–356. https://doi.org/10.1089/hum.2010.060

Merten O-W et al (2014) Manufacturing of viral vectors: part II. Downstream processing and safety aspects. Pharm Bioprocess 2(3):237–251. https://doi.org/10.4155/pbp.14.15

Merten OW, Hebben M, Bovolenta C (2016) Production of lentiviral vectors. Mol Ther Methods Clin Dev 3:16017–16017. https://doi.org/10.1038/mtm.2016.17

Mietzsch M et al (2020) Characterization of AAV-specific affinity ligands: consequences for vector purification and development strategies. Mol Ther Methods Clin Dev 19:362–373. https://doi.org/10.1016/j.omtm.2020.10.001

Moleirinho MG et al (2020) Current challenges in biotherapeutic particles manufacturing. Expert Opin Biol Ther 20(5):451–465. https://doi.org/10.1080/14712598.2020.1693541

Moreira AS et al (2021) Advances in lentivirus purification. Biotechnol J 16(1):2000019. https://doi.org/10.1002/biot.202000019

Nass SA et al (2018) Universal method for the purification of recombinant AAV vectors of differing serotypes. Mol Ther Methods Clin Dev 9:33–46. https://doi.org/10.1016/j.omtm.2017.12.004

National Center for Biotechnology Information (2022) Ammonium sulfate | (NH4)2SO4 - PubChem. https://pubchem.ncbi.nlm.nih.gov/compound/Ammonium-Sulfate. Accessed 12 Oct 2022

Negrete A, Pai A, Shiloach J (2014) Use of hollow fiber tangential flow filtration for the recovery and concentration of HIV virus-like particles produced in insect cells. J Virol Methods 195:240–246. https://doi.org/10.1016/j.jviromet.2013.10.017

Okada T et al (2009) Scalable purification of adeno-associated virus serotype 1 (AAV1) and AAV8 vectors, using dual ion-exchange adsorptive membranes. Hum Gene Ther 20(9):1013–1021. https://doi.org/10.1089/hum.2009.006

Olgun HB et al (2019) High-grade purification of third-generation HIV-based lentiviral vectors by anion exchange chromatography for experimental gene and stem cell therapy applications. Methods Mol Biol 1879:347–365. https://doi.org/10.1007/7651_2018_154

OxfordBiomedica (2021) Annual report. https://www.oxb.com/financial-reports/2021. Accessed 14 Oct 2022

Pei X et al (2018) Efficient capsid antigen presentation from adeno-associated virus empty virions in vivo. Front Immunol 9:844. https://doi.org/10.3389/fimmu.2018.00844

Perry C, Rayat ACME (2021) Lentiviral vector bioprocessing. Viruses 13(2):268. https://doi.org/10.3390/v13020268
Qu W et al (2015) Scalable downstream strategies for purification of recombinant adeno-associated virus vectors in light of the properties. Curr Pharm Biotechnol 16(8):684–695. https://doi.org/10.2174/1389201016666150505122228
Raghavan B et al (2019) Optimizing the clarification of industrial scale viral vector culture for gene therapy. Cell Gene Ther Insights 5(9):1311–1322. https://doi.org/10.18609/cgti.2019.137
Raj P, Vishal Gupta N (2016) Process analytical technology (PAT): a real time quality assurance. Int J Pharm Sci Rev Res 37(2):67–72
Rieser R et al (2021) Comparison of different liquid chromatography-based purification strategies for adeno-associated virus vectors. Pharmaceutics 13(5):748. https://doi.org/10.3390/pharmaceutics13050748
Ronzitti G, Gross DA, Mingozzi F (2020) Human immune responses to adeno-associated virus (AAV) vectors. Front Immunol 11:670. https://doi.org/10.3389/fimmu.2020.00670
Ruscic J et al (2019) Lentiviral vector purification using nanofiber ion-exchange chromatography. Mol Ther Methods Clin Dev 15:52–62. https://doi.org/10.1016/j.omtm.2019.08.007
Scanlon I et al (2022) A novel purification technology to increase processing efficiency purity and recovery of lentiviral particles for viral vector development [poster]. https://www.bioprocessonline.com/doc/a-novel-purification-technology-to-increase-processing-efficiency-purity-and-recovery-of-lentiviral-particles-for-viral-vector-development-0001. Accessed 14 Oct 2022
Schweizer M, Merten O-W (2010) Large-scale production means for the manufacturing of lentiviral vectors. Curr Gene Ther 10(6):474–486. https://doi.org/10.2174/156652310793797748
Segura MDLM et al (2005) A novel purification strategy for retrovirus gene therapy vectors using heparin affinity chromatography. Biotechnol Bioeng 90(4):391–404. https://doi.org/10.1002/bit.20301
Segura MM et al (2007) Production of lentiviral vectors by large-scale transient transfection of suspension cultures and affinity chromatography purification. Biotechnol Bioeng 98(4):789–799. https://doi.org/10.1002/bit.21467
Segura MM et al (2010) New protocol for lentiviral vector mass production. Methods Mol Biol 614:39–52. https://doi.org/10.1007/978-1-60761-533-0_2
Segura MM, Kamen AA, Garnier A (2011) Overview of current scalable methods for purification of viral vectors. Methods Mol Biol 737:89–116. https://doi.org/10.1007/978-1-61779-095-9_4
Soldi M et al (2020) Laboratory-scale lentiviral vector production and purification for enhanced ex vivo and in vivo genetic engineering. Mol Ther Methods Clin Dev 19:411–425. https://doi.org/10.1016/j.omtm.2020.10.009
Tanaka T, Hanaoka H, Sakurai S (2020) Optimization of the quality by design approach for gene therapy products: a case study for adeno-associated viral vectors. Eur J Pharm Biopharm 155:88–102. https://doi.org/10.1016/j.ejpb.2020.08.002
The European Commission (2017) Commission regulation (EU) 2017/1000 of 13 June 2017. Official Journal of the European Union, L 150(June), pp 14–18. http://eur-lex.europa.eu/legal-content/EN/TXT/PDF/?uri=CELEX:32017R1000&from=EN
ThermoFisher Scientific (2021) Viral safety in AAV production: how affinity chromatography effectively contributes. https://assets.thermofisher.com/TFS-Assets/BPD/Reference-Materials/viral-safety-in-aav-production-affinity-article.pdf
Toueille M et al (2018) Development of purification steps for several AAV serotypes using POROS™ CaptureSelect™ AAVX affinity chromatography. Cell Gene Ther Insights 4(7):637–645. https://doi.org/10.18609/cgti.2018.061
Tustian AD, Bak H (2021) Assessment of quality attributes for adeno-associated viral vectors. Biotechnol Bioeng 118(11):4186–4203. https://doi.org/10.1002/bit.27905
U.S. Food and Drug Administration (FDA) (2020) Chemistry, manufacturing, and control (CMC) information for human gene therapy investigational new drug applications (INDs) guidance for industry. https://www.fda.gov/regulatory-information/search-fda-guidance-documents/

chemistry-manufacturing-and-control-cmc-information-human-gene-therapy-investigational-new-drug. Accessed 12 Oct 2022

Urabe M et al (2006) Removal of empty capsids from type 1 adeno-associated virus vector stocks by anion-exchange chromatography potentiates transgene expression. Mol Ther 13(4):823–828. https://doi.org/10.1016/j.ymthe.2005.11.024

Valkama AJ et al (2020) Development of large-scale downstream processing for lentiviral vectors. Mol Ther Methods Clin Dev 17:717–730. https://doi.org/10.1016/j.omtm.2020.03.025

van der Loo JCM, Wright JF (2016) Progress and challenges in viral vector manufacturing. Hum Mol Genet 25(R1):R42–R52. https://doi.org/10.1093/hmg/ddv451

van Reis R, Zydney A (2007) Bioprocess membrane technology. J Membr Sci 297(1–2):16–50. https://doi.org/10.1016/j.memsci.2007.02.045

Venkatakrishnan B et al (2013) Structure and dynamics of adeno-associated virus serotype 1 VP1-unique N-terminal domain and its role in capsid trafficking. J Virol 87(9):4974–4984. https://doi.org/10.1128/jvi.02524-12

Venkiteshwaran A et al (2015) Mechanistic evaluation of virus clearance by depth filtration. Biotechnol Prog 31(2):431–437. https://doi.org/10.1002/btpr.2061

Viswanathan S, Puich M (2021) CMC obstacles in cell and gene therapy: four solutions to solve six challenges. Cell Gene Ther Insights 7(5):471–488. https://insights.bio/cell-and-gene-therapy-insights/journal/article/2049/CMC-obstacles-in-cell-and-gene-therapy-four-solutions-to-solve-six-challenges. Accessed 13 Oct 2022

Wang M (2011) Zeta plus™ VR filters for viral reduction. Bioprocess Int, p 62. https://www.bioprocessonline.com/doc/zeta-plus-vr-series-filters-for-viral-0003. Accessed 12 Oct 2022

Wang C et al (2019) Developing an anion exchange chromatography assay for determining empty and full capsid contents in AAV6.2. Mol Ther Methods Clin Dev 15:257–263. https://doi.org/10.1016/j.omtm.2019.09.006

Wasalathanthri DP et al (2020) Technology outlook for real-time quality attribute and process parameter monitoring in biopharmaceutical development—a review. Biotechnol Bioeng 117(10):3182–3198. https://doi.org/10.1002/bit.27461

Williams T et al (2020) Lentiviral vector manufacturing process enhancement utilizing TFDF™ technology. Cell Gene Ther Insights 6(3):455–467. https://doi.org/10.18609/cgti.2020.053

Winkler M et al (2021) Viral clearance in a downstream AAV process: case study using a model virus panel and a noninfectious surrogate. Bioprocess Int. https://www.biaseparations.com/en/library/publications/1112/viral-clearance-in-a-downstream-aav-process-case-study-using-a-model-virus-panel-and-a-noninfectious-surrogate. Accessed 1 July 2022

Wright JF (2014) Product-related impurities in clinical-grade recombinant AAV vectors: characterization and risk assessment. Biomedicines 2(1):80–97. https://doi.org/10.3390/biomedicines2010080

Yang S et al (2012) A simple and effective method to generate lentiviral vectors for ex vivo gene delivery to mature human peripheral blood lymphocytes. Hum Gene Ther Methods 23(2):73–83. https://doi.org/10.1089/hgtb.2011.199

Ye G-j et al (2014) Herpes simplex virus clearance during purification of a recombinant adeno-associated virus serotype 1 vector. Hum Gene Ther Clin Dev 25(4):212–217. https://doi.org/10.1089/humc.2014.060

Yu LX et al (2014) Understanding pharmaceutical quality by design. AAPS J 16(4):771–783. https://doi.org/10.1208/s12248-014-9598-3

Zurdo J et al (2015) Early implementation of QbD in biopharmaceutical development: a practical example. BioMed Res Int 2015:605427. https://doi.org/10.1155/2015/605427

Chapter 6
Filtration Principles and Techniques for Bioprocessing of Viral Vector-Based Therapeutics

Laurens Vergauwen, Claire Scanlan, Ratish Krishnan, Santosh Kumar Pandey, David Loong, and Amritanshu Bhushan

6.1 Introduction

The concept of gene therapy—altering or introducing genetic material to treat diseases—was first investigated more than 50 years ago. Clinical trials began in the early 2000s. Today, gene therapy products with enhanced safety profiles have been approved, and initial long-term efficacy data have been collected.

This success is driving aggressive development programs. There are over 1000 clinical trials currently active (Amarosi 2021). Because there is a pressing demand for a wide spectrum of gene therapy products in sufficient quantity, gene therapy has become one of the most heavily funded biotech sectors.

Viral vectors are commonly used to introduce genes for therapeutic applications. Although other viruses such as adenovirus and herpes simplex virus have also been used, the leading viral vectors for gene therapy are:

L. Vergauwen (✉) · S. K. Pandey
Merck KGaA, Darmstadt, Germany
e-mail: laurens.vergauwen@merckgroup.com

C. Scanlan · R. Krishnan
MilliporeSigma, Burlington, MA, USA

D. Loong
Merck Pte Ltd, Singapore, Singapore

A. Bhushan
Merck Life Science, Mumbai, India

S. Gautam et al. (eds.), *Bioprocess and Analytics Development for Virus-based Advanced Therapeutics and Medicinal Products (ATMPs)*,
https://doi.org/10.1007/978-3-031-28489-2_6

- *Adeno-associated viral vector (AAV)*: Commonly used for in vivo applications, for example gene-replacement therapy for spinal muscular atrophy.
- *Lentiviral vector (LVV)*: Commonly used for ex vivo applications, such as CAR-T therapy for immuno-oncology.

Gene therapy products must be manufactured following the Basics of Biosafety (BSL) Level 2, and in some regulatory jurisdictions, BSL 1.

At the time of publication, there were 11 approved drugs using these two viral vectors (FDA, 2023). This chapter focuses on these dominant vectors used for gene therapy.

6.1.1 Manufacturing Challenges for Midstream and Downstream Gene Therapy Production

6.1.1.1 Process Overview (Fig. 6.1)

There are four stages of gene therapy manufacturing:

- *Upstream production*: Viral vectors encoding the therapeutic gene of interest are produced in bioreactors using insect or mammalian cell culture. Transient transfection with plasmids and baculovirus infection are common techniques for vector production. Sometimes helper viruses are used to aid vector production, which must be subsequently removed.
- *Midstream processing*: Bioreactor material is harvested and typically concentrated. In the case of AAV, cells are typically lysed to increase vector recovery. Lysis releases large quantities of DNA, proteins, and cell debris, which complicates the process stream. Treatment with an endonuclease such as Benzonase® endonuclease (MilliporeSigma) simplifies processing and is followed by a clarification step. The characteristics of this harvest material have a tremendous impact on subsequent steps.
- *Downstream processing*: Viral vector harvest is purified using chromatography, ultrafiltration (UF), and diafiltration (DF) techniques. There may be additional bioburden removal steps or nanofiltration step to remove contaminants. Nanofiltration is applicable only for AAV process streams because LVV is too large to pass through the membrane pores.
- *Formulation*: Purified viral vector is formulated into the drug substance buffer using UF-DF techniques. Sterile filtration is a key quality assurance operation.

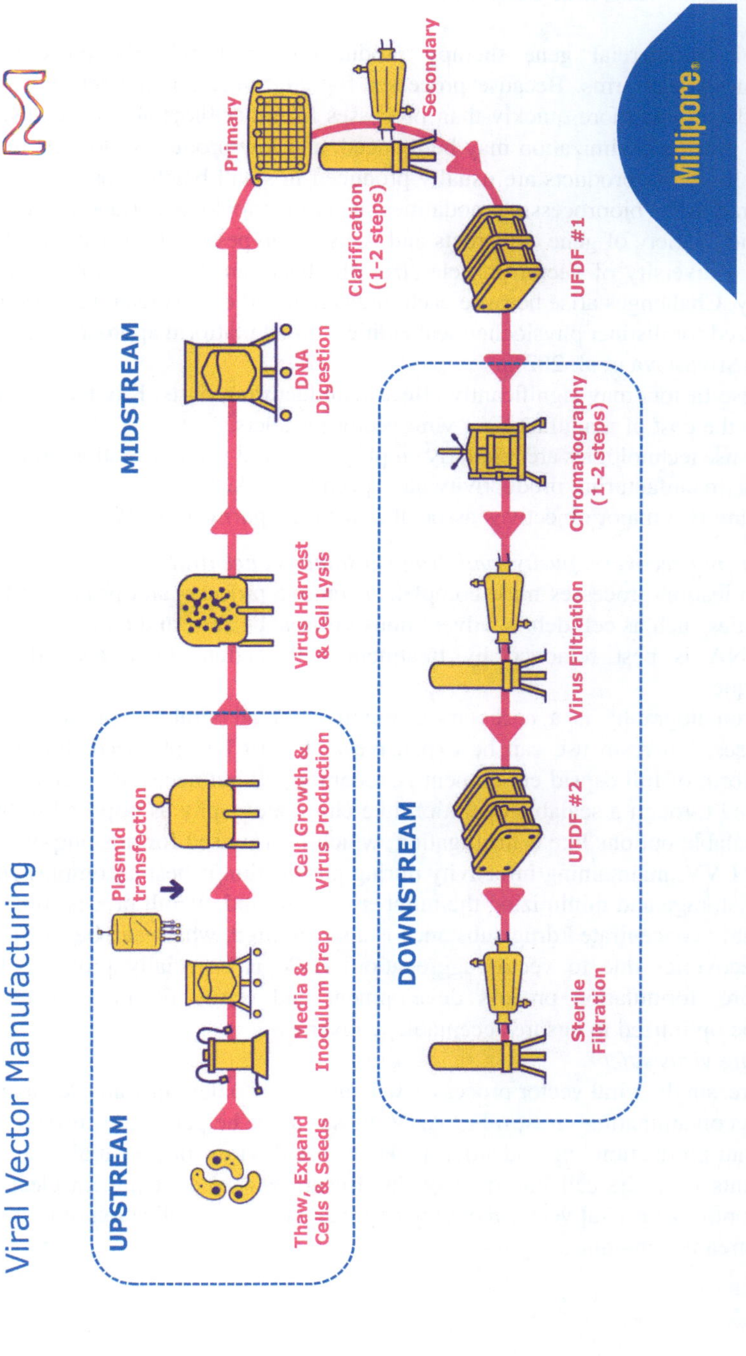

Fig. 6.1 Process overview detailing the typical process steps in gene therapy processes. Source: MilliporeSigma (2021) "Viral vector manufacturing process flow," MilliporeSigma Gene Therapy University Training

6.1.1.2 Strategies for Efficient and Cost-Effective Midstream and Downstream Gene Therapy Processing

Successful commercial gene therapy production programs rely on efficient manufacturing platforms. Because processes for viral vector manufacturing must often be developed more quickly than processes for other biopharmaceuticals, the degree of process optimization may be reduced, and low product yield may result. Additionally, these products are usually produced in small batch sizes when compared to traditional bioprocessing modalities such as monoclonal antibodies (mAbs).

The wide variety of gene constructs and AAV serotypes used for gene therapies results in a diversity of vector particle size, pI, shear sensitivity, and aggregation propensity. Challenges arise because each midstream and downstream process must be optimized for distinct physiochemical entities; broad platform approaches are not practical (Srivastava et al. 2021).

All these factors may significantly affect manufacturing costs. Key technologies can lower the cost of manufacturing viral vector products.

Single-use technologies are regularly employed in viral vector filtration processes to increase manufacturing productivity and speed to market.

There are two major objectives associated with the purification of viral vectors:

- *Maximizing recovery, purity, and drug substance concentration*

  Purification processes must completely remove process- and product-related impurities such as cell debris, adventitious viruses, buffers, and reagents. Residual DNA is best removed by treatment with endonuclease and filtration techniques.

  Chromatography is a necessary component of all viral vector purification processes, but resin use can be expensive and must be optimized. For AAV, some form of full capsid enrichment is becoming all but necessary. This is best achieved through a scalable operation like chromatography as opposed to other less-scalable options like centrifugation, which is also used for this purpose.

  For LVV, maintaining infectivity during purification is best accomplished by proper storage and minimizing the number of steps and overall process time.

  Highly concentrated drug substance is also required, which can be a cause of low recoveries due to vector aggregation. LVV is especially prone to this, therefore, formulation process development and sterile filtration operations must be optimized to ensure acceptable recoveries.

- *Assuring virus safety*

  Increasingly, viral vector processes will also have to demonstrate clearance of known contaminating viruses. For AAV, these can be helper viruses used during upstream production, e.g., adenovirus, HSV, baculovirus, or potential viral contaminants from the cell line or reagents. Ensuring viral contaminant clearance while optimizing viral vector recovery requires judicious application of mid- and downstream techniques.

6.2 Filtration Principles

6.2.1 Types of Filters Used in Gene Therapy Product Manufacture

Two different filtration methods are used to separate components in a feed stream during gene therapy production:

- Normal flow filtration (NFF).
- Tangential flow filtration (TFF) (also called cross-flow filtration).

6.2.1.1 Normal Flow Filtration

In normal flow filtration, or dead-end filtration, the feed stream moves perpendicular to the surface of the filter. Particulates and aggregates are retained in the pores or on the surface, forming a filter cake. NFF is frequently used for clarification, bioburden reduction, sterile filtration, and virus filtration. There are three types of NFF filters:

- *Depth filters*: Filter is a sponge-like material, some made of naturally derived materials (cellulose, diatomaceous earth, resin binder) and some of synthetic materials (polypropylene or glass fibers). Filters come in a variety of pore sizes and have increased solids retention capacity over membrane filters. In some depth filters, the media bears a charge for electrostatic retention. Diatomaceous earth provides improved feed capacity. Depth filters are generally used to process feed streams with a high particulate/solid load.
- *Surface filters*: Filter is usually a flat-sheet filter with permeable or semipermeable membrane. Filter relies on size exclusion for particle removal. Some are sterilizing grade (retention is correlated to bacterial retention). Non-sterilizing grade are used for bioburden reduction or prefiltration. This method is relatively low in cost; however, it is not suited for fine filtration due to a tendency to foul prematurely. This filter is generally used as a guard filter for the membrane filter. Also, this method does not offer full particle retention.
- *Membrane filters*: Filter is a flat or pleated sheet of inert polymer. Pores of a specific diameter in the filter determine the retention ratings. Membrane filters are generally used in sterile filtration applications.

6.2.1.2 Tangential Flow Filtration (TFF)

In tangential flow filtration, or cross-flow filtration, the feed stream flows parallel to the filter. One part goes through the membrane (permeate) and the rest (retentate) recirculates to the feed reservoir for further filtering. The cross-flow action minimizes particle buildup on the surface of the membrane.

TFF can be operated in batch, fed batch, and single-pass tangential flow filtration (SPTFF).

Two commonly used filtration devices are hollow-fiber units and cassettes (flat-sheet devices) composed of varying membranes such as mixed cellulose esters and polyethersulfone (PES).

These device types are available as multi-use and single-use.

TFF Devices
Two types of devices are primarily used in TFF applications: flat-sheet devices and hollow fiber devices.

- *Flat-sheet devices* are screened to aid in turbulent flow and comprise either regenerated cellulose or polyether sulfone membranes. Flat-sheet devices can be cleaned and reused, which improves process economics.

  A flat-sheet device TFF system cannot be considered "closed," and sterile filtration is necessary to ensure vector sterility. Single-use, pre-sterilized capsules that scale to the flat-sheet devices will soon be available. These will provide the advantage of flat-sheet devices and will be a comparable closed option for TFF.

  One benefit of a flat-sheet device is linear scalability. The feed channel length does not change across the multiple device sizes; instead, the membrane packets (consisting of the membrane, screen, and supports) increase without impacting the feed channel length.
- *Hollow fiber devices* are made up of fibers of multiple individual pore sizes that are cut to a specific length, sealed, and encapsulated in a module. The fluid flows on two sides, the lumen side, which is in the fibers, and the shell side, which is outside of the fibers.

  Unlike flat-sheet devices, the feed channel length does not scale linearly, so higher pump rates (and area) are needed for scale-up.

  On average, the permeate flux for hollow fibers is lower than that of flat-sheet devices, so more pump passes are usually required.

  Hollow fiber devices are considered closed systems, and they can be gamma sterilized. These systems are preferred for larger viruses that cannot be terminally sterilized.

6.2.1.3 Single-Use Vs. Multi-Use Systems

Single-use filter systems provide advantages that optimize facility efficiency and flexibility:

- Lower initial investment
- Decreased use of water
- Decreased risk of cross contamination
- Reduction in product turnover time (cleaning and cleaning validation are no longer necessary)
- Increased number of manufacturing batches possible

The use of multi-use systems requires qualification studies that document effective cleaning, regeneration, sanitizing, and consistency in performance for a given number of cycles.

6.3 Process Development

Small-scale optimization studies should be carried out on each filtration step prior to scale-up. The initial small-scale studies determine the appropriate filter and filter size for each process step based on hydraulic performance and product recovery. The small-scale sizing is then verified at the pilot scale (10–50 L bioreactor scale) before scaling up to full-scale manufacturing. The process development methods used for each of these steps are detailed in the following sections.

6.3.1 Clarification

Depth filtration is the most frequently employed technology for harvest and lysate clarification of both suspension and adherent systems. Depth filters are tested and sizing is provided using constant-flow Pmax™ and Tmax™ techniques. Small-scale filters, typically 20–23 cm^2, are tested at a constant flux (50–150 LMH). The differential pressure, filtrate turbidity, and filtrate volume are measured with respect to time in two- to three-minute intervals. The testing is stopped when the maximum operating pressure, Pmax™, is exceeded (typically 15–20 psi), or considerable turbidity breakthrough, Tmax™, is observed. A small-scale sterilizing-grade device is sometimes tested in-line with the depth filter to most accurately determine the point of turbidity breakthrough (the throughput at which the sterilizing-grade filter reaches Pmax™). The filtrates from these studies are assayed and tested to ensure adequate product recovery. The throughputs at which the filters reached capacity, either via pressure or turbidity breakthrough, are used to calculate approximate filter area for pilot- and large-scale operations. A safety factor of at least 1.5 is applied to account for process, feed, and filter variability.

6.3.2 Sterile and Virus Filtration

Sterilizing-grade and virus filters are often tested using constant pressure, Vmax™ techniques. A small-linearly scalable device, typically 3.5–17.7 cm^2, is used for this testing to provide filter area requirement with minimum available volume of representative feed. The pressure used for testing is the highest possible pressure drop (considered worst case) for that designated process step (typically 10–15 psi for most process intermediates). Sufficient volume of the test material is added to a pressure

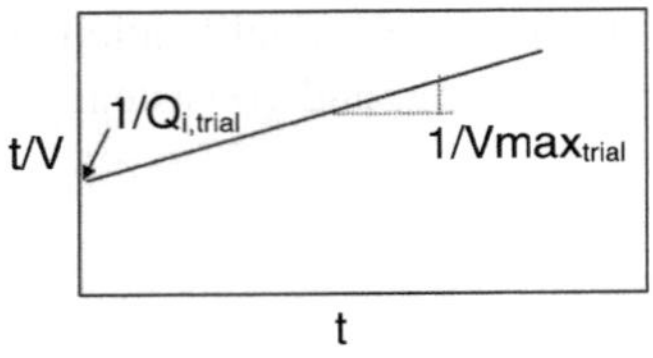

Fig. 6.2 Plotting time/volume vs. time for gradual pore plugging model Vmax™ sizing. Source: MilliporeSigma (2019) 'Vmax™, plotting time/volume vs. time', MilliporeSigma Normal Flow Filtration Program

vessel, and the vessel is pressurized to the desired test pressure using compressed air or nitrogen. The weight of the filtrate is measured and collected over time, and the testing is stopped when the flow stops (for highly plugging solutions) or at least 10–15% flux decay is achieved. By plotting time/volume vs. time (see Fig. 6.2), two essential parameters to determine the filter size can be identified: Vmax™ (L/m^2) and the initial flux (LMH). The inverse of the slope of this graph is the Vmax™ or the maximum volume that can be filtered by the test filter, and the initial flow rate is the inverse of the y-intercept.

Vmax™ is based on the gradual pore plugging model in which the equation, Amin = Vb/Vmax™ + Vb/(Qi × tb) where Vb refers to the batch volume (L), Qi is the initial flux (LMH), and tb is the process time (h). The Vmax™ model can be applied only if the correlation (r^2) in Fig. 6.2 is >0.99. If the correlation is <0.99, alternative fouling models (such as the combined model in which cake filtration is also occurring) should be used to accurately determine filter sizing (Giglia and Straeffer 2012).

In constant-flow Pmax™ and Tmax™ and constant pressure methodology of filter evalaution and sizing, a sufficient safety factor is applied to account for feed, process, and filter variability and scaling.

6.3.3 *Tangential Flow Filtration*

The molecular weight cutoff (MWCO) used in gene therapy applications is based on the vector size and the location of the step within the process (upstream, post-clarification, TFF1, or further downstream in the final formulation step, TFF2). For ultrafiltration and diafiltration, in which the product is concentrated on the upstream side of the membrane, the MWCO target is 1/3 to 1/5 the size of the product of interest. For AAV and lentivirus applications, 100 kDa and 300 kDa cutoffs are used for both steps, but in some AAV cases, a tighter MWCO is used for the final UF-DF step to ensure full product recovery. For process development, permeate control is recommended for MWCOs ≥100 kDa to prevent premature fouling of the membrane. There is not enough resistance from the membrane itself so permeate control enables a more efficient and stable process. Without permeate control, the permeate flux starts off high and decreases rapidly over time. Maintaining a constant permeate enables a more stable process and less polarization at the membrane surface (Valkama et al. 2020).

To determine the starting permeate flux, one needs to first find the "critical flux" i.e., the permeate flux at which the transmembrane pressure (TMP) is no longer stable and membrane fouling is starting to occur. This set point can be determined by setting the system in total recirculation mode and starting at a low permeate flow rate, measuring the TMP at the initial time and after ~5 min. If the TMP is stable, the permeate flow rate can be increased. This exercise is repeated until the TMP is no longer stable. The process will not be run at this point, but instead run at 20–50% of this critical flux. These TMP and permeate flow rate set points are used to initially concentrate the viral vector process feed to the concentration target. Samples should be analysed at various volumetric concentration factors (VCF) to ensure that there is no aggregation as the vector is concentrated (Raghunath et al. 2012).

The next critical process parameter is the target concentration to begin diafiltration. Often, this point is not at the target VCF, but instead is at some point in between. To minimize the amount of buffer and membrane area needed, the optimal diafiltration starting point is the point at the highest concentration. Permeate flux is determined by concentrating the vector to the target concentration and plotting the permeate flux times concentration (known as the diafiltration optimization parameter) versus product concentration. This will most likely yield an upside-down parabola. The maxima of this parabola is the optimal concentration to begin diafiltration. If this concentration is lower than the target concentration, the retentate is diluted (using process buffer) to the optimal concentration. Another critical flux experiment is performed to determine the starting permeate flux and the buffer exchange process is started (5–7 diavolumes is usually the target). The permeate pump may have to be adjusted to maintain a constant TMP (Raghunath et al. 2012).

Once the buffer exchange is complete, an additional critical flux study is done to determine the permeate pump set point for the final concentration step. The sizing for the ultrafiltration and diafiltration steps is determined based on the average flux and achieved trial loading (the scale-up loading should not exceed what was tested during the scale-down study) based on the target scale-up batch volume(s) and process time(s).

6.4 Clarification

6.4.1 Clarification Overview

The typical interfaces between the upstream and downstream processes are harvest and clarification. Clarification of viral vector feed stream harvest usually occurs early in the process to remove crude bioprocess material with a high load of cells and cell debris.

Several types of clarification methods are used either independently or in combination. The ideal clarification process results in high throughput for impurity removal as well as high product yield. Additionally, it is important that the choice of clarification method is consistent at different scales and easy to scale up.

Primary clarification is often carried out by centrifugation and/or filter-based technologies. Centrifugation as a primary clarification method prior to filtration (used as a secondary clarification method) is more economical for large-scale production (≥2000 L bioreactor scales); however, centrifuge scale-up and operation can be a hassle, and the capital expenditure to acquire a centrifuge is high. For bioreactor batch volumes ≤2000 L (which is where most viral vector harvest volumes lie), it is more economical to use disposable filter-based technologies. Some manufacturers also use tangential flow filtration (TFF) at low-shear conditions for clarification of the harvest, but implementation and production using TFF is more cumbersome for process development and for operators at the manufacturing scale. For this reason, single-use normal flow filtration is preferred, particularly for viral vector production.

Acoustic filters, which entrap suspended cells in soundwaves, can be also used. The cells can then be returned to the bioreactor during harvest. Disc stack continuous centrifugation is used for the processing of monoclonal antibodies and recombinant proteins but is not common in viral vector production because of low-volume processing (Perry and Rayat 2021).

The initial clarification step can be achieved using effective filtration options that are scalable and easy to implement.

Filter capacities depend on multiple process parameters such as cell culture, adherent or suspension cell culture, cell density at harvest, the degree of lysis, endonuclease use and activity, particle size distribution, and expression system. The typical viral vector lysate turbidity is >200 NTU. Similar to viral vaccine processes, depth filter followed by a bioburden reduction filter is commonly used in viral vector applications. TFF may also be an option in addition to the NFF options. Typical loading for TFF is 20–30 L/m^2 with low transmembrane pressures (TMP <0.5 bar) to reduce shear potential and to prevent premature membrane fouling (MillisporeSigma 2020).

The filter type has a direct impact on the efficiency of the clarification step:

- Disposable systems simplify cleaning and eliminate validation.
- Cascading filters consisting of a series of progressively finer filters prevent premature fouling of the sterilizing-grade filter. Fouling can result in reduced titer and can delay production (due to filter change outs) (Perry and Rayat 2021).

  Small-scale studies have shown that using multiple depth filter media types and grades (a combination of cellulose-based and synthetic media—polyacrylic fiber with silica filter aids) can reduce turbidity by 90% while maintaining high recovery.

  Using multiple depth filter grades also better enables a potential platform approach (across multiple serotypes) for the clarification step as it is more robust than using a single-media grade.

6.4.2 *Clarification Key Considerations*

- Is the filter positively charged? This especially impacts LVV production and some AAV serotypes.
- Is there a lysis step prior to clarification (can impact particle size distribution and impurity profile)?
- Is it a suspension or adherent system?
- Is the capacity sufficient? Do the filter requirements fit in the facility?

Examples of available depth filters:

- MilliporeSigma: Millistak® pod depth filters, Clarisolve® depth filters.
- Sartorius: Sartoclear® depth filters.
- Pall: Seitz® depth filters, Stax™ depth filters.
- 3M™: Zeta Plus™ filter capsules.

6.4.3 *Factors that Influence Clarification*

6.4.3.1 Cell Line

The cell line used for the production of viral vectors significantly impacts the choice of clarification methods. Adherent cell culture is cleaner than suspension cell culture. In some cases, only bioburden reduction and sterile filtration are required for an adherent system; depth filtration may not be necessary.

6.4.3.2 Lysis Conditions

Lysis, either mechanical or chemical, is used to break down the cell membrane.

Enveloped virus particles are generally released in the extracellular medium by budding through the cell membrane of the host cells. Non-enveloped viruses are assembled intracellularly. In the latter case, cell lysis is required to release the viral particles in the solution.

Lysis increases yield but also releases large quantities of DNA, proteins, culture media components, and cell debris into the bioprocess fluid. This impacts the clarification step. During harvest, these impurities are removed through centrifugation or filtration.

The fluid is then usually degraded with a nuclease such as Benzonase® endonuclease (MilliporeSigma).

Key Attributes and Considerations

The type of lysis performed can impact the downstream process:

- *Mechanical lysis* options are most effective at bench scale. Options include microfluidization, sonication, freeze–thaw cycles, and homogenization. These operations require specialized equipment, and parameters required for scale-up are not always evident (Srivastava et al. 2021).
- *Chemical lysis* options are most practical for large-scale production and are accomplished through the addition of low-concentrate surfactants (detergents). Choice of lysis detergent is based on effectiveness as well as environmental concerns and country regulations (Srivastava et al. 2021).

The type of surfactant used as well as the concentration impacts the capacity of the filters downstream. Optimum removal of such additives in the final product must be ensured.

A key contaminant in the viral vector process is carryover nucleic acid from lysed cells. These negatively charged large molecular components can interfere with virus purification. Regulations require carryover host cell DNA content to be <10 ng per dose, and the length to be <200 pb (Perry and Rayat 2021).

After the cell lysis, enzymatic treatment can degrade the released nucleic acids into smaller fragments before the next purification steps.

Benzonase® endonuclease (MilliporeSigma) is commonly used for the enzymatic degradation of nucleotides. The harvest can be treated with about 30 to about 50 U/mL of at 30–37 °C for four to 8 h (Adams et al. 2020).

Benzonase® endonuclease works optimally between pH 8.0 and 9.2 at 37 °C with a low level of $Mg2+$ (1–2 mM) needed for activity.

Recovery during the lysis step can be improved by the addition of a salt-tolerant nuclease (ArcticZymes Technologies SAN HQ). This enzyme is active at high salt concentration and helps in efficient digestion of DNA bound to the virus (Adams et al. 2020).

Studies have shown that the use of endonuclease can have a significant impact on clarification through the reduction of viscosity and/or breakdown of DNA.

6.5 Tangential Flow Filtration (TFF)

6.5.1 TFF Overview

Tangential, or cross-flow, filtration (TFF) is used in various process steps: clarification, concentration, buffer exchange, and/or additional impurity removal.

In TFF applications, fluid is pumped parallel to the surface of the filter membrane. A portion of the feed flow is recirculated through the retentate and back into the retentate vessel, and a portion is converted to permeate and is collected separately in a permeate vessel. Separation is size-based, so the selection of the molecular weight cutoff (MWCO) of the TFF device used is critical.

In microfiltration (MF), a molecular weight cutoff three to five times that of the product of interest enables the product to pass through to the permeate side.

In ultrafiltration (UF), a molecular weight cutoff that is one-third to one-fifth the size of the product of interest ensures concentration of the product of interest. The product is retained by the membrane and collected in the retentate vessel.

In microfiltration and some large virus ultrafiltration applications, because of the more open molecular weight cutoff used (0.1, 0.22, or 0.45 μm membranes are typically used), there is less resistance from the membrane so permeate control is recommended to prevent premature fouling and increased loading. A permeate pump or valve can be used to control the permeate. Permeate control is generally recommended when using a molecular weight cutoff ≥100 kDa.

6.5.2 TFF in AAV and LVV Processes

In AAV and LVV processes, there can be either one or two ultrafiltration and/or diafiltration steps.

The first TFF step (TFF1) typically occurs just prior to the first chromatography step and after the clarification step. The purpose of this step is to concentrate the vector to decrease load time and increase efficiency of the first chromatography step. In some cases, a diafiltration or buffer exchange is also accomplished in TFF1. This buffer exchange step may be necessary to change the composition or conductivity of the vector solution prior to loading on the column or it could also be for impurity or endonuclease removal. The selection of the molecular weight cutoff for TFF1 is dependent on the objective of the step. If impurity removal is desired, a more open cutoff ≥300 kilodalton (kDa) is recommended. If product recovery is the priority, a tighter cutoff (for AAV, 100 kDa) is preferred.

The purpose of the second TFF step (TFF2) is concentration, reaching the final desired vector titer, and exchanging into the final formulation buffer. The vector is purified prior to this step so product recovery as well as the use of non-adsorptive membrane material in the TFF device are of utmost importance.

6.5.2.1 AAV Retention vs. Molecular Weight Cut Off (MWCO)

There are many process considerations for the selection of a molecular weight cutoff for the concentration and buffer exchange of clarified AAV lysate (TFF1). If impurity and/or endonuclease removal is prioritized at this step, a more open 300 kDa MWCO is recommended. Models have shown that sieving of host cell proteins (HCPs) increases with increasing MWCO.

When chemical lysis is employed, the detergent used for lysis must be considered, especially whether the percentage of detergent used is higher than the critical micelle concentration (CMC) for that detergent. The most commonly used lysis detergents, Tween® 20 and Triton® X-100, create micelles in the 3.5–4.5 μm radius range. The approximate radius of a 300 kDa membrane is 13 μm, so these micelles fall within one-third of the pores and should sieve efficiently. But 100 kDa MWCOs and below could cause concentration of the lysis detergent.

Product recovery must also be considered. 100% AAV2 recovery has been seen with 30 kDa and 100 kDa Ultracel® membrane devices (MilliporeSigma) (using a feed flow rate of 4 LMM and 2.5% conversion), but a 10% loss was seen using a 300 kDa cutoff of the same membrane type. And even lower loss was seen with the 1000 kDa MWCO. These questions can also be answered if the priority is simply concentration at this upstream TFF1 step, or if ultrafiltration and diafiltration are both considered.

6.5.2.2 Processing Challenges

Many viral vectors have high aggregation potential, which increases with increasing concentration and titer. There are several ways to minimize aggregation: determining the maximum achievable concentration or virus titer, identifying the most effective combination of buffer excipients (salts, amino acids, and sugars are some examples), and adding an additional endonuclease step prior to or after concentration. This works when aggregation is caused by host cell DNA versus the vector itself (Wright et al. 2005; Adams et al. 2020).

Another challenge, particularly at the TFF1 step, is lower loading due to the higher impurity loads of the feed and the presence of the lysis detergent if greater than the critical micelle concentration (CMC).

6.5.2.3 Recovery Strategies

When the correct molecular weight cutoff (relative to the viral vector size) is used, high product recovery should be achieved during both TFF1 and TFF2 steps. A suitable post-use buffer recirculation and then a buffer push (to account for the system hold-up volume) is used to achieve this high product recovery. A high concentration of vector can be found at the surface of the membrane, particularly for "sticky," enveloped viral vectors, and this buffer recirculation and flush is necessary to resuspend it into solution. There are significant recovery gains with this buffer flush, but this also requires over-concentration to account for the post-use buffer flush dilution impact.

6.5.2.4 Shear Impact

In TFF applications, yield loss can sometimes be attributed to shear damage, which is caused by the recirculation of the product through the pump or within the device itself. For hollow-fiber and open-channel device formats, the highest shear is at the membrane wall. For screened, flat-sheet devices, the maximum shear is at the screen. The screen does act as a turbulence promoter, reducing the pumping requirements, but it is a complex flow where the screen both blocks and redirects flow, which then causes velocity changes at the screen that are higher than those seen at the wall (Perry and Rayat 2021) (Fig. 6.3).

The shear stress in a system is an important consideration, as some viral vectors are shear sensitive. There can be some advantages to the use of screen-containing devices, i.e., they can be operated at a lower feed flux, and the higher mass transfer leads to higher permeate fluxes which in turn leads to a lower amount of pump passes.

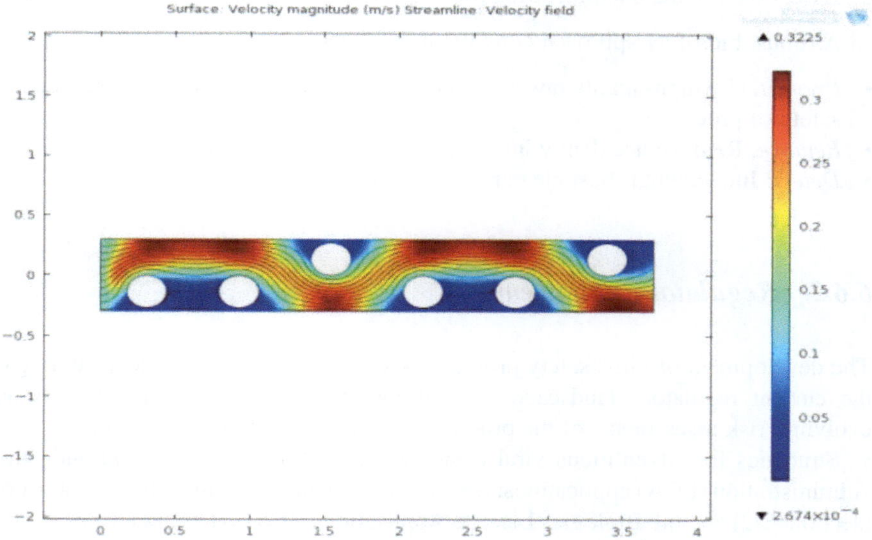

Fig. 6.3 Analysis of the surface velocity within a screened device showing the highest shear at the screen. Source: MilliporeSigma (2021) "Surface velocity within screened devices," MilliporeSigma Gene Therapy University Training

6.6 Viral Removal

6.6.1 Overview

Absolute viral safety in the manufacturing of biopharmaceuticals is essential to ensure patient safety.

An appropriate and robust viral clearance strategy must be developed and employed to ensure total viral clearance during production of vectors such as AAV and LVV for therapeutic purposes.

A manufacturing process designed with viral safety as a goal should include specific robust virus inactivation and removal steps, as well as purification unit operations that provide additional viral reduction.

A variety of GMP practices can mitigate risk:

- Careful choice and extensive testing of source materials for adventitious viruses.
- Rigorous testing procedures for raw materials, cell banks, virus seed stocks, bulk harvests, and drug substances.
- Assessment of the manufacturing process for inactivation and removal of viruses.

A robust biosafety approach consists of three components:

- *Prevent*: Use high-quality raw materials to prevent introduction of viruses. Ensure safety of processes.
- *Remove*: Remove accidently introduced adventitious agents.
- *Detect*: Implement robust clearance technologies.

6.6.2 Regulatory Requirements

The development of virus safety programs should be based on an understanding of the current regulatory landscape. In addition, because industry regulations are evolving, risk assessments of the production process should be conducted.

Strategies for adventitious viral clearance are outlined in two Food and Drug Administration (FDA) applications: the Investigational New Drug (IND) application (section A-2) and the Biologics License Application (BLA). These strategies should be taken into consideration prior to filing. The agency encourages open dialogue with gene therapy companies and offers several avenues for interaction including the INitial Targeted Engagement for Regulatory Advice on CBER producTs (INTER-ACT) meeting. This meeting with the Center for Biologics Evaluation and Research (CBER) is nonbinding and provides information about new products before the pre-IND meeting.

FDA's guidance document, "Chemistry, Manufacturing, And Control (CMC) Information for Human Gene Therapy Investigational New Drug Applications (INDs)," includes a section describing safety evaluation of adventitious agents (3.2.A.2) that should be reviewed carefully when establishing a strategy. The

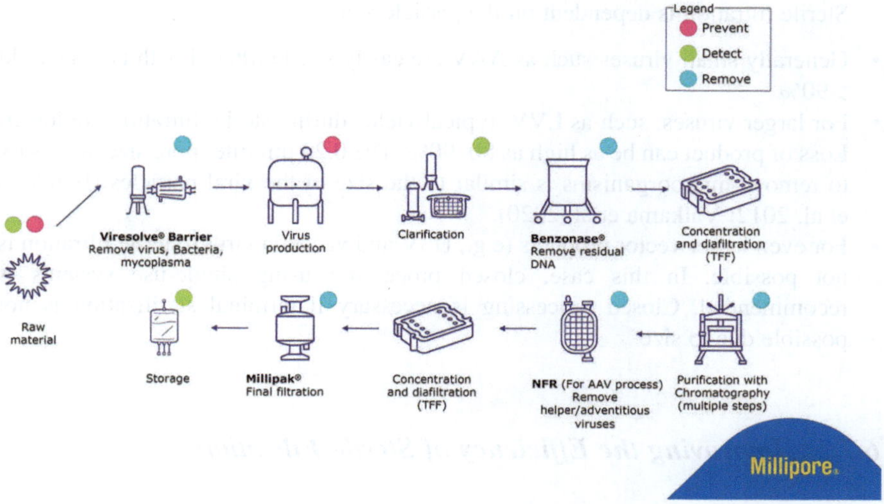

Fig. 6.4 Application of "prevent, detect, remove" strategy for virus safety as applied to the overall vector production workflow. Source: MilliporeSigma (2021) "Application of 'Prevent, detect, remove' strategy for virus safety as applied to the overall vector production workflow," MilliporeSigma Gene Therapy University Training

guideline also includes recommendations to assess the risk of contamination with adventitious agents and to provide information on viral safety studies (Fig. 6.4).

6.6.3 Nanofiltration as a Strategy for Efficient Viral Removal

Nanofiltration is effective for removing adventitious viruses. Filters such as Viresolve® normal flow retrovirus (NFR) filters with Retropore® membranes have been shown to quickly and efficiently remove retroviruses from media and protein feed streams. The pore size is approximately 50 nm; AAV pass through while larger viruses are retained. High recoveries with more than a six-log removal of retroviruses can be achieved (Hebben 2018).

6.7 Sterile Filtration

Bioburden control is imperative to mitigate the risk of microbial contamination in products produced under GMP regulatory requirements.

Standard regulatory requirement is sterile filtration using a 0.22 μm filter to remove any bacterial contamination. This is usually the final step of the downstream process before vialing and after purification, concentration, and formulation.

Sterile filtration is dependent on the particle size:

- Generally small viruses such as AAV are easily sterile filtered with typical yield >90%.
- For larger viruses, such as LVV, typical yields during sterile filtration are lower. Loss of product can be as high as 80–90%. The 0.22 μm filter pore size necessary to remove microorganisms is similar to the size of the viral particles (Bandeira et al. 2012; Valkama et al. 2020).
- For even larger vector products (e.g., HSV and vaccinia virus), sterile filtration is not possible. In this case, closed processing using single-use systems is recommended. Closed processing is necessary if terminal sterilization is not possible due to size.

6.7.1 *Improving the Efficiency of Sterile Filtration*

Sterile filtration can be optimized through process development, considering the following points:

- Pre-conditioning of the membrane (buffer flush).
- Membrane composition:
  - Millipore Express® SHC filter: polyethersulfone (PES) upstream 0.5 micron prefilter and downstream 0.2 micron sterilizing-grade filter.
  - MilliporeSigma Durapore® 0.22 μm hydrophilic polyvinylidene difluoride (PVDF) filters.
  - Sartorius Sartopore® capsule 0.2 μm (PES).
  - Pall Supor® filters (EKV–PES, Fluorodyne, PVDF).
- Membrane area (loading and capacity).
- Processing conditions: flow rate and/or pressure.
- Product formulation and aggregate control (some additives can be used to control aggregation) (Wright et al. 2005).
- The stage at which sterile filtration is performed.

6.7.1.1 Membrane Area Considerations

Appropriate filter sizing with optimal processing conditions (apt. pressure limit or flow rate) ensures minimal plugging and fouling during the sterile filtration process step. Due to the size and aggregate potential of these viral vectors, there is an increased incidence of product loss at high flux decays and/or pressure drops.

6.7.1.2 Final Filling Considerations

Shear stress during final filling steps can cause viral vectors to aggregate and form particulates.

Semi-automated filling processes for manufacturing small-volume gene therapy products can be subject to errors associated with crimping and sealing.

Peristaltic pump technology in a closed system provides low-shear filling. Closed systems coupled with real-time sterility testing can ensure sterility in cases where sterile filtration cannot be performed due to product loss.

References

Adams B, Bak H, Tustian AD (2020) Moving from the bench towards a large scale, industrial platform process for adeno-associated viral vector purification. Biotechnol Bioeng 117(10): 3199–3211. https://doi.org/10.1002/bit.27472

Amarosi D (2021) After challenging year, cell and gene therapy poised for further growth in 2021. HemOnc Today, 2021.02.03

Bandeira V et al (2012) Downstream processing of lentiviral vectors: releasing bottlenecks. Hum Gene Ther Methods 23(4):255–263. https://doi.org/10.1089/hgtb.2012.059

FDA (2023) https://www.fda.gov/vaccines-blood-biologics/cellular-gene-therapyproducts/approved-cellular-and-gene-therapy-products

Giglia S, Straeffer G (2012) Combined mechanism fouling model and method for optimization of series microfiltration performance. J Membr Sci 417–418:144–153. https://doi.org/10.1016/j.memsci.2012.06.026

Hebben M (2018) Downstream bioprocessing of AAV vectors: industrial challenges & regulatory requirements. Cell Gene Ther Insights 4(2):131–146. https://doi.org/10.18609/cgti.2018.016

MillisporeSigma (2020) Vaccine bioprocessing handbook, pp 1–22

Perry C, Rayat ACME (2021) Lentiviral vector bioprocessing. Viruses 13(2):1–46. https://doi.org/10.3390/v13020268

Raghunath B et al (2012) Best practices for optimization and scale-up of microfiltration TFF processes. Bioprocess J 11(1):30–40. https://doi.org/10.12665/j111.raghunath

Srivastava A et al (2021) Manufacturing challenges and rational formulation development for AAV viral vectors. J Pharm Sci 110(7):2609–2624. https://doi.org/10.1016/j.xphs.2021.03.024

Valkama AJ et al (2020) Development of large-scale downstream processing for lentiviral vectors. Mol Ther Methods Clin Dev 17:717–730. https://doi.org/10.1016/j.omtm.2020.03.025

Wright JF et al (2005) Identification of factors that contribute to recombinant AAV2 particle aggregation and methods to prevent its occurrence during vector purification and formulation. Mol Ther 12(1):171–178. https://doi.org/10.1016/j.ymthe.2005.02.021

Chapter 7
Chromatographic Purification of Viruses: State of the Art and Current Trends

Keven Lothert and Michael W. Wolff

7.1 Introduction

Viruses and virus-like particles (VLPs) play an important role for a variety of pharmaceutical applications. These include vaccines against infectious diseases, such as the annual influenza or the current coronavirus disease (COVID-19), as well as tumor therapy, immunomodulation, or viral vector applications, e.g., for gene therapy (Chen et al. 2018b, 2020; Hemminki et al. 2020). Although most of these applications are mainly relevant for human treatments, viral vaccines also play a crucial role in animal healthcare (Jorge and Dellagostin 2017). Efficient production processes are needed to meet the global demand for these various treatments.

In the past, a wide variety of expression systems have been used for vaccine production, resulting in a broad diversity of the product portfolio. Generally, the processes are organized into an upstream and a downstream process (USP and DSP). In the first part, USP, sufficient amounts of the target component are produced in various cells (Genzel 2015). In the second part, DSP, the target component is released from the cells, if necessary, and purified by eliminating product-related contaminants. Main constituents of these contaminants are proteins and DNA, derived from the host cells as well as from medium components. Maximum

K. Lothert
Institute of Bioprocess Engineering and Pharmaceutical Technology, Department Life Science Engineering, University of Applied Sciences Mittelhessen (THM), Giessen, Germany

M. W. Wolff (✉)
Institute of Bioprocess Engineering and Pharmaceutical Technology, Department Life Science Engineering, University of Applied Sciences Mittelhessen (THM), Giessen, Germany

Bioresources, Fraunhofer Institute for Molecular Biology and Applied Ecology (IME), Giessen, Germany
e-mail: michael.wolff@lse.thm.de

S. Gautam et al. (eds.), *Bioprocess and Analytics Development for Virus-based Advanced Therapeutics and Medicinal Products (ATMPs)*,
https://doi.org/10.1007/978-3-031-28489-2_7

acceptable limits are defined by the regulatory authorities, depending on the pharmaceutical aim and product type. For example, virus vaccines derived from non-tumorigenic host cells, such as human embryonic kidney (HEK) or Vero cells, may not exceed levels of residual host cell DNA of 10 ng_{DNA} per dose (Aubrit et al. 2015).

In order to meet these requirements, DSP commonly comprises different purification steps with varying operating mechanisms. In general, these can include cell disruption, precipitation (ammonium sulfate, calcium chloride, polyethylene glycol, etc.), clarification (filtration and centrifugation), to remove the particulate contaminants, followed by a concentration (tangential flow filtration) and, if necessary, a buffer exchange step. Subsequently, the actual purification and polishing takes place, most often by chromatography, as well as the final buffer exchange for the formulation (Wolff and Reichl 2011).

Depending on the therapeutic application, the product viruses need to be either intact and infective, e.g., for gene therapy, or inactive, e.g., for the immunization with inactivated pathogens, VLPs, or subunit and split vaccines. The requirements for a robust process management to produce infectious virus particles are considerably higher than those for inactive viral drug products, as the design space of the critical process parameters is smaller, and the transfer of process conditions to other viral products is often a greater challenge, even with platform technologies. This comprises the entire production process, with DSP usually being the bottleneck for the manufacturing process. Hence, the applied techniques have to be adapted for each new virus, and even for varying viral subtypes (Buckland 2015). In general, the different unit operations of DSP are optimized individually and in the context of the overall process chain, with regard to yield and purity as part of the process intensification. In the past, several reviews have been published, focusing on the general approaches for the DSP of cell culture-derived viruses (Wolff and Reichl 2011; Nestola et al. 2015a), with a special focus on chromatographic applications (Junter and Lebrun 2020; Zhao et al. 2019; Kramberger et al. 2015) or on specific viruses (Chen et al. 2020; Wolff and Reichl 2008; Loewe et al. 2020).

Here, we focus on liquid chromatographic techniques, which are generally an integral part of a DSP train for virus purification, and on a broad spectrum of chromatographic methods, as well as strategies of combined chromatographic steps, which are currently applied for the purification of cell culture-derived biological nanoplexes. Depending on the separation principle of the stationary phases, the methods belong to either the group of adsorption chromatography or the group of partition chromatography. During adsorption chromatography, the individual components of a feed material, the product of interest, or the contaminants are selectively retained on the stationary phase. In reference to the ligands used, adsorption chromatography can be further subdivided into ion-exchange chromatography (IEX), hydrophobic interaction chromatography (HIC), mixed-mode chromatography (MMC), and affinity chromatography, with an additional special form, the steric exclusion chromatography (SXC), gaining in importance recently.

Partition chromatography, on the contrary, works with an isocratic passage of the feed material through the column, mainly with size-exclusion chromatography (SEC).

Further key distinctions are based on the backbone used for the stationary phase (e.g., bead-based resins, monoliths, and membranes), the specific surface modification of the stationary phases, and the applied chromatography mode (product or contaminant adsorption).

The chapter describes the current and emerging chromatography methods for the purification of viruses, including the advantages and disadvantages of individual methods, also indicating the application potential for different production scales. Also, advantages and shortcomings of particular stationary phase backbones are discussed as well as continuous approaches by multi-column approaches. Finally, the possibilities of modeling and simulation are presented.

7.2 Choice of the Chromatographic Backbone

Traditionally, chromatographic DSP processes were mainly performed by using functionalized beads of different materials, particle diameters, and interparticle pore sizes. Apart from SEC and restricted access materials (see later in this chapter), however, the use of bead-based resins is not ideal for larger virus particles for the following reasons: (1) due to the size of the product, most of the resin surface is not accessible, particularly the inner particle surface area, which highly reduces the overall capacity of the stationary phase, and (2) the pressure drop across the column limits the maximum applicable flow rate and, thus, the optimization of processing times.

Consequently, convective flow materials, such as monoliths and membranes, are efficient alternatives (Curling and Gottschalk 2007; Gottschalk 2013; Krajacic et al. 2017). Monolithic columns consist of one piece, and chromatographic membrane modules are based on stacked individual membrane layers. Both comprise porous material, where the size distribution of the membrane pores is generally narrower. The structure of monoliths is often characterized by a bimodal pore size distribution of meso- and macropores. The macropores of monoliths are highly interconnected with multiple avenues of convective flow through the column. Hence, the mass transport of solutes through the column is relatively unaffected by the flow rate, which is also achieved with stacked membranes. Generally, for convective media, the mass transfer kinetics, and the column efficiency are affected by the structure, tortuosity, and the constriction of the end-to-end channels (Guiochon 2007).

Compared to bead-based columns, for convective matrices, pore diffusion limitations can be eliminated and the mass transfer is barely limited. The mass transfer of solutes is only affected by the film diffusion, which is relatively insignificant for larger particles, allowing the operation of processes at a higher flow rate (Roque et al. 2020). Other advantages of convective flow materials, as compared to bead-based resins, include a higher binding capacity for larger particles since, in relation to the outer surface of the beads, a larger area is available for interaction, and a greater column-to-column reproducibility is possible. Monolithic in situ polymerization avoids the primary source of column-to-column variability, which is the packing procedure for bead-based resins. Furthermore, recent developments in 3D-printing technologies enhance the reproducibility of the channel network inside the monoliths (Simon and Dimartino 2019).

The use of monoliths from some leading manufacturers, however, displays challenges when it comes to sterilization, e.g., for aseptic processing, and re-use, especially on larger scales. Industrial process-sized monolithic columns might need to be sterilized in place in a separate stainless-steel housing and, unfortunately, it may not be possible to re-fill the housing on site. Furthermore, cleaning in place is not trivial, irrespective of the scale, and fouling and channel blockade are an issue (Rajamanickam et al. 2015), especially due to lipids (Burden et al. 2012). For larger scale applications, disposable applications may be beneficial, as the necessity for cleaning validation is omitted. However, sterile units certified for good manufacturing (GMP) applications are not commercially available in most cases. In this regard, several studies focus on the reduction of the ecological footprint of the single-use process materials. In 2015, Fernandez and coworkers presented a method to prepare and functionalize biodegradable monolithic columns for adenovirus purification (Fernandes et al. 2015). They were able to obtain recoveries of up to 100% of the loaded virus in an anionic bind-and-elute approach. Nevertheless, the application of monoliths in a single-use manner is economically questionable, considering the alternative options.

These alternative convective flow materials are the previously mentioned membranes, which, in many cases, enable an easier handling, with regard to, e.g., the column weight. As a result, membranes are frequently applied for the chromatographic purification of viruses and VLPs (Hoffmann et al. 2019).

Similar to membranes, however, worth to be mentioned separately, is a stationary phase with nanofibers (Zhang et al. 2020). In contrast to membrane modules, which have stacked membrane layers with discrete pores across the area, the stationary phase of nanofibers resembles a loose network of filamentous strings. This results in larger pores between the fibers, thus further reducing the occurring pressure drop across the column and enabling higher flow rates. Additionally, the surface area is increased, leading to higher binding capacities. Nanofibers have been applied in the past for many different applications, such as tissue scaffolding, drug delivery, and filtration, to give some examples (Yoon et al. 2008; Leung and Ko 2011). For the chromatographic purification of biopharmaceuticals, several recent examples have led to an increased attention for this kind of matrix. Turnbull and coworkers applied cellulose nanofibers, modified with quaternary amine ligands, for the purification of adenovirus-5 particles (Turnbull et al. 2019). They described a productivity increase by 29 times compared to conventional bed chromatography. Furthermore, modified nanofiber materials have been used for the purification of lentiviral vectors, enabling a 100-fold product concentration and a 90% yield (Ruscic et al. 2019). Following these promising results, it can be expected that the importance of nanofibers will increase in future.

7.3 Adsorption Chromatography

Chromatographic purifications can be divided into adsorption and partition approaches. The latter is discussed in a later section of this chapter.

Adsorption chromatography utilizes the retention of either the target product [positive/bind-elute mode (Pereira Aguilar et al. 2020)] or the contaminants [negative/flow-through mode (Trabelsi et al. 2019)] on the surface of the stationary phase. Different surface chemistries and the resulting binding mechanisms used for virus purification by adsorption chromatography are described in the following.

7.3.1 Ion-Exchange Chromatography: Backbone of Most Chromatography-Based Viral Production Processes

A chromatographic approach for the purification of viruses, which has been used for decades, is the ion-exchange chromatography (IEX). IEX relies on charge interactions between the particles or dissolved compounds in the mobile phase and the ligands on the solid support. These interactions are controlled by the type and density of the ligands on the solid support, and by the type of ions and the ionic strength of the applied buffers. The overall surface charge of the particles in the mobile phase depends on their isoelectric point and is thus affected by the pH of the surrounding buffer. Positively charged molecules bind to a negatively charged surface (cation exchange, CEX), whereas negatively charged molecules bind to a positive surface [anion exchange, AEX (Cummins et al. 2017)]. For viruses and other biological nanoplexes, however, the overall charge is only partially relevant. Although charge estimations can be used to indicatively predict the performance of AEX and CEX methods, the charge density on the viral surface is multifaceted. Due to a complex surface composition, e.g., made of various (glyco-) proteins, clusters, or patches of different charges, may be present (Fig. 7.1). Accordingly, in some cases both AEX and CEX stationary phases can retain the same target virus at similar buffer conditions.

This opens a variety of possibilities to develop and optimize IEX processes for individual products. Consequently, IEX is a viable and frequently applied unit operation for different stages of the DSP process, e.g., as a capture, or as a polishing step. AEX and CEX methods have been applied for the purification of numerous viruses in the past and were summarized in previous reviews (Nestola et al. 2015a; Effio and Hubbuch 2015; Loewe et al. 2020). Examples of recent years are the application of IEX for the processing of the oncolytic measles virus (Eckhardt et al. 2021), human rhinoviruses (Allmaier et al. 2018), and adenoviruses (Turnbull et al. 2019), to name a few.

A striking benefit of the IEX is its generally wide acceptance in the pharmaceutical industry due to extensive experience with it and the large available data volume. Furthermore, chromatography materials are commercially available from various vendors for different scales of biopharmaceutical processing. Despite these

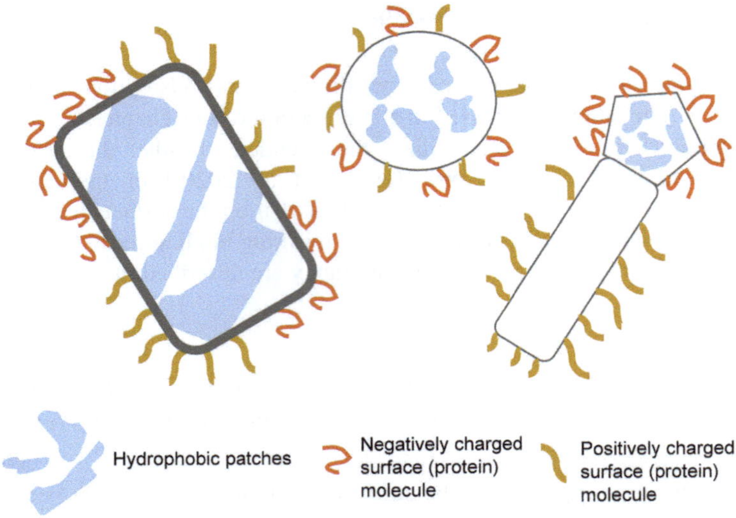

Fig. 7.1 Distribution of charges and hydrophobic areas on the surface of viral particles. The surface composition of viruses is complex, e.g., due to various (glyco-) proteins. Hence, the overall hydrophobicity and the isoelectric point of the whole particle might be different compared to individual areas, due to the presence of hydrophobic patches, or distinct clusters of specific net charges. In return, the retention behavior of viruses on a chromatographic stationary phase is not trivial, as multiple effects contribute to the interaction

advantages, IEX processes involve certain drawbacks, which will be shortly discussed in the following. The weaknesses include (1) the dependence on the isoelectric point and charge distribution of the product, (2) the low specificity and, thus, the contamination of the product fractions with contaminants, and (3) the elution conditions, which often require elevated ion concentrations. Regarding the isoelectric point (1), measurements for viruses are often challenging (Heffron and Mayer 2021) and require pure samples. Additionally, the charge depends on the surrounding buffer, of which at least the composition changes continuously during the production process. Moreover, changes in the production conditions (cell type, culture conditions, harvesting point, etc.) and the viral genotype or subtypes often come along with variations in the surface composition and alter the charge on the viral surface. This potentially leads to necessary adjustments in the process control of the IEX even for small changes in the USP. As a result, the IEX can only be used to a limited extent as a platform technology. This is directly related to the limited method specificity (2), as often the separation of the target viruses from the contaminating DNA is not trivial. Due to this circumstance, a careful adjustment of the salt ion or pH conditions during binding and elution is required (Lothert et al. 2020c). Variations in the pH, however, are usually difficult to accomplish, due to the narrow stability range of most of the viruses, additionally complicating the product-DNA resolution. Product elution is often accomplished by increasing the ion concentrations (3), rather than changing the pH. As this potentially affects the product configuration or activity as well, IEX approaches usually involve additional buffer exchanges to enable a successful chromatographic performance and to reduce the osmotic stress for the product (Olgun et al. 2019).

Despite all disadvantages, however, AEX and CEX are still the workhorses for the chromatographic purification of cell culture-derived nanoplexes, i.e., viruses as indicated, by frequent reports on IEX applications.

7.3.2 Hydrophobic Interaction Chromatography

A technique less frequently used for the purification of viruses is the hydrophobic interaction chromatography (HIC). It involves the separation of the target viral particle from process contaminants by a differential interaction of these components with hydrophobic sites on the surface of a solid support. Here, the interactions depend on the type and the density of the ligands as usually for adsorption chromatography. However, a broader repertoire of ligands is available for the HIC than for the IEX adsorber. Moreover, the types of ions used in the mobile phase for the adsorption and elution of the target compounds generally follow the Hofmeister series of lyotropic salts. This renders the process optimization more demanding than other adsorption chromatographies, as three process parameters (ligand, type, and concentration of buffer ions) need to be optimized. Furthermore, the adsorption of the target molecules requires an increased ion concentration (Tomaz 2017), which might affect the infectivity of the product viruses. Similar to the charge distribution, the hydrophobicity on the viral surface varies too, which also affects process optimization (Fig. 7.1). The HIC technique has a history in protein purification (Wang et al. 2012; Vailaya and Horváth 1996), but there are also several examples for virus purification processes. Current examples include the Orf virus (Lothert et al. 2020b), the influenza A and B viruses (Weigel et al. 2019), and the mumps and measles viruses (Sviben et al. 2017). For the latter, a combination with IEX was applied in the purification process, resulting in virus recoveries of 20% after IEX, and about 60% after HIC.

7.3.3 Selective Adsorption Chromatography: Affinity and Pseudo-Affinity Purification of Viruses

A possibility for an improvement of the purity in chromatography processes is the use of specific affinity ligands. However, the classical use of tags, as for recombinant proteins, is only possible to a very limited extent for nanoplexes. Therefore, specific ligands have to be identified via molecular modeling approaches, via screens of distinct ligand libraries, or via natural ligands described in the literature. Examples of potential ligands include all areas of chemical, biological, and biochemical components, such as antibodies, antibody fragments, specific binding proteins, aptamers, lectins, and (sulfated) carbohydrates. Affinity-based chromatography methods enable a highly specific capture of a target, even from a complex feed solution, such as the clarified cell culture harvest. Recently, Zhao et al. published a comprehensive overview on affinity chromatography for vaccine applications (Zhao et al. 2019). Here, we focus on a few promising platform examples.

A large number of viruses interact with heparin, a sulfated carbohydrate polymer (Liu and Thorp 2002), which is mainly used in small-scale laboratory experiments. Within these processes, excellent yields can be achieved, as described for cell culture-derived baculoviruses (85% recovery) via a commercially available POROS heparin column (Nasimuzzaman et al. 2016), or for the cell culture-derived enterovirus 71 (89% recovery) by custom-made hydrogel monoliths with heparin ligands (Gu et al. 2020). The reason why heparin is mainly used on a laboratory scale is that, currently, heparin cannot be synthesized chemically, and the use of animal components in production processes of human pharmaceuticals is critical for safety reasons (Wolff and Reichl 2011). However, other sulfated carbohydrate polymers, such as sulfated cellulose and dextran, show similar interactions with viruses, thus maintaining the platform character of the stationary phase, but omitting the requirement for using heparin itself. This improves the economic efficiency as well as avoiding the use of animal-derived products during the production process. Sulfated carbohydrate polymers were applied for virus purification, using resins, monoliths, and membranes with commercially available products, such as the Celufine® sulfate (AMSBIO), and Capto™ DeVirS (Cytiva) resins, as well as the Sartobind® sulfated cellulose membrane adsorber (Sartorius Stedim Biotech GmbH). All three stationary phases have already been successfully applied for virus purifications as shown in the examples hereafter. Sulfated cellulose resin column chromatography was applied to purify and concentrate the dengue virus and its performance, compared to conventional sucrose gradient centrifugation. Although the overall recovery of the infective virus was low for chromatography (~20%), an improvement compared to the density gradient method, in terms of recovery, purity, and, in particular, processing time, was reported (Kanlaya and Thongboonkerd 2016). Overall product recoveries of up to 81% (see Table 7.1) were achieved using sulfated cellulose membrane adsorbers, as was reported for the purification of cell culture-derived influenza A and B viruses (Fortuna et al. 2018, 2019; Opitz et al. 2009b), for influenza VLPs (Carvalho et al. 2018), and for the Modified Vaccinia Ankara virus (Wolff et al. 2010a, b).

Despite these promising reports, both approaches display certain drawbacks, which should be mentioned here. The resin-based approach suffers from the previously mentioned disadvantages of chromatographic resins, which are comparably low flow rates and a lower accessible surface area (see also the earlier section). The currently available membrane adsorbers, however, are characterized by a rather small pore size of approximately 0.8 μm. While this increases the binding area for smaller viruses, such as the influenza viruses, it impairs the recovery of larger viruses, as recently described for the Orf virus (Lothert et al. 2020c). In these experiments, recoveries of only 34% (from a clarified feed) and up to 54% (from a pre-purified feed) were possible. In order to overcome this, sulfated cellulose

Table 7.1 Recent chromatographic purification approaches of virus particles based on affinity interactions

Purified target	Affinity ligand and matrix	Max. virus recovery (%)	Impurity depletion (%)		Reference
			DNA	Protein	
Orf virus (enveloped)	Sulfated cellulose (membrane)	54	95	>99	Lothert et al. (2020c)
Adeno-associated virus	Immunoaffinity (resin)	93	Not reported		Wang et al. (2015)
Hepatitis C virus (enveloped)	Sulfated cellulose (membrane)	50	90	>99	Lothert et al. (2020a)
Influenza A/B (enveloped)	Sulfated cellulose (membrane)	81	96	74	Fortuna et al. (2018, 2019)
Influenza VLPs	Sulfated cellulose (membrane)	80	80	89	Carvalho et al. (2018)
Baculovirus (enveloped)	Heparin (resin)	85	Not reported		Nasimuzzaman et al. (2016)
Foot-and-mouth disease virus	Immobilized metal (Ni) affinity (beads)	Not reported	>99	>99	Biswal et al. (2015)
Mumps virus (enveloped)	Immunoaffinity (monoliths)	80	Not reported		Brgles et al. (2016)

membranes with larger pore sizes need to be commercially available for different scales and certified for GMP applications.

Besides the use of sulfated cellulose as a stationary phase, other affinity ligands, such as lectins (Chen et al. 2009; Opitz et al. 2007), or immobilized metal ions (Biswal et al. 2015; Opitz et al. 2009a) have been applied for some viruses, but are rarely used by comparison. Examples of recent applications using affinity ligands for the chromatographic purification of viruses and VLPs are depicted in Table 7.1. Additionally, the interested reader is referred to extensive reviews on affinity chromatography for vaccine manufacturing (Zhao et al. 2019) or, more specifically, on polysaccharide-based adsorbents (Junter and Lebrun 2020).

7.3.4 *Mixed-Mode Chromatography*

Mixed-mode chromatography (MMC) or multimodal chromatography refers to chromatographic methods that utilize a combination of binding modes, including ion exchange, hydrogen bonding, and hydrophobic interactions, between the ligands on the stationary phase and the target compounds (Zhang and Liu 2016; Junter and Lebrun 2020). MMC methods have been described in the literature since the 1950s, with the first methods mainly using hydroxyapatite (HA) as the stationary phase. HA is a naturally occurring mineral of calcium apatite, which enables the combination of CEX and metal affinity via its structure (Gagnon et al. 2009, 2010). Due to a limited understanding of multimodal interactions, the process optimization for viruses with complex surface structures of different proteins was challenging. However, after a careful adjustment of the process conditions, ceramic hydroxyapatite columns were successfully used as a DSP unit operation for gene therapy vectors (Kuiper et al. 2002), as well as a one-step purification for dengue virus particles (Kurosawa et al. 2012). The potential to design an efficient and robust unit operation with this method and with a deeper understanding of multimodal interactions prompted the development of new generations of mixed-mode stationary phases for biopharmaceutical applications (Zhang and Liu 2016; Gagnon et al. 2010). For the purification of viruses and VLPs, mainly negative mode MMC methods with resins are used today. The particles in these resins consist, in many of the applications, of an inert (ligand-free) outer coating and a highly cross-linked center with multimodal ligands. The interaction of macromolecules, viruses, and VLPs with the center ligands is controlled by the pore size and the thickness of the shell (Sanchez-Trasvina et al. 2020; Junter and Lebrun 2020). This special case of multimodal interaction will be discussed in a separate section on restricted access materials (RAM) later in this chapter.

7.3.5 Selective Size-Dependent Retention: Steric Exclusion Chromatography

The salting-out of proteins and macromolecules has not only been utilized for the application of HIC mentioned in an earlier section. Macromolecular crowding effects, induced by the addition of non-ionic polymers or ammonium sulfate, have also been applied for the precipitation and the subsequent purification of proteins and viruses since the 1970s (Ingham 1978). For more than five decades this approach was employed in batch mode, using centrifugal sedimentation and high polymer amounts primarily for proteins (Ingham 1978; Hammerling et al. 2017; Polson 1977; Atha and Ingham 1981). The method is also possible for viruses, enabling a simultaneous concentration of the product (Schagen et al. 2000), but in general the purification efficiency is poor and the recoveries are low, especially regarding infectious titers. Furthermore, the scale-up is difficult and requires high investment costs, due to appropriate devices for the batch centrifugation of large volumes.

The principle of sterically excluding macromolecules from a solution eventually found application in a chromatographic procedure and was initially described in 2012 by Peter Gagnon and his group (Gagnon 2012). They used monoliths as a stationary phase to purify large proteins and bacteriophages (Lee et al. 2012; Gagnon 2013). Later, the SXC method was optimized for large virus particles by applying cellulose membranes to purify Influenza A viruses (Marichal-Gallardo et al. 2017). This allowed higher flow rates by reducing the pressure drops across the column as well as the overall cost of goods in a single-use manner. For the SXC as a chromatography technology, usually polyethylene glycol (PEG) is used as a crowding agent. In general, the applied PEG concentrations are lower than those

Table 7.2 SXC applications for viruses with the achieved product recoveries and impurity depletions

Purified target	Stationary phase and pore size	Virus recovery (%)	Impurity depletion (%)		Reference
			DNA	Protein	
Bacteriophages	CIM-OH monoliths (2 μm)	>90	93	99.8	Lee et al. (2012)
Adeno-associated virus	Cellulose membranes (1 μm)	>95	94–98	80–85	Marichal-Gallardo et al. (2021)
Influenza A virus	Cellulose membranes (1 μm)	>95	99.7	92.4	Marichal-Gallardo et al. (2017)
Yellow fever virus	Cellulose membranes (1 μm)	>99.5	65–75	92	Marichal-Gallardo (2019)
Modified vaccinia Ankara virus	Cellulose membranes (1 μm)	>95	90	>90	Granicher et al. (2021)
Lentivirus	Cellulose membranes (2.5–3 μm)	88	80	80	Labisch et al. (2022)
Baculovirus	Cellulose membranes (3-5 μm)	91	85	99	Lothert et al. (2020d)
	Polyamide (5 μm)	81	81	>99.9	
	Glass microfiber (5 μm)	73	45	93	
Orf virus (two genotypes)	Cellulose membranes (1 μm)	>91	76–81	98–99	Lothert et al. (2020c, b)
Hepatitis C virus (two genotypes)	Cellulose membranes (1 μm)	>99	>99	84	Lothert et al. (2020a)

for precipitation techniques, hence reducing the impact on product integrity and increasing the overall recovery (Ingham 1978; Marichal-Gallardo et al. 2017). The general procedure for the SXC purification has been described elsewhere and is not discussed in detail here (Lee et al. 2012). With the use of cellulose membranes in recent years, the SXC was successfully applied for the purification of baculoviruses (Lothert et al. 2020d), hepatitis C viruses (Lothert et al. 2020a), different genotypes of the *parapoxvirus ovis* (Lothert et al. 2020b), adeno-associated viruses (Marichal-Gallardo et al. 2021), and the lentivirus (Labisch et al. 2022). The published literature indicates that the main driver of the SXC is the product size, which can be targeted by carefully adjusting the PEG size and concentration. However, also the isoelectric point of the virus is a critical process parameter, and consequently the pH and salt content (Lothert et al. 2020a). By a proper selection of adequate running conditions, usually between 90% and 100% virus recovery is reported (see Table 7.2). As the SXC additionally enables product concentration and supports buffer exchange during the elution step, it is described as a convenient option for a product capture after an initial clarification.

Although SXC displays a promising tool for the DSP of viruses delivering a nearly complete product recovery and a high impurity removal, its application is currently limited to laboratory-scale purifications. At present, industrialization and large-scale processing is restricted mainly by the limited availability of stationary phase material. The membranes used in the laboratories are custom-made. However, for pharmaceutical applications, GMP units are currently not available. Using monoliths, as initially described by Gagnon et al., would be an option, however, with limitations due to costs, possibilities for sterilization, and an increased fouling (Rajamanickam et al. 2015; Burden et al. 2012). Hence, future developments or product releases with adequate membranes could potentially boost the implementation of the SXC in biopharmaceutical processing. Furthermore, the recent revival of nanofibers (see earlier section) might also prove to be beneficial for SXC applications.

7.4 Partition Chromatography

In contrast to adsorption chromatography, partition chromatography is based on a varying speed at which the components in a mixture pass through the column. Usually, no binding interaction occurs. Instead of performing the individual load, wash, and elution steps, the separation is done isocratically. The most prominent example of partition chromatography is size-exclusion chromatography (SEC), which is described in more detail below.

7.4.1 Gentle Buffer Exchange and Polishing: Flow-Through Size-Exclusion Chromatography

In contrast to the SXC, the SEC is already a well-established purification tool for proteins as well as for viruses. SEC matrices usually consist of (cross-linked) beads, such as dextran or agarose, which do not allow binding. Instead of adsorbing to the stationary phase, molecules are separated by the different velocities required for the migration through the column (Striegel 2017). While smaller molecules can access nearly the entire column volume, larger particles are restricted by the pore size, leading to pore exclusion and diffusion effects. Hence, viruses elute first in SEC, followed by proteins of decreasing size and small nucleic acid fractions after nuclease treatment. An advantage of the size-exclusion chromatography is its gentle processing, as the entire product suspension is passed through the column isocratically. Furthermore, the SEC is well characterized, due to its long history in protein and virus applications; hence, mathematical modeling and resolution predictions are available (Gu 2015b; Gritti et al. 2020; Chen et al. 2018a). The recovery during the SEC is directly dependent on the desired purity, as it corelates with the size of the collected fractions and the resolution of the column. With an appropriate column resolution and by accurate peak fractionation, recoveries above 90% can be achieved (Hossienizadeh et al. 2021).

However, several disadvantages limit the application of the SEC in scalable virus purification processes. Firstly, the actual resolution is poor, and viruses cannot be separated well from other larger size species, such as extracellular vesicles or proteins with a high molecular weight. Secondly, the productivity of the SEC is low. The reason for this is mainly due to the low operating flow rates and the limited loading volume, because a maximum of 30% (usually rather less than 10% are used) of the column volume should be applied as a sample (Burgess 2018). While this is feasible for laboratory-scale applications, it would require extraordinarily large columns on an industrial scale. Hence, SEC is most often performed with already concentrated products at a later processing step, e.g., for polishing (Peixoto et al. 2007; Hossienizadeh et al. 2021) or for veterinary products with higher purity requirements, such as equine vaccines. It is also frequently applied for desalting and buffer exchange (Boix-Besora et al. 2022). Kalbfuss and coworkers applied the SEC as a primary chromatography unit, prior to an AEX polishing; however, they initially performed a sample concentration by tangential flow filtration (Kalbfuss et al. 2007).

7.5 New and Emerging Chromatography Approaches for Virus Purification

Research in recent years greatly advanced the performance of chromatographic virus purifications. This includes improved stationary phases, such as restricted access media (RAM), which are exceptionally suitable for the contaminant depletion without affecting product integrity. Additionally, yields are excellent with optimal process design, as the virus particles do not interact with the matrix. Moreover, the implementation of multi-column approaches and mathematic process models led to an improved throughput, a reduced buffer and processing time expenditure, and an easier method optimization. Hence, these topics shall be shortly discussed in the following sections.

7.5.1 Mixed-Mode Restricted Access Media: Efficient Tools for Contaminant Depletion

Another possibility to take advantage of the distinct size difference between viral products and contaminating proteins or DNA is the use of RAM, often also referred to as core-shell particle resins. These are MMC matrices that usually combine size exclusion on inert particles, with some kind of binding ligand, such as IEX, HIC, or mixed mode, within the core. This setup has proved to be extraordinarily suitable for the separation of viruses from surrounding smaller molecules, without the requirement of virus diffusion into the pore, as it is operated in the flow-through mode

(Sanchez-Trasvina et al. 2020). Thus, the contaminating protein and DNA can be easily removed from suspensions containing viruses as a product. Accordingly, as during SEC, the processing for the target virus is smooth, reducing losses, and resulting in high yields of active product (James et al. 2016) without comparable loading restrictions as for the SEC. A striking benefit of these applications is the platform character, enabling the use for various viral targets as depicted by the following examples.

For cell culture-derived Orf viruses, the advantage of the RAM material, applied in flow-through mode, was shown in comparison to various bind-and-elute methods (Lothert et al. 2020c). Here, the Capto™ Core 700 resin outperformed commonly used membrane adsorbers regarding infective virus recovery (>90%) and impurity depletion (>70% residual DNA removal).

However, these RAM materials are not only suitable for final product application but may also be applied for an efficient initial protein and DNA depletion. To give an example, Steppert et al. reported on a DSP process for the purification of a recombinant measles virus using RAM chromatography as a main purification step earlier this year (Steppert et al. 2022). Supported by filtration unit operations for clarification and concentration/diafiltration, the applied setup was scalable for up to 9 L loading volume. Further examples from recent years include the purification of the rabies virus [66% recovery (Trabelsi et al. 2019)], the yellow fever virus [84% recovery (Pato et al. 2019)], and influenza viruses [100% recovery (Tseng et al. 2018)].

The named literature descriptions mainly focus on the use of the Capto™ Core 700 (and 400) resin. This is due to the fact that this product family was the first commercially available RAM material. Recently, however, several other vendors entered the market offering a broad portfolio of RAM resins. These vary in the molecular weight cutoff in the outer shell (e.g., between 150 kDa and 5000 kDa), as well as in the chemistry of the core (e.g., anion and cation exchangers or mixed-mode ligands). Due to these modifications, the purification can be tailored to the targeted product and the contaminants to be removed. The easy process implementation and the broad application range will make the use of RAM materials an important feature of future virus DSP developments.

7.6 Continuous Chromatography for Vaccine Production Processes

Continuous processing has been a major focus of research in recent years in order to increase the throughput of DSP unit operations and to reduce the process footprint of individual batches, while exploiting the full potential of column capacity and minimizing buffer consumption. Applications usually involve the use of multi-column chromatography (MCC) operated, e.g., in the periodic counter-current

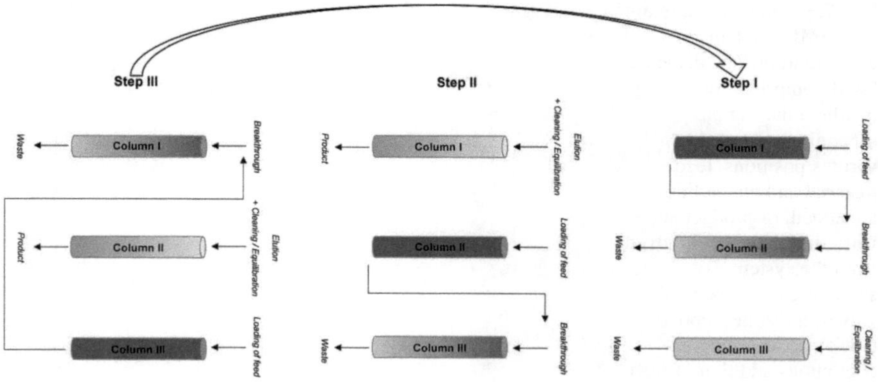

Fig. 7.2 Periodic counter-current chromatography (PCC) as an example for multi-column chromatography. The depicted PCC configuration comprises three columns. Column I is loaded in the beginning. To avoid sample loss, the breakthrough of column I is collected on column II and the third column is equilibrated in the meantime. Once the capacity of the first column is reached, the loading continues on column II with the breakthrough now being collected on column III. Simultaneously, the product is eluted from column one, and the column is regenerated and re-equilibrated to further continue the cycling between the columns. Once the capacity of column III is reached, the loading of the feed proceeds on column I in a new cycle until the required feed amount is processed. More than three columns can be integrated, and the column wash, prior to the elution, might be collected as well on one of the other columns

(PCC) or simulated moving bed (SMB) mode (Nestola et al. 2015a; Gerstweiler et al. 2021; Fortuna et al. 2021; Fischer et al. 2018).

The PCC is well suited for adsorption chromatography and utilizes at least three columns that are operated in parallel (Godawat et al. 2012). While the first column is loaded with the feed, its breakthrough is immediately directed onto the second column, allowing a continued loading until the defined utilization of the first column is reached (Fig. 7.2). In the meantime, the third column is equilibrated and prepared to receive the breakthrough of column 2 in the next stage. Once the loading of the feed and the collection of the breakthrough continues on columns 2 and 3, respectively, the first column is washed and the product eluted. In the last step, the loading of the feed proceeds on column 3, with its breakthrough being collected on column 1, and column 2 proceeding with wash and elution. Afterward, the cycle starts again from the beginning until the desired amount of feed material has been processed. This configuration reduces the required column volume and the loss of sample. Furthermore, the overall processing time is shortened, as the wash and elution steps can proceed simultaneously while the loading continues on the subsequent columns. The additional interconnection of further columns increases the intermediate time, and facilitates the sanitization and re-equilibration of readily eluted columns. In some applications, the system configuration allows the collection of the wash fraction on the column next in line to receive loading, in order to reduce product losses during the wash phase (Godawat et al. 2012).

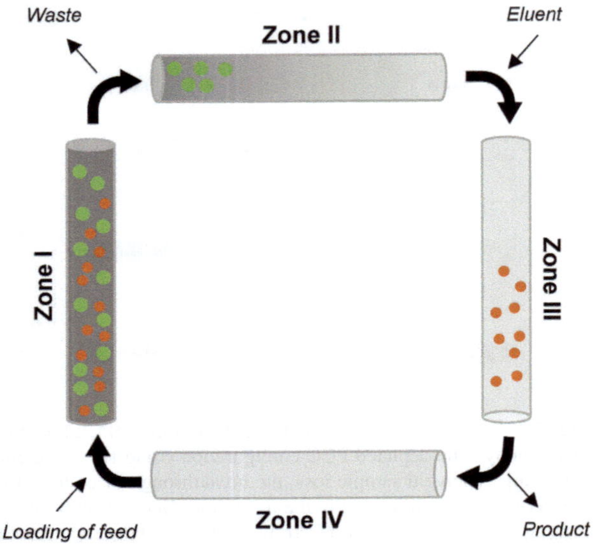

Fig. 7.3 Simulated moving bed (SMB) system configuration. The depicted SMB comprises four zones, in which the sample separation occurs. At various positions, feed material and eluent buffers are added, or product and waste fractions are removed from the system. By switching the respective valves, the zones "move" clockwise, enabling a continuous addition of feed and thus the generation of the purified product. Green and orange dots represent the solutes A and B, which should be separated

The SMB, on the other hand, utilizes a column setup that is interconnected in series (Fig. 7.3). Classically, the SMB system comprises four zones with each zone having at least two subsections (Harrison et al. 2015). Every subsection is an individual column segment, and the movement of the solid phase is simulated in a circular direction. This is achieved by periodically altering the points where the feed and buffer solutions are added, or the product and the waste are removed. In this way, the processes within individual zones shift, e.g., to the different conditions of (1) strongly binding analyte A, (2) binding of analyte B, (3) elution of analyte A, and (4) removal of analyte B. Due to this system configuration, the SMB is well suited for performing partition chromatography, as the serial interconnection of columns increases the overall length and thus the separation potential, which could probably not be achieved by using only one column. In order to maintain a stable system and to perform the process on steady-state conditions, an accurate knowledge of the separation principle, as well as a careful adjustment of buffers, valve switching, and flow rates, is necessary. This complexity requires an extensive method development and optimization, which needs to be done by modeling and by simulation approaches (see the next section).

To give a few examples, the use of PCC or SMB for the purification of viruses has been described for oncolytic adenoviruses [AEX PCC, 86% recovery (Mendes et al. 2021)], adeno-associated viruses [affinity PCC, 82% recovery (Mendes et al. 2022)], influenza viruses [pseudo-affinity PCC, 67% recovery (Fortuna et al. 2021), AEX SMB, 89% recovery (Fischer et al. 2018), SEC SMB, 70% recovery (Kröber et al. 2013)], and murine polyomavirus VLPs [AEX PCC, 89% recovery (Gerstweiler et al. 2022)]. Nestola et al. described the development of a purification process for adenoviruses by interconnecting two SEC columns by a simulated counter-current (Nestola et al. 2014) or SMB (Nestola et al. 2015b) technique, in order to avoid the limitations of the low loading volumes during SEC. With this continuous chromatography approach, the group was able to avoid bind-and-elute chromatography methods, yet could still remove up to 90% of the contaminating DNA. Furthermore, Silva and coworkers presented a detailed protocol for the implementation of (semi-) continuous processing for VLPs by IEX (Silva et al. 2020).

In addition, the interested reader shall be referred to recent reviews, covering the state-of-the-art and current developments of continuous processing of VLPs (Mittal et al. 2021) and continuous DSP of biopharmaceuticals in general (Gerstweiler et al. 2021).

7.7 Mathematic Models for Chromatographic Virus Purifications

Process development and optimization for the chromatographic purification of viruses is often elaborate, as a variety of parameters, e.g., buffer composition, product concentration, stationary phase, etc., highly affect the recovery of both the product and the contained impurities. Hence, these experiments require large amounts of consumables, feed material (i.e., virus harvests), and time. In order to reduce costs and effort, design of experiments (DoE) approaches are implemented (Neubauer et al. 2013). Another alternative is in silico simulations of the chromatographic process performance by modeling techniques (Shekhawat and Rathore 2019). This includes the characterization of the stationary phase material (Chen et al. 2018a; Ghosh et al. 2014), as well as an estimation of binding capacities, recovery yields, and purities. Most of the available literature in the field of chromatography modeling currently focuses on protein purifications (Gu 2015a). Here, a broad range of approaches is described to simulate the product interaction with the stationary phase for IEX (Schmidt et al. 2014; Rischawy et al. 2019; Benedini et al. 2020), HIC (Muca et al. 2010; Wang et al. 2016; Galeotti et al. 2020), SEC (Zelic and Nesek 2006), or affinity (Pfister et al. 2017; Dimartino et al. 2011) applications. Especially mass transfer parameters and binding behavior, according to the underlying adsorption and desorption isotherms, were extensively studied. The quality of the respective isotherm, for describing the retention characteristic of the specific compounds mathematically, depends on the actual type of physical or chemical interactions with the stationary phase. The most prominent examples, suitable for a variety of applications, are the Langmuir and the steric mass action isotherms. Additionally, the application of continuous chromatography modes (see previous section) has been described by modeling approaches (Zobel-Roos et al. 2018).

For viruses, however, literature on adsorption chromatography simulation is scarce. Recently, for a foot-and-mouth disease vaccine, a two-dimensional SEC-AEX method was described, and the experimental data fitted against mathematical modeling data to facilitate a later scale-up (Hossienizadeh et al. 2021). The governing model equations included several assumptions, such as plug flow with axial dispersion, isothermal conditions across the column, and ideal mixing in the aqueous phase. The model included the material transport in the bulk solvent and from the bulk solvent to the stationary phase, as well as occurring mass transfer resistances. The group applied the Yamamoto isotherm, which is useful to express the equilibrium of certain protein and ion exchange resin interactions (Hossienizadeh et al. 2021). Similarly, a modeling framework was presented for optimizing the SEC purification of influenza A viruses (Kalbfuss et al. 2008) and for AEX membrane chromatography using *Sf9* cell-derived VLPs (Effio et al. 2016).

In contrast to adsorption chromatography, the modeling of SEC processes is more straightforward, as no binding interaction occurs and the migration through the column depends on the individual properties of the analytes and not on other components in the mixture (Nestola et al. 2014). This is particularly beneficial in the case of SMB applications. Here, the use of the Langmuir isotherm has proven to be very effective, disregarding dispersion and mass transfer resistance (Migliorini et al. 1999). In this context, Nestola and coworkers designed a SEC step to later successfully set up a two-column SMB application for an adenovirus purification (Nestola et al. 2014, 2015b). They simplified the mixture of their feed material into three components, which are the product virus (i.e., the largest component), the last impurities exiting the column (i.e., the smallest ones), and thirdly everything eluting in between. By recording retention volumes, peak areas, and peak broadening, the duration of one process cycle was determined. Finally, the curve fitting was done by a deconvolution procedure to ultimately simulate the axial concentration of the three individual components.

Concerning adsorption chromatography, although most of the principles described for modeling protein purifications might be applicable to viruses as well, there are some fundamental differences that challenge a direct application of the existing models. These include (1) the complex binding behavior of viruses due to their multifaceted surface composition, (2) the large and varying size of different viruses and the resulting occupation of stationary phase ligands by individual particles, and (3) the interaction of viruses with other components in the suspension, e.g., DNA, which might shield possible binding partners on the virus' surface. Hence, future evaluations should focus on the verification and the adjustments of adsorption isotherms to viruses, which are already known from protein purifications. This might accelerate the development of suitable models, eliminating the need for rigorous experimental method optimization. While this is feasible for well-known and characterized methods such as IEX, for other methods, such as the SXC, the mechanistic understanding must be improved prior to identifying suitable mathematical descriptions. Especially in the case of the SXC, existing isotherms might not adequately describe the separation problem, as no specific ligand–target interaction takes place, but a multi-layered product association is possible. As the basis of the SXC principle resembles commonly used precipitation, modeling approaches should include this aspect as well and combine it with a chromatographic separation. Until present, however, the simulation of precipitation has been described exclusively for the quality by design optimization for protein processes, e.g., IgGs (Lohmann and Strube 2020). An adaptation to virus purification has yet to be described.

7.8 Conclusions and Outlook

Despite ongoing efforts to remove chromatography from virus purification protocols by a series of depth and crossflow filtration steps, as well as precipitation and extraction methods, chromatography is still an integral part of most viral DSP processes. The possibilities for the chromatographic purification of viruses are as versatile as the large number of different virus species suggests. Some approaches target a highly selective retention of the product, e.g., affinity and pseudo-affinity techniques. Others rather aim to separate entire groups of components from each other, based on general characteristics, such as charge in IEX, and size in SEC and SXC. However, none of these approaches alone enable the complete removal of contaminants. Hence, to allow a pharmaceutical usage of the derived products, a combination of at least two chromatographic methods, or the backing up of one chromatography mode by alternative technologies, is necessary. Additionally, a nuclease treatment, to reduce the numbers of base pairs of the host cell DNA to a safe fragment size according to the regulatory specifications, is usually required.

Developments in recent years clearly demonstrate the benefits of platform technologies, as these enable fast in-place processes, and facilitate a timely reaction to new pandemic outbreaks, e.g., annual influenza. Size-depending methods, such as SXC or pseudo-affinity using sulfated cellulose as well as RAM applications, are important representations of these platform technologies, as they target specific product properties (size and heparan-sulfate affinity, respectively), allowing a distinction of a broad range of viruses from process contaminants. Future research will most probably focus on developing and optimizing platform approaches. Another driver for future developments will be the wish to increase the productivity of chromatographic unit operations. For that reason, approaches for continuous chromatography operations are a regular focus in research. Similarly, stationary phase materials, as well as the design of their housings, are permanently optimized to allow a maximum capacity for the product (e.g., membranes and nanofibers), or to enable the gentle removal of contaminants without affecting the product (e.g., RAM materials). Ultimately, a better process understanding and a straightforward method optimization can be achieved by improving mathematical descriptions and modeling simulations of chromatographic virus purification processes.

Acknowledgments We want to thank Catharine Meckel-Oschmann for critical reading of the manuscript.

Funding This work was written within the framework of the industrial collective research program (IGF/Project no. 21631 BG/FE 2). It was supported by the Federal Ministry for Economic Affairs and Energy (BMWi) through the AiF (German Federation of Industrial Research Associations eV) based on a decision taken by the German Bundestag.

References

Allmaier G, Blaas D, Bliem C, Dechat T, Fedosyuk S, Gosler I, Kowalski H, Weiss VU (2018) Monolithic anion-exchange chromatography yields rhinovirus of high purity. J Virol Methods 251:15–21

Atha DH, Ingham KC (1981) Mechanism of precipitation of proteins by polyethylene glycols. Analysis in terms of excluded volume. J Biol Chem 256:12108–12117

Aubrit F, Perugi F, Leon A, Guehenneux F, Champion-Arnaud P, Lahmar M, Schwamborn K (2015) Cell substrates for the production of viral vaccines. Vaccine 33:5905–5912

Benedini LJ, Figueiredo D, Cabrera-Crespo J, Goncalves VM, Silva GG, Campani G, Zangirolami TC, Furlan FF (2020) Modeling and simulation of anion exchange chromatography for purification of proteins in complex mixtures. J Chromatogr A 1613:460685

Biswal JK, Bisht P, Subramaniam S, Ranjan R, Sharma GK, Pattnaik B (2015) Engineering foot-and-mouth disease virus serotype O IND R2/1975 for one-step purification by immobilized metal affinity chromatography. Biologicals 43:390–398

Boix-Besora A, Lorenzo E, Lavado-Garcia J, Godia F, Cervera L (2022) Optimization, production, purification and characterization of HIV-1 GAG-based virus-like particles functionalized with SARS-CoV-2. Vaccine 10:250

Brgles M, Sviben D, Forcic D, Halassy B (2016) Nonspecific native elution of proteins and mumps virus in immunoaffinity chromatography. J Chromatogr A 1447:107–114

Buckland BC (2015) The development and manufacture of influenza vaccines. Hum Vaccin Immunother 11:1357–1360

Burden CS, Jin J, Podgornik A, Bracewell DG (2012) A monolith purification process for virus-like particles from yeast homogenate. J Chromatogr B Analyt Technol Biomed Life Sci 880:82–89

Burgess RR (2018) A brief practical review of size exclusion chromatography: rules of thumb, limitations, and troubleshooting. Protein Expr Purif 150:81–85

Carvalho SB, Fortuna AR, Wolff MW, Peixoto C, Alves MP, Reichl U, Jt Carrondo M (2018) Purification of influenza virus-like particles using sulfated cellulose membrane adsorbers. J Chem Technol Biotechnol 93:1988–1996

Chen GY, Chen CY, Chang MD, Matsuura Y, Hu YC (2009) Concanavalin A affinity chromatography for efficient baculovirus purification. Biotechnol Prog 25:1669–1677

Chen QP, Schure MR, Siepmann JI (2018a) Using molecular simulations to probe pore structures and polymer partitioning in size exclusion chromatography. J Chromatogr A 1573:78–86

Chen YH, Keiser MS, Davidson BL (2018b) Viral vectors for gene transfer. Curr Protoc Mouse Biol 8(e58):1–7

Chen J-R, Liu Y-M, Tseng Y-C, Ma C (2020) Better influenza vaccines: an industry perspective. J Biomed Sci 27:33

Cummins PM, Rochfort KD, O'Connor BF (2017) Ion-exchange chromatography: basic principles and application. Methods Mol Biol 1485:209–223

Curling JM, Gottschalk U (2007) Process chromatography: five decades of innovation. Biopharm Int 20

Dimartino S, Boi C, Sarti GC (2011) A validated model for the simulation of protein purification through affinity membrane chromatography. J Chromatogr A 1218:1677–1690

Eckhardt D, Dieken H, Loewe D, Grein TA, Salzig D, Czermak P (2021) Purification of oncolytic measles virus by cation-exchange chromatography using resin-based stationary phases. Sep Sci Technol 57:886–896

Effio CL, Hubbuch J (2015) Next generation vaccines and vectors: designing downstream processes for recombinant protein-based virus-like particles. Biotechnol J 10:715–727

Effio CL, Hahn T, Seiler J, Oelmeier SA, Asen I, Silberer C, Villain L, Hubbuch J (2016) Modeling and simulation of anion-exchange membrane chromatography for purification of Sf9 insect cell-derived virus-like particles. J Chromatogr A 1429:142–154

Fernandes CS, Goncalves B, Sousa M, Martins DL, Barroso T, Pina AS, Peixoto C, Aguiar-Ricardo A, Roque AC (2015) Biobased monoliths for adenovirus purification. ACS Appl Mater Interfaces 7:6605–6612

Fischer LM, Wolff MW, Reichl U (2018) Purification of cell culture-derived influenza A virus via continuous anion exchange chromatography on monoliths. Vaccine 36:3153–3160

Fortuna AR, Taft F, Villain L, Wolff MW, Reichl U (2018) Optimization of cell culture-derived influenza A virus particles purification using sulfated cellulose membrane adsorbers. Eng Life Sci 18:29–39

Fortuna AR, Van Teeffelen S, Ley A, Fischer LM, Taft F, Genzel Y, Villain L, Wolff MW, Reichl U (2019) Use of sulfated cellulose membrane adsorbers for chromatographic purification of cell cultured-derived influenza A and B viruses. Sep Purif Technol 226:350–358

Fortuna AR, Taft F, Villain L, Wolff MW, Reichl U (2021) Continuous purification of influenza A virus particles using pseudo-affinity membrane chromatography. J Biotechnol 342:139–148

Gagnon P (2012) Purification of biological products by constrained cohydration chromatography. PCT/SG2012/000199

Gagnon P (2013) Steric exclusion chromatography. Genet Eng Biotechnol News 33:30–31

Gagnon P, Cheung CW, Yazaki PJ (2009) Cooperative multimodal retention of IgG, fragments, and aggregates on hydroxyapatite. J Sep Sci 32:3857–3865

Gagnon P, Cheung C-W, Lepin EJ, Wu AM, Sherman MA, Raubitschek AA, Yazaki PJ (2010) Minibodies and multimodal chromatography methods: a convergence of challenge and opportunity. Bioprocess Int 8:26–35

Galeotti N, Hackemann E, Jirasek F, Hasse H (2020) Prediction of the elution profiles of proteins in mixed salt systems in hydrophobic interaction chromatography. Sep Purif Technol 233:116006

Genzel Y (2015) Designing cell lines for viral vaccine production: where do we stand? Biotechnol J 10:728–740

Gerstweiler L, Bi J, Middelberg APJ (2021) Continuous downstream bioprocessing for intensified manufacture of biopharmaceuticals and antibodies. Chem Eng Sci 231:116272

Gerstweiler L, Billakanti J, Bi J, Middelberg APJ (2022) An integrated and continuous downstream process for microbial virus-like particle vaccine biomanufacture. Biotechnol Bioeng 119:2122–2133

Ghosh P, Vahedipour K, Leuthold M, Von Lieres E (2014) Model-based analysis and quantitative prediction of membrane chromatography: extreme scale-up from 0.08 ml to 1200 ml. J Chromatogr A 1332:8–13

Godawat R, Brower K, Jain S, Konstantinov K, Riske F, Warikoo V (2012) Periodic counter-current chromatography -- design and operational considerations for integrated and continuous purification of proteins. Biotechnol J 7:1496–1508

Gottschalk U (2013) The future of downstream processing. Biopharm Int:S39–S45

Granicher G, Babakhani M, Gobel S, Jordan I, Marichal-Gallardo P, Genzel Y, Reichl U (2021) A high cell density perfusion process for modified vaccinia virus Ankara production: process integration with inline DNA digestion and cost analysis. Biotechnol Bioeng 118:4720–4734

Gritti F, Hochstrasser J, Svidrytski A, Hlushkou D, Tallarek U (2020) Morphology-transport relationships in liquid chromatography: application to method development in size exclusion chromatography. J Chromatogr A 1620:460991

Gu T (2015a) Mathematical Modeling and scale-up of liquid chromatography

Gu T (2015b) Modeling and scale-up of size-exclusion chromatography. Mathematical modeling and scale-up of liquid. Chromatography

Gu H, Liu Y, Wang L, Zhang B, Yin D, Zhang Q (2020) Monolithic macroporous hydrogels prepared from oil-in-water high internal phase emulsions for high-efficiency purification of Enterovirus 71. Chem Eng J 401:126051

Guiochon G (2007) Monolithic columns in high-performance liquid chromatography. J Chromatogr A 1168:101–168. discussion 100

Hammerling F, Ladd Effio C, Andris S, Kittelmann J, Hubbuch J (2017) Investigation and prediction of protein precipitation by polyethylene glycol using quantitative structure-activity relationship models. J Biotechnol 241:87–97

Harrison RG, Todd PW, Rudge SR, Petrides DP (2015) Bioseparations science and engineering

Heffron J, Mayer BK (2021) Virus isoelectric point estimation: theories and methods. Appl Environ Microbiol 87:e02319–e02320

Hemminki O, Dos Santos JM, Hemminki A (2020) Oncolytic viruses for cancer immunotherapy. J Hematol Oncol 13(e84):1–15

Hoffmann D, Leber J, Loewe D, Lothert K, Oppermann T, Zitzmann J, Weidner T, Salzig D, Wolff M, Czermak P (2019) Purification of new biologicals using membrane-based processes. Current trends and future developments on (bio-) membranes

Hossienizadeh SMJ, Bagheri M, Alizadeh M, Rahimi M, Azimi SM, Kamalzade M, Es-Haghi A, Ghassempour A (2021) Two dimensional anion exchange-size exclusion chromatography combined with mathematical modeling for downstream processing of foot and mouth disease vaccine. J Chromatogr A 1643:462070

Ingham KC (1978) Precipitation of proteins with polyethylene glycol: characterization of albumin. Arch Biochem Biophys 186:106–113

James KT, Cooney B, Agopsowicz K, Trevors MA, Mohamed A, Stoltz D, Hitt M, Shmulevitz M (2016) Novel high-throughput approach for purification of infectious virions. Sci Rep 6:36826

Jorge S, Dellagostin OA (2017) The development of veterinary vaccines: a review of traditional methods and modern biotechnology approaches. Biotechnol Res Innov 1:6–13

Junter GA, Lebrun L (2020) Polysaccharide-based chromatographic adsorbents for virus purification and viral clearance. J Pharm Anal 10:291–312

Kalbfuss B, Wolff MW, Morenweiser R, Reichl U (2007) Purification of cell culture-derived human influenza A virus by size-exclusion and anion-exchange chromatography. Biotechnol Bioeng 96:932–944

Kalbfuss B, Flockerzi D, Seidel-Morgenstern A, Reichl U (2008) Size-exclusion chromatography as a linear transfer system: purification of human influenza virus as an example. J Chromatogr B Analyt Technol Biomed Life Sci 873:102–112

Kanlaya R, Thongboonkerd V (2016) Cellufine sulfate column chromatography as a simple, rapid, and effective method to purify dengue virus. J Virol Methods 234:174–177

Krajacic M, Ravnikar M, Strancar A, Gutierrez-Aguirre I (2017) Application of monolithic chromatographic supports in virus research. Electrophoresis 38:2827–2836

Kramberger P, Urbas L, Strancar A (2015) Downstream processing and chromatography based analytical methods for production of vaccines, gene therapy vectors, and bacteriophages. Hum Vaccin Immunother 11:1010–1021

Kröber T, Wolff MW, Hundt B, Seidel-Morgenstern A, Reichl U (2013) Continuous purification of influenza virus using simulated moving bed chromatography. J Chromatogr A 1307:99–110

Kuiper M, Sanches RM, Walford JA, Slater NK (2002) Purification of a functional gene therapy vector derived from Moloney murine leukaemia virus using membrane filtration and ceramic hydroxyapatite chromatography. Biotechnol Bioeng 80:445–453

Kurosawa Y, Saito M, Kobayashi S, Okuyama T (2012) Purification of dengue virus particles by one-step ceramic hydroxyapatite chromatography. World J Vaccines 02:155–160

Labisch JJ, Kassar M, Bollmann F, Valentic A, Hubbuch J, Pflanz K (2022) Steric exclusion chromatography of lentiviral vectors using hydrophilic cellulose membranes. J Chromatogr A 1674:463148

Lee J, Gan HT, Latiff SM, Chuah C, Lee WY, Yang YS, Loo B, Ng SK, Gagnon P (2012) Principles and applications of steric exclusion chromatography. J Chromatogr A 1270:162–170

Leung V, Ko F (2011) Biomedical applications of nanofibers. Polym Adv Technol 22:350–365

Liu J, Thorp SC (2002) Cell surface heparan sulfate and its roles in assisting viral infections. Med Res Rev 22:1–25

Loewe D, Dieken H, Grein TA, Weidner T, Salzig D, Czermak P (2020) Opportunities to debottleneck the downstream processing of the oncolytic measles virus. Crit Rev Biotechnol 40:247–264

Lohmann LJ, Strube J (2020) Accelerating biologics manufacturing by modeling: process integration of precipitation in mAb downstream processing. PRO 8

Lothert K, Offersgaard AF, Pihl AF, Mathiesen CK, Jensen TB, Alzua GP, Fahnoe U, Bukh J, Gottwein JM, Wolff MW (2020a) Development of a downstream process for the production of an inactivated whole hepatitis C virus vaccine. Sci Rep 10(e16261):1–13

Lothert K, Pagallies F, Eilts F, Sivanesapillai A, Hardt M, Moebus A, Feger T, Amann R, Wolff MW (2020b) A scalable downstream process for the purification of the cell culture-derived Orf virus for human or veterinary applications. J Biotechnol 323:221–230

Lothert K, Pagallies F, Feger T, Amann R, Wolff MW (2020c) Selection of chromatographic methods for the purification of cell culture-derived Orf virus for its application as a vaccine or viral vector. J Biotechnol 323:62–72

Lothert K, Sprick G, Beyer F, Lauria G, Czermak P, Wolff MW (2020d) Membrane-based steric exclusion chromatography for the purification of a recombinant baculovirus and its application for cell therapy. J Virol Methods 275:113756

Marichal-Gallardo P (2019) Chromatographic purification of biological macromolecules by their capture on hydrophilic surfaces with the aid of non-ionic polymers (PhD thesis). Otto-von-Guericke-Universität

Marichal-Gallardo P, Pieler MM, Wolff MW, Reichl U (2017) Steric exclusion chromatography for purification of cell culture-derived influenza A virus using regenerated cellulose membranes and polyethylene glycol. J Chromatogr A 1483:110–119

Marichal-Gallardo P, Borner K, Pieler MM, Sonntag-Buck V, Obr M, Bejarano D, Wolff MW, Krausslich HG, Reichl U, Grimm D (2021) Single-use capture purification of adeno-associated viral gene transfer vectors by membrane-based steric exclusion chromatography. Hum Gene Ther 32:959–974

Mendes JP, Silva RJS, Berg M, Mathiasson L, Peixoto C, Alves PM, Carrondo MJT (2021) Oncolytic virus purification with periodic counter-current chromatography. Biotechnol Bioeng 118:3522–3532

Mendes JP, Bergman M, Solbrand A, Peixoto C, Carrondo MJT, Silva RJS (2022) Continuous affinity purification of adeno-associated virus using periodic counter-current chromatography. Pharmaceutics 14:1346

Migliorini C, Gentilini A, Mazzotti M, Morbidelli M (1999) Design of simulated moving bed units under nonideal conditions. Ind Eng Chem Res 38:2400–2410

Mittal M, Banerjee M, Lua LHL, Rathore AS (2021) Current status and future challenges in transitioning to continuous bioprocessing of virus-like particles. J Chem Technol Biotechnol 97:2376–2385

Muca R, Marek W, Piatkowski W, Antos D (2010) Influence of the sample-solvent on protein retention, mass transfer and unfolding kinetics in hydrophobic interaction chromatography. J Chromatogr A 1217:2812–2820

Nasimuzzaman M, Lynn D, Van Der Loo JC, Malik P (2016) Purification of baculovirus vectors using heparin affinity chromatography. Mol Ther Methods Clin Dev 3:16071

Nestola P, Silva RJ, Peixoto C, Alves PM, Carrondo MJ, Mota JP (2014) Adenovirus purification by two-column, size-exclusion, simulated countercurrent chromatography. J Chromatogr A 1347:111–121

Nestola P, Peixoto C, Silva RRJS, Alves PM, Mota JPB, Carrondo MJT (2015a) Improved virus purification processes for vaccines and gene therapy. Biotechnol Bioeng 112:843–857

Nestola P, Silva RJ, Peixoto C, Alves PM, Carrondo MJ, Mota JP (2015b) Robust design of adenovirus purification by two-column, simulated moving-bed, size-exclusion chromatography. J Biotechnol 213:109–119

Neubauer P, Cruz N, Glauche F, Junne S, Knepper A, Raven M (2013) Consistent development of bioprocesses from microliter cultures to the industrial scale. Eng Life Sci 13:224–238

Olgun HB, Tasyurek HM, Sanlioglu AD, Sanlioglu S (2019) High-grade purification of third-generation HIV-based lentiviral vectors by anion exchange chromatography for experimental gene and stem cell therapy applications. Methods Mol Biol 1879:347–365
Opitz L, Salaklang J, Buttner H, Reichl U, Wolff MW (2007) Lectin-affinity chromatography for downstream processing of MDCK cell culture derived human influenza A viruses. Vaccine 25: 939–947
Opitz L, Hohlweg J, Reichl U, Wolff MW (2009a) Purification of cell culture-derived influenza virus A/Puerto Rico/8/34 by membrane-based immobilized metal affinity chromatography. J Virol Methods 161:312–316
Opitz L, Lehmann S, Reichl U, Wolff MW (2009b) Sulfated membrane adsorbers for economic pseudo-affinity capture of influenza virus particles. Biotechnol Bioeng 103:1144–1154
Pato TP, Souza MCO, Mattos DA, Caride E, Ferreira DF, Gaspar LP, Freire MS, Castilho LR (2019) Purification of yellow fever virus produced in Vero cells for inactivated vaccine manufacture. Vaccine 37:3214–3220
Peixoto C, Sousa MF, Silva AC, Carrondo MJ, Alves PM (2007) Downstream processing of triple layered rotavirus like particles. J Biotechnol 127:452–461
Pereira Aguilar P, Reiter K, Wetter V, Steppert P, Maresch D, Ling WL, Satzer P, Jungbauer A (2020) Capture and purification of human immunodeficiency virus-1 virus-like particles: convective media vs porous beads. J Chromatogr A 1627:461378
Pfister D, David L, Holzer M, Nicoud RM (2017) Designing affinity chromatographic processes for the capture of antibodies. Part I: A simplified approach. J Chromatogr A 1494:27–39
Polson A (1977) A theory for the displacement of proteins and viruses with polyethylene glycol. Prep Biochem 7:129–154
Rajamanickam V, Herwig C, Spadiut O (2015) Monoliths in bioprocess technology. Chromatography 2:195–212
Rischawy F, Saleh D, Hahn T, Oelmeier S, Spitz J, Kluters S (2019) Good modeling practice for industrial chromatography: mechanistic modeling of ion exchange chromatography of a bispecific antibody. Comput Chem Eng 130:106532
Roque ACA, Pina AS, Azevedo AM, Aires-Barros R, Jungbauer A, Di Profio G, Heng JYY, Haigh J, Ottens M (2020) Anything but conventional chromatography approaches in bioseparation. Biotechnol J 15:e1900274
Ruscic J, Perry C, Mukhopadhyay T, Takeuchi Y, Bracewell DG (2019) Lentiviral vector purification using nanofiber ion-exchange chromatography. Mol Ther Methods Clin Dev 15:52–62
Sanchez-Trasvina C, Fuks P, Mushagasha C, Kimerer L, Mayolo-Deloisa K, Rito-Palomares M, Carta G (2020) Structure and functional properties of Capto Core 700 core-shell particles. J Chromatogr A 1621:461079
Schagen FH, Rademaker HJ, Rabelink MJ, Van Ormondt H, Fallaux FJ, Van Der Eb AJ, Hoeben RC (2000) Ammonium sulphate precipitation of recombinant adenovirus from culture medium: an easy method to increase the total virus yield. Gene Ther 7:1570–1574
Schmidt M, Hafner M, Frech C (2014) Modeling of salt and pH gradient elution in ion-exchange chromatography. J Sep Sci 37:5–13
Shekhawat LK, Rathore AS (2019) An overview of mechanistic modeling of liquid chromatography. Prep Biochem Biotechnol 49:623–638
Silva RJS, Mendes JP, Carrondo MJT, Marques PM, Peixoto C (2020) Continuous chromatography purification of virus-based biopharmaceuticals: a shortcut design method. Methods Mol Biol 2095:367–384
Simon U, Dimartino S (2019) Direct 3D printing of monolithic ion exchange adsorbers. J Chromatogr A 1587:119–128
Steppert P, Mosor M, Stanek L, Burgstaller D, Palmberger D, Preinsperger S, Pereira Aguilar P, Mullner M, Csar P, Jungbauer A (2022) A scalable, integrated downstream process for production of a recombinant measles virus-vectored vaccine. Vaccine 40:1323–1333
Striegel AM (2017) Size-exclusion chromatography. Liquid chromatography

Sviben D, Forcic D, Ivancic-Jelecki J, Halassy B, Brgles M (2017) Recovery of infective virus particles in ion-exchange and hydrophobic interaction monolith chromatography is influenced by particle charge and total-to-infective particle ratio. J Chromatogr B Analyt Technol Biomed Life Sci 1054:10–19

Tomaz CT (2017) Hydrophobic interaction chromatography. Liquid chromatography

Trabelsi K, Ben Zakour M, Kallel H (2019) Purification of rabies virus produced in Vero cells grown in serum free medium. Vaccine 37:7052–7060

Tseng YF, Weng TC, Lai CC, Chen PL, Lee MS, Hu AY (2018) A fast and efficient purification platform for cell-based influenza viruses by flow-through chromatography. Vaccine 36:3146–3152

Turnbull J, Wright B, Green NK, Tarrant R, Roberts I, Hardick O, Bracewell DG (2019) Adenovirus 5 recovery using nanofiber ion-exchange adsorbents. Biotechnol Bioeng 116:1698–1709

Vailaya A, Horváth C (1996) Retention thermodynamics in hydrophobic interaction chromatography. Ind Eng Chem Res 35:2964–2981

Wang F, Min Y, Geng X (2012) Fast separations of intact proteins by liquid chromatography. J Sep Sci 35:3033–3045

Wang Q, Lock M, Prongay AJ, Alvira MR, Petkov B, Wilson JM (2015) Identification of an adeno-associated virus binding epitope for AVB sepharose affinity resin. Mol Ther Methods Clin Dev 2:15040

Wang G, Hahn T, Hubbuch J (2016) Water on hydrophobic surfaces: mechanistic modeling of hydrophobic interaction chromatography. J Chromatogr A 1465:71–78

Weigel T, Soliman R, Wolff MW, Reichl U (2019) Hydrophobic-interaction chromatography for purification of influenza A and B virus. J Chromatogr B Analyt Technol Biomed Life Sci 1117: 103–117

Wolff MW, Reichl U (2008) Downstream processing: from egg to cell culture-derived influenza virus particles. Chem Eng Technol 31:846–857

Wolff MW, Reichl U (2011) Downstream processing of cell culture-derived virus particles. Expert Rev Vaccines 10:1451–1475

Wolff MW, Siewert C, Hansen SP, Faber R, Reichl U (2010a) Purification of cell culture-derived modified vaccinia Ankara virus by pseudo-affinity membrane adsorbers and hydrophobic interaction chromatography. Biotechnol Bioeng 107:312–320

Wolff MW, Siewert C, Lehmann S, Hansen SP, Djurup R, Faber R, Reichl U (2010b) Capturing of cell culture-derived modified vaccinia Ankara virus by ion exchange and pseudo-affinity membrane adsorbers. Biotechnol Bioeng 105:761–769

Yoon K, Hsiao BS, Chu B (2008) Functional nanofibers for environmental applications. J Mater Chem 18:5326–5334

Zelic B, Nesek B (2006) Mathematical Modeling of size exclusion chromatography. Eng Life Sci 6: 163–169

Zhang K, Liu X (2016) Reprint of "Mixed-mode chromatography in pharmaceutical and biopharmaceutical applications". J Pharm Biomed Anal 130:19–34

Zhang B-T, Liu H, Liu Y, Teng Y (2020) Application trends of nanofibers in analytical chemistry. TrAC – Trends Anal Chem 131:115992

Zhao M, Vandersluis M, Stout J, Haupts U, Sanders M, Jacquemart R (2019) Affinity chromatography for vaccines manufacturing: finally ready for prime time? Vaccine 37:5491–5503

Zobel-Roos S, Stein D, Strube J (2018) Evaluation of continuous membrane chromatography concepts with an enhanced process simulation approach. Antibodies 7:13

Chapter 8
Purifying Viral Vectors: A Review of Chromatography Solutions

Louis Crowley, Jennifer J. Labisch, Maja Leskovec, Mojca Tajnik Sbaizero, Katy Mclaughlin, Piergiuseppe Nestola, and Amelie Boulais

8.1 Introduction

Downstream purification steps are typically among the most challenging in a bioprocess. Ongoing innovations in the biopharmaceutical industry deliver diverse modalities and increased titers, while stiff competition demands speed and productivity be at the forefront for any bioprocess. A 2022 survey of biotherapeutic developers and CMOs, viral and gene therapies were identified as the second most important biomanufacturing trend or operational area on which the industry must focus its efforts (BioPlan Associates Inc 2022). These findings demonstrate the spotlight that is currently on viral vectors as biotherapeutic tools, as well as the demand for increased efficiency across downstream processes.

Louis Crowley, Jennifer J. Labisch, Maja Leskovec and Mojca Tajnik Sbaizero contributed equally with all other contributors.

L. Crowley
Sartorius North America Inc, Bohemia, NY, USA

J. J. Labisch
Sartorius Lab Instruments GmbH & Co. KG, Göttingen, Germany

M. Leskovec · M. Tajnik Sbaizero
Sartorius BIA Separations d.o.o., Ajdovscina, Slovenia

K. Mclaughlin
Sartorius Stedim UK Ltd., Epsom, Surrey, UK

P. Nestola
Sartorius Stedim Switzerland AG, Tagelswangen, Switzerland

A. Boulais (✉)
Sartorius Stedim France FMT, Aubagne, France
e-mail: Amelie.Boulais@Sartorius.com

S. Gautam et al. (eds.), *Bioprocess and Analytics Development for Virus-based Advanced Therapeutics and Medicinal Products (ATMPs)*,
https://doi.org/10.1007/978-3-031-28489-2_8

More viral vector therapies continue to gain their way toward FDA approval. At the time of writing, 8 have been approved by the FDA, 25 are in late-stage development, and 120 are in Phase Two clinical trials (U.S. Food and Drug Administration (FDA) 2021). A wide range of viral vectors are being investigated as potential vectors for gene transfer. Different viruses have different biological properties, making them suitable for distinct gene therapy applications (Bulcha et al. 2021). Adeno-associated viruses (AAVs) and lentiviruses (LVs) are among the most commonly used viral vectors and will be the focus of this chapter.

Alongside the diversification of viral-based therapies are the ever-growing improvements in upstream cell culture technologies, generating increased cell densities and higher viral titers (Vellinga et al. 2014). As a result, there are constant demands on the downstream process to maximize productivity, shifting the production bottleneck to viral vector purification steps.

Compared to manufacturing protein-based therapeutics, developing a robust downstream process for viral vector production requires a special set of considerations. Viral particles are typically larger than proteins and may be more sensitive to shear stress, requiring developers to explore new chromatography matrices. Viral vectors are also highly diverse, making developing a platform approach for their purification complex. Finally, scalability remains a significant challenge; developing production approaches that support the efficient purification of large quantities of viral vectors that are not cost-prohibitive can often be a challenge.

8.1.1 Elements to Consider During Viral Vector Purification

8.1.1.1 Diverse Features

Across different viral vector types, a wide array of physical characteristics exist, including differences in isoelectric point, surface hydrophobicity, enveloped or non-enveloped, hydrodynamic diameter, and lability (Braas et al. 1996) and thus, each viral vector may be different based on the components of the viral surface. For instance, AAV vector therapies within research, preclinical and clinical studies currently consist of about 15 commonly used serotypes (AAV1, 2, 3b, 4-11, MYO, 7m8, DJ, and DJ/8) (Saksule 2022). These factors illustrate just one aspect of the overall complexity that must be examined when developing a downstream purification process.

8.1.1.2 Impurity Removal

Several factors relating to impurities found within the process must be considered in a viral vector purification scheme. These include protein-related impurities, process-related impurities, and contaminants. If the viral vector is intracellular (such as AAVs), the first step involves cellular lysis, generating a crude lysate consisting of numerous impurities. In fact, viral vectors may represent only 0.1% of the total proteins and nucleic acids produced by HEK293 cells (Goldfarb and Terova 2020). This has significant implications for downstream purification processes.

Product-related impurities include product variants that arise during manufacture or storage that have undesirable properties (e.g., aggregates, fragments, oxidized or deamidated species, and misfolded, mispacked, or empty viral particles). Process-related impurities are derived from the manufacturing process (e.g., reagents, starting materials, leachates). Contaminants are materials not intended to be part of the manufacturing process (e.g., adventitious viral and microbial agents, endotoxins, and mycoplasma).

Finally, unlike other modalities, AAVs pose a unique impurity profile challenge: separating full capsids (containing the proper genetic information) and empty capsids. The clearance of these empty capsids is of critical importance within a purification process, as they have been shown to pose a significant health risk in both animals and humans through undesired immunological responses even at low doses (Shirley et al. 2020; Dickerson et al. 2021).

Typically, a secondary chromatography step is employed to remove empty capsids. In this so-called polishing step, high resolution is as important as high dynamic binding capacity.

While the capsid structure is largely identical between empty and full particles, there is a difference in the isoelectric point, creating a charge differential of 0.4–0.5 pH units that can be exploited using charge-based chromatography.

8.1.1.3 Scalability

Purification strategies have been well established to satisfy all factors described toward removing process-related impurities, product-related impurities, and empty capsids (Burova and Ioffe 2005; Qu et al. 2007; Ayuso et al. 2010a, b; Burnham et al. 2015). These methods traditionally involved the utilization of density gradient centrifugation with cesium chloride. As viral vector gene-based therapy interest grows and FDA approvals expand, the demand for larger-scale solutions increases. The scalability of a purification scheme involving centrifugation is met with limitations in terms of operation, safety, and cost. To be incorporated into a manufacturing process, ultracentrifugation would be implemented at running at high speeds and long processing times. In addition, this would require the incorporation of cesium chloride to generate the proper density gradient. This dramatically increases the overall cost and safety at larger scales due to raw material cost and cesium chloride toxicity. While density gradient centrifugation is a viable technique for purifying viral vectors, it is typically not used in commercial manufacturing owing to its complexity and lack of scalability. Cesium chloride density gradients are most valuable in research and process development applications (Liddell 2022).

In this chapter, we introduce traditional and emerging technologies for the chromatographic separation of viral vectors before discussing—in detail—the value and potential drawbacks of each.

8.2 Chromatographic Matrices

There are four basic chromatography matrices with distinct features supporting different purification needs (Fig. 8.1).

8.2.1 Resins

The classical chromatography matrix is resins in a packed-bed column. Resins are spherical particles (beads) with a defined diameter, pore size, and pore size distribution (Barut et al. 2005). A wide variety of resin materials, such as agarose, dextran, cellulose, polystyrene, and polyacrylamide, are available (Jungbauer 2005). Mass transfer is based on diffusion into the pores of the resin, where a large proportion of the functional groups are located. As the mass transfer is diffusion-based, slow flow rates are applied when working with beads. Due to the small pore size of the beads, resin chromatography is typically used to purify small molecules. Large molecules, such as enveloped-viral vectors, will only reach a fraction of the binding sites (Perry and Rayat 2021).

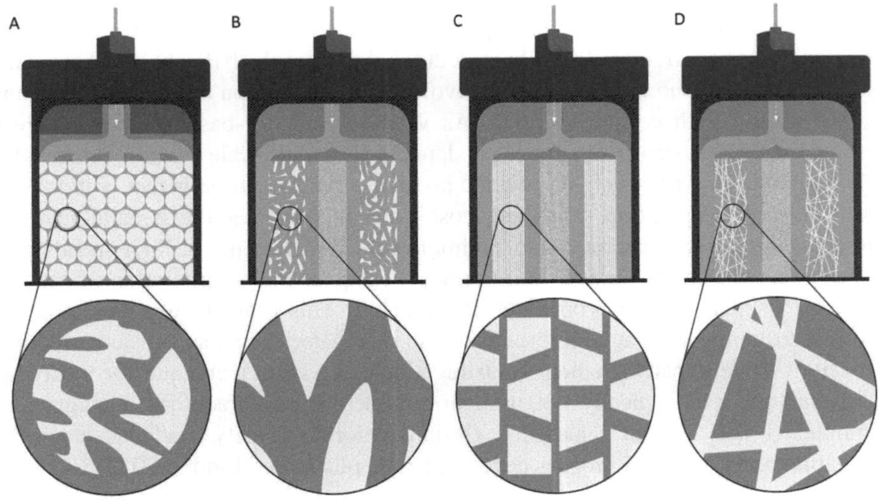

Fig. 8.1 Schematic structure of chromatography matrices. The chromatography matrix is shown in light gray. Matrices shown are resins (**a**), monoliths (**b**), membranes (**c**), and nanofibers (**d**)

8.2.2 Monoliths

Monoliths are a continuous block of polymer with a network of large, interconnected channels. The functional groups for chromatography are located on the surface of the walls of these connected channels. With their wide channels and no pores in the matrix, mass transport in monoliths is achieved by convection and not limited by diffusion, resulting in low back pressure and a high mass transfer (Barut et al. 2005; Jungbauer 2005). The large channel diameter is beneficial for the separation of large biomolecules and, therefore, particularly applicable for virus purification (Krajacic et al. 2017). Monolith columns are packed as ready to use, and do not require packing validation. Operation mode can be either single-use or multi-use, as monolith and housing can withstand several cleaning cycles.

8.2.3 Membranes

Membranes are polymers of different materials, like polyacrylamide hydrogel, modified polyethersulfone, or regenerated cellulose (Orr et al. 2013). They have large pore sizes, and the functional groups are located on the surface of the pores. Layered stacks of porous membranes create a column of around a few millimeters in thickness. One of the main characteristics of membrane chromatography is the very fast mass transfer due to convective mass transport and a low-pressure drop across the membrane (Barut et al. 2005; Zydney 2021). Due to the fast mass transport, the required residence time is short, and high flow rates can be achieved. Membrane chromatography can be operated in a single-use mode, eliminating the need for column packing operations, cleaning, regeneration, and validation, reducing labor and costs (Orr et al. 2013; Zydney 2021).

8.2.4 Nanofibers

Membrane nanofibers can be synthesized from different materials, like nylon, glass, or cellulose, randomly arranged to create a consistent stationary phase. The nanofiber stationary phase has a large surface that is entirely functionalized and has a convective mass transport (Turnbull et al. 2019). Nanofibers have been used to successfully purify lentiviral vectors (Ruscic et al. 2019) and adenoviral vectors (Turnbull et al. 2019) by anion-exchange chromatography (AEX).

The following sections will discuss the application of resins, monoliths, and membranes in viral vector purification.

8.3 Resins for the Purification of Viral Vectors

Downstream facilities worldwide, from process development to manufacturing, are well-equipped for modalities such as recombinant enzymes and monoclonal antibodies. A variety of chromatography-based equipment and resins have met regulatory requirements for various therapeutics. Therefore, resin-based chromatographic methodologies for the purification of adenovirus (AV), adeno-associated virus (AAV), and lentivirus (LV) have been developed to provide a scalable and robust process from small-scale process development to manufacturing. While resin-based chromatography is very well understood, the unique features of viral vectors challenge their separation capabilities.

Examining the physical properties of viral vectors, the property of size becomes an important aspect (Table 8.1). The smallest viral vector is AAV, with an average diameter of 20–26 nm. Most common resins possess pore sizes of 20–40 nm, meaning binding of even the smallest viral vectors would largely be restricted to the surface area of the resin beads, offering significantly reduced dynamic binding capacity, reduced flow rates, poor mass transfer, and an increase in pressure (Nestola et al. 2015). Therefore, resin-based chromatography is predominantly used in the purification of AAV and not for LV or AV.

A second issue is viral vector diversity. For instance, more than 100 AAV serotypes, hybrid constructs, or variants have been reported in the literature (Gao et al. 2004; Zincarelli et al. 2008; Schmit et al. 2020). Therefore, unlike Protein A-mediated purification of monoclonal antibodies, there is currently not a one-size-fits-all platform approach for purifying the wide range of AVs, AAVs, or LVs. Batch-to-batch variations within each AAV type change the production ratio of empty capsids versus full AAV particles, meaning the same purification process is not always applicable (Saksule 2022).

8.3.1 *Affinity Chromatography*

In the biopharmaceutical industry, a platform-based approach for downstream operations is essential for commercial feasibility and to supply a product that meets all regulatory requirements. As previously mentioned, methodologies involving density gradient centrifugation have been implemented as a purification strategy for viral

Table 8.1 Physical characteristics of viral vectors

Type	Diameter (nm)	Genome	Exogenous gene	Enveloped vs. non
AAV	20–26	ssDNA	5 kb	Non
AV	70–100	dsDNA	>8 kb	Non
LV	80–120	RNA	8–10 kb	Enveloped

vector production. This approach has been used in parallel with various chromatographic processes, including ion exchange, heparin affinity, hydrophobic interaction, and immobilized metal affinity chromatography (Smith et al. 2009; Lock et al. 2010; Vance et al. 2015; Clément and Grieger 2016).

A highly selective affinity-based capture step is typically necessary for establishing a truly platform-based approach for downstream operations. Ideally, this should provide high specificity separation of crude harvest material and recover the products at high yield and purity in a single step. An affinity capture step is typically followed by polishing steps to remove residual impurities such as DNA, leachables, and host cell proteins (HCPs). Implementing an affinity-based capture step affords numerous advantages:

- Simplified process development efforts
- Reduced number of process steps
- Decreased processing time
- Enhanced process robustness
- Increased product yields
- Improved scalability from process development into tech transfer for manufacturing

Additionally, while the overall cost of goods (COGs) appears to be a limiting factor in affinity-based resins due to the matrix price, it has been established that they can reduce the COGs by 1.5-fold (compared to a non-affinity process) (Łącki and Riske 2020). Affinity resins are not widely used in LV manufacturing processes—aside from a few research examples—because the desorption reagents have been shown to reduce LV infectivity (Yu and Schaffer 2006; Segura et al. 2007; Chen et al. 2010; Mekkaoui et al. 2018; Zhao et al. 2019).

8.3.1.1 Affinity Ligands

Affinity ligands have been developed for AAV using domain antibody fragments found to bind AAV capsid proteins. Three such resins include AAV8, AAV9, and AAVX from Thermo Fisher, which possess affinity toward AAV8, AAV9, and, in the case of AAVX, affinity toward a broad range of serotypes (1–9 as well as recombinant and chimeric serotypes tested). Cytiva has also produced AVB resin compatible with the purification of multiple AAV serotypes, including AAV1, AAV2, AAV5, AAV6, and AAVrh10.

Viral vector size presents a significant issue in resin chromatography. However, AAV8, AAV9, and AAVX resins utilize cross-linked polystyrene-divinylbenzene POROS™ beads with a Capture Select AAVX affinity ligand, a large throughpore matrix that provides a greater dynamic binding capacity as viral vectors can gain access to the interior bead pores.

AVB resin is based on a Capto high-flow agarose base matrix which has a rigid nature capable of tolerating increased pressures. For efficient viral vector binding, the ligand is attached to the base matrix by a hydrophilic spacer arm with high binding specificity (Oranje et al. 2004). The structural attributes and average resin particle size of 50 μm contribute to increased mass transfer, dynamic binding capacity, and resolution compared to traditional bead-based resins. These features are achieved at flow rates more amenable to production timelines (Nass et al. 2018; Toueille et al. 2018; Goldfarb and Terova 2020).

In some studies, binding capacities of $>1 \times 10^{13}$ viral genome per mL were obtained using AAV9 resins, with recovery rates of ≥70% and upward of 95% for both natural and synthetic peptides (Terova et al. 2017, 2018). Thermo Fisher claims a dynamic binding capacity of $1 \times 10^{13} - 1 \times 10^{14}$ vg/mL, which is in good agreement with published data. Studies have shown that AAVX resins possess a greater binding affinity toward AAV8 than AAV8 resin itself (Ravault and Laroudie 2022). Residence times had no significant impact on the binding capacity while providing recoveries close to 100% with scalability upward of 20-fold (Ravault and Laroudie 2022).

Regarding potential impurities, studies have demonstrated that DNA and other contaminants are largely isolated to the flowthrough when viral vector affinity resins are employed (Rieser et al. 2021). Thermo Fisher AAVX resins can clear four contaminating viruses to acceptable log reduction levels and two to moderate levels.

While affinity resins have improved viral vector production processes, the number of recombinant viral vector serotypes and chimeras is also expanding, keeping pace with the increased interest in gene therapy. New recombinant serotypes and variants may not possess the same binding affinity as currently available affinity resins.

A potential solution to this issue lies within custom affinity ligands. There have been several custom-designed affinity ligands for modalities such as recombinant enzymes and proteins from companies including Affilogics, Astrea, Avitide, Prometics, Navigo, and TaG CxY. However, developing ligands with high specificity toward the molecule of interest is time-consuming, taking an average of 3–6 months.

Additionally, while acid-stable, the overall caustic tolerance of affinity ligands is limited, and cleaning procedures may involve a wide array of acids, denaturants, and alcohols. Manufacturers of viral vector affinity resins recommend sodium hydroxide concentrations to be within the range of only 10–25 mM. This can lead to inefficient cleaning and reduced resin lifetime. However, when developing custom affinity ligands, the resin matrix backbone can be selected to be more amenable to production and enhanced resin lifetime. A proper COGs evaluation and risk assessment should be performed to measure the viability of such a technique and its affordability for smaller start-up organizations, larger CDMOs, and manufacturing facilities.

8.3.2 Anion-Exchange Chromatography

Affinity chromatography cannot fully remove impurities such as DNA. Therefore, a final polishing step is usually required during viral vector purification. More

Table 8.2 Characteristics of anion exchange resins

Resin	Type of AEX resin	Pore size (Angstrom)
D50	Weak	1100
PI	Weak	2000
HQ50	Strong	2000
XQ	Strong	1100

importantly, affinity resins lack a fundamental characteristic in the purification of viral vectors: their inability to separate full and empty capsids. The distinct isoelectric points of empty and full capsids are exploited to enrich the sample for full viral particles using anion exchange (Urabe et al. 2006; Qu et al. 2007, 2015; Lock et al. 2012; Venkatakrishnan et al. 2013; Joshi et al. 2021).

Currently, DEAE and Q resins are the most widely utilized anion-exchange ligands within the industry. They are typically operated in a bind and elute mode to separate full versus empty capsids and perform the final removal of DNA and any other impurities that remain from the initial purification steps. HIC resins are ineffective as proper desorption requires high kosmotropic salt, which can impact the overall integrity of the viral capsid (Rieser et al. 2021). It should be noted here that while widely utilized, care should be taken with Q-based anion-exchange resins as viral binding may be inhibited at higher NaCl concentrations.

Traditional anion-exchange resins utilized in viral vector purification include Fractogel EMD DEAE-650 (EMD Millipore) (Kamen and Henry 2004). Streamline Q XL (Cytiva) (Peixoto et al. 2006), Q Sepharose XL (Cytiva) (Qu et al. 2007), and Q Sepharose FF (Rodrigues et al. 2006). Published data indicates that elution requires a long shallow gradient (which may be in the range of 60–100 column volumes at low flow rates) due to the marginal difference in isoelectric points between empty and full capsids (Lock et al. 2015, 2016). Considering the operational flow rates of anion-exchange resins, this proves to be a highly time-consuming process.

To address these challenges utilizing resin chromatography, studies purifying plasmid DNA for viral vectors have used Thermo Fisher POROS™ anion-exchange resins, taking advantage of the higher throughpore structure (Becerra and Buyel 2021). In one study, four anion-exchange resins consisting of two weak and two strong anion-exchangers were evaluated, each with pore sizes more amenable to viral vector binding (Table 8.2).

D50 resin is comparable to DEAE resin as it possesses a dimethylethyaminopropyl functional group with slightly different chemistry. HQ50 resin serves as a strong anion-exchanger as it contains primary, secondary, tertiary, and quaternary amines. Combined with increased porosity, these chemistries may represent a viable alternative to traditional anion-exchange resins to achieve a higher mass transfer, performance at higher flow rates, and reduced pressure, allowing for the efficient separation of full versus empty capsids.

8.4 Membrane Chromatography for Viral Vector Purification

The most commonly employed membrane-based chromatography mode is ion-exchange chromatography. Membranes are functionalized with quaternary amines (Q), resulting in strong anion-exchangers or diethylaminoethyl (DEAE) ligands. Cation exchange membranes are functionalized with, for example, sulfonic acid (S) groups (Ladisch 2001). Commercial ion exchange membranes are available from Sartorius (Sartobind® Q, D, and S), Pall (Mustang® Q and S), and Merck Millipore (Natrix®), among others.

Q functionalized membranes, particularly Mustang® Q and Sartobind® Q membranes, are frequently used in LV applications (Table 8.3). Membrane chromatography for AAVs has been performed with anion and cation exchange chromatography, hydrophobic interaction chromatography, and steric exclusion chromatography (Table 8.4).

8.4.1 Membrane Module Design

Small-scale ion-exchange membrane chromatography units consist of several stacked membrane layers assembled in their housing so that the flow is directed frontally from above. This results in a dead-end flow (Orr et al. 2013). For mid-scale and large-scale membrane chromatography, capsules are used. In Sartobind® capsules, the membrane is rolled up. Within these capsules, the feed flows radially from the external fluid channel through the membrane to the internal fluid channel. Mustang® Q XT140 and XT5000 capsules have the membrane incorporated in a pleating technology (Mustang Q XT chromatography capsules 2009).

Sartorius enabled further scaling capabilities in the Sartobind® cassette system, which consists of a modular concept and achieves the same flow pattern as the capsules. Two membrane stacks are incorporated into a cassette with a central inlet flow channel. The fluid flows from the internal flow channel perpendicular to the external flow channel and then to the outlet. These cassettes can be assembled in a pilot filter holder or process holder, resulting in a membrane volume of 20.8 L and ~100 L, respectively (Fischer-Frühholz et al. 2018).

The membrane can be thought of as a very flat column with a low bed height (the height of the superimposed membrane layers). For instance, for Sartobind®, the membrane layers are 4 or 8 mm. During scale-up, the bed height is kept constant while the volume of the membrane is increased by, for example, making the capsule longer and wider. As a result, the flow rate is scaled up proportionally to the membrane volume. The buffer volume of each step is usually expressed in membrane volumes and remains the same during scale-up. Therefore, the actual buffer volume scales proportionally with the membrane volume. Similarly, the load of viral particles per membrane volume is kept constant during scale-up.

Table 8.3 Membrane chromatography purification strategies for lentiviral vectors (LVs)

Purification method	Viral vector, envelope protein	Stationary phase	Vector recovery (%)	Impurity clearance	References
AEX	LV, VSV-G	Mustang Q, quaternary amine	Final titer after whole DSP process 10^8–10^9 TU/mL	n.a.	Tinch et al. (2019)
AEX	LV, VSV-G	Mustang Q, quaternary amine	56% [a]	n.a.	Kutner et al. (2009)
AEX	LV, VSV-G	Sartobind Q, quaternary amine	33.1%	DNA and protein clearance 95–100% [b]	Valkama et al. (2020)
AEX	LV, VSV-G	Sartobind D, DEAE	29%	>99% protein and 95% dsDNA removal	Lesch et al. (2011)
AEX	LV, VSV-G	Sartobind D, DEAE	28 ± 4%	n.a.	Bandeira et al. (2012)
AEX	LV, VSV-G	LentiSELECT40, LentiSELECT500, LentiSELECT1000 (Contains Sartobind Q)	55.5 ± 3.4%, 39.6 ± 5.5%, 43.6 ± 9.4%	n.a.	Zimmermann et al. (2011)
AEX	LV, VSV-G	Mustang Q, Sartobind Q, quaternary amine	90%, 63%	DNA removal >70%, protein removal 99% for Mustang Q	Moreira et al. (2021)
AEX	LV, measles virus H and F glycoproteins	Mustang Q, quaternary amine	80%, 65%	n.a.	Marino et al. (2015)
AEX	LV, VSV-G, and BaEV-R-less	Mustang Q, quaternary amine	55 ± 14% [c] 33 ± 2.9% [c]	n.a.	Bauler et al. (2020)
AEX	LV, VSV-G	Mustang Q, quaternary amine	29.6–33.4% [c]	n.a.	
AEX	LV, VSV-G	Mustang Q, quaternary amine	>60% [b]	n.a.	Slepushkin et al. (2003)
SXC	LV, VSV-G	Hydrosart 10242, Hydrophilic stabilized cellulose membrane	86%	80% total protein and	Labisch et al. (2022)

(continued)

Table 8.3 (continued)

Purification method	Viral vector, envelope protein	Stationary phase	Vector recovery (%)	Impurity clearance	References
				dsDNA removal	

n.a., data not available/not mentioned
[a]Calculated from text or table
[b]Obtained or calculated from a graph
[c]Recovery of several downstream processing steps, including chromatography

Table 8.4 Membrane chromatography purification strategies for adeno-associated vectors (AAVs)

Purification method	Viral vector, envelope protein	Stationary phase	Vector recovery (%)	Impurity clearance	Refs
SXC	AAV2, AAV6, AAV-1P5, and AAV-9A2	Whatman regenerated cellulose membrane	126.1 ± 10.9%	Total protein 80–85%, DNA 94–98%	Marichal-Gallardo et al. (2021)
CEX, AEX	AAV8	Mustang S, Mustang Q, sulfonic acid, quaternary amine	49 ± 14.42%[a]	n.a.	Okada et al. (2009)
CEX, AEX	AAV2/8, AAV5	Mustang S, Mustang Q, sulfonic acid, quaternary amine	41%, 67%	80%[b]	Davidoff et al. (2004)
HIC	AAV1, AAV-MutC, AAV5, AAV8	Sartobind Phenyl Nano	87.5%, 89.6%, 76.8%, 101.4%, (93.6%, 97.3%, 81.2%, 91.7%)[c]	>90% (HCP), >80% dsDNA	McNally et al. (2020)
AEX	AAV5	Mustang Q	60% of full capsids and 13% of total capsids	< 0.2 ng/mL of DNA and < 10 ng/mL of HCP (final concentrations after AEX)	Hejmowski et al. (2022)

n.a., data not available/not mentioned
[a]Calculated from text or table
[b]Based on silver-stained SDS PAGE gel
[c]Values without brackets are for clarified media supernatant, values in brackets are for lysed and clarified material, Order corresponds to the listing of serotypes

It is possible to transfer a resin-based purification process to a membrane-based purification process. For instance, a 2019 publication described the intensification of an AV anion-exchange chromatography purification process by replacing the resin-based column with a membrane adsorber (Abrecht et al. 2019). As well as a capacity increase of over tenfold, significant time and cost savings were achieved. The smaller facility footprint and higher operating flow rates of membrane adsorbers reduced buffer consumption by 58% and reduced processing time. The requirement for column-packing, validation, and cleaning is also eliminated, further speeding up the process (Abrecht et al. 2019).

8.4.2 LV Vector Purification with Ion-Exchange Membrane Chromatography

Typically, clarified viral vector material is used as a starting material for LV purification with IEX (Table 8.3), although in a large-scale LV purification process, 11 kg of concentrated material (after TFF) was used as the starting material (Valkama et al. 2020). Large-scale LV downstream processing protocols use Q or DEAE AEX chromatography after clarification (Merten et al. 2016). Some protocols include adjusting the starting material with NaCl to reduce protein binding to the membrane, increasing membrane capacity, maximizing final purity (McNally et al. 2014; Valkama et al. 2020), and reducing the amount of DNA in the eluate (McNally et al. 2014). Elution is carried out by increasing the salt concentration, which typically falls within 0.4 and 2 M NaCl. This can be performed using an isocratic, stepwise, or linear gradient.

Enveloped viral vectors, like LVs, are susceptible to high salt concentrations, leading to a loss of infectivity (Zimmermann et al. 2011; Labisch et al. 2021). To maintain infectivity, the salt concentration of the LV sample is often immediately reduced by dilution or an ultrafiltration and diafiltration step (Bandeira et al. 2012; Tinch et al. 2019; Bauler et al. 2020; Moreira et al. 2021). Depending on the purification mode and viral vector, pre- and post-treatment of chromatography runs may be required. These may include adjusting conductivity for optimal binding to a stationary phase or diluting fractions to reduce salt concentrations to preserve vector infectivity. Pre- and post-chromatography treatments increase buffer consumption and can attenuate the concentration aspect of anion-exchange chromatography (Perry and Rayat 2021). In general, anion-exchange chromatography results in functional LV particle recovery yields ranging from 28 to 90%, with most publications reporting recoveries between 30 and 60% (Table 8.3). Impurity clearance of LV purification using AEX often yielded above 90% (Lesch et al. 2011; Valkama et al. 2020; Moreira et al. 2021).

8.4.3 *AAV Vector Purification with Ion-Exchange Membrane Chromatography*

In contrast to LVs, AAVs are robust non-enveloped viral particles capable of handling harsher conditions such as a pH change or high ionic strengths (Segura et al. 2011). As a result, a post-elution treatment is not necessary when purifying AAV by ion-exchange chromatography, but a pre-treatment may be required depending on the previous processing step (Hejmowski et al. 2022). AAV purification by ion-exchange chromatography has been performed with cation exchange followed by anion-exchange chromatography (Davidoff et al. 2004; Okada et al. 2009). The exchange chromatography step was performed in a flowthrough mode, meaning AAVs do not adsorb to the membrane and are found in the flowthrough, while positively charged impurities adsorb to the membrane. The collected flowthrough is then loaded onto an anion-exchanger to separate empty AAV capsids from full capsids (Davidoff et al. 2004).

As described previously, ultracentrifugation was initially performed to enrich the sample for full capsids. Anion-exchange chromatography has been the method of choice for avoiding scale-up limitations of density gradient purification, and membrane-based anion-exchange chromatography can be implemented as a polishing chromatography step in AAV production (Adams et al. 2020). A high dilution of the affinity purification eluate is likely to be required, as harsh elution conditions are necessary for affinity chromatography, and dilution of the eluate is typically performed to adjust pH and conductivity for the subsequent AEX step (Hejmowski et al. 2022). An elution gradient is useful when fractionating closely related biomolecules, for example, empty and full viral vector particles (Do Minh and Kamen 2021).

With dual ion-exchange chromatography, purified AAVs were obtained with only 0.8% (Okada et al. 2009) to $<10\%$ empty particles (Davidoff et al. 2004) and a total recovery of 41–67%, depending on the AAV serotype (Table 8.4). Full capsid recovery of 60% has been reported for membrane-based anion exchange chromatography following affinity purification (Hejmowski et al. 2022). Several serotypes were successfully purified, meaning anion-exchange membrane chromatography could represent a platform-able polishing step (Boenning et al. 2020). Employing anion-exchange membranes for polishing can intensify the AAV purification process, as membrane chromatography devices can be operated at high flow rates. A comparison of membrane and resin chromatography for anion-exchange-mediated AAV purification found that membrane loading was 40× faster (Boenning et al. 2020).

8.4.3.1 Developing an Ion-Exchange Chromatography Process

For anion-exchange chromatography process development, a variety of parameters can be screened on a small scale, including:

- Different membranes, including strong and weak anion-exchangers
- Fresh or frozen starting material loaded directly or after clarification, concentration, or adjustment to a specific NaCl concentration
- Different elution patterns, either stepwise or linear gradient elution to yield higher vector recoveries or better resolution of impurities from the vector-containing fraction, e.g., for empty/full capsid separation
- Flow rate
- Additives like sugars (e.g., sucrose) and salts (e.g., $MgCl_2$) in buffers, which may increase viral vector stability
- Buffer type (Tris, HEPES, PBS), including salt concentration and pH
- Amount of viral particles loaded
- The implementation of a DNA digestion step before chromatography

Using a design of experiments (DoE) approach can be extremely useful to vary multiple parameters and reduce the number of experiments required. Factors of particular significance can be obtained from the model, and the optimal process parameters can be predicted. Ideally, small-scale devices should be used, as these tools enable the performance of a considerable number of experiments with minimal starting material. Several suppliers offer 96-well chromatographic membrane plates that facilitate high-throughput process development and screening, as demonstrated by Moreira et al. (Moreira et al. 2021). In this study, the researchers first performed a DoE in 96-well plates with two different stationary phases (Mustang® Q and Sartobind® Q) varying three factors: (1) the amount of virus loaded, (2) elution pH, and (3) NaCl concentration for elution. After evaluating the DoE, the predicted ideal process conditions were validated through small-scale chromatography runs with devices comprising membrane volumes of up to 1 mL. Here, some minor parameter variations can be implemented before further scale-up. For instance, Moreira et al. performed a stepwise elution, and the effect of sucrose addition was tested on small-scale Mustang® Q and Sartobind® Q devices.

8.4.4 Viral Vector Purification with Steric Exclusion Membrane Chromatography

Steric exclusion chromatography has also been proposed for the purification of LV, AAV (Marichal-Gallardo et al. 2017; Labisch et al. 2022), and other viruses, including baculovirus, influenza A, and Orf (Marichal-Gallardo et al. 2017; Lothert et al. 2020a, b). Exclusion chromatography was first described by Lee et al. for the purification of IgG using nanoparticles as a stationary phase (Lee et al. 2012). Later, Marichal-Gallardo et al. successfully used cellulose membranes as a stationary phase (Marichal-Gallardo et al. 2017).

The steric exclusion chromatography principle is based on depletion attraction using polyethylene glycol (PEG). Upon PEG addition to the viral particle solution, depletion zones are formed as PEG molecules are sterically excluded from this area.

The thermodynamically unfavorable increase in free energy promotes the physical reorganization of the viral vectors. (Lekkerkerker and Tuinier 2011; He 2014) When the depletion zones of a viral particle and the stationary phase overlap, the excluded volume is reduced, thereby decreasing the free energy. The negative osmotic pressure leads to weak attraction (Lekkerkerker and Tuinier 2011; Lee et al. 2012; He 2014). The optimal PEG size and concentration depend on the size of the target molecule. Viral particles are eluted with a PEG-free buffer, e.g., Tris buffer with a low salt concentration, reversing the association of the viral particles with the membrane. No pre- or post-treatment of the viral vector material is required.

High recoveries of approximately 86% for LV and 126% for AAV have been reported for steric exclusion chromatography-based purification, with an impurity removal of $\geq$ 80% (Marichal-Gallardo et al. 2017; Labisch et al. 2022). It does not require any chemical interaction between the target species and the stationary phase, meaning steric exclusion chromatography facilitates milder elution conditions. This feature might be of particular interest for fragile enveloped viral vectors.

8.4.4.1 Developing a Steric Exclusion Chromatography Process

When developing a steric exclusion-based membrane chromatography process, a variety of process parameters can first be screened, including:

- Hydrophilic stationary phases (preferably stabilized cellulose membranes)
- Fresh or frozen starting material loaded after clarification or concentration
- Flow rate
- PEG size and concentration
- Amount of viral particles loaded
- The implementation of a DNA digestion step before chromatography

So far, steric exclusion chromatography has mainly been applied at a small scale and a scale-up approach (Labisch et al. 2023) was recently published which allowed purification of approximately 1 L LV feed solution.

8.4.5 Hydrophobic Interaction Membrane Chromatography

For hydrophobic interaction chromatography, buffers with high ionic strengths are used (e.g., ammonium sulfate), as these buffers reduce the solvation of the viral vector in the sample and thereby promote the adsorption of the viral vector on the hydrophobic matrix. The vectors are eluted by decreasing the salt concentration (Cazes 2009).

To date, only one study demonstrating the performance of membrane-based hydrophobic interaction chromatography for the purification of AAVs has been published (McNally et al. 2020). In this study, a membrane adsorber with phenyl ligands served as a stationary phase. Some AAV serotypes require cell disruption to release cell-associated AAVs. Both disruption by repeated freeze-thaw cycles and chemical lysis are feasible methods to release viral particles (Adams et al. 2020). McNally et al. performed an integrated chemical lysis and primary capture technique with hydrophobic interaction chromatography to purify AAVs. The cells were lysed with an ammonium sulfate buffer, in which the subsequent chromatography step was also performed. Thus, the required pre-treatment for hydrophobic interaction chromatography was combined with cell lysis. After incubation with ammonium sulfate, the formed non-aqueous phase containing high concentrations of DNA was removed. The subsequent hydrophobic interaction chromatography removed over 90% of host cell proteins (HCPs) and over 80% of DNA contaminants (the DNA removal does not include the DNA physically removed from the lysate before chromatography) and achieved recoveries of 81–97%, depending on the serotype. Without cell lysis before chromatography, recoveries of 77–101% were obtained, depending on the serotype, but with lower AAV titers as cell-associated AAVs were not released. A subsequent polishing step with anion-exchange chromatography to separate empty and full particles might be necessary.

8.5 Monolith Chromatography for Viral Vector Purification

One of the main challenges associated with viral vector processing is how existing technologies can support the purification of the large industrial volumes required for commercial manufacturing. Monolithic chromatography is widely accepted as an advantageous approach for purifying viral vector-based therapeutics and is ideally suited to solve speed and scalability challenges. The structure of the interconnected channels in the monolith matrix is designed to have high porosity and lack dead ends, supporting convective mass transport. The surface of the large channels contains active binding sites easily accessible to nanoparticles, such as viral vectors (Podgornik and Štrancar 2005; Podgornik et al. 2013; Kramberger et al. 2015). Standard monolith channel sizes for preparative chromatography are 2 and 6 μm. Sartorius is a producer of commercially available CIM™ (Convective Interaction Media) monoliths.

Convective mass transport is independent of the flow rate, i.e., faster flow rates do not influence column capacity or resolution. As a result, chromatography processes using monoliths can be operated with high speed, even more than ten times faster than porous-based media (Kramberger et al. 2015). This enables the purification of large volumes of material in a relatively short time and increases process productivity. The high porosity of monolithic material (60–65%) results in lower pressure drops. This results in high operating flow rates on a large industrial scale using single-use chromatographic equipment, which can otherwise represent a bottleneck when transferring a process to an industrial scale.

The high dynamic binding capacity (DBC) of monoliths for large biomolecules, such as viral vectors, drives a significant purification process improvement since smaller columns and reduced buffer volumes can be employed. This translates to increased recovery and significant time and cost savings. The reason for the high dynamic binding capacity is the high adsorption area of the porous structure (Jungbauer and Hahn 2008). The monolith's interconnected channels are highly accessible to the molecules which pass through. Conversely, with porous matrices, molecules must fit into dead-ends, often too small for the viral vectors. On the other hand, convective mass transport in monoliths does not support size exclusion chromatography (SEC) principles, since SEC is based on the diffusion of particles into the pores, which is a characteristic monoliths cannot achieve. Importantly, the DBC is unaffected by the high speed of monoliths (Podgornik and Štrancar 2005). In the case of AVs, process development resulted in at least double DBC compared to alternative matrices (Fernandes et al. 2013). Similar results were also reported for LV vectors (Bandeira et al. 2012).

The availability of a wide range of column sizes makes separating viral vectors by monolithic chromatography highly scalable, addressing one of the more significant challenges of their production. The CIMmultus™ line is suitable for preparative applications, with column sizes ranging from 1 mL to 40 L. For screening purposes, CIM™ 96-well monolithic plates and 0.1 mL CIMmic™ discs are available. The Sartorius portfolio also contains various chemistries of analytical HPLC CIMac™ 0.1 mL columns. Smaller screening units are important for initial process development studies, DoE experiments, or studies where only a limited amount of sample is available.

Ion exchangers are the primary chromatography tool of choice for purifying viral vectors by monolithic chromatography (Tables 8.5 and 8.6). The presence of protein subunits on the surface of the viral vector creates complex patterns of ion charges that interact uniquely with ion exchange media, serving as an ideal characteristic for a robust purification process. By changing the pH of the sample or chromatographic buffers, either cation or anion-exchange media can be implemented for the same viral vector (Banjac et al. 2014). Other chemistries such as hydrophobic interactions and multi-modal ligands are also used to purify viral vectors, described repeatedly for AAV vectors (Table 8.5). In the case of LV vectors, affinity monolith chromatography has been used in some studies, although most of the published processes used weak anion exchangers, such as DEAE (Table 8.6).

Table 8.5 Monolith chromatography purification strategies for adeno-associated viruses (AAVs)

Purification method	Viral vector	Stationary phase	Vector recovery (%)	Impurity clearance	References
HIC	rAAV2/8	CIMmultus OH	23% (TFF) 3% (PEG precipitate) 0.3% (cell pellet)	n.a.	Rieser et al. (2021)
CEX	rAAV2/8	CIMmultus SO3	65%, 64%, 86%, 55% (cell pellet) 92%, 80% (PEG precipitate)	Total copy number of residual hcDNA in range of 10^7	Rieser et al. (2021)
AEX	rAAV2/8	CIMmultus QA	59% (cell pellet lysate, CEX capture)	Good full / empty separation. Total copy number of residual hcDNA in range of 10^7	Rieser et al. (2021)
AEX	rAAV2/8	CIMmultus PrimaS	70%, 70%, 78% (cell pellet) 63%, 31% (PEG precipitate)	17% (SV-AUC) or 12% (analytical AEX) of remaining empty capsids in full fraction. Total copy number of residual hcDNA $6*10^6$ to $5*10^7$	Rieser et al. (2021)
CEX	wt rAAV2/2, rAAV2/2.NN, rAAV2/2.GL	CIMmultus SO3	93–99%	n.a.	Rieser et al. (2021)
AEX	AAV5, AAV6, AAV8	CIMmultus QA	73% (AAV5) 80–83% (AAV6 and AAV8)	8.8-fold vector capsid (full) enrichment (AAV5), 2.6x (AAV8), 2.6x and 2.4x (AAV6). Final purity: 80% full (AAV5), 97 % full (AAV8), >90% full (AAV6).	Joshi et al. (2021)
Multi-modal, and AEX	AAV8 in AAV9	CIMmultus PrimaT, QA and PrimaS	69% (PrimaT, AAV8, Magnesium gradient elution) 100% (PrimaT, AAV9, Magnesium gradient elution)	100% reduction of empty capsids (PrimaT, AAV9, Magnesium gradient elution) 75% reduction of empty capsids (PrimaS)	Gagnon et al. (2021a)

(continued)

Table 8.5 (continued)

Purification method	Viral vector	Stationary phase	Vector recovery (%)	Impurity clearance	References
AEX	AAV8	CIMmultus QA	n.a.	Viral reduction in QA elution: ≥5.1 log XMuLV, ≥ 4.6 log HAV, ≥ 3.8 log HSV-1, ≥ 5.1 log PRV, ≥ 5.6 log Reo-3, ≥ 5.4 log MVM, ≥ 3.9 log MVM-MVP	Winkler et al. (2021)
AEX	AAV2	CIMmultus QA	More than 70%	4.6-fold enrichment of full AAV, more than 90% full in elution fraction	Dickerson et al. (2021)
CEX	AAV2	CIMmultus SO3	n.a.	>90% (AVB Sepharose + SO3 column): low levels of residual NHK, HSV, BSA protein; low levels of BHK, HSV DNA, undetectable benzonase and AVB ligand	Ye et al. (2015)
AEX	AAV/Anc80	CIMmultus QA	100%	70% full AAV (AUC)	Gu et al. (2018)

n.a., data not available/ not mentioned

8.5.1 Considerations for Designing a Monolithic Chromatography Process

The key factors that influence the selection process for monolithic chromatography and process conditions are outlined below.

The stability and infectivity of viral vectors should not be compromised during the sample preparation or chromatography. This principle is particularly important when designing chromatographic conditions, such as pH and conductivity of the process. For example, LVs are more sensitive than AAVs. Therefore, acceptable parameters must be developed within a narrow window. In contrast, AAVs are less pH- and conductivity-sensitive, so a relatively large range of parameters can be used for purification. Additives and stabilizers like sugars, amino acids, surfactants, and salts, are often added to keep viruses in favorable conditions throughout the downstream process (Srivastava et al. 2021).

To reduce development timelines using multi-factor screening and increase comparability during downstream process development, scientists can perform trial experiments using CIM® 0.05 mL monolithic 96-well plates and 0.1 mL CIMmic™ discs, which are small units designed specifically for small sample volumes and fast operating when developing the downstream process. The obtained

Table 8.6 Monolith chromatography purification strategies for lentiviral vectors (LVs)

Purification method	Viral vector	Stationary phase	Vector recovery (%)	Impurity clearance	References
Affinity	cTag-modified LV	CIMac high-performance streptavidin	20	n.a.	Mekkaoui et al. (2018)
AEX	GALV-TR	CIM DEAE	64	n.a.	Boudeffa et al. (2019)
AEX	LV	CIMmultus DEAE	65	99.9% total protein removal, 96.5% total DNA removal	Lesch et al. (2011)
AEX	LV	CIMmultus DEAE	80	99.9% total DNA removal	Bandeira et al. (2012)
Affinity	LV	CIM-IDA-Ni2+	69	99.8% total protein removal, 97.5 % total DNA removal	Cheeks et al. (2009)

results can be easily applied to the CIMmultus™ line, which is scalable to large industrial volumes (BIA Separations 2022). CIM™ 0.05 mL monolithic 96-well plates are designed to quickly screen many possible options, such as performing large DoE studies without using a chromatographic system. CIMmic™ discs can be installed in standard chromatographic systems and are of particular use when determining dynamic binding capacity or other critical process parameters when a small sample volume is available.

When choosing a suitable chromatography ligand, defining the dynamic binding capacity is an important criterion, as the same virus material can behave differently under the same conditions where an alternative ligand is used (Kramberger et al. 2015). Strong ion-exchangers, such as CIM™ QA (quaternary amine, anion-exchange) or CIM™ SO3 (sulfonate, cation exchanger), are typically used for AAV purification. Regarding LV vectors, CIM™ DEAE was used in several studies for LV purification (Table 8.6). Recently, multi-modal monolithic columns, such as PrimaT and PrimaS, were introduced due to different binding selectivity, which can be exploited especially for fine-tune separation of viral vectors, e.g., AAV empty and full capsid separation (Gagnon et al. 2021a). For example, PrimaS is a multi-modal chromatography ligand that combines elements of hydrogen bonding with anion exchange chromatography. Hydrophobic interaction chromatography (CIM™ OH column) also has value in the purification of viruses, especially with its power to eliminate host-cell DNA from the samples (Sviben et al. 2017). When using HIC, high salt concentrations are required to bind viral vectors to the matrix (Rieser et al. 2021).

8.5.1.1 Pre-chromatography Sample Preparation

Viral vectors are usually purified from a complex starting material due to the media and host-cell remains in the harvest. In the case of cell-secreted products, the complexity of the sample is decreased but still includes host-cell contaminants. These impurities can present a patient safety risk and must be removed to minimal values defined by regulatory agencies. Removing as much DNA and protein contamination as possible before the first chromatography step is also important for increasing dynamic binding capacity and overall process productivity. For lysed AAV harvest, several options of sample pre-treatment are possible, including clarification, tangential flow filtration (TFF), and nuclease treatment. TFF can be performed also coupled with salt-tolerant nuclease (Gagnon et al. 2020). For LV processes, clarified LV supernatant is usually loaded onto monolithic columns.

Before loading the viral sample onto the chromatography matrix, pH and conductivity must be kept constant either by employing a robust pre-chromatography step or titration and dilution to achieve target conditions used for chromatography. Samples should also be filtered to prevent column clogging.

8.5.1.2 Chromatography Strategies

The first chromatographic step in an industrial platform using monoliths is a capture step. During capture, most host cell contaminants are removed, and the viral vector is bound to the monolith matrix. The productivity of the capture step is determined by the dynamic binding capacity, which is influenced by the sample preparation and the availability of binding sites on the chromatographic matrix. Usually, a bind-elute chromatography step is performed either with a pH or salt gradient.

Since most viruses have an isoelectric point below pH 6 (or for AAV, around pH 6) (Venkatakrishnan et al. 2013; Michen and Graule 2010). they efficiently bind to anion-exchangers at neutral pH (AAV at slightly basic pH). It is worth mentioning that the isoelectric points of viruses are not useful as an absolute value, since viruses are large macromolecular structures with non-uniform surface charges. This means that viruses can bind to both anion and cation exchangers at a certain pH (Banjac et al. 2014). However, most host-cell-derived impurities also bind to anion-exchangers at neutral pH in case of lysed harvest. Therefore, anion-exchangers are used mainly for capture step when supernatant is used instead of lysed harvest and purification is performed under neutral pH, such as in the case of LV purification.

In contrast, when lowering pH using acidic buffers, viruses can bind selectively to cation exchangers (Kramberger et al. 2015). However, this is typically only appropriate for AAVs, which tolerate higher pH changes than LVs. For the capture step of AAV vectors, CIM™ SO3 chemistry (cation exchanger) or CIM™ OH (hydrophobic interaction) is the chromatography method of choice. Capture step does not separate empty and full AAV capsids, so a second step (polishing) is needed.

8.5.1.3 Separation of Empty and Full AAV Capsids

An advantage of monoliths is that resolution is independent of the flow rate (Kramberger et al. 2015). Lock et al. demonstrated that monoliths deliver excellent resolution for the separation of empty and full AAV capsids in a method that is now widely used (Lock et al. 2012). Separating full and empty particles with monoliths is done primarily using anion-exchange columns, such as the CIM™ QA (Table 8.5). Moreover, researchers studying anion-exchange monolithic chromatography for empty/full AAV separation have observed that additives in chromatography buffers, such as the presence of divalent metal ions—e.g., magnesium—changed the separation profile (Gagnon et al. 2021b). Following these results, empty and full AAV separation was also investigated using a multi-modal matrix, including multi-modal metal affinity chromatography (CIM™ PrimaT). A robust recovery of full capsids and complete elimination of empty capsids was achieved (Gagnon et al. 2021a). Similar results were obtained using a weak anion-exchanger combined with hydrogen bonding (CIM™ PrimaS) (Gagnon et al. 2021b).

Since AAV polishing is a fine-tuned process, empty and full capsid separation is more serotype-dependent than the capture step. Removal of full capsids and other AAV species requires further research and development to be able to achieve complete separation for all serotypes, with developing new surface chemistries as well as optimization of existing methods.

8.5.2 Recommendations for Scale-Up

Scaling up chromatographic processes from laboratory or process development scale to pilot and manufacturing scale is an important and challenging task that has been dealt with by several strategies. Scaling up monolithic chromatography is straightforward due to the flow-independent separation and resolution, which provide consistent separation quality. Moreover, the radial format delivers a short chromatographic path, and the high porosity results in a low-pressure drop with a linear connection to the flow rate. When approaching scale up of monolithic chromatography processes, key factors to take into account include:

- Determining column loading capacity during process development is critical since it provides crucial information for process economics, optimizing efficiency, and calculating the correct column sizes at each scale.
- The gradient can be shortened to decrease buffer consumption (especially after product elution) by taking into account the same gradient length ratio as during process development.
- In general, it is recommended that the conductivity of elution is maintained during scale-up by transferring the same gradient duration in column volumes, using the same flow rate in column volumes/min (when possible), and applying the same loading amount per mL of column volume.

When changing certain parts of a process or building a new viral vector purification process, all steps must be harmonized to reach the best outcome. Owing to their easy operation and wide range of working conditions, monoliths enable the flexible performance of consecutive and adjacent steps during the manufacturing process.

8.6 Continuous Chromatography

Constant innovations in the chemistries, composition, and structure of chromatographic media will improve viral vector production. These include developments within membrane adsorption, convective media in monolithic chromatography, and bead-based affinity resins. As progress chromatographic separation technologies continue to improve, multiple other factors can also be optimized to boost productivity. These include increasing resin lifetime, increasing resin utilization, decreasing processing time, increasing laboratory productivity while reducing the overall footprint of equipment, and decreasing the risk of contamination.

In traditional fed-batch chromatography following harvest clarification, the material is processed via chromatographic means based upon the dynamic binding capacities of the matrix. Thus, harvest material may require storage before further processing if titers exceed the overall chromatography capacity. Moreover, it has been well established that in traditional fed-batch resin-based chromatography, not all of the resin is fully utilized due to mass transfer limitations (Silva et al. 2015). Continuous or semicontinuous chromatography helps overcome these issues and offers significant benefits in a downstream process.

Rather than the single packed column employed for traditional chromatography, in continuous chromatography, multiple smaller columns are placed in series. This arrangement decreases the amount of matrix material required for processing. In 2012, several researchers demonstrated the application of this principle for the purification of monoclonal antibodies and recombinant enzymes, maximizing the resin binding occupancy where harvest material was overloaded onto a chromatographic medium (Godawat et al. 2012; Mahajan et al. 2012; Warikoo et al. 2012). Upon detection of a breakthrough by UV, the flowthrough of this first column is then directed to the second column in series, and during this, the first column in series can begin the subsequent steps of washing, elution, and regeneration. This allows the resin to be utilized more efficiently while also increasing productivity.

Continuous chromatography has slowly worked its way into the biopharmaceutical landscape but remains primarily confined to monoclonal antibody production. This is a logical progression, as monoclonal antibody production benefits from a relatively standardized downstream platform approach with Protein A affinity as the initial capture step. Protein A resin is more costly than other resins, and upstream production yields have continued to increase, requiring downstream operations to respond. Therefore, it is of great interest to fully maximize the use of Protein A resin while minimizing cost by reducing resin and buffer volume requirements.

Initial studies using continuous chromatography required utilizing traditional FPLC systems, e.g., AKTA Explorer. However, with the ever-increasing interest and needs of industry, multiple systems are now available to allow for continuous chromatography multi-column operation from process development to manufacturing scale.

Thus far, the application of continuous chromatography to viral vector production has been limited. As viral vectors continue to gain approval—and with upstream improvements in production—it stands to reason that continuous chromatography will become more in demand. Notably, two studies have demonstrated a multi-column approach through size exclusion chromatography (Kröber et al. 2013; Nestola et al. 2015). These studies both took advantage of the size of viral vectors and operated in an isocratic negative or flowthrough mode. While both studies used a size-exclusion method for purification, the molecule of interest differed along with the overall strategy. In the first study, Influenza (H1N1) was purified utilizing a 3-column continuous feed strategy (Kröber et al. 2013). In the second study, Adenovirus (Ad5) was purified via a semicontinuous two-column feed strategy (Gagnon et al. 2020). For the purification of H1N1, a recovery of 70% and a 3.8-fold productivity increase were observed (Kröber et al. 2013) while the purification of Ad5 yielded a recovery of 86% and a 6.1-fold productivity increase (Gagnon et al. 2020).

With the advent of viral vector affinity bead-based resins, monolithic chromatography, and membrane-based chromatography taken in conjunction with these findings, improved yields along with increased productivity and decreased costs by continuous chromatography can help accelerate the field and meet patient needs. This is particularly valuable as drug developers move toward addressing larger indications with their AAV pipeline, increasing the therapeutic doses; continuous chromatography can lead the way to the implementation of more cost-effective and high-throughput downstream trains.

8.7 Conclusion

While most virus-based therapeutics were designed for the treatment of rare diseases, there is a growing interest in their application to more common disorders. Continued improvements in manufacturing strategies are necessary to deliver the increased yields and reduced costs required to make these biotherapeutics commercially viable.

Keeping up with the demands of the dynamic biopharmaceutical landscape requires a comprehensive understanding of production challenges and opportunities for innovation in the downstream purification of viral vectors. This includes the potential for standardization (production platforms), process simplification (reducing purification steps), and intensification (continuous processing) to accelerate patient access to supply the demand of pharmaceuticals globally to patients in need.

Downstream processing remains the bottleneck in viral vector production. The relative infancy of the field coupled with the diversity of viral vectors means a one-size-fits-all approach for virus purification does not yet exist. While density gradient ultracentrifugation is a valid method for isolating viral vector particles, it is not practical at large scales. Thus, chromatography remains the method of choice for virus purification.

Chromatography resins have often been the workhorse for the large-scale purification of biotherapeutics. However, as made clear in this chapter, they are certainly not the only option worth exploring. Convective media (membranes and monoliths) should be considered viable when screening potential purification methods.

References

Abrecht H et al (2019) Sartobind Q membrane adsorbers for adenovirus downstream processing. BioProcess Int. Available at: https://bioprocessintl.com/analytical/downstream-development/adenovirus-downstream-process-intensification-implementation-of-a-sartobind-q-membrane-adsorber/. Accessed 28 June 2022

Adams B, Bak H, Tustian AD (2020) Moving from the bench towards a large scale, industrial platform process for adeno-associated viral vector purification. Biotechnol Bioeng 117(10): 3199–3211. https://doi.org/10.1002/bit.27472. Wiley

Ayuso E et al (2010a) High AAV vector purity results in serotype- and tissue-independent enhancement of transduction efficiency. Gene Ther 17(4):503–510. https://doi.org/10.1038/gt.2009.157

Ayuso E, Mingozzi F, Bosch F (2010b) Production, purification and characterization of adeno-associated vectors. Curr Gene Ther 10(6):423–436. https://doi.org/10.2174/156652310793797685

Bandeira V et al (2012) Downstream processing of lentiviral vectors: Releasing bottlenecks. Hum Gene Ther Methods 23(4):255–263. https://doi.org/10.1089/hgtb.2012.059

Banjac M et al (2014) Purification of Vero cell derived live replication deficient influenza A and B virus by ion exchange monolith chromatography. Vaccine 32(21):2487–2492. https://doi.org/10.1016/j.vaccine.2014.02.086

Barut M et al (2005) Convective Interaction Media short monolithic columns: Enabling chromatographic supports for the separation and purification of large biomolecules. J Sep Sci 1876–1892. https://doi.org/10.1002/jssc.200500246

Bauler M et al (2020) Production of lentiviral vectors using suspension cells grown in serum-free media. Mol Ther Methods Clin Dev 17:58–68. https://doi.org/10.1016/j.omtm.2019.11.011

Becerra A, Buyel JF (2021) Optimizing downstream purification of high-quality plasmid DNA for gene therapy and vaccine production. Cell Gene Ther Insights 7(11):1217–1230. https://doi.org/10.18609/cgti.2021.162

BIA Separations (2022) Fast and semi-automatic optimization of capture step for rAAV purification using CIM® SO3 96-well monolithic plate. Available at: https://www.biaseparations.com/en/library/application-notes/1155/fast-and-semi-automatic-optimization-of-capture-step-for-raav-purification-using-cimr-so3-96-well-monolithic-plate. Accessed 5 July 2022

BioPlan Associates Inc (2022) 19th annual report and survey of biopharmaceutical manufacturing capacity and production biotherapeutic developers and contract manufacturing organizations

Boenning K et al (2020) Platform approach for the AAV purification process including membrane chromatography for polishing. Cytotherapy 22(5):S186. https://doi.org/10.1016/j.jcyt.2020.03.393

Boudeffa D et al (2019) Toward a scalable purification protocol of GaLV-TR-pseudotyped lentiviral vectors. Hum Gene Ther Methods 30(5):153–171. https://doi.org/10.1089/hgtb.2019.076

Braas G et al (1996) Strategies for the isolation and purification of retroviral vectors for gene therapy. Bioseparation 6(4):211–228. Available at: https://pubmed.ncbi.nlm.nih.gov/9032984/. Accessed 30 June 2022

Bulcha JT et al (2021) Viral vector platforms within the gene therapy landscape. Signal Transduct Targeted Ther. Springer Nature. https://doi.org/10.1038/s41392-021-00487-6

Burnham B et al (2015) Analytical ultracentrifugation as an approach to characterize recombinant adeno-associated viral vectors. Hum Gene Ther Methods 26(6):228–242. https://doi.org/10.1089/hgtb.2015.048

Burova E, Ioffe E (2005) Chromatographic purification of recombinant adenoviral and adeno-associated viral vectors: Methods and implications. Gene Ther 12(1):S5–S17. https://doi.org/10.1038/sj.gt.3302611

Cazes J (2009) Encyclopedia of chromatography. CRC Press, Boca Raton, FL, p 733. Available at: https://pubs.acs.org/doi/10.1021/ja9099467. Accessed 28 June 2022

Cheeks MC et al (2009) Immobilized metal affinity chromatography of histidine-tagged lentiviral vectors using monolithic adsorbents. J Chromatogr A 1216(13):2705–2711. https://doi.org/10.1016/j.chroma.2008.08.029

Chen R et al (2010) Affinity recovery of lentivirus by diaminopelargonic acid mediated desthiobiotin labelling. J Chromatogr B Analyt Technol Biomed Life Sci 878(22):1939–1945. https://doi.org/10.1016/j.jchromb.2010.05.019

Clément N, Grieger JC (2016) Manufacturing of recombinant adeno-associated viral vectors for clinical trials. Mol Ther Methods Clin Dev 3:16002. https://doi.org/10.1038/mtm.2016.2. Springer

Davidoff AM et al (2004) Purification of recombinant adeno-associated virus type 8 vectors by ion exchange chromatography generates clinical grade vector stock. J Virol Methods 121(2): 209–215. https://doi.org/10.1016/j.jviromet.2004.07.001

Dickerson R et al (2021) Separating empty and full recombinant adeno-associated virus particles using isocratic anion exchange chromatography. Biotechnol J 16(1). https://doi.org/10.1002/biot.202000015

Do Minh A, Kamen AA (2021) Critical assessment of purification and analytical technologies for enveloped viral vector and vaccine processing and their current limitations in resolving co-expressed extracellular vesicles. Vaccines 823. MDPI AG. https://doi.org/10.3390/vaccines9080823

Fernandes P et al (2013) Bioprocess development for canine adenovirus type 2 vectors. Gene Ther 20(4):353–360. https://doi.org/10.1038/gt.2012.52

Fischer-Frühholz S, Leuthold M et al (2018) Membrane chromatography cassettes for bind & elute applications of viruses and large proteins. Vaccine Technol VII 1600(April):37079. Available at: https://www.researchgate.net/publication/344512978_Membrane_chromatography_cassettes_for_bind_and_elute_applications_of_viruses_and_large_proteins. Accessed 28 June 2022

Gagnon P et al (2020) Streamlining industrial purification of adeno-associated virus. BioProcess Int 18(11–12):1–4. Available at: https://www.biaseparations.com/en/library/publications/1106/streamlining-industrial-purification-of-adeno-associated-virus. Accessed 1 July 2022

Gagnon P, Leskovec M et al (2021a) Removal of empty capsids from adeno-associated virus preparations by multimodal metal affinity chromatography. J Chromatogr A 1649:462210. https://doi.org/10.1016/j.chroma.2021.462210

Gagnon P, Goričar B et al (2021b) Separation of empty and full adeno-associated virus capsids from a weak anion exchanger by elution with an ascending pH gradient at low ionic strength. BioProcess J. Available at: https://www.biaseparations.com/en/library/publications/1134/separation-of-empty-and-full-adeno-associated-virus-capsids-from-a-weak-anion-exchanger-by-elution-with-an-ascending-ph-gradient-at-low-ionic-strength. Accessed 1 July 2022

Gao G et al (2004) Clades of adeno-associated viruses are widely disseminated in human tissues. J Virol 78(12):6381–6388. https://doi.org/10.1128/jvi.78.12.6381-6388.2004
Godawat R et al (2012) Periodic counter-current chromatography—design and operational considerations for integrated and continuous purification of proteins. Biotechnol J 7(12):1496–1508. https://doi.org/10.1002/biot.201200068
Goldfarb M, Terova O (2020) Downstream purification of adeno-associated virus for large-scale manufacturing of gene therapies. Cell Gene Ther Insights 6(7):95–963. https://doi.org/10.18609/cgti.2020.106
Gu B et al (2018) Establishment of a scalable manufacturing platform for in-silico-derived ancestral adeno-associated virus vectors. Cell Gene Ther Insights 4(S1):753–769. https://doi.org/10.18609/cgti.2018.080
He Q (2014) Investigation of stabilization mechanisms for colloidal suspension using nanoparticles. University of Louisville. https://doi.org/10.18297/etd/593
Hejmowski AL et al (2022) Novel anion exchange membrane chromatography method for the separation of empty and full adeno-associated virus. Biotechnol J 17(2). https://doi.org/10.1002/biot.202100219
Joshi PRH et al (2021) Development of a scalable and robust AEX method for enriched rAAV preparations in genome-containing VCs of serotypes 5, 6, 8, and 9. Mol Ther Methods Clin Dev 21:341–356. https://doi.org/10.1016/j.omtm.2021.03.016
Jungbauer A (2005) Chromatographic media for bioseparation. J Chromatogr A 3–12. https://doi.org/10.1016/j.chroma.2004.08.162
Jungbauer A, Hahn R (2008) Polymethacrylate monoliths for preparative and industrial separation of biomolecular assemblies. J Chromatogr A. Elsevier 62–79. https://doi.org/10.1016/j.chroma.2007.12.087
Kamen A, Henry O (2004) Development and optimization of an adenovirus production process. J Gene Med. https://doi.org/10.1002/jgm.503
Krajacic M et al (2017) Application of monolithic chromatographic supports in virus research. Electrophoresis 2827–2836. Wiley-VCH. https://doi.org/10.1002/elps.201700152.
Kramberger P, Urbas L, Štrancar A (2015) Downstream processing and chromatography based analytical methods for production of vaccines, gene therapy vectors, and bacteriophages. Hum Vaccines Immunother 1010–1021. Taylor and Francis. https://doi.org/10.1080/21645515.2015.1009817
Kröber T et al (2013) Continuous purification of influenza virus using simulated moving bed chromatography. J Chromatogr A 1307:99–110. https://doi.org/10.1016/j.chroma.2013.07.081
Kutner RH et al (2009) Simplified production and concentration of HIV-1-based lentiviral vectors using HYPERFlask vessels and anion exchange membrane chromatography. BMC Biotechnol 9. https://doi.org/10.1186/1472-6750-9-10
Labisch JJ et al (2021) Infectious titer determination of lentiviral vectors using a temporal immunological real-time imaging approach. PLoS One 16(7 July):e0254739. Edited by B Lu. https://doi.org/10.1371/journal.pone.0254739.
Labisch JJ et al (2022) Steric exclusion chromatography of lentiviral vectors using hydrophilic cellulose membranes. J Chromatogr A 1674:463148. https://doi.org/10.1016/j.chroma.2022.463148
Labisch JJ, Paul R, Wiese GP, Pflanz K (2023) Scaling up of steric exclusion membrane chromatography for lentiviral vector purification. Membranes 13(2):149
Łącki KM, Riske FJ (2020) Affinity chromatography: an enabling technology for large-scale bioprocessing. Biotechnol J 1800397. Wiley-VCH. https://doi.org/10.1002/biot.201800397
Ladisch MR (2001) Bioseparations engineering: principles, practice , and economics. Wiley, New York
Lee J et al (2012) Principles and applications of steric exclusion chromatography. J Chromatogr A 1270:162–170. https://doi.org/10.1016/j.chroma.2012.10.062
Lekkerkerker HNW, Tuinier R (2011) Colloids and the depletion interaction, Lecture notes in physics. Springer, Dordrecht. https://doi.org/10.1007/978-94-007-1223-2

Lesch HP et al (2011) Production and purification of lentiviral vectors generated in 293T suspension cells with baculoviral vectors. Gene Ther 18(6):531–538. https://doi.org/10.1038/gt.2010.162

Liddell J (2022) Challenges in downstream purification of advanced therapies. Eur Pharm Rev 27(1):36–41

Lock M et al (2010) Rapid, simple, and versatile manufacturing of recombinant adeno-associated viral vectors at scale. Hum Gene Ther 21(10):1259–1271. https://doi.org/10.1089/hum.2010.055

Lock M, Alvira MR, Wilson JM (2012) Analysis of particle content of recombinant adeno-associated virus serotype 8 vectors by ion-exchange chromatography. Hum Gene Ther Methods 23(1):56–64. https://doi.org/10.1089/hgtb.2011.217

Lock M, Vandenberghe LH, Wilson JM (2015) Scalable production method for AAV. United States. Available at: https://patents.google.com/patent/US9198984B2/en. Accessed 4 July 2022

Lock M, Vandenberghe LH, Wilson JM (2016) Scalable production method for AAV. United States. Available at: https://patents.google.com/patent/US20160040137A1/en. Accessed 4 July 2022

Lothert K, Sprick G et al (2020a) Membrane-based steric exclusion chromatography for the purification of a recombinant baculovirus and its application for cell therapy. J Virol Methods 275. https://doi.org/10.1016/j.jviromet.2019.113756

Lothert K, Pagallies F et al (2020b) Selection of chromatographic methods for the purification of cell culture-derived Orf virus for its application as a vaccine or viral vector. J Biotechnol 323: 62–72. https://doi.org/10.1016/j.jbiotec.2020.07.023

Mahajan E, George A, Wolk B (2012) Improving affinity chromatography resin efficiency using semi-continuous chromatography. J Chromatogr A 1227:154–162. https://doi.org/10.1016/j.chroma.2011.12.106

Marichal-Gallardo P et al (2017) Steric exclusion chromatography for purification of cell culture-derived influenza A virus using regenerated cellulose membranes and polyethylene glycol. J Chromatogr A 1483:110–119. https://doi.org/10.1016/j.chroma.2016.12.076

Marichal-Gallardo P et al (2021) Single-use capture purification of adeno-associated viral gene transfer vectors by membrane-based steric exclusion chromatography. Hum Gene Ther 32(17–18):959–974. https://doi.org/10.1089/hum.2019.284

Marino MP et al (2015) A scalable method to concentrate lentiviral vectors pseudotyped with measles virus glycoproteins. Gene Ther 22(3):280–285. https://doi.org/10.1038/gt.2014.125

McNally DJ et al (2014) Optimised concentration and purification of retroviruses using membrane chromatography. J Chromatogr A 1340:24–32. https://doi.org/10.1016/j.chroma.2014.03.023

McNally DJ et al (2020) Development and optimization of a hydrophobic interaction chromatography-based method of AAV harvest, capture, and recovery. Mol Ther Methods Clin Dev 19:275–284. https://doi.org/10.1016/j.omtm.2020.09.015

Mekkaoui L et al (2018) Lentiviral vector purification using genetically encoded biotin mimic in packaging cell. Mol Ther Methods Clin Dev 11:155–165. https://doi.org/10.1016/j.omtm.2018.10.008

Merten OW, Hebben M, Bovolenta C (2016) Production of lentiviral vectors. Mol Ther Methods Clin Dev 16017. Elsevier. https://doi.org/10.1038/mtm.2016.17

Michen B, Graule T (2010) Isoelectric points of viruses. J Appl Microbiol 388–397. https://doi.org/10.1111/j.1365-2672.2010.04663.x

Moreira AS et al (2021) Enhancing the purification of Lentiviral vectors for clinical applications. Sep Purif Technol 274:118598. https://doi.org/10.1016/j.seppur.2021.118598

Mustang Q XT chromatography capsules (2009) Filtration- separation. Available at: https://shop.pall.co.uk/uk/en/biotech/chromatography/membranes/zidgri78lhe. Accessed 29 June 2022

Nass SA et al (2018) Universal method for the purification of recombinant AAV vectors of differing serotypes. Mol Ther Methods Clin Dev 9:33–46. https://doi.org/10.1016/j.omtm.2017.12.004

Nestola P et al (2015) Improved virus purification processes for vaccines and gene therapy. Biotechnol Bioeng 843–857. Wiley. https://doi.org/10.1002/bit.25545

Okada T et al (2009) Scalable purification of adeno-associated virus serotype 1 (AAV1) and AAV8 vectors, using dual ion-exchange adsorptive membranes. Hum Gene Ther 20(9):1013–1021. https://doi.org/10.1089/hum.2009.006

Oranje P et al (2004) 423. Isolation of an adeno-associated virus (AAV)-specific camelid-derived single chain antibody fragment: a novel tool for purification of AAV vectors of different serotypes. Mol Ther 9:S162. https://doi.org/10.1016/j.ymthe.2004.06.344

Orr V et al (2013) Recent advances in bioprocessing application of membrane chromatography. Biotechnol Adv 450–465. Elsevier. https://doi.org/10.1016/j.biotechadv.2013.01.007

Peixoto C et al (2006) Purification of adenoviral vectors using expanded bed chromatography. J Virol Methods 132(1–2):121–126. https://doi.org/10.1016/j.jviromet.2005.10.002

Perry C, Rayat ACME (2021) Lentiviral vector bioprocessing. Viruses 13(2):268. https://doi.org/10.3390/v13020268

Podgornik A, Štrancar A (2005) Convective Interaction Media® (CIM)—Short layer monolithic chromatographic stationary phases. Biotechnol Annu Rev 281–333. https://doi.org/10.1016/S1387-2656(05)11009-6. Elsevier

Podgornik AŠ et al (2013) Fast separation of large biomolecules using short monolithic columns. J Chromatogr B Analyt Technol Biomed Life Sci 80–89. https://doi.org/10.1016/j.jchromb.2013.02.004

Qu G et al (2007) Separation of adeno-associated virus type 2 empty particles from genome containing vectors by anion-exchange column chromatography. J Virol Methods 140(1–2):183–192. https://doi.org/10.1016/j.jviromet.2006.11.019

Qu W et al (2015) Scalable downstream strategies for purification of recombinant adeno-associated virus vectors in light of the properties. Curr Pharm Biotechnol 16(8):684–695. https://doi.org/10.2174/1389201016666150505122228

Ravault V, Laroudie N (2022) Platform optimization for efficient AAV purification: insights from a CDMO. Cell Gene Ther Insights 08(01):145–158. https://doi.org/10.18609/cgti.2022.001

Rieser R et al (2021) Comparison of different liquid chromatography-based purification strategies for adeno-associated virus vectors. Pharmaceutics 13(5). https://doi.org/10.3390/pharmaceutics13050748

Rodrigues DS, Salomão R, Kallas EG (2006) Profound peripheral T-lymphocyte depletion and activation in disseminated tuberculosis. Braz J Infect Dis 10(1):59–61. https://doi.org/10.1016/j.jchromb.2006.03.061

Ruscic J et al (2019) Lentiviral vector purification using nanofiber ion-exchange chromatography. Mol Ther Methods Clin Dev 15:52–62. https://doi.org/10.1016/j.omtm.2019.08.007

Saksule A (2022) Adeno-associated viral vector process development challenges in early research & preclinical study. Cell Gene Ther Insights 08(03):411–420. https://doi.org/10.18609/cgti.2022.060

Schmit PF et al (2020) Cross-packaging and capsid mosaic formation in multiplexed AAV libraries. Mol Ther Methods Clin Dev 17:107–121. https://doi.org/10.1016/j.omtm.2019.11.014

Segura MM et al (2007) Production of lentiviral vectors by large-scale transient transfection of suspension cultures and affinity chromatography purification. Biotechnol Bioeng 98(4):789–799. https://doi.org/10.1002/bit.21467

Segura MM, Kamen AA, Garnier A (2011) Overview of current scalable methods for purification of viral vectors. Methods Mol Biol (Clifton, N.J.) 89–116. https://doi.org/10.1007/978-1-61779-095-9_4

Shirley JL et al (2020) Immune responses to viral gene therapy vectors. Mol Ther 709–722. Cell Press. https://doi.org/10.1016/j.ymthe.2020.01.001

Silva RJ et al (2015) Improving the downstream processing of vaccine and gene therapy vectors with continuous chromatography. Pharm Bioprocess 3(8):489–505. https://doi.org/10.4155/pbp.15.29

Slepushkin V et al (2003) Large-scale purification of a lentiviral vector by size exclusion chromatography or mustang Q ion exchange capsule. BioProcess J 2(5):89–95. https://doi.org/10.12665/j25.dropulic

Smith RH, Levy JR, Kotin RM (2009) A simplified baculovirus-AAV expression vector system coupled with one-step affinity purification yields high-titer rAAV stocks from insect cells. Mol Ther 17(11):1888–1896. https://doi.org/10.1038/mt.2009.128

Srivastava A et al (2021) Manufacturing challenges and rational formulation development for AAV viral vectors. J Pharm Sci 2609–2624. Elsevier BV. https://doi.org/10.1016/j.xphs.2021.03.024

Sviben D et al (2017) Recovery of infective virus particles in ion-exchange and hydrophobic interaction monolith chromatography is influenced by particle charge and total-to-infective particle ratio. J Chromatogr B Analyt Technol Biomed Life Sci 1054:10–19. https://doi.org/10.1016/j.jchromb.2017.04.015

Terova O, Hermans, P, Soltys, S, Detmers F (2017) Affinity chromatography accelerates viral vector purification for gene therapies. BioPharm Int 27–35. MJH Life Sciences

Terova O et al (2018) Overcoming downstream purification challenges for viral vector manufacturing: enabling advancement of gene therapies in the clinic. Cell Gene Ther Insights 4(2): 101–111. https://doi.org/10.18609/cgti.2018.017

Tinch S et al (2019) A scalable lentiviral vector production and purification method using mustang Q chromatography and tangential flow filtration. Methods Mol Biol 135–153. Humana. https://doi.org/10.1007/978-1-4939-9065-8_8

Toueille M et al (2018) Development of purification steps for several AAV serotypes using POROS™ CaptureSelect™ AAVX affinity chromatography. Cell Gene Ther Insights 4(7): 637–645. https://doi.org/10.18609/cgti.2018.061

Turnbull J et al (2019) Adenovirus 5 recovery using nanofiber ion-exchange adsorbents. Biotechnol Bioeng 116(7):1698–1709. https://doi.org/10.1002/bit.26972

Urabe M et al (2006) Removal of empty capsids from type 1 adeno-associated virus vector stocks by anion-exchange chromatography potentiates transgene expression. Mol Ther 13(4):823–828. https://doi.org/10.1016/j.ymthe.2005.11.024

U.S. Food and Drug Administration (FDA) (2021) Approved cellular and gene therapy products. U.S Food and Drug Administration, pp 14–16. Available at https://www.fda.gov/vaccines-blood-biologics/cellular-gene-therapy-products/approved-cellular-and-gene-therapy-products. Accessed 30 June 2022

Valkama AJ et al (2020) Development of large-scale downstream processing for lentiviral vectors. Mol Ther Methods Clin Dev 17:717–730. https://doi.org/10.1016/j.omtm.2020.03.025

Vance MA, Mitchell A, Samulski RJ (2015) AAV biology, infectivity and therapeutic use from bench to clinic. In: Gene therapy—principles and challenges. InTech. https://doi.org/10.5772/61988

Vellinga J et al (2014) Challenges in manufacturing adenoviral vectors for global vaccine product deployment. Hum Gene Ther 318–327. Mary Ann Liebert Inc. https://doi.org/10.1089/hum.2014.007.

Venkatakrishnan B et al (2013) Structure and dynamics of adeno-associated virus serotype 1 VP1-unique N-terminal domain and its role in capsid trafficking. J Virol 87(9):4974–4984. https://doi.org/10.1128/jvi.02524-12

Warikoo V et al (2012) Integrated continuous production of recombinant therapeutic proteins. Biotechnol Bioeng 109(12):3018–3029. https://doi.org/10.1002/bit.24584

Winkler M et al (2021) Viral clearance in a downstream AAV process: case study using a model virus panel and a noninfectious surrogate. BioProcess Int. Available at: https://www.biaseparations.com/en/library/publications/1112/viral-clearance-in-a-downstream-aav-process-case-study-using-a-model-virus-panel-and-a-noninfectious-surrogate. Accessed 1 July 2022

Ye GJ et al (2015) Safety and biodistribution evaluation in cynomolgus macaques of rAAV2tYF-CB-hRS1, a recombinant adeno-associated virus vector expressing retinoschisin. Hum Gene Ther Clin Dev 26(3):165–176. https://doi.org/10.1089/humc.2015.076

Yu JH, Schaffer DV (2006) Selection of novel vesicular stomatitis virus glycoprotein variants from a peptide insertion library for enhanced purification of retroviral and lentiviral vectors. J Virol 80(7):3285–3292. https://doi.org/10.1128/jvi.80.7.3285-3292.2006

Zhao M et al (2019) Affinity chromatography for vaccines manufacturing: Finally ready for prime time? Vaccine 37(36):5491–5503. https://doi.org/10.1016/j.vaccine.2018.02.090
Zimmermann K et al (2011) Highly efficient concentration of lenti- and retroviral vector preparations by membrane adsorbers and ultrafiltration. BMC Biotechnol 11(1):55. https://doi.org/10.1186/1472-6750-11-55
Zincarelli C et al (2008) Analysis of AAV serotypes 1-9 mediated gene expression and tropism in mice after systemic injection. Mol Ther 16(6):1073–1080. https://doi.org/10.1038/mt.2008.76
Zydney AL (2021) New developments in membranes for bioprocessing—a review. J Membr Sci. Elsevier BV. https://doi.org/10.1016/j.memsci.2020.118804.

Chapter 9
Particle Analytics: Comparative Approaches for Analysis of Viral Vectors

Irini Skaripa-Koukelli, Aurelia Raducanu, Joost van Haasteren, Juliana Albrecht, Philippe Ringler, and Alexandra H. E. Machado

9.1 Introduction

Recombinant adeno-associated virus (rAAV) has become the leading platform for ongoing in vivo gene therapy efforts, thanks to its favorable safety profile and ability to infect a broad spectrum of cell types. The development of engineered capsids with clinically desirable features and optimized genome designs, along with the recent approval of rAAV-based gene therapies in the United States and Europe have proved the tremendous potential of these therapies and generated optimism in the field (Dunbar et al. 2018; Wang et al. 2019b). Although rAAV is a relatively simple virus, it is still more complex not only in comparison with small molecule therapeutics but even when compared with protein-based therapeutics, as it is composed of two elements, namely the protein capsid and a DNA genome (Wang et al. 2019b;

I. Skaripa-Koukelli · J. Albrecht · A. H. E. Machado (✉)
Pharma Technical Development Biologics Europe (PTDE), F. Hoffmann-La Roche Ltd., Basel, Switzerland
e-mail: irini.skaripa_koukelli@roche.com; juliana.albrecht@roche.com; alexandra.machado@roche.com

A. Raducanu
Pharma Technical Development Cell & Gene Therapy (PTC), Roche Diagnostics GmbH, Penzberg, Germany

J. van Haasteren
Pharma Technical Development Cell & Gene Therapy (PTC), F. Hoffmann-La Roche Ltd., Basel, Switzerland
e-mail: joost.van_haasteren@roche.com

P. Ringler
Biozentrum, University of Basel, Basel, Switzerland
e-mail: philippe.ringler@unibas.ch

S. Gautam et al. (eds.), *Bioprocess and Analytics Development for Virus-based Advanced Therapeutics and Medicinal Products (ATMPs)*,
https://doi.org/10.1007/978-3-031-28489-2_9

Ramsey et al. 2021). This results in an increased complexity of the analytical requirements to assess whether the rAAV products meet the required specifications.

Physicochemical and biological characterization of rAAVs is required over the course of the therapeutic development to test the impact of various modifications, e.g., differentially engineered capsids or alternative formulations during early development, as well as post-approval to compare between batches, different manufacturing sites, and production processes (Cockroft and Wilson 2021). Properties that must be monitored and controlled include the capsid and genomic titer or the preparation, the empty capsid content (full-to-empty ratio), genome and capsid integrity and, finally, the extent of aggregation. For example, empty capsids generate the need for an increased dose to achieve the expected therapeutic effect, and can also give rise to undesirable immune reactions, which raise safety concerns (Wright 2014). At least two scenarios are possible in the case of partially filled capsids: the genome might not be intact or, alternatively, the incorrect DNA might be encapsulated. As is the case for empty capsids, the former can cause immunogenicity along with the need for an increased dose. The latter has the potential to induce the expression of an undesirable protein, which could also be immunogenic. Additionally, aggregated particles can, at best, result in a dampened therapeutic effect—which can still be detrimental if only a single administration is possible, but can also trigger an immune reaction (Wright et al. 2005; Mohammad 2021). Thus, a robust set of analytical tools, including orthogonal methods, is needed to guarantee that all the quality criteria are met uniformly to ensure the safety and efficacy of gene therapy products and to harness the full potential of these immensely promising therapeutics.

At present, a variety of methods are used to assess capsid and genomic titer and integrity, the full-to-empty (F/E) ratio, and aggregation status, each with its own advantages and limitations, including their suitability for good manufacturing practice (GMP) procedures, which require qualified instrumentation, method validation, trained users and good documentation practices among others. In this chapter, we limit our discussion to analytical methods associated with particle characterization, while we refer the reader to the relevant chapters for the characterization of rAAV formulations in terms of other properties, e.g., potency and capsid identification. While not exhaustive, this chapter describes the main analytical methods currently used by the biopharmaceutical industry at the various stages of rAAV development for this purpose. We provide the theoretical basis for each technique (although in-depth analysis of the basic principles is out of the scope of this chapter) and focus on the advantages, limitations, and practical details related to the analysis of rAAV samples.

9.2 Analytical Ultracentrifugation (AUC)

Ultracentrifugation is employed for the purification and characterization of rAAVs, and it is classified as preparative and analytical ultracentrifugation (AUC), respectively. Like centrifugation, ultracentrifugation separates molecules or particles based on their densities. Particles with densities higher than the density of the medium sediment, while those with densities lower than the density of the medium float, and centrifugation accelerates that sedimentation or floating process. In AUC, a similar instrument to preparative ultracentrifugation is used (e.g., Beckman Coulter Optima XPN and Optima AUC, for preparative and analytical ultracentrifugation, respectively), the difference being that in AUC the movement of concentration gradient of the analyte in the centrifugation cell is monitored over time and that information is then processed using appropriate software to derive information about the molar mass, shape, size, and aggregation state of the analyte.

The birth of AUC dates back to the beginning of the twentieth century, when T. Svedberg conducted the first AUC experiments to study colloid systems (Nobel prize in chemistry, 1926). AUC has been a powerful tool in biological and chemical research throughout the twentieth century, as exemplified by its use for major discoveries such as the mechanism of DNA replication. Gradient density ultracentrifugation using cesium chloride (CsCl) or iodixanol gradients is traditionally used to separate empty, full, and partially filled capsids, but it is now more commonly used during downstream processing of rAAV. Today, AUC is a versatile technique used for the separation and biophysical characterization of various types of particles and biomolecules, including rAAVs, and modern AUC for rAAVs is being optimized for use with the native rAAV medium, without a CsCl or iodixanol gradient.

The sedimentation coefficient (S) measures the rate of sedimentation of an analyte, and is defined as the ratio of the velocity per unit centrifugal acceleration, based on the Svedberg equation (Eq. 9.1):

$$S = \frac{\upsilon}{\omega^2 r} = \frac{D \times M_{\mathrm{B}}}{RT} = \frac{m_o(1 - \nu_0 \rho_w)}{6\pi\eta R_o \frac{f}{f_o}} \tag{9.1}$$

where υ is the sedimentation speed, ω is the angular velocity of the rotor and r is the distance of a particle to the rotor axis ($\omega^2 r$ is referred to as centrifugal acceleration). From this equation, we can derive the following insights:

- S is proportional to the analyte's buoyant mass (M_{B}), hence heavier particles (e.g. full rAAV capsids) sediment faster than less dense particles (e.g., empty rAAV capsids)
- S is proportional to the analyte's density (ρ), hence denser particles (e.g. iron nanoparticles) sediment faster than less dense particles (e.g., polystyrene nanoparticles)
- S is inversely proportional to the shape factor of the analyte ($\frac{f}{f_o}$), hence, isotropic particles (such as supercoiled DNA) will sediment faster than anisotropic/spread-out particles (such as linearized DNA) (Bhattacharya 2019).

The sedimentation coefficient is measured in Svedberg units (S) which is essentially a measure of time (1 S = 10–13 s) but it also offers a measure of particle size.

Two types of analysis can be performed using AUC, namely, sedimentation velocity (SV-AUC) and sedimentation equilibrium (SE-AUC), with the former being the method of choice to characterize rAAVs. In SV-AUC, high rotational speed is employed (up to 289,848 rcf or 60,000 rpm with the Beckman Coulter Optima AUC system) (Demeler 2022), which gradually depletes the analyte from the center of the rotor. A detector (UV-Vis, Rayleigh interference (dRI), fluorescence) monitors the resulting concentration gradient of the analyte across the length of the cell over time, from which, in turn, the sedimentation speed can be deduced. The sedimentation speed of the analyte in a specific medium is determined by its size, molecular weight, and shape. The resulting readout is an absorbance versus radius plot repeated at multiple time points during the centrifugation time. Initially ($t = 0$), the absorbance versus radius plot is expected to be a flat line at maximum absorbance (total analyte concentration) indicating a homogeneous distribution of the analyte across the length of the centrifugation cell. The absorbance signal—hence the analyte concentration—lowers close to the meniscus (small radius), and increases toward the bottom of the cell (large radius) as the centrifugation proceeds, and the analyte eventually forms a pellet at the bottom of the centrifugation cell. Multiple plots are collected over time and their analysis provides the sedimentation coefficient and the diffusion coefficient of the analyte, according to O. Lamm's Hydrodynamic theory (1929). This allows an estimation of the molar mass, and shape (anisotropy) and molecular weight (MW) of the studied particle (Cole et al. 2008; Bhattacharya 2019; Sucato 2020; Haberland et al. 2021).

In contrast, the SE-AUC experiment is performed at a lower rotor speed as the aim is to find the point at which diffusion and sedimentation forces balance each other out to create an equilibrium radial concentration gradient. Accordingly, a pellet is not expected in a SE-AUC experiment. By analyzing the concentration gradient at equilibrium, an accurate measurement of the molecular weight of the analyte can be obtained. Molar mass measurements obtained by SE-AUC are surpassed in accuracy by mass spectrometry measurements, however, SE-AUC has two major advantages over mass spectrometry (MS) and liquid chromatography–mass spectrometry (LC-MS) for macromolecules and large constructs. First, SE-AUC has a very wide range of MW that it can measure and, second, the macromolecule or construct can be measured in essentially any buffer, and therefore it can be in a biologically relevant state. The tradeoff for the better accuracy of SE-AUC over SV-AUC is that the experiment is longer (a few hours for SV vs a few days for SE-AUC) (Cole et al. 2008; Bhattacharya 2019; Sucato 2020).

9.2.1 Practical Considerations for rAAV Analysis

- **Determining the Required rAAV Titer and Sample Volume**
  One of the first questions arising when considering the introduction of a new characterization method is what concentration (or titer, in the case of viral vectors) is needed to acquire reliable data. Optima AUC, the gold standard AUC instrument, is equipped with UV-Vis, fluorescence and differential refractive index (dRI) detectors, and the absorbance or emission wavelengths can be

recorded in one experiment. The generic detection wavelength is 230 nm and the appropriate concentration can be determined by measuring the optical density (absorbance) of serial dilutions of a rAAV sample with a known titer. Generally, optical density of approximately 0.5 ABU is appropriate for measurement with AUC, but the acceptable range is 0.1–1 ABU for a path length of 10 mm. This can be optimized using a NanoDrop prior to the AUC measurement. Optical density of 0.5 ABU roughly corresponds to a titer of 1×10^{12} viral particles per mL (vp/mL) for rAAV samples. It is important to consider the required sample volume, as it is higher than what is required for most other characterization techniques used for the quality control (QC) of rAAV samples. At the time of writing, the minimum volume for Optima AUC is 0.1 mL (maximum volume 0.5 mL). The resolution of the measurement, however, increases with an increase in sample volume (minimization of the air gap) because the air gap in the cell resulting from smaller sample volumes obviously does not provide any sedimentation information. Therefore, higher sample volumes are preferable.

- **Experimental Parameters and Throughput**

  When using absorbance to measure the sedimentation coefficient, the extinction coefficients of the samples are in general required to determine the molar concentration and the full and empty capsids. More recently, a new analysis method has been developed, where the F/E capsid ratio can be determined without the extinction coefficient at the isosbestic wavelength (Maruno et al. 2021). Three to seven samples can be analyzed in a single run using four or eight-slot rotors, respectively, as one of the slots is needed for balance. Experiments can run up to 6–8 h per round/sample.

- **Full-to-Empty (F/E) Ratio Determination**

  Analysis of rAAVs can be performed with monitoring the sedimentation coefficient at multiple wavelengths, which allows for the identification and quantitation of both the protein (280 nm) and the DNA (260 nm) components. By performing simultaneous detection at these two wavelengths, AUC can provide information on the F/E capsid ratio of rAAVs. Additional measurements at 230 nm can provide higher sensitivity at low concentrations (Haberland et al. 2021). The multiple wavelength analysis also allows the identification of partially filled, aggregated, or capsids overloaded with DNA.

- **Data Analysis**

  SEDFIT/SEDPHAT and ULTRASCAN are the most commonly used, user community-developed software for the analysis of AUC data. Both programs, along with their source code, are available on GitHub (https://sedfitsedphat.github.io/download.htm for SEDFIT, and https://github.com/ehb54/ultrascan3 for Ultrascan) and benefit from strong support by the AUC community. The methods used for the analysis of AUC data can be clustered based on whether they are based on a model or not. In "model-free" approaches, the minimum assumptions are made and are therefore suitable for the initial analysis of an unknown system. Further, more detailed analysis of particle interactions may require a model-based method. Numerical fitting algorithms from the software deconvolve the raw data into peak distributions (Cole et al. 2008). Although

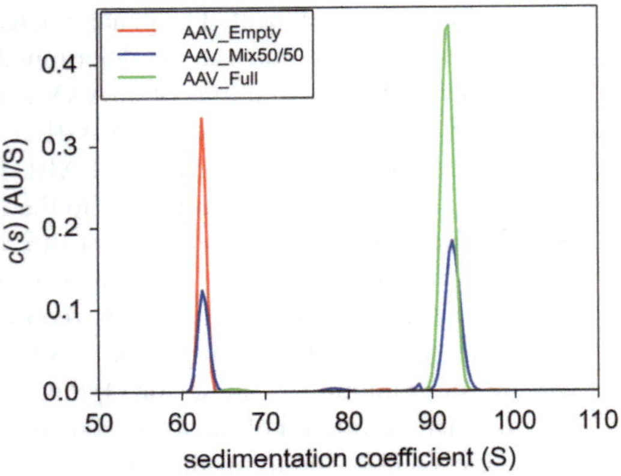

Fig. 9.1 Overlays of distribution plots of the sedimentation coefficient of rAAV8 empty (red), rAAV8 full (green), and a mixture of full and empty rAAV8 (red) obtained by SV-AUC measurements

variations may happen for different rAAV serotypes, empty and full particles appear as peaks around 60–70 S and 90–100 S, respectively (Fig. 9.1). Partially filled capsids appear in between, whereas overfilled capsids and aggregates appear at higher and fragments at lower sedimentation coefficients (Burnham et al. 2015; Maruno et al. 2021).

- **Other Analytical Parameters**

  AUC has very good resolution, comparable precision, and repeatability to chromatographic techniques and shows good linearity. The limit of detection (LOD) and limit of quantification (LOQ) are defined by the noise of the data curves (the raw curves, not the deconvoluted ones). The LOQ has been found to be in the range of 3×10^{11} vp/mL (Maruno et al. 2021).

AUC is considered the gold standard for the determination of capsid loading and product-specific impurity analysis of rAAVs. By allowing simultaneous multi-wavelength analysis (Optima AUC), it can provide identification and reliable quantitation of multiple species present in samples (e.g., empty, partially and fully filled capsids, overfilled capsids) as well as an estimation of the size of DNA contained in each of the species. It can measure the samples in their native state and particle separation can be monitored in situ and in real time (Burnham et al. 2015). Unlike chromatographic methods, it does not require the use of standards, is not impacted by peak broadening due to dilution nor does it rely on stationary phase, hence excluding preferential interactions with certain species. Additionally, it does not require staining as, e.g., negative staining electron microscopy and it is a non-destructive method, i.e., the sample can be recovered after the analysis. AUC is a very valuable technique in evaluating different rAAV production processes (Burnham et al. 2015; Nass et al. 2017).

However, AUC implementation as a routine analytical method is limited by several factors. Its low throughput, lack of scalability, long runs, and the requirement to use large volumes of sample at relatively high concentrations can make it difficult to implement to monitor processes or in large screenings. Furthermore, it requires specialized equipment as well as technical expertise to operate the equipment and analyze the data, and it may be challenging to validate in a GMP setting due to concerns with data analysis software compliance (Nass et al. 2017). Nevertheless, AUC often outperforms other methods in providing accurate and precise rAAV F/E ratios as seen in various studies and it can be a valuable orthogonal method to facilitate validation of other methods with lower resolution but shorter run times (Fu et al. 2019; Haberland et al. 2021; Werle et al. 2021). It should be part of the extended characterization analytical panel for rAAV characterization.

9.3 Digital Polymerase Chain Reaction/Enzyme-Linked Immunosorbent Assay (dPCR/ELISA)

9.3.1 Background

A common method to determine the F/E ratio of rAAV is by comparing the capsid titer with the genomic titer. Here, two independent methods are combined where one queries the amount of intact viral capsids and the other the amount of encapsulated viral genomes. In the case of a viral preparation in which 50% of the capsids do not contain a viral genome and thus the F/E ratio is 0.5, the capsid titer would be expected to be twice as high as the genomic titer. Genomic titers, as discussed in an earlier chapter, are customarily measured through quantitative or digital polymerase chain reaction (PCR) methods whereas capsid titer is measured through enzyme-linked immunosorbent assay (ELISA). These methods are commonplace in many (molecular biology) labs that work with rAAV vectors and both genome and capsid titers are routinely measured, making F/E ratio a "free" data point in the appraisal of rAAV CQAs. However, since the genome:capsid titer ratio approach is based on two independent methods, it compounds the potential drawbacks of both those methods.

9.3.2 Capsid Titer

There are multiple methods that can, as discussed in this chapter, measure the amount of (intact) rAAV particles in a preparation, including SEC(-MALS), SLS, and EM. However, these all require costly specialist equipment that may not be readily available in many labs. ELISA on the contrary is a commonly used assay that can also be employed to measure the amount of intact viral capsids. It works by capturing rAAV particles with a first antibody that is attached to the bottom of a well. After a washing step, a second antibody is then used to bind to the captured rAAV, after which an enzyme (commonly horseradish peroxidase) that is conjugated to the second antibody or attached via a biotin-streptavidin bond is activated to metabolize a chemical that can be absorbed in a common plate reader. The amount of measured

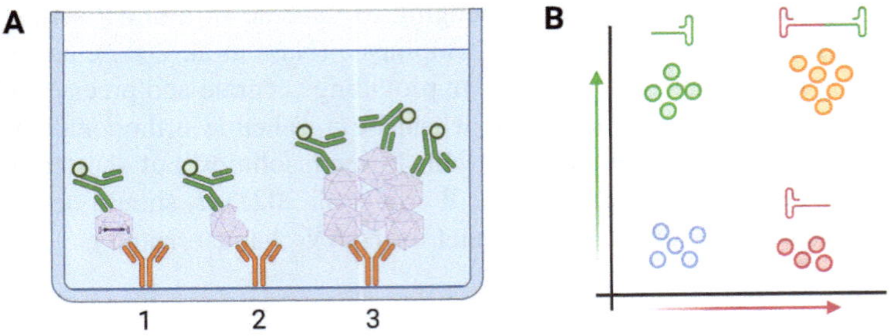

Fig. 9.2 rAAV direct ELISA (**a**). rAAV particles are captured by an immobilized anti-rAAV antibody (brown). Capsids are then detected by the second anti-rAAV antibody that is conjugated to a detection enzyme (e.g., horseradish peroxidase). This method suffers from the possibility of detecting broken capsids (2) or aggregates (3) which overestimates and underestimates the particle titer respectively. Genome integrity dPCR (**b**) relies on the amplification of two targets (here indicated with red and green halves of the rAAV genome). Double positive droplets (in orange, top right) suggest the presence of an intact genome, whereas green and red droplets represent single positive droplets and hence truncated genomes. Created with BioRender.com

metabolized chemical correlates with the amount of rAAV that is captured by the first and bound by the second antibody (Fig. 9.2a).

9.3.2.1 Considerations

The use of antibodies in the ELISA assay makes most of these assays serotype specific, with some exceptions where epitopes are detected that are common between different serotypes. As a result, these types of ELISA can be used for the identification of rAAV gene therapy products for instance during manufacturing. On the other hand, this specificity could hamper the pharmaceutical development of novel engineered capsids when no suitable ELISA is available commercially and may need to be developed in-house. Perhaps because commercially available assays are provided with an internal standard curve, the variability between replicates and between labs is small compared to qPCR-based genome titer (Lock et al. 2010; Ayuso et al. 2014). To our knowledge, no direct, quantitative, multi-lab comparison is made between ELISA and other methods of viral particle analysis. Compared to other titration methods such as quantitative PCR (qPCR) or digital PCR (dPCR), ELISA is time-consuming when performed manually but does allow for automation through the use of liquid handling robots or specialized equipment (ProteinSimple 2022). While the commonly used ELISA assay by manufacturer ProGen is advertised to be specific to intact rAAV particles by means of recognition of a structural epitope (i.e., denatured rAAV viral proteins, VP, will not be recognized) the multifaceted nature of the rAAV dodecahedron capsid suggests that a capsid particle that is broken in half still may have ~30 recognition sites for the capture and detection antibody to bind and show up as a false-positive "intact" particle (see Fig. 9.3 for an example of broken rAAV particles and Fig. 9.2a2). Furthermore,

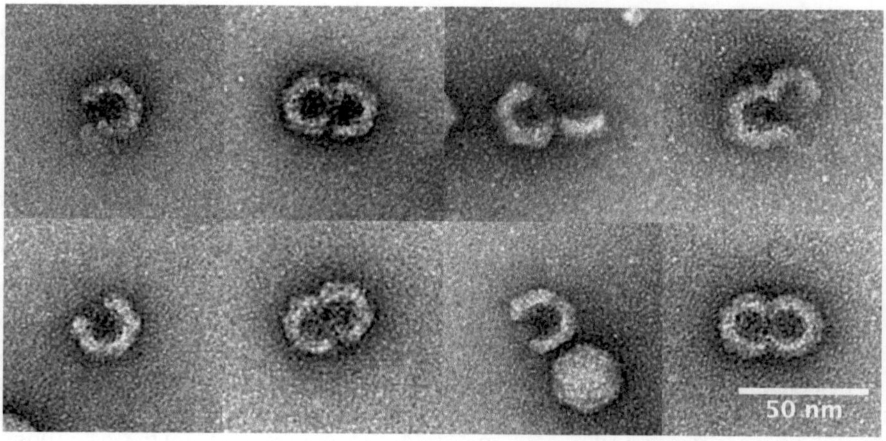

Fig. 9.3 Transmission electron micrographs depicting rAAV2 broken capsids after negative staining with uranyl acetate

aggregation may cause an underestimation of particle titer as particles in the middle of a large aggregate will not be measured (Fig. 9.2a3).

9.3.3 Genomic Titer

The genomic titer of a rAAV preparation commonly relies on the amplification of a specific region of the viral genome through PCR. The concentration of viral genomes is then quantified in relative terms by comparing the amplification curves with that of a standard curve of a known concentration (termed quantitative PCR or qPCR) or in absolute terms by the detection of amplification and thus the presence of the genome in small nanoliter size droplets or compartments (termed digital PCR or dPCR). In case the amplified target is in a region of a potential contaminating sequence (e.g., the plasmid backbone of the viral genome used in virus production), of which the sequence is known a priori, PCR methods can also effectively be used to determine the contamination in a sample. As expected, the ability to accurately quantify the genomic titer is paramount for an accurate F/E ratio determination.

9.3.3.1 Considerations

There are several rAAV-specific common pitfalls that should be avoided to achieve accurate genome titration results using dPCR, which are highlighted below. Considerations common to all dPCR applications have been comprehensively described in the Minimum Information for Publication of Digital PCR Experiments (dMIQE) guidelines (The dMIQE Group et al. 2020).

- **DNAse Digestion.** Contaminating and genomic DNA that is present outside of the capsid is commonly removed by DNAse digestion prior to partitioning/droplet generation. This ensures that only encapsulated (and thus active) genomes are measured. Care must be taken to inactivate the DNAse to prevent the enzyme from digesting the amplification primers or detection probe. Doing this at an early stage in the dilution series common for dPCR rAAV titration (see below) allows for dilution of the DNAse buffer components that may interfere with PCR efficacy.
- **Decapsidation.** To allow efficient amplification of the target on the rAAV genome, the genome must be released from the capsid shell. This is routinely done by application of heat, commonly in the first enzyme activation step of the thermocycling protocol (Lock et al. 2010; Dobnik et al. 2019); however, there is evidence that digestion with proteinase K increases the measured genome titer (Suoranta et al. 2021). There is some evidence that Rep proteins are bound covalently to the rAAV genome (Snyder et al. 1990; Prasad and Trempe 1995), which may interfere with the processing of the DNA polymerase enzyme during amplification, something which a proteinase K treatment would prevent.
- **Detection Range and Loss.** dPCR is a very sensitive assay and can theoretically be as little as a single copy per well. However, the upper limit of detection is determined by the amount of partitions or droplets that are formed during partitioning, which is roughly 10,000–20,000 for the varying machines that are on the market. This means that the range is quite limited at max four orders of magnitude. Thus, for many vector preparations, which are in the 1×10^{13}–1×10^{14} vector genomes per mL (vg/mL) range, a high dilution factor has to be pipetted to hit this range. As a result, a relatively small error or unexpected loss can accumulate and have a major influence on the accuracy of the assay. To overcome this, the use of surfactants, tubes that avoid binding DNA and carrier RNA have been proposed to avoid the adsorption of DNA to the walls of the used consumables (Sanmiguel et al. 2019).
- **Controls.** Given the many processing steps, appropriate controls are required. Plasmid DNA can be used to determine the efficacy of the DNAse enzyme and a no-template control is useful to address contamination. A positive rAAV control of known concentration (with a given assay) whose result must be between a predetermined range is advised to determine whether the current experiment is within the limits set as well as to address variability between experiments and users.

9.3.4 Genome Integrity dPCR

While F/E ratio is commonly based on the ratio between genomic and capsid titer, one might argue that the use of the amount of intact rAAV genomes instead of genomic titer is more predictive of the potency of the tested viral preparation. Genomic titer measured through qPCR or dPCR only reflects the concentration of a specific part of the rAAV genome, namely the target of PCR amplification, which is normally 100–150 bp in size. The presence of the remaining potential 4.6 kb of sequence is therefore an open question. The percentage of intact viral genomes encapsulated into rAAV capsids is a quality attribute that has a major impact on the potency of rAAV preparations as truncated or fragmented genomic DNA molecules that are loaded into the rAAV capsids are less likely to lead to successful transduction. Truncated rAAV genomes showed substantially lower transgene expression capacity in vitro in an experiment in which an oversized genome was attempted to be packaged into rAAV capsids (Wu et al. 2010). Any residual expression was determined to likely be a result of genome recombination (Wu et al. 2010). Third-generation sequencing methods such as GP-seq (Tai et al. 2018) measure genome integrity by exploiting the long-range sequencing capabilities of platforms such as PacBio and Oxford Nanopore and are able to sequence the entire length of the genome, provided the necessary sequencing adapters can be readily ligated to the ends of the extracted rAAV genomes. dPCR-mediated genome integrity assays rely on the ability of digital PCR to amplify and detect two or more targets on the rAAV genome simultaneously (Fig. 9.2b). A correlation was found between the percentage of intact rAAV genomes measured through dPCR and its potency in an in vitro fluorescent reporter gene assay (Furuta-Hanawa et al. 2019). An intact genome will show up in the analysis as a double (or triple or quadruple depending on the amount of targets used) positive droplet or compartment whereas a genome that is truncated and has lost one of the two targets on the genome will be measured being positive for only one of the fluorophores associated with the amplified target. Since the dPCR approach relies on relatively small (~100–150 bp) amplicons, any truncations that have occurred outside of the most 3′ and 5′ amplicons will not be picked up. This can partially be overcome by the use of multiple targets and target combinations and designing them toward the ends of the genome. It is likely unwise to rely on amplicon targets on the rAAV ITR sequence (which are the outermost sequences of the rAAV genome) since these amplicons will be present on both ends of the genome, thus creating a false-positive when one of the two ITRs is intact. Furthermore, care must be taken to prevent a concentration that is too high, as this may inflate the amount of perceived intact genomes by virtue of multiple truncated genomes in the same droplet or compartment being measured as an intact genome. Nonetheless, we argue that using the amount of intact genomes, a percentage of the genome titer will likely create a more representative idea of the F/E ratio as an estimate of potency as partial genomes are omitted from the equation.

9.4 Dynamic Light Scattering (DLS)

Dynamic light scattering (DLS), also known as Photon Correlation Spectroscopy or Quasi-Elastic Light Scattering, is a well-established technique used to calculate the size and size distribution of sub-micron particles (0.5 nm–1μm) based on the measurement of their Brownian motion. Despite certain limitations mainly related to its low resolution, DLS has gained popularity as a technique to estimate the extent of aggregation in rAAV formulations. When combined with static light scattering (SLS) or when scattered light is measured at multiple angles (multi-angle dynamic light scattering, MADLS), this technique can also provide an estimation of the titer. In the following subsections, we summarize the theoretical basis of the DLS measurement, followed by certain rAAV-specific considerations and experimental details.

9.4.1 *Theoretical Background*

9.4.1.1 Basic Principle

Dynamic light scattering analysis is based on the Brownian motion, while the diffusion rate, defined as the translational diffusion coefficient (D), is determined by the Stokes–Einstein equation (Eq. 9.2) and affected by the size, temperature, and the viscosity of the medium.

$$D = \frac{kT}{6\pi\eta R} \tag{9.2}$$

where k is the Boltzmann constant, T is the absolute temperature, η is the viscosity of the medium, and R (or R_h) is the hydrodynamic radius of the particle. Hydrodynamic size (expressed as diameter or radius) is the size of a sphere that diffuses at the same rate as the particle in question. The size of this sphere is not limited to the core particle, but it includes the surrounding ions and surfactants or adsorbed polymers bound to its surface. Further, the hydrodynamic radius of the particle is affected by the type and concentration of ions in the medium by dictating the thickness of the electric double layer surrounding the particle. When the movement of particles dispersed in liquid is recorded over time, information on the size of the particles can be derived, as large particles diffuse slowly, resulting in similar positions over time, while smaller particles diffuse faster, thus their spatial arrangement changes quickly.

9.4.1.2 Autocorrelation Function: From Diffusion Rate to Size

If particles are in motion, the pattern of scattered light intensity will change over time. The fluctuating intensity signal, generated by the simultaneous diffusion of many dispersed particles, is recorded over time. Snapshots of these intensity signals are taken, and a digital autocorrelator compares them back with the original signal recorded at the start of the measurement. Between consecutive snapshots, taken within the time frame of a few nano- or microseconds, the intensity signals are very similar to each other (well correlated). The correlation begins to decrease over the course of the measurement, as later snapshots are compared with the original one. Eventually, there is no correlation between the measured intensity signal and the original one. This process of comparing measured intensity snapshots to the originally measured intensity is called autocorrelation and gives rise to the autocorrelation function. The attenuation time of each measurement should be adapted according to the autocorrelation function and depends on the size and size distribution (polydispersity) of each sample. A proper autocorrelation function is an important quality attribute of acceptable measurements, since the extinction coefficient is derived from it.

9.4.1.3 Small Particles and Large Particles

For particles that are small compared to the wavelength of the incident laser beam ($d < \lambda/10$, where d is the particle diameter and λ the wavelength of the laser), scattering is considered isotropic, meaning that it is equal in all directions. According to the Rayleigh approximation, the intensity of the scattered light, I, is proportional to R^6, and inversely proportional to λ^4, the wavelength of the incident light. It can then be deduced that at a given wavelength, a difference in particle size of one order of magnitude will result in an intensity difference of six orders of magnitude. Therefore, it would be difficult to measure a mix of particles with vastly different sizes, for example, 10 and 1000 nm, as the contribution of the 10 nm particles to the overall intensity of the scattered light would be negligible. Further, a shorter wavelength results in higher signal.

When the particle size becomes comparable to the wavelength of the laser beam (often 633 nm), light scattering is no longer homogeneous and becomes angle-dependent, producing a complex function of maxima and minima. In this case, the Rayleigh approximation cannot be used for the calculation of the diffusion coefficient and the generalized theory for light scattering (Mie theory) is used by the software to obtain the correct size across all measurement angles. The conversion of the intensity signal to volume is also based on the generalized theory of light scattering (Malvern Instruments 2010; Malvern Panalytical 2021).

9.4.1.4 Important Considerations

DLS is used in rAAV applications to give an indication of the aggregation state of the viral particles. An example of the use of DLS to assess how different parameters (serotype, capsid loading, buffer type) affect the aggregation state of AAV samples is presented in Fig. 9.4. As for other nanoparticle applications, its popularity relies on the fact that it is a quick and facile, non-destructive technique that requires none or minimal sample preparation. It is also particularly relevant for the measurement of particles that are intended to be formulated in liquid media as the measurement takes place in the native environment of the particle—as opposed to electron microscopy techniques, for example, where the sample is typically dry, under vacuum, and

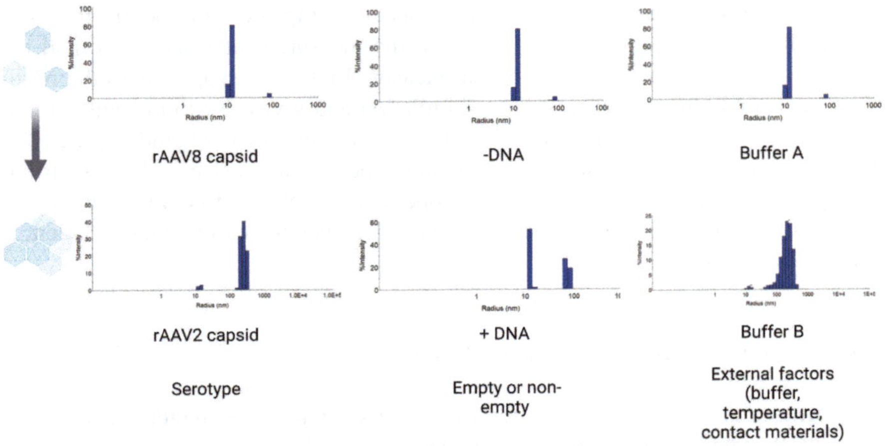

Fig. 9.4 Size distribution graphs of rAAV samples measured by dynamic light scattering (DLS). From left to right, rAAV samples of different serotypes (rAAV2 and rAAV8), empty and full capsids (–DNA and +DNA, respectively), and different media (Buffer A and B) were measured using the DynaPro plate reader (Wyatt). The output of each measurement is presented as a size distribution (regularization) graph of the intensity signal. rAAV titer was 5×10^{12} vp/mL, data were acquired for 5 s, and shown as the average of 20 acquisitions per sample. The data were generated using Dynamics (Wyatt Technology) and the figure created with BioRender.com

otherwise manipulated. However, DLS has significant limitations, which should be borne in mind when using it to draw conclusions about particle size distributions.

- DLS is a spectroscopic technique, in which the size and size distribution of the particles in the measured samples is based on numerical transformations of the light intensity measurements. Its resolution is therefore, in principle, limited compared to techniques involving direct particle counting (e.g., electron microscopy, Coulter counter) or fractionation (e.g., ultracentrifugation, chromatography). Unlike the aforementioned techniques, DLS provides relative distributions rather than absolute numbers.
- DLS is highly sensitive to temperature fluctuations and buffer parameters such as ionic strength and viscosity. Therefore, these parameters must be controlled during a DLS measurement.
- Scattered light intensity is proportional to the sixth power of the particle radius; therefore, even a small proportion of large particles will skew the intensity distribution significantly. As such, it is important that dust particles are carefully removed from the buffers and the sample containers. For the same reason, DLS is not the optimal method for the measurement of highly polydisperse samples.
- Unlike fractionation techniques and electron microscopy, DLS does not separate between empty and full capsids.
- The refractive index of particles of interest needs to be known in advance in order to obtain accurate results, as well as the refractive index of the buffer solution.

– The respective buffer of each sample needs to be set as blank as soon as the buffer shows a difference in refractive index.

9.4.2 Experimental Requirements

9.4.2.1 Sample Preparation

Optimizing DLS Measurements

The DLS measurement is sensitive to contaminants such as dust particles from the environment, or particulate contamination of laboratory equipment, host cell-derived contaminants such as nucleic acids and proteins. This is particularly important when measuring relatively low concentrations of small particles, such as in the case of rAAVs.

Although the diameter of the rAAV capsid is approximately ~25–26 nm (Wang et al. 2019b; Large et al. 2021), slightly higher values have been reported for the hydrodynamic radius (R_h) as measured by DLS. For example, Zhang et al. measured rAAV9 samples using the DynaPro Plate Reader or DynaPro NanoStar, and reported R_h values of 14.7 ± 0.7 nm and 15.6 ± 0.1 nm for empty and full capsids, respectively (Zhang et al. 2019). It is important that all consumables are particle-free, so as not to skew the size distribution. Samples are handled in particle-free conditions in a laminar flow (LAF) cabinet using sterile consumables. Small sample volumes are preferred and it is common that several experimental conditions are tested at once, therefore the 96-well or 384-well plate formats are preferred over the cuvette format. If the sample is not exposed to elevated temperatures for the measurement (e.g., temperature ramp to test aggregation onset), the sample can be recovered after the measurement and used for other analyses. Pressurized air can be used to remove dust particles from the wells before adding the samples. It is advisable that the buffers used to disperse the rAAVs and the blank samples are also filtered (0.02μm pore cut-off is appropriate). Air bubbles should also be removed, e.g., by centrifugation. The plate can be sealed with a transparent film before being transferred outside the LAF cabinet in order to protect the samples from contamination and for biosafety reasons.

To confirm that the instrument is functioning properly and that the buffers are free from particles, a blank water or buffer sample should be included. The autocorrelation function of the blank should be visually inspected (random pattern of peaks and troughs) and it should display a low count rate (<500 kCnts/s). A positive control of known particle size can also be included, e.g., polystyrene beads 60 nm in 10 mM NaCl in water. The theoretical sample concentration range is 1×10^{11}–1×10^{15} vp/mL for rAAVs, but the lower limit may be higher depending on the sample. To obtain reliable results the sample should show a three to five times higher absolute particle count than the respective buffer.

9.4.2.2 Equipment Requirements

DLS instruments employ either a detector at 90° (e.g., DynaPro® NanoStar®, Wyatt Technology, or Zetasizer Nano S90®, Malvern Instruments) or a backscatter detection system at 173° (e.g., Zetasizer Nano S®, Malvern Instruments) and at 158° (DynaPro Plate Reader®, Malvern Instruments) close to the incident light of 180°. Backscattering (as opposed to forward scatter) limits the impact of the multiple scattering phenomenon (scattering of the same photon by more than one particle), which is particularly relevant in highly concentrated samples. Furthermore, the backscatter detection system is more appropriate for the measurement of small particles, especially in the case of polydisperse samples. Anisotropic scatterers, like dust particles, scatter more light in the forward direction as their scattering becomes wavelength-dependent compared to smaller-size particles (Rayleigh scattering), which have nearly equal scattering in both directions.

9.4.2.3 Data Analysis and Evaluation

Typically, the cumulant analysis (monomodal distribution) method is often employed for the analysis of rAAV samples and this is set as the default analysis in the system software. This method is suitable for samples with the polydispersity index (PDI) of ~0.1, i.e., monodisperse system. The PDI value of 0.1–0.7 represents nearly monodisperse preparation, whereas PDI >0.7 suggests broad distribution of macromolecular sizes in solution and non-monomodal distribution methods should be considered for data analysis. PDI of <0.05 is rarely observed in biological systems.

Prior to assessing the particle size distribution in each sample, it is important to establish that the particle concentration is significantly higher than that in the blank (buffer only) samples (3–5×). For unknown samples, it is important to confirm that the obtained R_h is repeatable. If it changes over time, then the acquisition time and number of acquisitions should be adjusted until a consistent measurement is obtained. As an example, the authors use 20 acquisitions in a timeframe of 5 s, which is satisfactory for most rAAV samples. In a similar vein, if the shape of the autocorrelation function is not as expected (should be sigmoidal, returning to baseline), optimization steps may be required. It is possible to extend the acquisition time to allow for the autocorrelation function to return to baseline or filter the samples to remove large aggregates prior to measurement. However, the aim of the experiment might be to assess the aggregation state of the samples, hence removing the larger particles might not be desirable. Measurement of polydisperse samples from which removal of large particles is not desired may be improved by decreasing the acquisition time (e.g., 3 s) and increasing the number of acquisitions. If the sample is monodisperse, only one inflection point is observed in the autocorrelation function, and the PDI is small. More often than not, there is at least a second peak, which shows aggregation and it does not affect data quality (given that the autocorrelation function still returns to baseline). In this case, where more than one particle populations are detected, the regularization fit more accurately describes the data than the cumulant fit. With low sample concentration, the polydispersity index will increase even with a monodisperse sample because the noise of the autocorrelation function increases. Intensity-weighted size distribution is the primary readout of the DLS measurement. Intensity can be mathematically converted to mass, volume, and number, which is useful if the sample shows monodispersity and a simple unimodal distribution, and the shape of the particles is known. Mass, volume, and number values can also provide a good comparison with readouts from different techniques such as laser diffraction, which is based on the mass/volume of particles, or microscopy, which is based on the number of particles.

9.5 Size Exclusion Chromatography Coupled with Multi-angle Light Scattering (SEC-MALS)

Size exclusion chromatography (SEC) coupled with multi-angle light scattering (MALS) or multi-angle laser light scattering (MALLS) has been used for the characterization of rAAV particles in solution allowing the determination of capsid molecular weight, capsid titer, encapsulated DNA molar mass, size-distribution and oligomerization state or aggregation profile (McIntosh et al. 2021).

SEC is traditionally used for protein purification and to estimate the molecular weight of the proteins. Separation is based on size differences of the analytes or, more specifically, their hydrodynamic volume, using a stationary phase with defined pore size. Large analytes do not fit into the pores and move between the stationary phase particles, thus eluting faster, while small particles enter the pores and take longer to elute. A monomeric virus fraction from a viral suspension can be similarly separated from small sub-viral debris but also from multimeric fractions corresponding to dimers or multimers (Uliyanchenko 2014). After separation, each fraction exiting the column is subjected to a coherent light beam and the subsequent scattered light is analyzed by a MALS detector. The information provided by the MALS detector together with UV absorbance and/or differential refractive index is analyzed and can be used to calculate the molecular weight and concentration of the molecules within a peak (Sahin and Roberts 2012).

9.5.1 *Theoretical Background*

MALS is a form of static light scattering (SLS) in which the intensity of the scattered light is collected by an array of detectors and measured as a function of different angles dependent on the position of detectors. Static light scattering is used to determine the weight averaged molar mass (M_w), radius of gyration (r_g) or the root mean square radius (rms) and the second virial coefficient (A_2). When a laser light encounters a macromolecule the electric field of the oscillating light wave induces the formation of a dipole particle which in turn scatters light. The intensity of radiated light depends on how strong the polarity of the dipole is. The more polarizable a molecule is, the stronger is the intensity of scattered light. The polarizability of the molecule in contrast to the surrounding medium can be determined by measuring the change in the refraction index (Δn) of a solution with the change in molecular concentration (Δc) by measuring dn/dc (=$\Delta n/\Delta c$) using a refractometer (Barron 2004; Xu 2015).

Each of the analytes in the mobile phase scatters light due to the induced dipole by the laser beam. The intensity of the signal is proportional to the concentration of the analyte. In a dimer, each of the monomers scatters light which oscillates in a phase relationship with each other. Thus, the scattered light of a dimer is two times more intense than the one scattered from individual monomers. Since the intensity of scattered light at each angle is proportional to the molar mass and the concentration of the molecules, light scattering can be thus used to determine molar mass, concentration, and the formation of aggregates. For macromolecules whose radius is approximately 20 times smaller than the wavelength of the incident light, there is no angular dependence of the scattered light. The intensity is equal at every angle and thus the molecules scatter light isotropically. When the macromolecules are larger, the light scattered by different regions of the particle varies with angle thus providing information about the shape of the molecule (e.g., sphere, rod, or random coil) (Barron 2004; Xu 2015).

According to Zimm's calculations based on Rayleigh–Debye–Gans theory, the following equation represents the foundation of calculating the wanted parameters upon measurements from a MALS experiment (Eq. 9.3):

$$\frac{K^* \times c}{R(\theta, c)} = \frac{1}{M_w \times P(\theta)} + 2A_2 \times c \tag{9.3}$$

where $R(\theta,c)$ is the ratio of the excess intensity of light scattered by the solution and the pure solvent as a function of scattering angle θ and concentration of the sample c; M_w is the weight-averaged molecular weight; A_2 is the second virial coefficient; K^* is the optical parameter equal to $4\pi^2(dn/dc)^2 n_0^2/N_a\lambda_0^4$; N_a is the Avogadro's number; $P(\theta)$ is a complex function that describes the angular dependence of the scattered light, and can be related to the rms; n_0 is the refraction index of the solvent and λ_0 is the vacuum wavelength of the laser. Expanding the function $P(\theta)$ one obtains:

$$P(\theta) = 1 - \frac{16\pi^{2\sim}\mathrm{n}_0^2}{3\,\lambda_0^2}\left\langle r_g^2 \right\rangle \sin^2\left(\frac{\theta}{2}\right) + O\left[\sin^4\left(\frac{\theta}{2}\right)\right] \tag{9.4}$$

where r_g is the rms (Wen et al. 1996; Minton 2016).

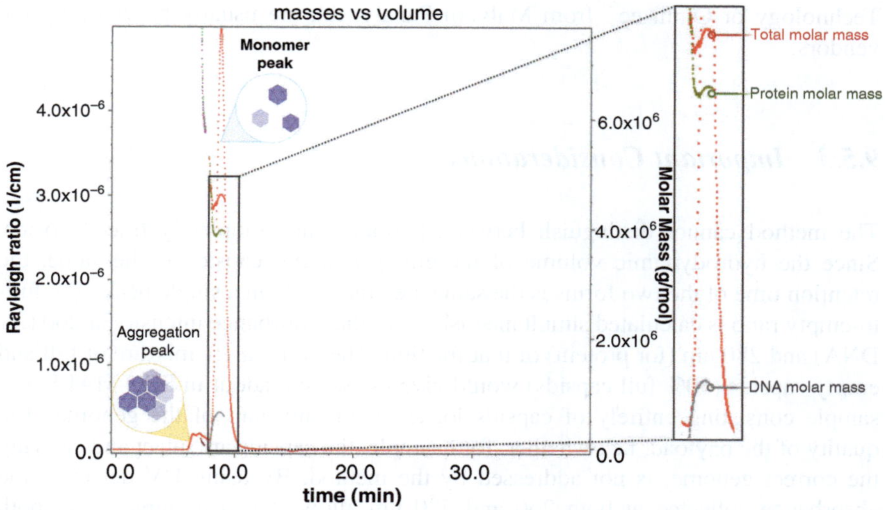

Fig. 9.5 SEC-MALS mass vs. volume graph for a rAAV sample. The red trace corresponds to the signal obtained by the light-scattering detector. Two peaks are highlighted in yellow and blue and correspond to the aggregates and monomers, respectively. The molar mass for the protein and DNA component of the rAAV sample, as well as the total molar mass of the construct are indicated for the monomer peak. The concentration of the rAAV sample used to generate this graph was 5×10^{12} vg/mL and the injection volume was 25μL. The graph was generated using the ASTRA software (Wyatt Technology) and the figure created with BioRender.com

The data collected from all angles are combined and a fitting method is applied in order to determine the size, molar mass, and concentration of the molecules. An example of a SEC-MALS plot obtained for rAAV can be seen in Fig. 9.5.

9.5.2 Equipment Requirements

Required equipment for an SEC-MALS experiment includes an HPLC unit and an SEC column with the appropriate pore size for AAV separation, selected to be compatible with a downstream MALS system, which requires minimal particle shedding and low flow rates, typically between 0.1 and 0.35 mL/min. In addition, an isotropic scatterer is required as reference standard, e.g., BSA (Bovine Serum Albumin), which is used to test the performance of the instrument. The system is also equipped with UV detectors capable of measuring absorbance at both 260 and 280 nm.

Examples of MALS detectors are DAWN® (18 angles available) or miniDAWN® (3 angles) from Wyatt Technology or Viscotek SEC-MALS 20® (up to 20 angles) from Malvern Panalytical. Data analysis software, such as Astra® from Wyatt

Technology or Omnisec® from Malvern Panalytical, are usually provided by the vendors.

9.5.3 *Important Considerations*

The method cannot distinguish between partially and completely filled capsids. Since the hydrodynamic volume of the empty and full capsids is the same, the retention time of the two forms is the same, i.e., they elute in a single peak. The full-to-empty ratio is calculated simultaneously from the absorbance intensity at 260 (for DNA) and 280 nm (for protein) of that fraction. Therefore, a 1:1 mixture of full and empty capsids (50% full capsids) would give the same readout in SEC-MALS as a sample consisting entirely of capsids loaded with only half of the genome. The quality of the payload, i.e., whether, for example, the capsids are intact and carrying the correct genome, is not addressed by the method. By using UV detectors the absorbance collected at both 260 and 280 nm allows the determination of both protein and DNA content. The calculations generally assume a complete recovery from the SEC column, which is never 100%. Total sample amount per injection should be relatively high when using a refractive index detector, at least 1×10^{12} vp/injection in the authors' experience

Large aggregates cannot be analyzed since MALS is restricted to a radius range between 10 and 600 nm. In addition, a guard column, which is often used with SEC, the column frit or the actual SEC column might shear the large aggregates thus modifying the sample. The number of angles used has an impact on the accuracy of the measurement. The more angles are used the more precise the measured parameters will be, however making the data analysis more complex. In fact, one major disadvantage of the method is that it requires intensive and time-consuming data analysis.

The advantages of using SEC-MALS are nevertheless important: relatively simple sample preparation, it does not require a calibration curve, multiple attributes are simultaneously determined (capsid and genomic titer, full/empty ratio, impurities and aggregation profile, estimation of non-encapsulated DNA) and further sample characteristics, e.g., molecular weight and hydrodynamic volume of the molecules, can be measured.

9.6 Electron Microscopy (EM)

9.6.1 *Background*

Electron microscopy (EM) is a high-resolution imaging technique, which has become a very valuable tool in determining the F/E capsid ratio of rAAVs (Gimpel et al. 2021; Werle et al. 2021). Furthermore, it is commonly used to visualize structural details, morphology, size distribution, and integrity of rAAVs (as seen in Fig. 9.3) as well as for quantification of viral capsids and impurities in the samples (Horowitz et al. 2013; Tan et al. 2018; Dobnik et al. 2019; Mietzsch et al. 2020; Rumachik et al. 2020). Electron microscopy will be discussed in more detail in the next chapter and therefore here we will focus more on the F/E capsid ratio determination.

In electron microscopy, samples are irradiated with a high-energy electron beam. The generated electron images have nanometer resolution and depict the differences in electron density and mass-thickness between the sample and the surrounding layer. The most commonly used EM techniques for the determination of F/E capsid ratio of rAAVs are negative stain EM and cryogenic EM (cryo-EM), both of which are validated to be used as release testing techniques in a GMP environment.

9.6.2 Considerations

In negative stain EM, particles are adsorbed onto an electron microscope surface, dried and stained with a heavy-metal salt such as uranyl acetate or phosphotungstic acid to increase contrast. The stain is not applied to the viral particles but to the surrounding space and full, empty, and partially filled rAAV capsids can be quantified by manual visual identification. In an electron image, full rAAV capsids appear opaque/bright due to stain exclusion, empty capsids show a darker electron-dense center since they are permeable to the stain and partially filled capsids show a dimmer center than the full capsids (Horowitz et al. 2013; Dobnik et al. 2019; Fu et al. 2019). While the sample preparation is simple, negative stain EM has certain drawbacks. For example, the use of a stain during sample preparation can alter the state and even damage the capsids. Furthermore, the process of counting and assigning empty and full rAAV particles is done manually and requires many images to be acquired to achieve statistical significance. Besides being time-consuming, the accurate identification of the capsids can be ambiguous due to a lack of contrast or to the presence of impurities such as proteins and cell debris, which can interfere with the results. To overcome some of these drawbacks, Vironova has recently introduced a low voltage transmission electron microscopy with an automated image acquisition and analysis system (miniTEM) and additional improvements on sample preparation have been reported (Vironova 2018; Hauser et al. 2020).

In cryogenic electron microscopy (cryo-EM), samples are vitrified by rapidly freezing them in the hydrated state at cryogenic temperatures. The specimen is visualized in the frozen state, with the capsids embedded in a layer of amorphous ice. With the images obtained by cryo-EM, a three-dimensional (3D) capsid reconstruction with high resolution can be obtained by averaging multiple two-dimensional (2D) images of homogeneous samples. Similarly to negative stain EM, this method also allows the detection of full, partially filled, and empty rAAV capsids (Subramanian et al. 2019). Full capsids observed with cryo-EM show an overall gray staining (no distinction between shell and core), whereas empty

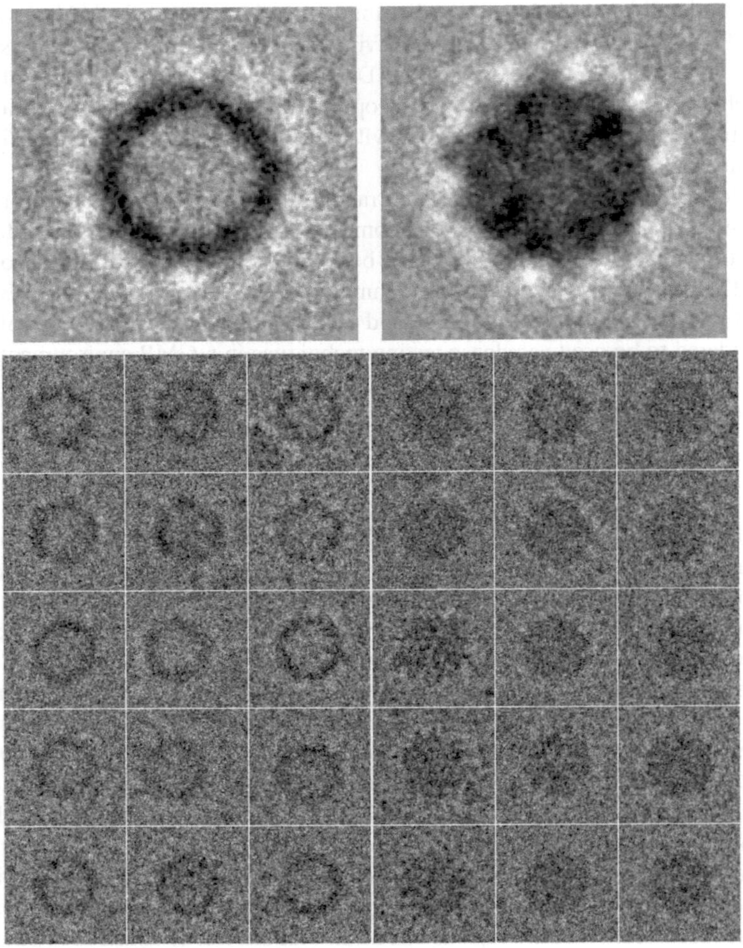

Fig. 9.6 Transmission electron micrographs of frozen-hydrated rAAV2 empty and full capsids (all box sizes 40.7 nm). Top-left panel: empty capsid, class averaging of 2D projections from 22 particles (18 representative raw images are shown in the bottom left panel). The darker ring corresponds to the icosahedral assembly of capsid proteins and the central inner part of the empty capsid appears in lighter gray. Top-right panel: full capsid, class averaging of 2D projections of 62 particles (18 representative raw images are shown in the bottom right panel). The darker disk corresponds to the packed genetic material and capsid proteins

capsids display a darker ring surrounding a light internal center (Fig. 9.6). Radial density profiles and internal density histograms stemming from 2D classification as well as further 3D classification analyses of samples can complement these results. Furthermore, these analyses also allow to identify partially filled capsids, which are characterized by an intermediate genome density, and can detect differences resulting from different preparation methods (Subramanian et al. 2019).

One study has compared negative stain and cryo-EM and showed that the ratios of full and empty particles obtained by the two methods varied by less than 5% (Besir et al. 2019). However, cryo-EM has several advantages over negative stain EM (Adrian et al. 1984). Cryo-EM employs a limited electron dose and allows the preservation of the capsids native or near-native states, improving the determination of F/E capsid ratio. Moreover, the quantification in cryo-EM is not impacted by the use of heavy metal contrast that can lead to uneven staining of capsids and potential artifacts, nor by additional impurities, which can be seen in negative stain images, as these are eliminated during data processing. However, to ensure reliable results, more than 10,000 particles should be sampled, which is reflected in the amount of time required for the complete analysis (~7–8 h).

Negative stain and cryo-EM have been compared to other techniques as F/E capsid ratio determination orthogonal techniques and have shown several degrees of agreement. For example, negative stain EM has shown to be a useful tool when used in combination with genomic titer methods (dPCR and qPCR) to allow a better understanding of the molecular results obtained for rAAVrh.10hCLN2 (Dobnik et al. 2019). A new method developed for AEX-HPLC has also been compared to AUC and negative stain EM, showing good comparability between the three assays with regard to F/E capsid ratio for different samples of rAAV9 (Fu et al. 2019). This was not the case for another comparison of rAAV9 samples, where discrepancies were obtained between negative stain EM and AUC, likely due to the presence of partially filled capsids (Tomono et al. 2018). Subramanian and colleagues have compared cryo-EM with AUC and showed good agreement between the two methods for the full and empty quantification of rAAV3 samples (Subramanian et al. 2019). Another study has compared SV-AUC, charge detection mass spectrometry (CDMS), cryo-EM, A260/A280 ratio, SEC-MALS, and capsid titer-to-genome ratio. The cryo-EM results have shown a general overestimation of empty rAAV capsids when compared with the theoretical expected percentage of empty capsids and the other methods (Werle et al. 2021). Overestimation may be minimized by increasing the amount of particles sampled, by improving the detection of partially filled capsids and by re-sorting particles incorrectly classified with iterative rounds of 3D classification.

In summary, EM techniques are the only ones that allow direct visualization of the capsids with powerful magnification. Nevertheless, it should be highlighted that other microscopic techniques, such as atomic force microscopy (AFM), can also be useful in visualization of the capsids and, e.g., quantification of aggregates (Chen 2007; Bernaud et al. 2018). EM techniques are widely applicable, straightforward to interpret, require low sample volume (<25µL for negative stain and ~3µL for cryo-EM) and are often recommended to be included in the QC analytical panel. However, many capsids need to be imaged to provide accurate quantification and throughput is limited, hindering its use in for instance large formulation screens. Furthermore, the equipment is expensive, requires special maintenance, its operation entails some degree of qualification by operators and, even if it can be validated, it may be more challenging than other methods to be implemented in a GMP environment. Nevertheless, further developments can be made to improve the methods, increase the throughput and reduce costs. Particularly cryo-EM has wide potential and should continue to be compared with other orthogonal methods to better understand its advantages and limitations.

9.7 Optical Density (OD)/Absorbance

Optical density (OD) or absorbance is a relatively simple, fast, and cheap method to estimate the capsid titer and F/E ratio based on the ratio between absorbance at 260 and 280 nm. The absorbance of light occurs via the chromophores present both in capsid proteins as well as in the genomic single-stranded DNA. While both components absorb light in the UV range, the maximum of absorption for DNA is approximately 260 nm, whereas for the proteins is 280 nm. For DNA the chromophores are present in the stacked purines and pyrimidines, whereas for proteins in aromatic amino acids such as tryptophan and tyrosine as well as disulfide bonds (Glasel 1995).

Values for extinction coefficients of proteins at 280 nm can be predicted and calculated from their primary sequence according to first principles using a standard formula (Gill and Hippel 1989; Pace et al. 1995). However, empirical determination of extinction coefficient of proteins in different buffers might slightly differ from the theoretical value due to sensitivity of various amino acids to pH or solvent polarity. Similarly, pH and ionic strength influence the extinction coefficient of nucleic acids at 260 nm.

According to Lambert–Beer's law, measuring the absorbance of both protein and nucleic acid at both wavelengths and given the known either predicted or measured extinction coefficient (ε) of the two components at the two different wavelengths ($\lambda 1$ and $\lambda 2$) the two equations can be resolved in order to determine the ratio of full capsids (Eqs. 9.5 and 9.6).

$$A_{\lambda 1} = \varepsilon_{\text{DNA};\lambda 1} C_{\text{DNA}} + \varepsilon_{\text{protein};\lambda 1} C_{\text{protein}} \tag{9.5}$$

$$A_{\lambda 2} = \varepsilon_{\text{DNA};\lambda 2} C_{\text{DNA}} + \varepsilon_{\text{protein};\lambda 2} C_{\text{protein}} \tag{9.6}$$

A simple OD-based method was developed by Sommer et al. to quantify the full and empty capsids in rAAV2 virus preparations. To minimize Rayleigh scattering, a denaturation step using 0.1% sodium dodecyl sulfate (SDS) and heat is necessary for accurate absorbance measurement. Experimental determination of the extinction coefficient of the empty capsids from a pure preparation was in concordance with predicted theoretical value when denaturation was performed (Sommer et al. 2003). Similarly, Porterfield and Zlotnick described a method to determine the concentration of protein and nucleic acid components of two ssRNA viruses, hepatitis B virus, and brome mosaic virus. The method can be virtually applied to any type of virus and does not require sample denaturation but it relies on estimating and subtracting the Rayleigh scattering from the spectra (Porterfield and Zlotnick 2010).

9.7.1 *Considerations*

The method does not discriminate between different types of proteins and nucleic acids, including potential contaminants. Therefore, it can be used only for highly pure viral preparations. Similarly, the buffer and excipient components present in the samples that can absorb in the UV range might interfere with the method. In addition, OD analysis requires a relatively high concentration of pure samples ($>5 \times 10^{11}$ vg/ml), which cannot be recovered due to the requirement for sample denaturation prior to measurement.

9.8 Ion Exchange Chromatography (IEC)

Ion-exchange chromatography (IEC, IEX) and more specifically AEX (anion exchange chromatography) have been widely used to separate and characterize full and empty AAV capsids. Encapsulation of negatively charged ssDNA genome lowers the isoelectric point (pI) of the viral particle in comparison to an empty one. This difference is exploited in AEX where a positively charged stationary phase binds negatively charged AAV particles, which can be subsequently separated into full and empty AAV capsids based on their relative charge by applying a gradient for example with two suitable mobile phases.

9.8.1 *Theoretical Background*

High-performance liquid chromatography (HPLC) is an essential method in protein analytics. It is a fast method, which offers high robustness and reproducibility with a possibility for automation for high-throughput analysis. The capability of coupling an HPLC device with additional systems, for example, a mass spectrometer or a MALS detector, allows to derive further information for the obtained peaks regarding their molecular weight and size.

The principle of separation in ion exchange chromatography is based on the differential partition of sample components between a stationary phase consisting of the chromatographic medium packed into a vertical column and a mobile phase consisting mainly of a buffer solution that is pumped through the column. The separation begins already with the injection of the sample into the column and, depending on the buffer system used each analyte elutes at a different time point. A detector, for example, an UV detector or a fluorescence detector, measures the signal given by each analyte that elutes from the column resulting in a chromatogram plot (Lottspeich and Engels 2018).

In ion-exchange chromatography the charge of the analyte interacts electrostatically with the oppositely charged ligand which is immobilized on the stationary phase. While the strength of this interaction depends mainly on the charges of the protein and the ion exchanger, it is also influenced by several factors such as: other ions present in the buffer that can compete with the analyte for the binding site of the stationary phase, the dielectric constant of the medium, non-electrostatic and/or hydrophobic interactions with the column, pH or temperature. Protein charge varies with the pH and hence its binding can be influenced. At a pH lower than its pI, the protein will be positive and will bind to negatively charged cation exchangers as it happens in cation exchange chromatography (CEX). On the contrary, at a pH higher than its pI, the protein charge will be negative and it will bind to positively charged negative exchangers as is the case in anion exchange chromatography (AEX). Elution of the bound molecules is usually achieved by a change in the pH or most commonly in the ionic strength by increasing the salt concentration (Janson 2011).

9.8.2 Considerations

Empty AAV capsids (AAV1-12) have a mean calculated pI value of 6.3 whereas the full capsids packaged with a 4.7 kb genome have an average calculated pI value of 5.9 (Venkatakrishnan et al. 2013). This difference allows the separation of most of the serotypes into empty and full capsids. Both preparative and analytical IEX methods have been developed to purify and enrich full AAV particles or to determine F/E ratio. Several studies have reported chromatographic methods for the purification of full AAVs after ultracentrifugation (Tamayose et al. 1996; Zolotukhin et al. 1999; Debelak et al. 2000). Brument et al. developed a two-step chromatographic method based on ion exchange resins to separate rAAV2 and 5 without the need for pre-purification from the cell lysate. In a first step, a cation exchanger resin was used and binding conditions were achieved by adjusting the lysate to pH 6.5 to separate the AAV particles from the cellular proteins. Subsequently, an anion exchanger resin was employed to enrich the full particle fractions, and binding was achieved at pH 8.2. In both steps, a salt gradient was used for eluting the particles. While in the second step full rAAV5 particles were almost completely separated from the empty ones, for the full rAAV2 this step was unsuccessful with no enrichment of the full fraction. This was one of the first studies that succeeded in the enrichment of a full fraction solely by chromatographic methods and at the same time suggested the necessity of IEX method optimization for each serotype separately (Brument et al. 2002). A report from Urabe et al. described several buffer systems used to separate rAAV1 particles. The study showed that different elution buffers containing commonly used ions such as NH^{4+}, $(CH_3)_4N^+$, PO_4^{3-}, SO_4^{2-} were more suitable of resolving rAAV from empty capsids than using NaCl gradients, while trimethylammonium sulfate ($C_6H_{20}N_2O_4S$) was most effective in separating full from empty rAAV particles (Urabe et al. 2006). To improve the sensitivity of the method Fu et al. developed an AEX method to determine the F/E ratio of rAAV2 in which the detection was based on the intrinsic native fluorescence of the capsid proteins. Aromatic amino acids such as tryptophan, tyrosine, and phenylalanine manifest intrinsic fluorescence, which allows their detection at low concentrations.

Table 9.1 Summary of IEX methods that successfully separate full and empty capsid of different rAAV serotypes

Serotype	LC system	Elution system	Detection system	Publication
rAAV2 and rAAV5 No separation between full and empty for rAAV2; only for rAAV5	Two-step purification on: (I) SP Sepharose High Performance resin - Cation exchanger (Amersham Biosciences) (II) Source 15Q - Anion exchanger/ FPLC (Amersham Biosciences)	NaCl gradient	UV	Brument et al. (2002)
rAAV1	POROS HQ (Applied Biosystems)	Trimethylammonium sulfate gradient	UV	Urabe et al. (2006)
rAAV8	CIM-QA monolith Anion exchanger/ FPLC	NaCl gradient	UV	Lock et al. (2012)
	(I) CIMac SO3 monolith Cation exchanger (II) CIMac rAAV Anion exchanger	NaCl gradient	UV, fluorescence, light scattering	Gagnon et al. (2021)
rAAV2	CIMac rAAV full/empty-0.1 analytical column Anion exchanger HPLC	0–200 mM NaCl gradient	Fluorescence	Fu et al. (2019)
rAAV6.2	CIMac rAAV full/ empty Agilent HPLC	Combination of salt gradient with a 7.5–9 pH gradient	Fluorescence	Wang et al. (2019a)

The use of intrinsic fluorescence enables the measurement of viral samples of almost one order of magnitude more diluted than when using UV detection (from 5×10^{10} vg/injection as reported with UV detection down to 6×10^{9} vg/injection). In addition, the DNA molecules do not manifest native fluorescence thus eliminating the interfering signal observed when detecting UV absorbance and increasing the accuracy of the measurement. It has been observed that when using UV absorbance at 280 the relative quantity of full capsid tends to be overestimated (Fu et al. 2019). Table 9.1 summarizes the methods reported so far aiming to separate the full capsids.

IEC is a promising method to separate full from empty capsids allowing high-throughput analysis and, with appropriate equipment and software, fulfilling regulatory requirements. Moreover, the possibility of coupling it with detectors such as MALS can offer a more precise quantification of particle concentration in the separated peaks. Similar to other chromatographic methods, IEC is however non-discriminatory regarding protein or DNA identity. Additionally, the sample

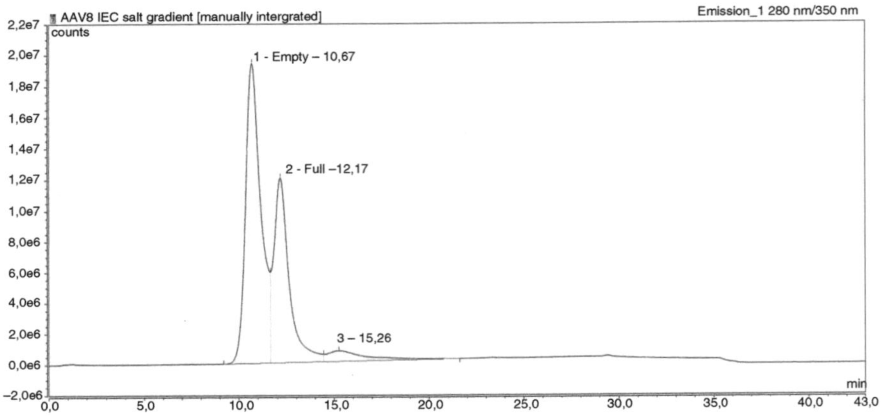

Fig. 9.7 Representative AEX-HPLC Chromatogram of a rAAV8 commercial sample. Fluorescence signals were collected at 350 nm upon excitation at 280 nm. The two peaks correspond to the empty and full particles, respectively, and their integration allows the relative quantification of empty and full capsids. The third peak eluting at ~15.26 min represents impurities

must be relatively clean and the buffer must be devoid of high salt concentration making it difficult to use for serotypes prone to aggregation such as rAAV2. In Fig. 9.7 a typical chromatogram showing a successful separation between full and empty rAAV8 particles is depicted.

9.9 Mass Photometry (MP)

Mass photometry (MP) is a light scattering-based technique introduced by Young and colleagues in 2018 (Young et al. 2018), which measures the mass of single macromolecules or particles in solution by capturing the interference of scattered and reflected light. MP is a novel technique, originally developed for the measurement of protein samples to overcome the limitations of fluorescence-based methods, for example, for the label-free assessment of protein–protein interactions (Foley et al. 2021). Over the last couple of years, it has been applied on the measurement of the heterogeneity of rAAV samples, as a simpler and more economical alternative to SV-AUC, the current gold standard method to assess the sample content in full, empty, and partially filled capsids (Wu et al. 2022).

For an MP measurement, the sample solution containing the rAAVs or proteins is applied on a microscopy slide and the sample is illuminated through a light source under the slide. Both reflected and scattered light are captured by the detector. The intensity of the scattered light is proportional to the molecular mass of the illuminated particle, while the reflected light arises from the surface of the slide and is constant. The primary output of the MP measurement is the distribution of interferometric contrast signals (or simply, "mass photometry contrast") of different analytes detected in the sample plotted as histograms. The contrast signal is directly correlated to the mass of the macromolecule (Young et al. 2018), and a molecular mass distribution can be derived with the use of an appropriate calibrant to convert the contrast signals into mass signals. The dynamic range of MP is 40 kDa–5 MDa, with a reported less than 2% mass error (Young et al. 2018; Wu et al. 2022). A standard protocol for MP measurements has been published by Wu and Piszczek (Wu and Piszczek 2021) and, to the best of the authors' knowledge, Refeyn is the only manufacturer of a commercially available instrument, specifically developed for the characterization of AAV samples (Samux^{MP}, (Refeyn 2022a)).

MP presents major advantages directly relevant to rAAV characterization for research and clinical purposes. In principle, MP is a label-free technique measuring light scattered by a particulate analyte in solution; therefore, it is able to characterize unmodified rAAV capsids in their native environment. This is in contrast to techniques such as TEM, which require negative staining of the sample, or cryo-EM, which avoids the need for negative staining but, by definition, requires ultra-low temperatures for measurement. Additionally, the presently available instrument only requires 10μL of sample volume with a concentration in the range of 1×10^{11} vp/mL, although optimization might be required depending on sample purity and the purpose of the measurement. The data acquisition and analysis is complete within a few minutes and it can be performed on the software that accompanies the instrument (both developed by Refeyn) with minimal requirements for user training. According to Wu et al., who directly compared MP and SV-AUC characterization of the same AAV8 sample, the most up-to-date AUC instrument (Optima AUC, Beckman Coulter) requires 400μL of similar concentration (as determined by prior measurement of optical density, e.g., by NanoDrop) and the SV-AUC run alone is 60 min with additional time required for data analysis and extensive training of the operator (Wu et al. 2022). As MP is a very recently developed technique, and its application for rAAV characterization is at its infancy with fewer than a handful of publications available at the time of writing, the reliability and robustness of the technique, especially regarding protein/DNA contaminants and other sources of noise, and performance characteristics, such as its accuracy and precision are yet to be fully confirmed (Gimpel et al. 2021).

Potential limitations of the method include its sensitivity to impurities that would decrease the resolution of the analysis, and the requirement for a low-scattering medium, although rAAVs would generally be suspended in aqueous media so this might not be relevant. Also, the latest MP instrument available is optimized for rAAV measurements, and, as such, it is less sensitive to low molecular mass contaminants. The upper limit of molecular mass that MP is able to measure is 5–10 MDa; therefore, MP cannot be used to assess the aggregation status of the rAAV sample or for the measurement of samples of larger viruses. Furthermore, MP measures molecular mass and is therefore incapable of identifying whether the identity of the encapsulated genome is correct. As MP counts the particles of different molecular mass at a single molecule level, it could be expected that it would also provide the titer. However, this is not possible at present, and only information on the relative populations of full, empty, and partially filled capsids can be provided. This is because the data acquisition is a video recording starting when the sample is applied by the operator. The speed by which the sample is applied and the measurement started can introduce variability, which can, however, be overcome if the measurement is automated, so titer measurements could become available in the near future (Refeyn 2022b). MP, similarly to DLS and MALS, is a light scattering technique and hence, the mass is not directly measured but rather derived from the scattering intensity based on appropriate calibration, in contrast to AUC, which directly measures the density of the analyte. Lastly, MP is not currently compliant with GMP requirements but this is expected to be resolved in the near future. Overall, MP is a novel method that holds great potential for facilitating and accelerating the characterization of rAAV samples, but further investigation is needed before it can be broadly adopted.

9.10 Charge Detection Mass Spectrometry (CDMS)

9.10.1 Background

Charge detection mass spectrometry (CDMS) can determine the mass of charged particles or ions by measuring both the mass-to-charge ratio (m/z) and charge of individual particles. Advances in sensitivity now allow the use of CDMS to detect full, empty, and partially packaged rAAV genomes as the mass of the rAAV particle differs between full (highest mass), partial (intermediate mass) and empty (lowest mass) particles (Pierson et al. 2016). This single-particle approach has yielded impressive results when it comes to determining accurate F/E ratios (Werle et al. 2021) and has also deepened the understanding of how rAAVs are packaged. For instance, one curious result that arose from the CDMS method's excellent ability to distinguish genome sizes is that there is a high proportion of capsids in self-complementary rAAV (scrAAV) preparations that have a mass almost exactly in between full and empty (Pierson et al. 2016). This suggests that scrAAV genomes are truncated at the mutated ITR (mITR) hairpin that connects the two complementary strands of the genome. These packaged truncated genomes, which can reach up to 42% of the total amount of capsids (Pierson et al. 2016), may lead to unexplained reduction in transduction efficacy as they will be readily detected and judged as "normal" by for instance dPCR genomic titer determination. Interestingly, undersized viral genomes (i.e., genomes less than half of the rAAV packaging capacity) that do not contain the mITR and thus are expected to produce ssDNA genomes, can package scrAAV genomes in ~2–3% of the times as measured by CDMS (Pierson et al. 2016; Werle et al. 2021; Wörner et al. 2021).

9.10.2 Considerations

As is the case for AUC, CDMS was reported to provide accurate and reproducible reports for F/E determination (Werle et al. 2021) but requires highly specialized equipment and skilled users to operate them. Additionally, since CDMS is a relatively new method for rAAV analysis and thus commercial equipment is few and far in between, question marks are raised when it comes to the implementation of this method in GMP grade facilities, particularly when it comes to acquisition and analysis software (Werle et al. 2021).

9.11 Differential Scanning Fluorimetry (DSF)

The melting temperature (T_m) is a characteristic of the different rAAV serotypes and serves as a surrogate for the thermal stability of the capsid. Differential scanning fluorimetry (DSF), also known as ThermoFluor or Thermal Shift Assay, is commonly used to measure protein unfolding. In the context of rAAV characterization, DSF is employed to identify the rAAV serotype, as first introduced by Rayaprolu and colleagues in 2013 (Rayaprolu et al. 2013). T_m has been found to be characteristic of the different rAAV serotypes independently of the genome payload and hence DSF is used in the characterization of a rAAV sample, specifically for serotype identification. Further, VP3, the most abundant capsid protein, has been identified as the sole determinant of thermal stability (Bennett et al. 2017). Increasingly, DSF is also being used to measure the stabilizing or destabilizing effects of formulation excipients, such as surfactants, manifested as an increase or decrease in the T_m, respectively. DSF can be used to measure T_m either in combination with a fluorescent dye, SYPRO Orange (SO-DSF), or, as more recently shown, for the measurement of the intrinsic fluorescence of tryptophan (iDSF) (Bennett et al. 2017; Rieser et al. 2020).

SYPRO Orange (SO) is a sensitive, ready-to-use, fluorescent dye, developed for protein detection and quantification of protein bands in sodium dodecyl sulfate polyacrylamide gel electrophoresis (SDS-PAGE) gels as an alternative to Coomassie™ Blue. SO binds to hydrophobic regions of proteins. The fluorescence signal sharply increases when the capsid denatures and the normally inaccessible hydrophobic regions are exposed and available for interaction with SO.

For the calculation of T_m the fluorescence emission is recorded over a temperature ramp, typically starting at room temperature and reaching 99 °C in a linear temperature gradient (e.g., 1 °C/min). The primary output is a plot (thermogram) of the fluorescence signal (*y*-axis) against the temperature (*x*-axis). The change rate of fluorescence emission is represented by the first derivative of the function (slope). T_m is the temperature at which the first derivative reaches its maximum value (steepest slope in the thermogram) and is defined as the temperature at which 50% of the protein is denatured. An example of a plot showing different rAAV profiles is depicted in Fig. 9.8.

In a similar vein, the intrinsic tryptophan fluorescence intensity is tightly dependent on the protein conformation (3D structure). Tryptophan is a hydrophobic amino acid and is therefore exposed upon protein denaturation or unfolding. Tryptophan emission wavelength undergoes a red shift (decrease in the λ_{max}) when the amino acid is in an aqueous environment. The sample emission is scanned at 330 and

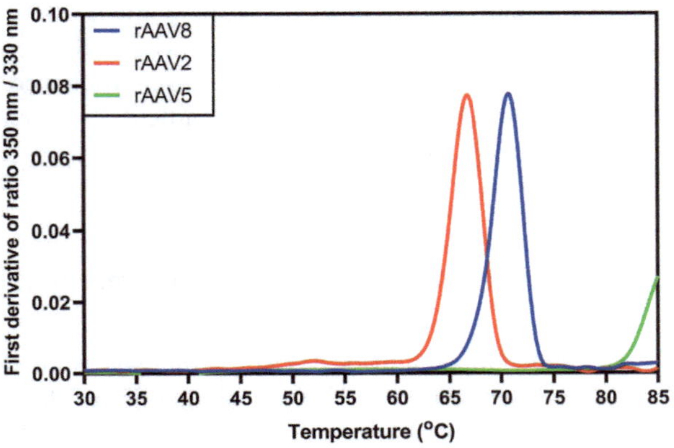

Fig. 9.8 DSF measurements of intrinsic fluorescence signals of different rAAV serotypes (rAAV2 in blue, rAAV5 in red, and rAAV8 in blue) with increasing temperature. The peak is indicative of the melting temperature, T_m

350 nm simultaneously and the 350/330 ratio is plotted against the temperature gradient. Thus, when denaturation occurs typically the emission at 330 nm changes more drastically than the 350 nm (red shift). As in the case of SO, the first derivative transformation gives the temperature when the slope (dEmission/dT) is maximum, which is the reported T_m (Andari and Grimm 2021; 2bind 2022).

DSF is a quick, cost-effective, facile, and robust method requiring only a small amount of sample. For intrinsic fluorescence determination, it requires none or minimal sample preparation. Importantly, DSF measurement is carried out in solution, unlike crystallographic techniques employed for in-depth characterization of the protein 3D structure. DSF can generally distinguish serotypes that differ by a single amino acid as long as the change impacts the charge and pI. Thus, it is more sensitive to changes between acidic-basic amino acids. rAAV serotypes with very similar T_m in a specific buffer cannot be reliably distinguished by a measurement in that buffer alone. For example, Bennett and colleagues found that rAAV7, rAAV9, and rAAVrh.10 displayed a similar T_m in PBS (76.5–77 °C). Nevertheless, identification of these serotypes by means of SO-DSF was still possible by measuring the samples in different buffers (Bennett et al. 2017). Surfactants such as poloxamer 188 (e.g., PluronicTM F68), polysorbate 20 or 80 (e.g., Tween 20 or 80) are increasingly used to prevent rAAV aggregation and adsorption to the contact materials, e.g., vials and syringes. Surfactants interfere with the SO-DSF measurement due to their interaction with SO, which elevates the fluorescence signal baseline. Although this interaction does not shift the T_m, an elevated baseline would cause a lower sensitivity, as SO-DSF measurement is concentration-dependent. Therefore, a label-free measurement, such as iDSF is advantageous. Of note, iDSF signal is not concentration-dependent, although the sample concentration must fall within a certain range for a reliable measurement. Further, studies involving rAAVs are generally limited by sample availability. rAAV production remains expensive and analytical methods that can generate reliable results with the minimum sample quantity are favored. Therefore, iDSF is preferred as it requires 10μL of sample, compared to 22.5μL of sample for SO-DSF, while a similar concentration range is still appropriate (1×10^{12}–1×10^{13} vp/mL).

Tryptophan residues are generally conserved in rAAVs, which underscores the relevance of this technique (Andari and Grimm 2021). The major limitation of iDSF is that specialized, expensive equipment is required for measurement, e.g., Prometheus NanoDSF (NanoTemper). A limitation of both SO-DSF and iDSF is that the measurement is destructive because by definition the rAAV capsid must be denatured and a temperature ramp (e.g., 25–99 °C) is employed for that purpose. However, as T_m measurement is commonly combined with other analytical techniques to characterize a particular sample, the samples used for other non-destructive analytical techniques (e.g., DLS) can be retained and used for the DSF measurement. Additionally, more recent equipment allows measuring the intrinsic fluorescence together with back-reflection, DLS and/or SLS combining several measurements in a single run, e.g., the newer version of the aforementioned Prometheus (Panta) and Uncle (Unchained Labs). The latter equipment can also detect the DNA ejection from the capsid by using, e.g., SYBR Gold as a dye (Unchained Labs 2020).

The use of DSF for the identification of rAAV serotypes is well established. Alternative techniques for capsid identification include mass spectrometry, protein sequencing, and antibody-based techniques such as ELISA (Pacouret et al. 2017). These techniques are laborious and time-consuming, while ELISA in particular is also limited by the availability and selectivity of antibodies against existing and newly engineered serotypes. DSF presents as a solution to the need for a quick and simple method to confirm capsid identity independently of its genome payload (Pacouret et al. 2017; 2bind 2022).

Apart from capsid identity, thermal stability, expressed as an increase in T_m, can be used as an indicator for overall stability to investigate, for example, the potentially stabilizing effects of excipients during formulation development. Such an interpretation of T_m has obvious practical advantages over the long and tedious traditional degradation studies, as its results can be obtained and analyzed easily and quickly, within a few hours. However, whether an increase in T_m can safely be interpreted as an increase in stability more broadly, e.g., resistance to aggregation, increased infectivity, is a matter of discussion. In particular, when examining the effect of surfactants, which have shown to increase the T_m in one study (Rieser et al. 2020), a question arises whether the apparent stabilization effect is real or simply the result of an increase in the non-adsorbed—thus measured—particles. SO-DSF is sensitive to concentration changes and therefore susceptible to this bias, which is not the case for iDSF, according to a study by Rieser and colleagues (Rieser et al. 2020).

9.11.1 Important Considerations

Sample

- A high concentration of $>1 \times 10^{12}$ vp/ml is required to obtain reliable results.
- Very stable serotypes (e.g., rAAV5) need a more extended temperature range. The stability of rAAV in a buffer depends on the particle concentration and the buffer composition, thus those parameters need to be taken in consideration.

Data Analysis and Evaluation

- The equipment software provides automatically not only the melting temperature (T_m), but also the onset temperature (T_{onset}), i.e., the temperature at which the protein start to unfold. Currently available instruments (e.g., Prometheus) are equipped with a light scattering detector, which allows the detection of the onset of aggregation (T_{agg}), thus assessing the colloidal stability of the sample along with the capsid stability.

9.12 Outlook

With the in vivo gene therapy field maturing and considering the plethora of rAAV-based gene therapy clinical trials planned or already underway, the need for reliable analytical methods to accurately assess relevant CQAs is more evident than ever. Indeed, there are several cases where the FDA has issued a CMC (partial) clinical hold (Solid Biosciences Inc 2018; Selecta Biosciences Inc 2021) to ask for clarification on product attributes to allow the clinical trial to go forward. This kind of hold-up affects the involved company negatively financially but also potentially prevents patients from timely access to the medicine.

Another aspect that is fast becoming key in pharmaceutical development of gene therapy products is the possibility of in-process monitoring during manufacturing. This would help identify problems or out-of-specification products earlier in the manufacturing process, saving valuable time and resources. To achieve this, speed is of the essence. Analytical methods that take little time to perform and analyze are thus favored. In line with this, automation of physical methods (e.g., liquid handling robots) and analysis can decrease hands-on time and may allow current methods to be implemented as in-process controls. For instance, ELISA for rAAV capsid titer is a cumbersome and time-consuming method when performed manually, but this is a prime example of a method that can be performed by a liquid handling robot or specialized automated ELISA device. Further, automation efforts can facilitate the miniaturization of analytical methods. This is imperative as the cost of rAAV production is high and reducing the fraction of the virus preparation used for CQA analysis can help keep this cost down both during development and clinical or commercial production.

A question that is still outstanding is which analytical methods related to a particular CQA should be used in preparation of, for instance, a regulatory filing. The lack of clear gold standards for many of these as well as the fact that most methods have clear drawbacks (Table 9.2) suggests that the "safe" tactic would be a matrix approach that combines orthogonal methods to cover one's bases (similar to what is described for potency assays as described in a previous chapter). However, a package of methods that is too large could hamper development as it interferes with the necessary agility in early-stage pharmaceutical development and it may place an undue strain on the financial capability of a small company and the (often limited)

Table 9.2 Summary of the discussed methods, including advantages and disadvantages and comparison with other methods

Method	Quality attribute	Advantages	Disadvantages	Examples of comparative studies with other methods
AUC	– F/E ratio – Aggregation	– No dependence on interactions with stationary/mobile phase – Minimum assumptions – Broad spectrum of concentration and matrices can be used – Detects partially filled capsids	– Not currently validated for GMP – Relatively high sample volume required – Time-consuming – High cost, specialized instruments and highly trained operators required	– TEM and AEX (Fu et al. 2019) – SEC-MALS and AEX (Haberland et al. 2021) – CDMS, cryo-EM, OD A260/A20 ratio, SEC-MALS and capsid titer-to-genome titer ratio (Werle et al. 2021) – OD 260/280 ratio (Maruno et al. 2021)
ELISA	– Capsid titer	– Routine method, readily implemented in many labs – Can be GMP validated – Commercial kits available	– Serotype-specific, novel engineered capsids may need to be custom developed – May struggle to accurately detect aggregates and broken capsids – Little to no sample preparation necessary	
dPCR	– Genomic titer	– Readily implemented in many labs – Can be GMP validated	– Only detects small subset of total genome – Results depend strongly on sample preparation steps	
dPCR/ ELISA	– F/E ratio	– Standard methods – Methods readily implemented; data is 'free'	– Combining two methods increases variability and pitfalls of both	– OD A260/A20 ratio (Sommer et al. 2003)
DLS	– Aggregation	– Rapid and facile – Minimal sample preparation – Non-destructive (sample can be re-used assuming measurement at room temperature)	– More appropriate for screening rather than in-depth analysis – Low resolution (bulk method) – Heavily impacted by large particle contamination	– MADLS, DSC and SEC-MALS (Cole et al. 2021)
SEC-MALS	– Capsid titer	– Accurate analysis – Relatively small	– Time-consuming – Analysis can be	– AEX and AUC (Haberland et al.

(continued)

Table 9.2 (continued)

Method	Quality attribute	Advantages	Disadvantages	Examples of comparative studies with other methods
	– Aggregation – F/E ratio	volume required for analysis – Minimal sample preparation	impacted by changes in sample medium, choice of mobile and stationary phase – Compatibility with, e.g., sample medium and mobile phase needs to be checked – 100% recovery assumed	2021) – DLS, MADLS and DSC (Cole et al. 2021) – AUC, CDMS, cryo-EM, OD A260/A20 ratio and capsid titer-to-genome titer ratio (Werle et al. 2021)
EM	– F/E ratio	– Direct visualization of capsids – Low sample volume required – Detects partially filled capsids – Can be GMP validated	– May require staining; tedious sample preparation for Cryo-EM – Limited throughput and long analysis time – Requires specialized equipment	Negative stain EM: – AUC (Tomono et al. 2018) – AUC and AEX-HPLC (Fu et al. 2019) Cryo-EM: – AUC (Subramanian et al. 2019) – SV-AUC, CDMS, A260/A20 ratio, SEC-MALS and capsid titer-to-genome titer ratio (Werle et al. 2021)
OD/ Absorbance	– F/E ratio	– Cheap and fast routine method – Allows fast screen of F/E ratio	– Accuracy is dependent on sample purity – Does not discriminate between different contaminating protein or DNA molecules present – Unsuitable for in-process control samples unless coupled with a purification technique	– qPCR/ELISA (Sommer et al. 2003) – AUC (Maruno et al. 2021) – AUC, CDMS, cryo-EM, SEC-MALS and capsid titer-to-genome titer ratio (Werle et al. 2021)
IEC/AEX	– F/E ratio	– Determination of F/E ratio during process development possible with higher accuracy in pure samples – Relative small	– Incompatibility with some buffers containing high salt – Resolution of separated peaks is serotype dependent – Method	– AUC and TEM (Fu et al. 2019) – AUC and SEC-MALS (Haberland et al. 2021)

(continued)

Table 9.2 (continued)

Method	Quality attribute	Advantages	Disadvantages	Examples of comparative studies with other methods
		volume required – High reproducibility – Possibility of high throughput and automation – GMP compliant	development for each serotype – Relatively low dynamic range dependent on the differences in pI – Does not separate abnormally packed capsids	
MP	– F/E ratio	– Low sample volume required – Very short analysis time – Detects partially filled capsids	– Very new method – Not currently validated for GMP	– AUC (Wu et al. 2022)
CDMS	– F/E ratio	– Accurate, repeatable method – Detects partially filled capsids	– Very new method – No GMP validated software from vendors	– AUC, cryo-EM, OD A260/A20 ratio, SEC-MALS and capsid titer-to-genome titer ratio (Werle et al. 2021)
DSF	– Identity – Thermal stability	– Rapid and facile – Minimal sample preparation – Label-free (intrinsic fluorimetry)	– May not be able to detect minor changes	– SDS-PAGE and intrinsic fluorescence spectroscopy (IFS) (Pacouret et al. 2017)

supply of material. The answer to this quandary may come from the Bespoke Gene Therapy Consortium (BGTC) (NIH 2021), which the NIH and FDA have spearheaded to help accelerate the development of gene therapy programs. This public-private collaboration, which involves many pharmaceutical companies involved in the gene therapy field, aims to do so by setting up a standard set of analytical methods. Currently, not all equipment used in the methods outlined in this chapter are GMP compliant when supplied by the vendors, meaning that any methods need to be validated for reproducibility and precision as well as data integrity. The hope is that standardized methods will help remove these hurdles in GMP validation.

In this chapter, we have described methods used in rAAV particle characterization where it pertains to CQAs regarding aggregation and virus genome content. While this list is not exhaustive by any means, we believe it provides a comprehensive insight into key aspects that can make or break rAAV gene therapy development and can provide a way to overcome the current challenges in clinical development and commercialization.

Acknowledgments We would like to acknowledge Matthias Lauer and Sylwia Huber (Roche Pharma Research and Early Development, Therapeutic Modalities, Roche Innovation Center Basel) for support with the electron micrographs and AUC data, respectively. Additionally, Carsten Elger and Dominik Kochardt (Pharma Technical Development Cell & Gene Therapy, Roche Diagnostics GmbH) are acknowledged for supporting the SEC-MALS and IEC sections. Lastly, we would like to thank Elaine Peters (Pharma Technical Development Cell & Gene Therapy, Genentech) and Tobias Graf (Pharma Technical Development Cell & Gene Therapy, Roche Diagnostics GmbH) for reviewing the manuscript.

References

2bind (2022) The gold standard for challenging protein stability analysis. https://2bind.com/nanodsf/. Accessed 5 Jan 2022

Adrian M, Dubochet J, Lepault J, McDowall AW (1984) Cryo-electron microscopy of viruses. Nature 308:32–36. https://doi.org/10.1038/308032a0

Andari JE, Grimm D (2021) Production, processing, and characterization of synthetic AAV gene therapy vectors. Biotechnol J 16:2000025. https://doi.org/10.1002/biot.202000025

Ayuso E, Blouin V, Lock M et al (2014) Manufacturing and characterization of a recombinant adeno-associated virus type 8 reference standard material. Hum Gene Ther 25:977–987. https://doi.org/10.1089/hum.2014.057

Barron LD (2004) Molecular light scattering and optical activity, 2nd edn. Cambridge University Press, Cambridge

Bennett A, Patel S, Mietzsch M et al (2017) Thermal stability as a determinant of AAV serotype identity. Mol Ther Methods Clin Dev 6:171–182. https://doi.org/10.1016/j.omtm.2017.07.003

Bernaud J, Rossi A, Fis A et al (2018) Characterization of AAV vector particle stability at the single-capsid level. J Biol Phys 44:181–194. https://doi.org/10.1007/s10867-018-9488-5

Besir H, Odenwald C, Hinner I, Betts K (2019) Cryo and negative staining EM—a comparison study with AAV capsids. Progen Biotechnik

Bhattacharya A (2019) An introduction to characterization of biomolecules using analytical ultracentrifugation. https://www.youtube.com/watch?v=Ilo84HN6UEU. Accessed 22 Feb 2022

Brument N, Morenweiser R, Blouin V et al (2002) A versatile and scalable two-step ion-exchange chromatography process for the purification of recombinant adeno-associated virus serotypes-2 and -5. Mol Ther 6:678–686. https://doi.org/10.1006/mthe.2002.0719

Burnham B, Nass S, Kong E et al (2015) Analytical ultracentrifugation as an approach to characterize recombinant adeno-associated viral vectors. Hum Gene Ther Method 26:228–242. https://doi.org/10.1089/hgtb.2015.048

Chen H (2007) Atomic force microscopy of recombinant adeno-associated virus-2 prepared by centrifugation. Scanning 29:238–242. https://doi.org/10.1002/sca.20067

Cockroft A, Wilson A (2021) Comparability: what we can learn from the review of advanced therapy medicinal products. Regen Med 16:655–667. https://doi.org/10.2217/rme-2021-0026

Cole JL, Lary JW, Moody TP, Laue TM (2008) Analytical ultracentrifugation: sedimentation velocity and sedimentation equilibrium. Methods Cell Biol 84:143–179. https://doi.org/10.1016/s0091-679x(07)84006-4

Cole L, Fernandes D, Hussain MT et al (2021) Characterization of recombinant adeno-associated viruses (rAAVs) for gene therapy using orthogonal techniques. Pharmaceutics 13:586. https://doi.org/10.3390/pharmaceutics13040586

Debelak D, Fisher J, Iuliano S et al (2000) Cation-exchange high-performance liquid chromatography of recombinant adeno-associated virus type 2. J Chromatogr B Biomed Sci Appl 740:195–202. https://doi.org/10.1016/s0378-4347(00)00100-6

Demeler B (2022) Optima AUC analytical ultracentrifuge from Beckman Coulter. https://www.news-medical.net/Optima-AUC. Accessed 18 Feb 2022

Dobnik D, Kogovšek P, Jakomin T et al (2019) Accurate quantification and characterization of adeno-associated viral vectors. Front Microbiol 10:1570. https://doi.org/10.3389/fmicb.2019.01570

Dunbar CE, High KA, Joung JK et al (2018) Gene therapy comes of age. Science 359:eaan4672. https://doi.org/10.1126/science.aan4672

Foley EDB, Kushwah MS, Young G, Kukura P (2021) Mass photometry enables label-free tracking and mass measurement of single proteins on lipid bilayers. Nat Methods 18:1247–1252. https://doi.org/10.1038/s41592-021-01261-w

Fu X, Chen W-C, Argento C et al (2019) Analytical strategies for quantification of adeno-associated virus empty capsids to support process development. Hum Gene Ther Method 30:144–152. https://doi.org/10.1089/hgtb.2019.088

Furuta-Hanawa B, Yamaguchi T, Uchida E (2019) Two-dimensional droplet digital PCR as a tool for titration and integrity evaluation of recombinant adeno-associated viral vectors. Hum Gene Ther Method 30:127–136. https://doi.org/10.1089/hgtb.2019.031

Gagnon P, Goricar B, Mencin N et al (2021) Multiple-monitor HPLC assays for rapid process development, in-process monitoring, and validation of AAV production and purification. Pharm 13:113. https://doi.org/10.3390/pharmaceutics13010113

Gill SC, von Hippel PH (1989) Calculation of protein extinction coefficients from amino acid sequence data. Anal Biochem 182:319–326. https://doi.org/10.1016/0003-2697(89)90602-7

Gimpel AL, Katsikis G, Sha S et al (2021) Analytical methods for process and product characterization of recombinant adeno-associated virus-based gene therapies. Mol Ther Methods Clin Dev 20:740–754. https://doi.org/10.1016/j.omtm.2021.02.010

Glasel J (1995) Validity of nucleic acid purities monitored by 260nm/280nm absorbance ratios. Biotechniques 18:62–63

Haberland S, Hanelt I, Ettelt V et al (2021) Quality control of AAV capsid loading. Protagen

Hauser J, Kylberg G, Colomb-Delsuc M et al (2020) A microfluidic device for TEM sample preparation. Lab Chip 20:4186–4193. https://doi.org/10.1039/d0lc00724b

Horowitz ED, Rahman KS, Bower BD et al (2013) Biophysical and ultrastructural characterization of adeno-associated virus capsid uncoating and genome release. J Virol 87:2994–3002. https://doi.org/10.1128/jvi.03017-12

Janson J-C (2011) Protein purification: principles, high resolution methods, and applications, 3rd edn. Wiley, Hoboken, NJ

Large EE, Silveria MA, Zane GM et al (2021) Adeno-associated virus (AAV) gene delivery: dissecting molecular interactions upon cell entry. Viruses 13:1336. https://doi.org/10.3390/v13071336

Lock M, McGorray S, Auricchio A et al (2010) Characterization of a recombinant adeno-associated virus type 2 reference standard material. Hum Gene Ther 21:1273–1285. https://doi.org/10.1089/hum.2009.223

Lock M, Alvira MR, Wilson JM (2012) Analysis of particle content of recombinant adeno-associated virus serotype 8 vectors by ion-exchange chromatography. Hum Gene Ther Part B Methods 23:56–64. https://doi.org/10.1089/hgtb.2011.217

Lottspeich F, Engels JW (2018) Bioanalytics: analytical methods and concepts in biochemistry and molecular biology. Wiley, Weinheim, Germany

Malvern Instruments (2010) Dynamic light scattering: an introduction in 30 minutes. https://warwick.ac.uk/fac/cross_fac/sciencecity/programmes/internal/themes/am2/booking/particlesize/intro_to_dls.pdf. Accessed 10 Apr 2022

Malvern Panalytical (2021) Dynamic light scattering (DLS). https://www.malvernpanalytical.com/en/products/technology/light-scattering/dynamic-light-scattering. Accessed 30 Nov 2021

Maruno T, Usami K, Ishii K et al (2021) Comprehensive size distribution and composition analysis of adeno-associated virus vector by multiwavelength sedimentation velocity analytical ultracentrifugation. J Pharm Sci 110:3375–3384. https://doi.org/10.1016/j.xphs.2021.06.031

McIntosh NL, Berguig GY, Karim OA et al (2021) Comprehensive characterization and quantification of adeno associated vectors by size exclusion chromatography and multi angle light scattering. Sci Rep 11:3012. https://doi.org/10.1038/s41598-021-82599-1

Mietzsch M, Barnes C, Hull JA et al (2020) Comparative analysis of the capsid structures of AAVrh.10, AAVrh.39, and AAV8. J Virol 94:e01769-19. https://doi.org/10.1128/jvi.01769-19

Minton AP (2016) Recent applications of light scattering measurement in the biological and biopharmaceutical sciences. Anal Biochem 501:4–22. https://doi.org/10.1016/j.ab.2016.02.007

Mohammad R (2021) Key considerations in formulation development for gene therapy. Drug Discov Today 27:292–303. https://doi.org/10.1016/j.drudis.2021.08.013

Nass SA, Mattingly MA, Woodcock DA et al (2017) Universal method for the purification of recombinant AAV vectors of differing serotypes. Mol Ther Methods Clin Dev 9:33–46. https://doi.org/10.1016/j.omtm.2017.12.004

NIH (2021) NIH, FDA and 15 private organizations join forces to increase effective gene therapies for rare diseases. https://www.nih.gov/news-events/news-releases/nih-fda-15-private-organizations-join-forces-increase-effective-gene-therapies-rare-diseases. Accessed 12 May 2022

Pace CN, Vajdos F, Fee L et al (1995) How to measure and predict the molar absorption coefficient of a protein. Protein Sci 4:2411–2423. https://doi.org/10.1002/pro.5560041120

Pacouret S, Bouzelha M, Shelke R et al (2017) AAV-ID: a rapid and robust assay for batch-to-batch consistency evaluation of AAV preparations. Mol Ther 25:1375–1386. https://doi.org/10.1016/j.ymthe.2017.04.001

Pierson EE, Keifer DZ, Asokan A, Jarrold MF (2016) Resolving adeno-associated viral particle diversity with charge detection mass spectrometry. Anal Chem 88:6718–6725. https://doi.org/10.1021/acs.analchem.6b00883

Porterfield JZ, Zlotnick A (2010) A simple and general method for determining the protein and nucleic acid content of viruses by UV absorbance. Virology 407:281–288. https://doi.org/10.1016/j.virol.2010.08.015

Prasad K-MR, Trempe JP (1995) The adeno-associated virus Rep78 protein is covalently linked to viral DNA in a preformed virion. Virology 214:360–370. https://doi.org/10.1006/viro.1995.0045

ProteinSimple (2022) Ella: Your Next Generation ELISA. https://www.proteinsimple.com/ella.html. Accessed 29 Apr 2022

Ramsey JP, Khatwani SL, Lin M et al (2021) Overview of analytics needed to support a robust gene therapy manufacturing process. Curr Opin Biomed Eng 20:100339. https://doi.org/10.1016/j.cobme.2021.100339

Rayaprolu V, Kruse S, Kant R et al (2013) Comparative analysis of adeno-associated virus capsid stability and dynamics. J Virol 87:13150–13160. https://doi.org/10.1128/jvi.01415-13

Refeyn (2022a) SamuxMP: a mass photometer for AAV analytics. https://www.refeyn.com/samuxmp-mass-photometer. Accessed 15 May 2022

Refeyn (2022b) Are your AAVs filled? Determine rapidly with mass photometry. https://www.workcast.com/AuditoriumAuthenticator.aspx?cpak=1202283353645039&pak=4090471351879666. Accessed 15 May 2022

Rieser R, Penaud-Budloo M, Bouzelha M et al (2020) Intrinsic differential scanning fluorimetry for fast and easy identification of adeno-associated virus serotypes. J Pharm Sci 109:854–862. https://doi.org/10.1016/j.xphs.2019.10.031

Rumachik NG, Malaker SA, Poweleit N et al (2020) Methods matter: standard production platforms for recombinant AAV produce chemically and functionally distinct vectors. Mol Ther Methods Clin Dev 18:98–118. https://doi.org/10.1016/j.omtm.2020.05.018

Sahin E, Roberts CJ (2012) Size-exclusion chromatography with multi-angle light scattering for elucidating protein aggregation mechanisms. In: Voynov V, Caravella JA (eds) Therapeutic proteins: methods and protocols, 2nd edn. Humana, Totowa, NJ, pp 403–423

Sanmiguel J, Gao G, Vandenberghe LH (2019) Quantitative and digital droplet-based AAV genome titration. Methods Mol Biol 1950:51–83. https://doi.org/10.1007/978-1-4939-9139-6_4

Selecta Biosciences Inc (2021) Selecta biosciences provides update on phase 1/2 clinical trial of SEL-302 for the treatment of methylmalonic acidemia-U.S. FDA has issued a clinical hold on phase 1/2 clinical trial of SEL-302 due to CMC related questions. https://ir.selectabio.com/node/11211/pdf. Accessed 3 May 2022

Snyder RO, Im DS, Muzyczka N (1990) Evidence for covalent attachment of the adeno-associated virus (AAV) rep protein to the ends of the AAV genome. J Virol 64:6204–6213. https://doi.org/10.1128/jvi.64.12.6204-6213.1990

Solid Biosciences Inc (2018) Prospectus summary and business—overview. https://www.sec.gov/Archives/edgar/data/1707502/000119312518019022/d404258dfwp.htm. Accessed 3 May 2022

Sommer JM, Smith PH, Parthasarathy S et al (2003) Quantification of adeno-associated virus particles and empty capsids by optical density measurement. Mol Ther 7:122–128. https://doi.org/10.1016/s1525-0016(02)00019-9

Subramanian S, Maurer AC, Bator CM et al (2019) Filling adeno-associated virus capsids: estimating success by cryo-electron microscopy. Hum Gene Ther 30:1449–1460. https://doi.org/10.1089/hum.2019.041

Sucato C (2020) Considerations in the use of analytical ultracentrifugation for characterization and QC testing of AAV gene delivery vectors. https://www.youtube.com/watch?v=UclBt0dU8kE. Accessed 22 Feb 2022

Suoranta T, Laham-Karam N, Ylä-Herttuala S (2021) Optimized protocol for accurate titration of adeno-associated virus vectors. Hum Gene Ther 32:1270–1279. https://doi.org/10.1089/hum.2020.318

Tai PWL, Xie J, Fong K et al (2018) Adeno-associated virus genome population sequencing achieves full vector genome resolution and reveals human-vector chimeras. Mol Ther Methods Clin Dev 9:130–141. https://doi.org/10.1016/j.omtm.2018.02.002

Tamayose K, Hirai Y, Shimada T (1996) A new strategy for large-scale preparation of high-titer recombinant adeno-associated virus vectors by using packaging cell lines and sulfonated cellulose column chromatography. Hum Gene Ther 7:507–513. https://doi.org/10.1089/hum.1996.7.4-507

Tan YZ, Aiyer S, Mietzsch M et al (2018) Sub-2 Å Ewald curvature corrected structure of an AAV2 capsid variant. Nat Commun 9:3628. https://doi.org/10.1038/s41467-018-06076-6

The dMIQE Group, Whale AS, Spiegelaere WD et al (2020) The digital MIQE guidelines update: minimum information for publication of quantitative digital PCR experiments for 2020. Clin Chem 66:1012–1029. https://doi.org/10.1093/clinchem/hvaa125

Tomono T, Hirai Y, Okada H et al (2018) Highly efficient ultracentrifugation-free chromatographic purification of recombinant AAV serotype 9. Mol Ther Methods Clin Dev 11:180–190. https://doi.org/10.1016/j.omtm.2018.10.015

Uliyanchenko E (2014) Size-exclusion chromatography—from high-performance to ultra-performance. Anal Bioanal Chem 406:6087–6094. https://doi.org/10.1007/s00216-014-8041-z

Unchained Labs (2020) DNA leaks before capsids pop: AAV thermal stability on Uncle. https://www.unchainedlabs.com/wp-content/uploads/2021/11/DNA-leaks-before-capsids-pop-AAV-thermal-stability-on-Uncle.pdf. Accessed 8 Apr 2022

Urabe M, Xin K-Q, Obara Y et al (2006) Removal of empty capsids from type 1 adeno-associated virus vector stocks by anion-exchange chromatography potentiates transgene expression. Mol Ther 13:823–828. https://doi.org/10.1016/j.ymthe.2005.11.024

Venkatakrishnan B, Yarbrough J, Domsic J et al (2013) Structure and dynamics of adeno-associated virus serotype 1 VP1-unique N-terminal domain and its role in capsid trafficking. J Virol 87:4974–4984. https://doi.org/10.1128/jvi.02524-12

Vironova (2018) Automated integrity analysis of AAV and adenovirus particles using MiniTEM. Vironova

Wang C, Mulagapati SHR, Chen Z et al (2019a) Developing an anion exchange chromatography assay for determining empty and full capsid contents in AAV6.2. Mol Ther Methods Clin Dev 15:257–263. https://doi.org/10.1016/j.omtm.2019.09.006

Wang D, Tai PWL, Gao G (2019b) Adeno-associated virus vector as a platform for gene therapy delivery. Nat Rev Drug Discov 18:358–378. https://doi.org/10.1038/s41573-019-0012-9

Wen J, Arakawa T, Philo JS (1996) Size-exclusion chromatography with on-line light-scattering, absorbance, and refractive index detectors for studying proteins and their interactions. Anal Biochem 240:155–166. https://doi.org/10.1006/abio.1996.0345

Werle AK, Powers TW, Zobel JF et al (2021) Comparison of analytical techniques to quantitate the capsid content of adeno-associated viral vectors. Mol Ther Methods Clin Dev 23:254–262. https://doi.org/10.1016/j.omtm.2021.08.009

Wörner TP, Snijder J, Friese O et al (2021) Assessment of genome packaging in AAVs using Orbitrap-based charge-detection mass spectrometry. Mol Ther Methods Clin Dev 24:40–47. https://doi.org/10.1016/j.omtm.2021.11.013

Wright JF (2014) AAV empty capsids: for better or for worse? Mol Ther 22:1–2. https://doi.org/10.1038/mt.2013.268

Wright JF, Le T, Prado J et al (2005) Identification of factors that contribute to recombinant AAV2 particle aggregation and methods to prevent its occurrence during vector purification and formulation. Mol Ther 12:171–178. https://doi.org/10.1016/j.ymthe.2005.02.021

Wu D, Piszczek G (2021) Standard protocol for mass photometry experiments. Eur Biophys J 50: 403–409. https://doi.org/10.1007/s00249-021-01513-9

Wu Z, Yang H, Colosi P (2010) Effect of genome size on AAV vector packaging. Mol Ther 18:80–86. https://doi.org/10.1038/mt.2009.255

Wu D, Hwang P, Li T, Piszczek G (2022) Rapid characterization of adeno-associated virus (AAV) gene therapy vectors by mass photometry. Gene Ther. https://doi.org/10.1038/s41434-021-00311-4

Xu R (2015) Light scattering: a review of particle characterization applications. Particuology 18: 11–21. https://doi.org/10.1016/j.partic.2014.05.002

Young G, Hundt N, Cole D et al (2018) Quantitative mass imaging of single biological macromolecules. Science 360:423–427. https://doi.org/10.1126/science.aar5839

Zhang X, Wang W, Kenrick S (2019) Characterization of AAV-based viral vectors by DynaPro DLS/SLS instruments. Wyatt Technology

Zolotukhin S, Byrne BJ, Mason E et al (1999) Recombinant adeno-associated virus purification using novel methods improves infectious titer and yield. Gene Ther 6:973–985. https://doi.org/10.1038/sj.gt.3300938

Chapter 10
Phase-Appropriate Potency Assay Development for AAV-Based Gene Therapy Products

Elaine E. Peters, Roland Pach, and Shyamtanu Datta

10.1 Introduction

The number of clinical trials and approved marketed therapies with viral vector-based gene therapies continues to steadily increase. AAV vectors remain the major platform for in-vivo gene therapy due to their proven safety profile and high efficiency of transduction of a broad range of target tissues. However, like all ATMPs (Advanced therapy medicinal products), viral vector-based gene therapies are not considered to be well-characterized biological products. This makes it challenging to determine and control critical quality attributes (CQAs) and, therefore, characterization studies should be completed early in the development life cycle to gain an understanding of factors that can impact product efficacy and/or safety.

E. E. Peters (✉)
Pharma Technical Development Cell and Gene Therapy USA, Genentech, Inc., South San Francisco, CA, USA
e-mail: peters.elaine@gene.com

R. Pach
Pharma Technical Development Analytics Biologics, F. Hoffmann-La Roche Ltd., Basel, Switzerland

S. Datta
Pharma Technical Development Cell and Gene Therapy, F. Hoffmann-La Roche Ltd., Basel, Switzerland

S. Gautam et al. (eds.), *Bioprocess and Analytics Development for Virus-based Advanced Therapeutics and Medicinal Products (ATMPs)*,
https://doi.org/10.1007/978-3-031-28489-2_10

10.2 Phase Appropriate Potency Assay Development

Development and validation of an appropriate potency assay is particularly difficult. Most ATMPs have multiple mechanisms of action or have mechanisms of action that are part of a cascade of separate actions which are not necessarily serial in nature. For example, a gene therapy product has at least two mechanisms of action, the ability to transfer a genetic sequence to a cell, and the effect of the expressed genetic sequence.

In order to obtain marketing approval for an AAV product, a validated potency assay or assay matrix must be described and justified in the marketing application. The assay(s) must:

- Demonstrate product specificity and reflect the relevant mechanism of action
- Provide test results for the release of the product
- Provide quantitative data
- Meet predefined acceptance and/or rejection criteria
- Include appropriate reference materials, standards, and/or controls (to monitor assay performance)
- Establish and document the accuracy, sensitivity, specificity, linearity, repeatability, and reproducibility/precision of the test methods employed through validation
- Measure the identity and strength of all active ingredients
- Demonstrate stability indicating properties

In early phase clinical investigations, it may not be possible to meet all of the requirements described above for licensed biological products and thus an incremental approach is recommended (Fig. 10.1).

In preclinical, Phase 1 and early Phase 2 studies, limited quantitative information on relevant biological attributes may be sufficient. Quantitative assay acceptance criteria should be established and should be adjusted throughout the product development stages to reflect manufacturing and clinical experience.

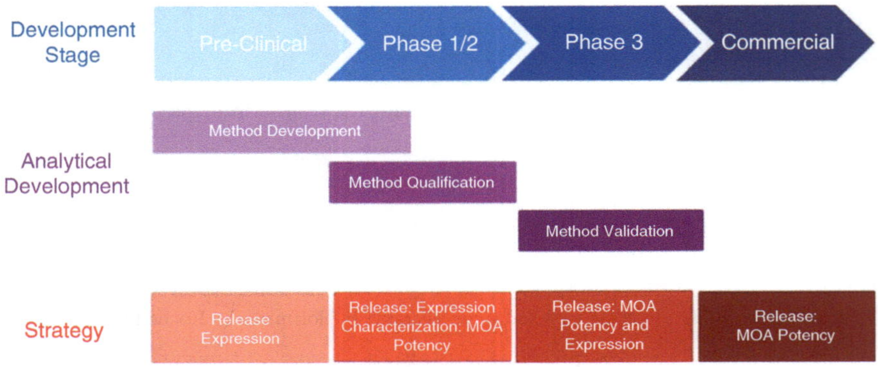

Fig. 10.1 Phase-appropriate assay development

As clinical development progresses into later phase studies (i.e., Phase 3) and product knowledge increases, improved potency measurement(s) that quantitatively assesses relevant biological product attribute(s) should be implemented.

The primary objective of later phase investigational studies is to gather meaningful data about product efficacy, which is determined by adequate and well-controlled clinical trial(s). One aspect of an adequate and well-controlled trial is administering product lots with similar potency. Therefore, the final potency assay or assay matrix design and acceptance criteria should establish appropriate limits for potency to assure that product lots are well-defined, biologically active, and consistently manufactured.

Regulators recommend considering the establishment of multiple potential potency assays for measuring potency as early as possible during early clinical development with the intention of eventually identifying and qualifying the most appropriate potency assay(s). Throughout all stages of the development life cycle, data should be acquired from these assays to aid the final assay selection process.

10.3 Connection Between Manufacturing Process and Product Potency

While an appropriate potency testing matrix is essential for drug product release and shelf life determination; it is equally important to have a suitable potency assay to support process development and process validation. There are a number of areas where there is a close connection between the manufacturing process and product potency including; process impurities, empty/full capsid ratio, post-translational modifications and structure.

For example, key to validating the manufacturing process is demonstrating consistent removal of process impurities. These impurities may occur across the entire manufacturing process of an AAV product. For example, impurities may be introduced during cell culture growth and harvest, purification processes or from the use of single-use technologies. Process related impurities may include cell-derived impurities, buffer components, antibiotics, surfactants, anti-foaming agents, process-enhancing agents, or compounds that leach from product contacting materials.

Process-related impurities can impact a number of critical quality attributes including potency. Ideally, process impurities should be completely removed. However, in practice, low levels of impurities may remain in the final drug product. It is therefore important to complete characterization studies early in product development early to understand which process and product impurities can impact potency. The data from these characterization studies can be used to inform process development and the establishment of critical process parameters to ensure effective control; removal, reduction to consistent low impurity levels.

Evaluation of AAV vector product-related impurities can be challenging as these may closely resemble the vector itself. The ratio of viral particles that are full, empty (do not contain genetic material), or partially full (only contain a fragment of the genetic material) is one such example. Therapeutic efficacy generally requires full capsids. Product dosing tends to be based on total viral particle load which is not limited to only full capsids and there is evidence that empty viral particles can compete with full viral particles for cellular uptake and intracellular trafficking as well as increasing vector-related adverse effects (Parker et al. 2003). Thus, empty or partially full capsids can impact both efficacy and safety.

The distribution of empty, full, and partially full capsids can vary significantly between manufactured lots/batches. Potency assays have multiple roles in understanding the impact of empty/full capsid ratios. Firstly, they can ensure batches with consistent efficacy are used in clinical studies and for approved products. The potency assay(s) can also play a vital role in characterizing the partially full capsids. Gaining an understanding of the functionality of partially full capsids can guide bioprocessing development.

Many of the common degradation pathways of AAV products may be impacted and controlled by the process used for manufacturing. These include: aggregation (Rodrigues et al. 2019; Wright et al. 2005), proteolysis (Salganik et al. 2012), unfolding/denaturing, oxidation (Ayuso et al. 2010), deamidation (Giles et al. 2018), disulfide formation (Pulicherla et al. 2012), isomerization and, phosphorylation. For example, low ionic strength, buffer and excipient pH can lead to aggregation. Sheering from fill/finish processes can result in aggregation, precipitation, and unfolding.

10.4 The Matrix Approach

With the described complexities of gene therapy products as well as processes, an incremental approach might be needed to the development and implementation of potency tests, applying the "matrix approach."

Depending on the therapeutic area, in discovery transgenic animal models might be designed and chosen for mirroring the very complex as well as multiple mechanisms of actions (MoAs). They serve often also as proof of concept. At the analytical method development phase during preclinical and early clinical development stage (Fig. 10.1), even such a transgenic animal model can be used as an in-vivo potency assay (Bishnu et al. 2018).

In comparison to in-vitro cell based potency assay, in-vivo potency assay shows less robustness, reproducibility and precision (Aronson et al. 2020) which are crucial elements of a release test in a GMP compliant environment. Therefore, from a generic compliance and quality perspective, the in-vivo potency assay is inappropriate as an element of the commercial control strategy.

It is crucial to design and select the adequate in-vitro potency assay methods along the intracellular processing path of the transgenic gene of interest (GOI) by application of the matrix approach as early as possible during the early development stage (Fig. 10.1). For gene therapy products besides 'transduction efficiency' it is recommended to show its functionality whose readout depends strongly on the intrinsic complexity and biological function of the GOI. Already during the early development stage, the life cycle and continuous improvement begin. A strong effort is to avoid any product-related in-vivo testing in the control system and reduce its usage to 'bioassay bridging' or to characterization studies only. Ideally, the selected in-vitro potency assay is consistent with in-vivo efficacy (Aronson et al. 2020).

Not every MoA has to be or can be shown for the potency assay, mainly the key MoA(s) is adequate and robust enough for the needs of the release-, stability- and comparability testing. For multiple MoAs, the matrix approach selects two or more complementary potential measures of potency. Deducing from the current understanding, the mode of actions (MoAs) and what product attributes contribute to biological effect(s), these analytical methods can be biological or non-biological ones. Although the in-vitro potency assays should reflect the predicted MoAs, in some cases there is no correlation to the in-vivo efficacy possible.

In its entirety, the matrix of complementary measures should cover the intracellular processing of the transgenic sequence (GOI) up to its functionality in the target cells. Experimentally this can be performed by e.g. transduction assay of representative cell lines or in-vivo, e.g. in adequate transgenic animal models.

At a minimum, one potency assay has a quantitative read-out (and therefore capable of being a stability indicating functional assay). This measure in units of activity is calibrated against an adequate qualified reference. In the case of the absence of a reference, the usage of a well characterized 'assay control' provides the monitoring of the assay performance (not used to assign potency values for the test sample).

Figure 10.2 is an overview of an in-vitro recombinant AAV (rAAV) based on the transduction of representative cells showing intracellular processing of transgenic information for a recombinant protein. This figure illustrates all process steps of a transduction starting by uptake of the viral vectors, transport to nucleus, over transcription to translation, followed by post-translational modifications and intracellular translocation; and the resultant intracellular products (primary product, e.g., mRNA) that can be analyzed. As shown, numerous methods can be used to characterize each cellular step allowing for the overall determination of potency. In the best case, the functionality of transgenic information can be measured in a quantitative and stability-indicating manner directly from the same experiment.

10.5 Analytical Tools to Measure Potency Assays

The potency assay for a viral-based gene therapy product should essentially demonstrate an evaluation of the efficiency of gene transfer (infectivity/transduction/delivery) and the level of expression of the therapeutic sequence or its direct activity. There are many read-out methods described to determine quantitatively and/or qualitatively the different analytes (virus, DNA/RNA, polypeptide and mature protein) at the different intracellular processing steps. As illustrated in Fig. 10.2

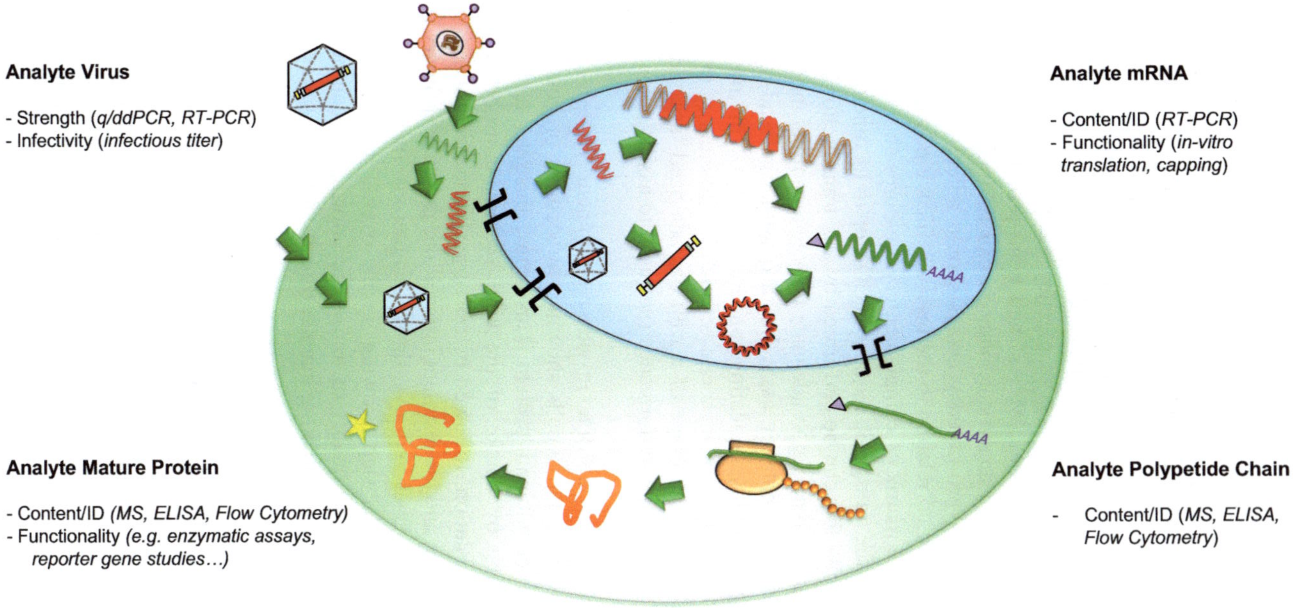

Fig. 10.2 In-vitro AAV/Lentivirus-based cell transduction.

for infectivity/delivery and primary product, molecular biological methods like qPCR, ddPCR, RT-PCR are mainly used. For the determination of expression level and potency biochemical/immunochemical based methods like ELISA, Flow Cytometry, LC-MS and SPR are used. The choice of analytical method determining the functionality depends on the predicted mechanism of action, complexity/higher order structure and post-translational modifications. ICH Q6B guideline recommends that functional assays of the transgene "measure an organism's biological response to a product, a biochemical or physiological response at the cellular level, enzymatic reaction rates, or ligand- and receptor-binding." The FDA issued a guidance for industry on potency tests for cellular and gene therapy products (FDA Guidelines 2011) and EMA issued a Guideline on the quality, non-clinical and clinical aspects of gene therapy medicinal products (EMA guidelines 2021). Both guidelines empathize the importance and necessity to include the determination of functionality of the therapeutic sequence in the potency assay strategy. Depending on the intrinsic complexity and multiple manners of the MoAs only a qualitative read-out might be feasible. In this context quantitative immunochemical methods are added complementary to the functional assay with the target to determine the expressed level and integrity of protein/polypeptide. If even surrogate assays based on molecular methods (nucleic acid amplification) are selected for potency release testing, a 'bioassay bridging' to the functionality assay or to in-vivo efficacy are needed. If not feasible, biological potency tests in animal tissues-maintained ex vivo or in whole animals can be considered. Transgenic animals or animals with transplanted human tissues or systems, or environmentally challenged animal models (e.g., laser-induction, stimulus-induced) may be suitable for this purpose. However, to reduce the use of animals in accordance with the 3R principles and high variability in animal experiments, a validated in-vitro method is generally preferred over animal testing wherever possible. Albeit discussing the transgene specific functional assay which will depend on the biochemical characteristic of the transgene is out of scope for this chapter, we have listed a few state-of-the-art technologies that are being used to analyze the transduction, infectivity, and level of gene or protein expression.

10.5.1 qPCR and ddPCR

Quantitative polymerase chain reaction (qPCR) and digital droplet PCR (ddPCR) quantify the cDNA vector via fluorescence (FS) detection during or after amplification in a thermocycler (Campomenosi et al. 2016). These two methods are currently employed to quantify the gene expression of the transgene for the potency assay. More recently, ddPCR emerged as a powerful alternative that offers excellent accuracy and precision. AAV genome or rAAV reference standard physical titer is traditionally determined by qPCR but large multi-site clinical studies highlighted its low precision, indicated by poor coefficient of variation (CV), and low reproducibility. Stochastic and systematic sources of this variability, such as low replication efficiency or choice of the primer's target sequence, have been identified and alleviated with additional DNA pre-treatments and optimized primer conditions improving repeatability and intermediate precision to miniscule level. However, such extensive efforts have not developed the qPCR assay to be optimal. Compared to qPCR, ddPCR does not require a standard curve and measures the endpoint of PCR cycles, exhibiting less sensitivity to variable replication efficiency. Thus, ddPCR is significantly more precise than qPCR, at repeatability and intermediate precision of very low CV, and more tolerant to matrix effects and inhibitors (Dobnik et al. 2019). Both qPCR and ddPCR exhibit a dynamic range of 2–4 orders of magnitude. It is also noteworthy that infectious titers determined by qPCR and ddPCR may differ, complicating the switch from qPCR to ddPCR during development or manufacturing. One-step RT-ddPCR method can serve as a framework for analytical and potency assays in many upcoming gene therapy programs, both gene augmentation and repression (Clarner et al. 2021). First, absolute quantification of transgene and a housekeeping gene in a two-dimensional reaction allows for AAV sample specific internal control eliminating the need for standard curves. This absolute quantification nature enables inter-study data comparison with a simplified data output format. Second, the one-step RT and PCR process, coupled with automated droplet reading, significantly shortens the hands-on manipulation in the workflow as well as the associated variability. In addition, as more industry laboratories are using ddPCR to titer AAV vectors, it serves as a good method to use the same platform technology for analytical assays to maintain consistency in potency assay. However, with ddPCR being one of the sophisticated techniques is not completely devoid of drawbacks. It has been demonstrated that poorly handled droplets, rough pipetting, interference caused by the phenomenon of "rain" in the adjusting of a threshold, the copy number of input DNA, adjusting the cycling conditions, and read-out for any cellular sample transduced with AAV being lower than 10,000 droplets and reduce the sensitivity and efficiency of the ddPCR-based potency assay. Moreover, an oversized transgene in AAV is truncated from the terminals, neither qPCR nor ddPCR, using a selected amplicon toward the end of the transgene, can accurately determine the titer of complete genomes, as the PCR replication cycle only considers the presence of the amplicon itself, i.e., the DNA flanked by the primer sequences. In such cases, long read NGS or Nanopore NGS must be used as an alternative technology.

10.5.2 Flow Cytometry

Flow cytometry is a versatile laser-based state-of-the-art platform that has the potential to be used in every stage of a viral based gene therapy development program, moving from discovery, through to preclinical and into clinical phases of development. With the potential for implementation in each of these stages, there is a need to develop robust flow cytometric methods and to validate the methods in alignment with the associated regulatory environments (GLP, US FDA, EMA,

International Organization for Standardization [ISO] and Standard for Clinical Laboratory Accreditation [ISO15189]). One way to determine transduction efficiency is possible by intracellular flow cytometric detection of transgene expression. Cells grown and transduced with viral-based gene therapy products are permeabilized and fixed. After washing, the cells can be labeled with specific primary and secondary antibodies conjugated with fluorescent dye that can be measured on a Flow cytometry machine. Flow cytometry allows for the quantification and characterization of protein translated from the transgene at the single-cell level. The flow assay distinguishes between transgene-positive and transgene-negative cells based on the specific staining of cells with antibodies coupled to another fluorescence dye to distinguish live and dead cells using cell viability markers. A single-cell population was selected based on SSC/FSC plots, and a gate was set with a maximum background of ~0.5% of transgene positive cells in stained un-transduced control cells. The assay optimization parameters included: selection of target cell, protein concentration, peptide-pulse incubation time, assay incubation time, multiplicity of infection of the certain AAV serotype, optimization of the voltage settings for each flow cytometer detector, titration of each antibody for optimal performance in the panel, the pairing of dyes with targets and minimize spillover between multiple dyes, and the number of events to collect. A gating strategy must be defined and standardized during assay development. A few manual adjustments to the gating might be necessary given assay and instrument variation. However, a well-defined and methodical approach should be followed. At present, flow cytometry methods are being evaluated for their capability to analyze AAV-mediated transgene expression at multi-site trials. For comparing results from multi-site trials standard flow cytometry protocols must be adopted to ensure identical processing and storage of AAV-based GTP to ensure that the functional assay is not compromised and the evaluation of capsid and gene products for cellular immunity. Despite being a high throughput and highly precise technology, one of the challenges of flow cytometry lies in finding the appropriate antibody for measuring the protein expression. The presence of fluorescent tags to the transgenes solves the issue to some extent for a proof-of-concept method development, but such fluorescent tags are not available in the final GTPs that go into patients.

10.5.3 ELISA and SPR

An enzyme-linked immunosorbent assay (ELISA) quantifies the capsid titer via the photometric activity of a substrate reaction catalyzed by an enzyme linked to monoclonal antibodies specific to a conformational epitope of the protein expressed from the cargo. In its most common and specific variant, sandwich ELISA, a capture antibody is used to immobilize the protein of interest on a solid support, and a detection antibody facilitates quantification with a biotin-conjugated antibody and streptavidin peroxidase catalyzing the biotin/streptavidin peroxidase color reaction. The assay is typically performed after a purification step but is sufficiently robust and specific to be used on samples with moderate matrix effects and free capsid proteins, like cell lysates. Disadvantage of ELISA includes multiple lengthy incubation and washing steps prior to analysis and akin to FACS finding an appropriate antibody remains to be a challenge. The principles of concentration or Surface Plasmon Resonance (SPR) resemble those applied to automated ELISA. Both are performed on a solid support in which one interaction partner is immobilized (ligand) and the other is free in solution (analyte). However, it is noteworthy that SPR assays increase operational efficiency by introducing automation and eliminating time-consuming washing steps. The ability to culminate several methods and run-in sequence to link each assay step increases the degree of automation and consumes less time. Depending on the assay format, the signal reported in the assay is either directly or inversely proportional to the amount of active analyte bound to the ligand. SPR assays and ELISA assays share many similarities and are based on the same experimental principles. The conversion of an ELISA assay into a SPR assay may only require the transfer of the antibody pairs used in the ELISA and the assigned buffer conditions. The advantages of using SPR assays include: (a) Results are obtained in less than half the time compared with ELISA; (b) Very low operating expenses thanks to the automated setup; (c) The ability to reuse the sensor surface significantly reduces consumables expenses; (d) Low sample consumption and high reproducibility that minimizes the need for assay reruns; (e) Easy-to-use: no biotinylated antibodies, streptavidin-HRP reagent, and other developing reagents nor wash buffer preparation and repetitive washing steps are needed; (f) Automation decreases inter- and intra-assay variability; (g) Suitable for more complex assays, such as bispecific antibodies; and (h) SPR assays not prone to edge effects or the hook effect as in ELISA assays. Nevertheless, it must be taken in account that the conversion of an ELISA to an SPR assay may involve re-qualification and validation of the conditions and configurations, not unlike the steps required for development and optimization of an ELISA method.

10.5.4 Mass Spectrometry (MS)

Another important tool that has been widely used in the development of many small-molecule pharmaceuticals is Mass Spectrometry. However, for large molecules, the loss of charge state resolution in the measured m/z (mass-to-charge ratio) spectrum has been a bottleneck. Charge detection MS (CDMS), where the m/z and charge are directly measured for each ion, however, has been able to overcome this limitation. It is worth mentioning that for the full characterization of rAAV samples, CDMS is very high throughput technology and minimal sample consumption (<20 mL) at a relatively low titer. From the recent development of the method and with detailed studies, this technology appears to be universally applicable to all natural and non-natural AAV serotypes measured to date and it provides an accurate and precise measurement of the empty/full ratio. Though MS has the potential to quantitate the potency of GTP, owing to the heterogeneous nature of the DNA packaged into the capsids, the method development is quite challenging and may have to be complemented with alternate technologies like NGS.

10.6 Future Directions and Challenges Ahead

- A major challenge in translating adeno-associated virus (AAV) vectors for the treatment of human disease is the challenge in demonstrating the product function in a potency assay, a requirement in the clinical translation to product. Standalone methods to analyze AAV potency suffer from poor efficacy of infectivity and do not yield the correct measures of potency. For viral-based gene therapy products, initial IND-enabling efficacy studies typically rely on established in-vitro interaction and binding studies. As process development advances, the need may arise to enhance potency to capture fully the product's potential in the intended disease.
- Additionally, these assays may require cell line development specific to each product and, sometimes, a parallelization of assay performance in-vivo and in-vitro. It is noteworthy that engineering cell lines for specific functional potency is one of the major challenges for gene therapy, as it requires knowledge of disease pathogenesis. The specific cells are quite complex to handle and not robust to analyze. Sometimes even a whole network of 3D cell cultures is needed to demonstrate the specific mechanism of action.
- The health authorities recommend that potency assays must include in-vitro and/or in-vivo tests as well as full characterization of the biological activities and that this be done early in development to assure identity, quality, purity, potency, and stability. Modes of action of viral-based gene therapy products are quite complex. It is necessary to have a thorough understanding of the mechanism of action involved in both the disease and treatment (e.g., infection, integration, and expression of transgene), and the also in-vivo fate of the product as it relates to migration from the administration site, cellular differentiation, viral infection, and transgene expression.
- Developing in-vitro potency assays for viral vector-based gene therapy products includes at least two steps: demonstrating the transfer of genetic material and showing that the transferred genetic material has the desired effect in patients. Potency assays should be quantitative, high throughput, and expressed in units of activity calibrated against recognized standard or reference standard from a reliable source. The functional potency assay matrix includes cell line and animal model selection and/or development specific to the gene therapy product and its promoter, viral transducibility and infectivity, tissue selectivity of AAV serotype (s), dose-ranging studies, bioactivity assays (e.g., enzymatic or non-enzymatic assays), the emergence of common impurities during sample processing, and assaying for relative potency against a reference material.
- Gene therapy products using AAV vector-based delivery currently undergo in-vitro and in-vivo potency assays, for the specific mechanisms of action under investigation, at the DNA, RNA, and protein levels. Although mostly standard

assays based on the technologies mentioned in the last sections are powerful tools that can be used for preclinical studies, no clear public guidelines exist that address the requirements for performing these tools in a regulated environment. It is also to be noted that while same methods can be used for measuring physicochemical characterization of the AAVs as well as the bioanalytical properties of the AAV vectors (e.g., ddPCR is used for both physical titer and infectivity or infectious titer), the acceptance criteria for bioanalytical assay may be different from that of the physicochemical analysis because of the complexity of bioanalytical assay. Therefore, here also one must take care to keep the acceptance criteria more based on the assay than the technology. Moreover, evolving regulatory expectations, new technologies, and the quality of AAV-based products present a major concern for the development of a potency assay.
- Another major challenge for bringing bioanalytical technologies in a regulated environment is data integrity of these technologies. FDA emphasizes on completeness, consistency, and accuracy of data. According to the health authorities, the bioanalytical data for release and stability must be attributable, legible, contemporaneous, original, and accurate (ALCOA). Verifying data integrity poses a significant challenge in method development. It is recommended to have data integrity checks built in throughout the various phases of gene therapy development.
- Apart from above mentioned, other challenges that exist in developing potency assays tests for viral-based gene therapy products are: (1) inherent variability of starting materials; (2) small lot size and materials for testing; (3) narrow window of stability; (4) lack of appropriate reference standards; (5) interactions between multiple active ingredients (vectors containing multiple genes); (6) Raw materials/Excipients; and (7) Product and process-related impurities/HCPs. These challenges have been covered in detail in the previous sections of this chapter.

Despite the significant number of challenges ahead, with the growing amount of knowledge generated each day and the collaborative kind of approach amongst industries and academia, one can be optimistic to develop better potency assays. This will eventually bring more gene therapies safely and efficiently to the market to cure diseases.

References

Aronson SJ, Bakker RS, Moenis S et al (2020) A quantitative in vitro potency assay for adeno-associated virus vectors encoding for the *UGT1A1* transgene. Mol Ther Methods Clin Dev 18:250–258. https://doi.org/10.1016/j.omtm.2020.06.002

Ayuso E, Mingozzi F, Bosch F (2010) Production, purification and characterization of AAV. Curr Gene Ther 10:423–436

Bishnu P, Chen A, Salami CO et al (2018) In vivo potency assay for adeno-associated virus–based gene therapy vectors using AAVrh.10 as an example. Hum Gene Ther Methods 29(3):146–155. https://doi.org/10.1089/hgtb.2017.246

Campomenosi P, Gini E, Noonan DM, Poli A, Rotolo N, Dominioni L, Imperatori A (2016) A comparison between quantitative PCR and droplet digital PCR technologies for circulating microRNA quantification in human lung cancer. BMC Biotechnol 16(1):60

Clarner P, Lau SK, Chowdhury T, Guilmette E, Trapa P, Lo SC, Shen S (2021) Development of a one-step RT-ddPCR method to determine the expression and potency of AAV vectors. Mol Ther Methods Clin Dev 23:68–77

Dobnik D, Kogovsek P, Jakomin T, Kosir N, Tusek Znidaric M, Leskovec M, Kaminsky SM, Mostrom J, Lee H, Ravnikar M (2019) Accurate quantification and characterization of adeno-associated viral vectors. Front Microbiol 10:1570

European Medicines Agency Guideline on quality, non-clinical and clinical aspects of medicinal products containing genetically modified cells, 2021

FDA Guidance for industry: potency tests for cellular and gene therapy products, 2011

Giles AR, Sims JJ, Turner KB et al (2018) Deamidation of amino acids on the surface of adeno-associated virus capsids leads to charge heterogeneity and altered vector function. Mol Ther 26(12):2848–2862

ICH Q6B specifications: Test procedures and acceptance criteria for biotechnological/biological products

Parker A, Nagy D, Vargas J, Anand V, Qu G, Wright JF, Couto L (2003) in-vivo performance of AAV2 vectors purified by CsCl gradient centrifugation or column chromatography. Mol Ther 7: S390

Pulicherla N, Kota P, Dokholyan NV, Asokan A (2012) Intra- and inter-subunit disulfide bond formation is nonessential in adeno-associated viral capsids. PLoS One 7(2):1–6

Rodrigues GA, Shalaev E, Karami TK, Cunningham J, Slater NKH, Rivers HM (2019) Pharmaceutical development of AAV-based gene therapy products for the eye. Pharm Res 36(2):1–20

Salganik M, Venkatakrishnan B, Bennett A et al (2012) Evidence for pH-dependent protease activity in the adeno-associated virus capsid. J Virol 86(21):11877–11885

Wright JF, Le T, Prado J et al (2005) Identification of factors that contribute to recombinant AAV2 particle aggregation and methods to prevent its occurrence during vector purification and formulation. Mol Ther 12(1):171–178

Chapter 11
Quantification of Virus Infectivity: The Key Assay for the Development of Viral Therapeutics

Ramona Businger, Daniel Hochdorfer, Dominik Hotter, and Johannes Solzin

11.1 Introduction

In the late nineteenth century, viruses were identified as infectious agents which could not be removed by filtration. Plant viruses, such as the tobacco mosaic virus, were among the first viruses to be studied (Lecoq 2001). In the following, bacterial viruses (bacteriophages) were discovered by Twort in 1915 and d'Herelle in 1917 (Duckworth 1976). For the quantification of bacteriophage infectivity, d'Herelle set up the first titration method in 1917, known as plaque assay (d'Herelle 1917). Other viruses such as the virus causing foot-and-mouth disease and the yellow fever virus, representing the first identified human pathogenic virus, were studied and titrated in animals since cell culture techniques were not available before the middle of the twentieth century (Anonymous 1925; Staples and Monath 2008; Reed et al. 1901).

The term infection can refer to quite different understandings of its outcomes, including the mere entry of viral particles into cells, amplification of viral genomes, expression of viral genes, release of viral progeny, or pathogenic effects. Likewise, assays to determine the infectious titer can be based on measuring different end-points, and thus, include or exclude certain aspects of the viral replication cycle. However, a common feature of all methods to quantify infectivity is that they involve infection of susceptible hosts, represented either by animals, cultivated cells, or model systems such as embryonated eggs. Of note, there are also other, host-independent methods available to quantify viruses. These include qPCR or ELISA, which quantify single viral components such as genomes or viral proteins,

R. Businger · D. Hochdorfer · D. Hotter · J. Solzin (✉)
Development Viral Therapeutics Center, Boehringer Ingelheim GmbH & Co KG, Ochsenhausen, Germany
e-mail: johannes.solzin@boehringer-ingelheim.com

S. Gautam et al. (eds.), *Bioprocess and Analytics Development for Virus-based Advanced Therapeutics and Medicinal Products (ATMPs)*,
https://doi.org/10.1007/978-3-031-28489-2_11

respectively. Nanoparticle tracking analysis (NTA) or virus counters, which are based on flow cytometry, detect and quantify virus particles as a whole. Although these methods are generally faster, more precise, and easier to handle, they do not evaluate the ability of the virus particles to infect a cell. In the context of quality control for virus-based therapeutics, an assay to determine the infectious titer is one of the most important assays, as this is directly linked to its potency.

11.2 Principles for Infectious Virus Quantification

There are many possibilities to measure the infectivity of a virus sample. However, all classical methods that are used to measure infectious titer are either based on the principle of endpoint dilutions with a binary evaluation of infected cultures or on plaque assays (i.e., counting of infection events). The next two subchapters will describe these assays and variations thereof.

11.2.1 Endpoint-Dilution Assay

In endpoint-dilution assays, the sample of interest is serially diluted and applied in multiple technical replicates to susceptible host cells at a large concentration range. The range includes a high concentration at which a maximal effect is expected to be observed and a concentration at which no effect is expected to be observed. Finally, the sample concentration with the "half-maximal" effect will be reported. The tissue culture infectious dose 50 ($TCID_{50}$) is commonly used as an endpoint assay to determine the infectious titer of virus-based therapeutics. $TCID_{50}$ describes the dilution at which half of the inoculated cell culture wells are infected.

In detail, replicates of the virus sample are serially diluted, often in $\log_{10}$ or 0.5 $\log_{10}$ steps. Then, wells containing a cell monolayer are inoculated with these virus dilutions (Fig. 11.1). After an incubation period, which, in the classical assay setup, depends on the time the virus needs to replicate and cause a detectable cytopathic effect (CPE, see Sect. 11.3.1), all inoculated cell culture wells are assessed for the presence of the virus-induced CPE. The titer of the sample is then calculated based on the number of infected wells using, e.g., the Spearman–Kärber formula (Kärber 1931; Spearman 1908), and reported as $TCID_{50}$/mL.

11.2.2 Plaque Assay

The original plaque assay was developed with bacteria cultures on agar plates that were infected with bacteriophages (Panec and Katz 2006) and was adapted to animal viruses in 1952 by Dulbecco (Dulbecco 1952). The underlying principle of the

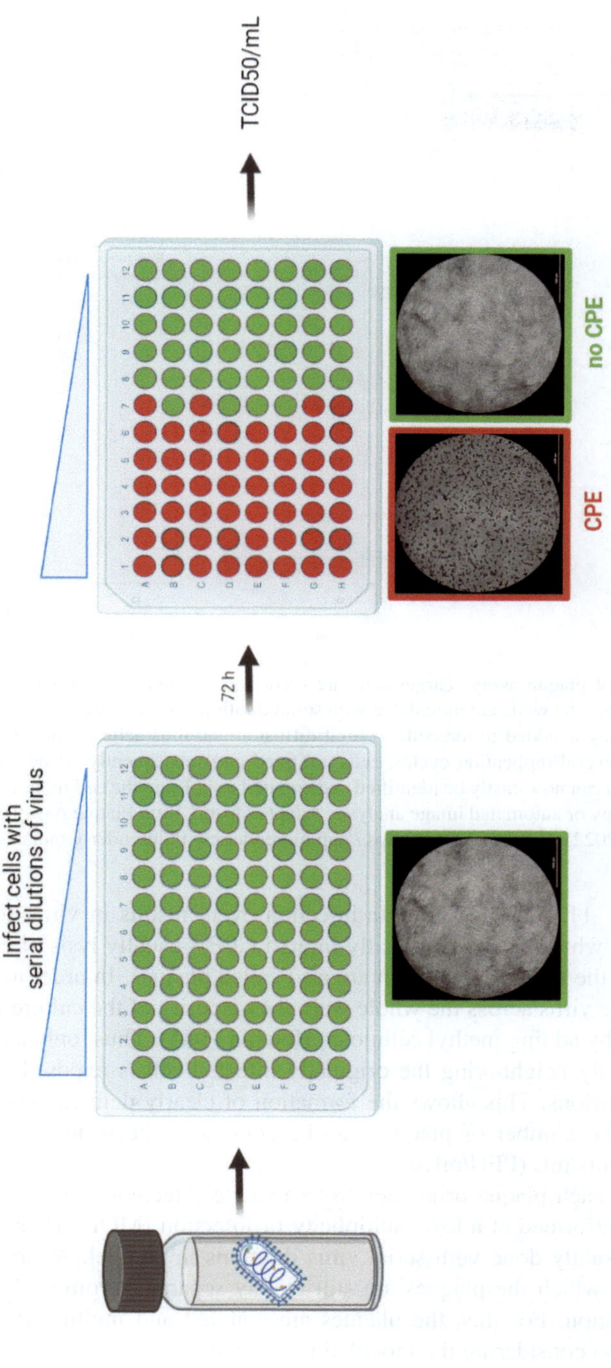

Fig. 11.1 Principle of the $TCID_{50}$ titration technique. Target cells are infected in multiple (e.g., eight) technical replicates with serial virus dilutions on a 96-well plate. After an incubation period covering several viral replication cycles, the wells are evaluated for cytopathic effects (CPE). The final titer of the virus ($TCID_{50}$/mL) is calculated, by, e.g., the Spearman–Kärber formula (created with Biorender.com and adapted from Hochdorfer et al. 2022, with permission from Elsevier)

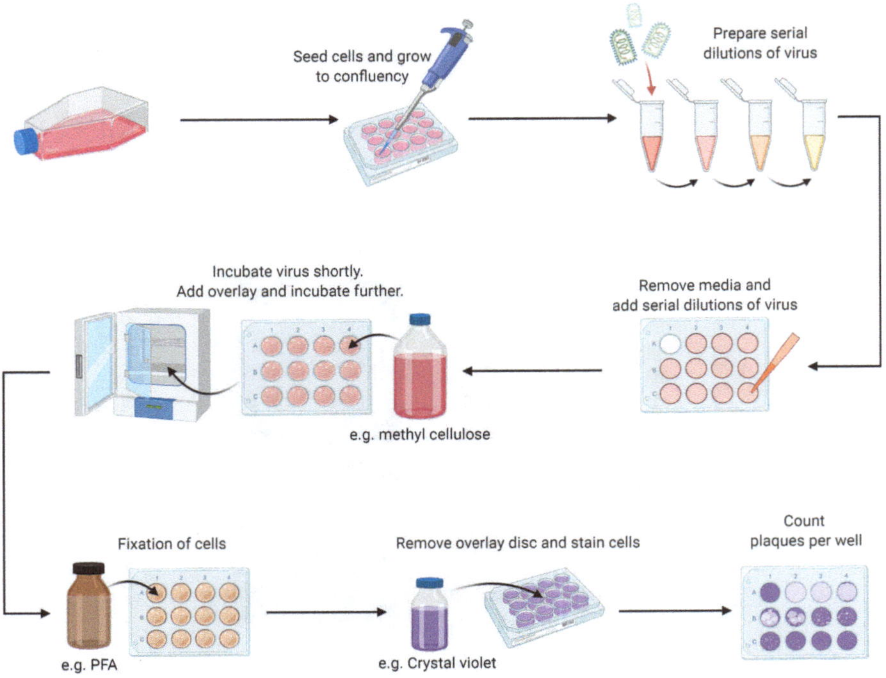

Fig. 11.2 Principle of plaque assay. Target cells are seeded and grown to confluency, typically done on 12-well plates. The wells are inoculated with serial dilutions of the virus. After a short time of infection, an overlay is added to the cells to limit diffusion and thus achieve locally restricted virus spread. After several replication cycles, cells are fixed and subsequently stained with, e.g., crystal violet. Plaques can now easily be identified and counted as holes in the cell monolayer either by classical microscopy or automated image analysis. Adapted from "Viral Plaque Assay Protocol," by BioRender.com (2022). Retrieved from https://app.biorender.com/biorender-templates

plaque assay (Fig. 11.2) is that a single infection event results in virus spread to neighboring cells, which causes a spatially limited CPE—usually lysis of cells that results in holes in the monolayer, which are also called plaques. In order to prevent the diffusion of the virus across the whole well, the viscosity of the culture medium is increased, e.g., by adding methyl cellulose (Hotchin 1955). Thus, only a confined area of cells directly neighboring the originally infected cell is exposed to newly formed progeny virions. This allows the formation of clearly defined and easy-to-detect plaques. The number of plaques can be counted to determine the titer as plaque-forming units/mL (PFU/mL).

To ensure that each plaque originates from a single infection event, the plaque assay has to be performed at a low multiplicity of infection (MOI). Therefore, the plaque assay is usually done with serial virus dilutions in parallel. Afterward, the lowest dilution at which the plaques are still clearly separated from each other is chosen for evaluation. For this, the plaques are counted and multiplied with the dilution factor, also considering the inoculation volume.

While cell lysis leading to plaques is the classical read-out of the plaque assay, the same principle also works for other replication-induced morphological changes, such as cell rounding. By staining infected cells, this principle can also be applied to viruses that do not cause a visible CPE or to assays in which infection needs to be evaluated prior to the formation of a visible CPE. These groups of infected, but still viable cells are called focus/foci instead of plaque/plaques. Therefore, the resulting titer from these assays is often reported as focus-forming units/mL (FFU/mL) instead of plaque-forming units/mL (PFU/mL). When applying virus-specific staining methods, individual infected cells can even be identified before they become a focus or plaque, respectively. There are several advantages in identifying single infected cells, rather than the fully formed foci/plaques: (1) for most viruses this significantly reduces the total assay duration, (2) there is no need to restrict diffusion as the readout is performed before a second replication cycle occurs, (3) smaller assay formats can be used (96-well instead of 24-well), (4) it is not necessary to infect at very low virus concentrations, (5) the assay can also be performed with suspension cells employing, e.g., flow cytometry as readout, (6) as the fraction of infected cells can be quantified at high virus concentrations, the titer can be extrapolated using the Poisson distribution on the results from a wide range of virus concentrations (Delwart 2011), and (7) automation of the assay readout is usually easier to implement.

It is worth mentioning that results from $TCID_{50}$ and plaque assays are not directly comparable. Theoretically, it is possible to convert titers from PFU/mL to $TCID_{50}$/mL and vice versa, as based on Poisson the relationship between these is $\frac{PFU}{mL} = -\ln(0.5) \times \frac{TCID_{50}}{mL}$, i.e., 1 PFU/mL ≈ 0.7 $TCID_{50}$/mL (Myers et al. 1994). However, for this relation to be true, the corresponding assays have to be performed under the exact same conditions, which is practically never the case. In this context, it is also important to mention that even though titers are expressed as PFU/mL or $TCID_{50}$/mL, these values should be considered relative to the assay system that was used to determine them. Specifically, the cell type, cell cultivation, inoculation volume, inoculation medium, incubation time, etc., significantly affect the measured titers.

11.3 Infectivity Readouts

Various readouts can be used for infectivity assays and need to be chosen according to the properties of the virus analyzed and the chosen target cell type.

11.3.1 *Morphological Changes of the Cells*

The classical readout of infectivity assays is to determine replication-induced cytopathic effects (CPE) to the target cells. These CPEs can be manifested by cell lysis but also by more subtle morphological changes, including changes in granularity, shrinking, swelling, or rounding of the cells as well as syncytia formation with neighboring cells. The type and extent of CPE depends on the virus' biology and the target cell type. Readouts that depend on a detectable CPE or even cell lysis can only be chosen when analyzing replicating viruses that cause a visible CPE.

VSV is an example of such a virus and is additionally clinically relevant, since it is a promising oncolytic virus (Muik et al. 2014) as well as an approved vector for vaccines such as ZEBOV against Ebola (Piszczatoski and Gums 2020; Saphire 2020). VSV is an actively replicating virus that causes a host cell shutdown and finally cell lysis of infected cells, resulting in a visible CPE (Wan et al. 2015; Redondo et al. 2015). Therefore, VSV qualifies to be analyzed by all morphological readouts presented below. Furthermore, both $TCID_{50}$ and plaque assay are suitable for VSV titration.

11.3.1.1 Visual CPE Evaluation

Historically, evaluation of CPE positive/negative wells ($TCID_{50}$) or counting of plaques (plaque assay) is done directly by visual evaluation or with the help of a light microscopy device. Lysis of the cells, represented by holes in the cell monolayer, can be easily interpreted by the operator either label-free or following additional staining steps of the remaining cell monolayer with, e.g., crystal violet (Baer and Kehn-Hall 2014). CPE evaluation can be done manually but this step can also be automated to reduce hands-on time and generate operator-independent results: Imaging readers take images from each well and analyze them automatically by an algorithm which determines the remaining cell area ($TCID_{50}$) (Hochdorfer et al. 2022) (Fig. 11.3) or count holes in the monolayer (plaque assay). In general, this readout requires adherent cells that form a confluent monolayer at the time of the readout.

When using cell lysis as a readout, the assay duration strongly depends on the replication kinetics of the virus.

Visual CPE evaluation is often chosen for $TCID_{50}$ or plaque assays because it is label-free with no or only few handling steps. The assay is usually performed in standard 96-well plates and can be analyzed with a regular light microscope which is available in all standard molecular biology or virology laboratories.

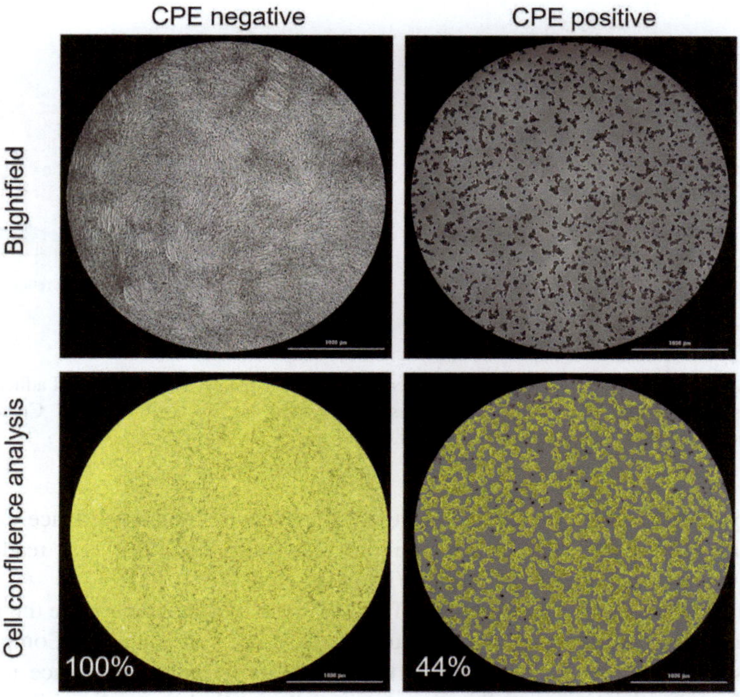

Fig. 11.3 Brightfield images of infected and non-infected BHK-21 cells ($TCID_{50}$). Yellow: automated analysis of cell area for the evaluation of cell lysis and therefore productive infection with numbers indicating % confluence. Reprinted from Hochdorfer et al. (2022), with permission from Elsevier

11.3.1.2 Metabolic Activity

Measuring the metabolic activity of cells is an alternative approach to determining whether a well is infected or not. Virus-induced cell lysis reduces the number of viable cells, and thus, the overall metabolic activity within a well (Filone et al. 2017). Depending on the substrate, the metabolic activity can be determined with a microtiter plate reader by measuring the absorption, fluorescence, or luminescence, respectively. Based on a pre-defined cutoff value, this approach categorizes a complete well as "infected" or "not infected" and can therefore only be used in an endpoint-dilution assay ($TCID_{50}$) rather than a plaque assay.

11.3.1.3 Impedance Measurement

Using special cell culture plates with integrated gold microelectrode sensor arrays (called E-plates) and the application of a low current, an electric field is generated which is impeded by the presence of cells (Fig. 11.4). Viral infection detaches or

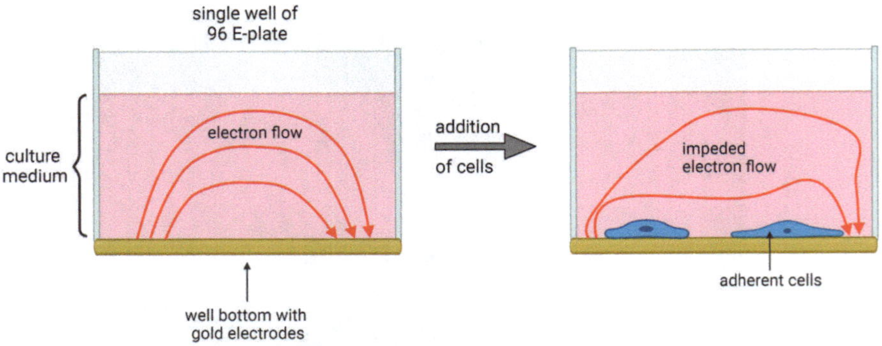

Fig. 11.4 Principle of impedance measurement in 96-well E-plates. The presence of adherent cells impedes the electron flow which can be used as a surrogate for cell viability. Created with BioRender.com and adapted from Şener et al. (2017)

even lyses the cells, reducing the electrical impedance. Thus, impedance measurement as a surrogate marker for viral infection also qualifies as a readout for infectivity assays (Fang et al. 2011; Witkowski et al. 2010).

This readout is only applicable for $TCID_{50}$, since it cannot determine the number of plaques per well, but rather measures the overall cell viability. Compared to microscopic readout methods, which typically require days, impedance measurement can already detect smaller changes in cell morphology and viability, which allows to obtain faster titration results within hours after infection. Furthermore, impedance measurement is label-free, easy to perform, generates objective results, and supports highly automated sample throughput. However, it also requires cost-intensive 96-well E-plates and specialized lab equipment.

11.3.1.4 Laser Force Cytology

Laser force cytology (LFC) was recently presented as a new readout for a fast and label-free determination of infectivity (Hebert et al. 2018). LFC uses a combination of optical and fluidic forces within a microfluidic channel to identify and quantify changes in intrinsic biophysical properties of single cells. Virus infection induces cellular changes in membranes, cytoplasmic organelles, and nuclear features, that manifest themselves as changes in velocity, size, shape, and orientation of cells as they pass through the region of laser photon pressure. These subtle infection-dependent changes can be measured already relatively early after infection, without the need for labels or antibodies. Depending on the combination of virus type and host cells, the user can choose from a variety of different infection metrics, determined by the LFC instrument, to find the best dependency between virus dose and change in cellular properties. For example, Hebert et al. correlated velocity-based LFC infection metrics of Vero cells infected with VSV with the corresponding

$TCID_{50}$ data, demonstrating the ability of LFC to measure infectivity (Hebert et al. 2018).

The ability of LFC to quantify cellular changes that occur early during the viral replication cycle is a large advantage compared to infectivity assays that rely merely on the visual evaluation of CPEs and can save multiple days until results are available. For example, classical brightfield image-based $TCID_{50}$ assays analyzing VSV infection take around 3 days, whereas LFC could be applied already 16 h post infection (Hebert et al. 2018). A downside of LFC is, however, the need of specialized equipment and the fact that the microfluidic cell analysis requires cells in suspension and therefore laborious sample preparation in the case of adherent cell lines.

11.3.2 Viral Antigen Expression

In contrast to changes in cellular properties, detection of viral antigens is a virus-specific readout, which quantifies infectivity at the level of protein expression, rather than induction of CPE. Typically, viral proteins are detected by specific antibodies. Therefore, this readout depends on the availability and performance of antibodies. Furthermore, the choice of the viral antigen is crucial since it must be expressed in a detectable amount. Staining of viral antigens helps in identifying, quantifying, and counting infected cells by microscopy or flow cytometry. Typically, secondary antibodies used for this readout are fluorescently labeled (Coons and Kaplan 1950; Noyes 1955). Identifying infected cells by staining of viral proteins and detected by microscopy can be used for both $TCID_{50}$ and plaque assay.

Genetically modified viruses that express a specific reporter protein, e.g., luciferase or fluorescent proteins, facilitate the direct identification of infected cells. Additionally, reporter cells are another possibility to show the presence of infectious virus. There are cell lines available that express a certain reporter upon the presence of specific viral proteins.

The staining of viral proteins as titration readout is suitable for replicating viruses that do not lyse the cells when releasing virus progeny (Rapp et al. 1959). Since the readout relies on an early step in the replication cycle, it can also be applied for lytic viruses before the onset of cell lysis. In conclusion, for most viruses, this is a possibility to measure the infectious titer much faster compared to the classical CPE analysis that often involves several replication cycles.

11.3.3 Viral Nucleic Acid Detection

Another possibility to quantify infection is the detection of viral nucleic acid. This is the favorable readout when analyzing non-replicating single-round infection virus vectors, e.g., adenoviruses (AdV), lentiviruses, or non-replicating vectors such as recombinant adeno-associated viruses (rAAV).

Nowadays, rAAV are most often used as vectors for gene therapy projects (Zhao et al. 2021). For this purpose, rAAV carry specific therapeutic transgenes and therefore lack the *rep* and *cap* genes. For titration of single-round virus vectors, the missing viral genes need to be provided in the cells by transcomplementation. This is either done by transient plasmid transfection or by using stably transduced cell lines carrying the specific genes. For rAAV specifically, titration is typically done in HELA RC32, which carry *rep* and *cap* of AAV2. Moreover, rAAV replication needs a helper gene which is often delivered by the coinfection with AdV5 or additional transient transfection of the helper gene. With all genes present, rAAV nucleic acids are amplified, extracted, and finally detected by qPCR or ddPCR (Lock et al. 2020; Zhen et al. 2004).

In this special case of an endpoint dilution $TCID_{50}$ assay, the presence of viral nucleic acids rather than morphological changes is used to detect infected wells. The principle of plaque assay is not suitable for combination with this readout.

Detection of viral nucleic acids is highly specific for the virus analyzed. This titration technique depends on specific primers and probes which need to be chosen and established appropriately to the virus. It is time and money consuming but has the advantage that it can be applied to any virus very early in its replication cycle even before the detection of viral antigens by antibodies.

11.3.4 Conclusion

There is a variety of possible readouts which can be combined with either the $TCID_{50}$ or the plaque principle (Table 11.1) to quantify virus infectivity. However, as there is no universal understanding of the term infectivity, it is crucial to choose an assay setup, which reflects the virus' mode of action, relevant for therapeutic use (delivery of nucleic acids, protein expression, or the ability to lyse cells). For some viruses, e.g., non-replicating rAAV, this guidance restricts the set of potential assay variants to only $TCID_{50}$ in combination with the detection of viral nucleic acids. For other viruses, e.g., VSV, the titration method and the readout principle can be chosen freely depending on individual assay requirements or prerequisites, such as acceptable assay duration, availability of virus-specific labels, feasible extent and complexity of handling steps, or available equipment. Irrespective of the readout principle, a common challenge to all infectivity assays is to find an appropriate cell line. The goal is to use a cell line which is straightforward to cultivate, susceptible to the virus analyzed, and ideally physiologically relevant for the indication of the viral therapeutic (e.g., appropriate cancer cell line for VSV as an oncolytic virus or organ-specific target cells for transduction with rAAV in gene therapy).

Table 11.1 Overview on titration readouts

Readout	Advantages	Disadvantages	Titration principle	Required from the infected cells	Special reagents required	Equipment required
CPE evaluation by microscopy	– Label-free is possible – Fast readout	– Non-specific readout	$TCID_{50}$ & Plaque	Visible CPE (e.g., rounding, cell lysis)	(Typically) none	Imaging reader or microscope
Impedance measurement	– Label-free – Fast and automated readout	– Non-specific readout – High costs of consumables	$TCID_{50}$	Cell detachment or cell lysis	None	– xCELLigence real-time cell analyzer – 96-well E-plates
Metabolic activity	– Fast and automated readout	– Non-specific readout – Additional handling step	$TCID_{50}$	Cell lysis	Viability substrates/kits	Micro titer plate reader
Laser force cytology	– Readout is possible early during replication	– Indirect readout – Expensive equipment required	$TCID_{50}$[a]	Visible CPE	None	Lumacyte Radiance instrument
Viral antigen expression	– Virus-specific readout – Readout is possible early during replication	– Dependent on antibodies or reporters	$TCID_{50}$ & Plaque	Viral proteins expressed	Specific antibodies	Micro titer plate reader, fluorescence microscope, or flow cytometer
Viral nucleic acid detection	– Virus-specific readout	– Dependent on primer/probes – Many handling steps	$TCID_{50}$	Amplification of viral nucleic acid	DNA/RNA extraction kit, specific primers/probes	qPCR or ddPCR instrument

[a]LFC determines an "infection metric" that can be correlated to a $TCID_{50}$ value (Hebert et al. 2018)

11.4 Development of Infectivity Assays

While the chapters above describe the underlying principles of infectivity assays and the corresponding readouts, the following chapter addresses different aspects that have to be considered during their development for industry. As guidelines such as USP <1032> (Pharmacopeia US 2016) and <1033> (Pharmacopeia US 2017) already cover general aspects of developing cell-based assays, the following chapter will be focused on special features that accompany infectivity assays that are not or insufficiently addressed in these guidelines.

11.4.1 Timelines for Assay Development and Validation

As the field is quickly evolving, there are to date no general regulatory guidelines available regarding expected timelines for the development and validation of infectivity assays of viral therapeutics. For most virus-based therapeutics, infectivity is a major part of the drug's potency, which represents the mode of action. Therefore, one can refer to the potency guidelines for virus-based gene therapy products, which request that a suitable potency assay should be in place before material for the first clinical trial is produced (EMA 2016). Validation of these assays should be performed prior to phase III clinical trials (EMA 2016). In the case of viral therapeutics, where the assay is also dose-defining, authorities expect the assay to be validated even in preparation for phase I. Furthermore, as the infectious titer is also the most critical quality parameter for process development, the assay should be established and proven to be scientifically sound even before starting the development of a product-manufacturing process (FDA 2011).

11.4.2 Choosing the Right Cell Model

For all cell-based assays, the cell line should support the mode of action. Moreover, in combination with an appropriate readout, the cell line should allow a sufficient signal-to-noise ratio, which is a prerequisite for an assay to be precise and robust. Accordingly, different cell lines need to be evaluated right at the beginning of the assay development. In addition, the cells for the infectivity assay must be susceptible to the virus of interest. The degree of the susceptibility of a cell can depend on various parameters, which need to be considered during later assay optimization (see below).

11.4.3 Assay Precision

Cell-based assays typically suffer from relatively high variability. As for most cell-based assays, the high variability in infectivity assays can be traced back to the cells themselves, due to, e.g., inhomogeneous cell population across a micro-titer plate or different behavior of the cells between experiments. Of course, these root causes of variability can be reduced via meticulous standardization of all assay steps that involve cells and cell cultivation (Bal-Price and Coecke 2011). Instead, using "ready-to-assay" cells might be a better approach. "Ready-to-assay" means that cells are aliquoted and frozen at the appropriate concentration and thawed just before seeding into the micro-titer plates (Wehmeier and Loa 2019). "Ready-to-assay" cells are becoming more and more popular in recent years as not only the laborious routine cell culture can be omitted, but especially the day-to-day variability caused by routine cell culture is, in principle, abolished. However, the preparation of a ready-to-assay cell bank might not be suitable for every cell line, as the viability of the cells might decrease significantly by the additional freeze and/or thaw step.

Day-to-day or inter-plate variability caused by cells can be mitigated by introducing reference standard material on each plate, and thereby, reporting relative infectious titers instead of absolute values. If no representative material is available, e.g., during early development phases, the assay performance could also be monitored via the introduction of an assay control. By trending the results of the assay control, already slight changes in assay performance can be detected early on, which allows appropriate countermeasures in time. Furthermore, an assay control is particularly advantageous, for example, for the transfer of an assay to a CMO or for monitoring the assay performance between several analytical laboratories, since small changes in the implementation of the assay can have significant effects on the infectious titer, particularly in the case of absolute assays.

In addition to cells being the culprits for the high variance, the many pipetting steps that usually come with biological assays can contribute to the variance due to error propagation. Here, semi-automated pipetting aids (e.g., Hochdorfer et al. 2022; Filone et al. 2017) or even fully automated robots can decrease the significance of this problem, as they increase the consistency of pipetting.

There are of course other technical aspects such as plate uniformity, or the number of replicates that influence the variability of an assay, which are all described in USP<1032> in detail (Pharmacopeia US 2016). However, especially for $TCID_{50}$ assays, the increment of sample concentration within the serial dilution steps (e.g., $\log_{10}$ vs. 0.5 $\log_{10}$ steps) affects the resolution and variability of the results. Therefore, it should be carefully balanced during assay development to find the right compromise between an increase in precision and lowering the working range.

An additional contribution to variability is the carry-over effect. In comparison to chemical or protein-based therapeutics, this effect has a much higher significance for infectivity assays, as, e.g., already a single virus attached to a pipette tip, can cause a well to be CPE-positive, even though wells at previous dilutions steps are

CPE-negative. For this reason, pipette tips should be changed after every step of the serial dilution.

11.4.4 Assay Optimization

Especially for cell-based assays, which have a high intrinsic variability, a proper assay optimization is mandatory. As different assay factors tend to interact with each other (e.g., effects of cell cultivation on receptor expression or cell cycle and thus on susceptibility), assays that include cells have a much higher complexity. It is very helpful for assay optimization to understand these interactions. Therefore, a classical "one factor at a time" (OFaT) approach is of limited value, because interactions can only be understood via statistical design of experiment (DoE) approaches. In addition, it is much easier and more likely to find the optimum of an assay and identify the conditions where the assay performs robustly using DoE (Draper 2012; Montgomery 2019). For this reason, guidelines like USP<1032> recommend using DoEs (Pharmacopeia US 2016). Yet, many labs still hesitate to use DoE approaches, often because they are not familiar with these procedures, or they believe that those approaches are difficult to employ (DOE 2009; Onyeogaziri and Papaneophytou 2019). Additionally, the hesitation is also likely caused by DoE approaches being considered very laborious and time-consuming, as up to four independent DoE approaches are necessary during assay development (i.e., screening, characterization, optimizing, verification/robustness) (Montgomery 2019). To circumvent this, Solzin et al. have recently developed a simplified DoE approach called ixDoE, which allows comparable output with a single DoE approach, and thus, takes significantly less time (Solzin et al. 2023).

11.4.5 Selectivity of the Assay

The ratio of infectious virus particles to total particles (i.e., infectious plus non-infectious particles) is usually much lower than one (Delwart 2011). This can be either caused by damaging the virus during USP/DSP or simply being intrinsic to the biology of the respective virus and its interaction with the host cell itself. The ratio is a critical quality attribute as virus lots with different ratios could for instance lead to significant differences in the innate immune response of the animal/patient (Timmers et al. 2020). However, as described above, most of the infectivity assays do detect only the functional virus particles but not virus particles or virus-like particles which are not able to replicate. Also, infectivity assays are not able to directly detect aggregates. Aggregates might also be of interest as "infectious units" should not in every case be interpreted as single infectious virions, because a single infection event might require multiple infectious virions. For example, baculoviruses form crystalline protein structures that contain multiple infectious virions (Slack and Arif 2006).

Finally, to detect the total number of infectious and non-infectious viruses as well as the number of aggregates, it is crucial to guarantee a constant quality of virus material in the production process. Especially for AAV-based approaches it is well known that the ratio of full/empty capsids (i.e., virus particles with or without DNA) is critical, e.g., in terms of causing a strong immunogenic response (Timmers et al. 2020). To detect those product-related impurities orthogonal methods are necessary to be employed but each of those methods has its own limits with regard to its significance. To determine aggregates or fragments typically methods such as analytical ultracentrifugation, asymmetric flow-field-flow fractionation, dynamic light scattering, or multi-angle light scattering are used (Gimpel et al. 2021) and the list of methods is constantly growing.

11.5 Conclusion

The development of infectivity assays is complex, and many points need to be considered. Here we provided an overview of important aspects when developing an infectivity assay. These included a variety of possible assay formats, read-outs, as well as how to address technical challenges. The infectivity assay for virus-based therapeutics does not only represent a main part of the mode of action but is also often the dose-defining, and therefore, the most crucial assay. In addition, the assay is already required early in pre-clinical phases, as it is key for process development. Finally, it pays off to invest in the development of a specific and robust infectivity assay right from the beginning, in order to avoid that this key assay is becoming a bottleneck, which can cause delays in project timelines.

References

Anonymous (1925) Research on foot-and-mouth disease. Nature 116:489–490. https://doi.org/10.1038/116489a0

Baer A, Kehn-Hall K (2014) Viral concentration determination through plaque assays: using traditional and novel overlay systems. J Vis Exp. https://doi.org/10.3791/52065

Bal-Price A, Coecke S (2011) Cell culture techniques. Neuromethods 1–25. https://doi.org/10.1007/978-1-61779-077-5_1

Coons AH, Kaplan MH (1950) Localization of antigen in tissue cells. J Exp Med 91:1–13. https://doi.org/10.1084/jem.91.1.1

Delwart E (2011) Principles of virology. Volume I: Molecular biology and Volume II: Pathogenesis and control. 3rd ed. By Flint SJ, Enquist LW, Racaniello VR, Skalka AM. ASM Press, Washington, DC. 2009. Q Rev Biol 86:144–144. https://doi.org/10.1086/659935

d'Herelle F (1917) Sur un microbe invisible anatagoniste des bacilles dysenterique. CR Acad Sci Paris 165:373–375

DOE in assay development trends 2009 report, published by HTStec Limited, Cambridge, UK, July 2009
Draper NR (2012) Optimal design of experiments: a case study approach by Peter Goos and Bradley Jones. Int Stat Rev 80:341–342. https://doi.org/10.1111/j.1751-5823.2012.00187_23.x
Duckworth DH (1976) Who discovered bacteriophage? Bacteriol Rev 40:793–802. https://doi.org/10.1128/br.40.4.793-802.1976
Dulbecco R (1952) Production of plaques in monolayer tissue cultures by single particles of an animal virus. Proc Natl Acad Sci 38:747–752. https://doi.org/10.1073/pnas.38.8.747
EMA (2016) Guideline on potency testing of cell based immunotherapy medicinal products for the treatment of cancer
Fang Y, Ye P, Wang X et al (2011) Real-time monitoring of flavivirus induced cytopathogenesis using cell electric impedance technology. J Virol Methods 173:251–258. https://doi.org/10.1016/j.jviromet.2011.02.013
FDA (2011) Guidance for industry potency tests for cellular and gene therapy products
Filone CM, Miller D, Wahl-Jensen V (2017) Ebolaviruses, methods and protocols. Methods Mol Biol 1628:163–175. https://doi.org/10.1007/978-1-4939-7116-9_12
Gimpel AL, Katsikis G, Sha S et al (2021) Analytical methods for process and product characterization of recombinant adeno-associated virus-based gene therapies. Mol Ther Methods Clin Dev 20:740–754. https://doi.org/10.1016/j.omtm.2021.02.010
Hebert CG, DiNardo N, Evans ZL et al (2018) Rapid quantification of vesicular stomatitis virus in Vero cells using Laser Force Cytology. Vaccine 36:6061–6069. https://doi.org/10.1016/j.vaccine.2018.09.002
Hochdorfer D, Businger R, Hotter D et al (2022) Automated, label-free TCID50 assay to determine the infectious titer of virus-based therapeutics. J Virol Methods 299:114318. https://doi.org/10.1016/j.jviromet.2021.114318
Hotchin JE (1955) Use of methyl cellulose gel as a substitute for agar in tissue-culture overlays. Nature 175:352–352. https://doi.org/10.1038/175352a0
Kärber G (1931) Beitrag zur kollektiven Behandlung pharmakologischer Reihenversuche. Naunyn-schmiedebergs Archiv Für Exp Pathologie Und Pharmakologie 162:480–483. https://doi.org/10.1007/bf01863914
Lecoq H (2001) Discovery of the first virus, Tobacco mosaic virus: 1892 or 1898? Comptes Rendus De L'académie Des Sci - Ser Iii - Sci De La Vie 324:929–933. https://doi.org/10.1016/s0764-4469(01)01368-3
Lock M, Wilson J, Sena-Esteves M, Gao G (2020) Sensitive determination of infectious titer of recombinant adeno-associated viruses (rAAVs) using TCID 50 end-point dilution and quantitative polymerase chain reaction (qPCR). Cold Spring Harb Protoc. https://doi.org/10.1101/pdb.prot095653
Montgomery DC (2019) Design and analysis of experiments, 10th edn. Wiley, Hoboken, NJ
Muik A, Stubbert LJ, Jahedi RZ et al (2014) Re-engineering vesicular stomatitis virus to abrogate neurotoxicity, circumvent humoral immunity, and enhance oncolytic potency. Cancer Res 74:3567–3578. https://doi.org/10.1158/0008-5472.can-13-3306
Myers L, McQuay LJ, Hollinger FB (1994) Dilution assay statistics
Noyes WF (1955) Visualization of Egypt 101 virus in the mouse's brain and in cultured human carcinoma cells by means of fluorescent antibody. J Exp Med 102:243–248. https://doi.org/10.1084/jem.102.3.243
Onyeogaziri FC, Papaneophytou C (2019) A general guide for the optimization of enzyme assay conditions using the design of experiments approach. Slas Discov 24:587–596. https://doi.org/10.1177/2472555219830084
Panec M, Katz DS (2006) Plaque assay protocols
Pharmacopeia US (2016) <1032> Design and development of biological assays
Pharmacopeia US (2017) <1033> Biological assay validation

Piszczatoski CR, Gums JG (2020) Ervebo (Ebola Zaire Vaccine, Live/rVSVΔG-ZEBOV-GP): the first licensed vaccine for the prevention of ebola virus disease. J Pharm Technology 36:243–250. https://doi.org/10.1177/8755122520950692

Rapp F, Seligman SJ, Jaross LB, Gordon I (1959) Quantitative determination of infectious units of measles virus by counts of immunofluorescent foci. Proc Soc Exp Biol Med 101:289–294. https://doi.org/10.3181/00379727-101-24915

Redondo N, Madan V, Alvarez E, Carrasco L (2015) Impact of vesicular stomatitis virus M proteins on different cellular functions. PLoS One 10:e0131137. https://doi.org/10.1371/journal.pone.0131137

Reed W, Carroll JAS, Agramonte A (1901) The etiology of yellow fever: An additional note. J Amer Med Assoc XXXVI:431–440. https://doi.org/10.1001/jama.1901.52470070017001f

Saphire EO (2020) A vaccine against ebola virus. Cell 181:6. https://doi.org/10.1016/j.cell.2020.03.011

Şener LT, Albeni̇z G, Di̇nç B, Albeni̇z I (2017) iCELLigence real-time cell analysis system for examining the cytotoxicity of drugs to cancer cell lines. Exp Ther Med 14:1866–1870. https://doi.org/10.3892/etm.2017.4781

Slack J, Arif BM (2006) The baculoviruses occlusion-derived virus: virion structure and function. Adv Virus Res 69:99–165. https://doi.org/10.1016/s0065-3527(06)69003-9

Solzin J, Eppler K, Knapp B, Buchner H, Bluhmki E (2023) Optimising cell-based bioassays via integrated design of experiments (ixDoE) - a practical guide. SLAS Discov 28(1):29–38. https://doi.org/10.1016/j.slasd.2022.10.004. Epub 2022 Oct 28. PMID: 36415004

Spearman C (1908) The method of 'right and wrong cases' ('constant stimuli') without Gauss's formulae. Br J Psychol 1904-1920 2:227–242. https://doi.org/10.1111/j.2044-8295.1908.tb00176.x

Staples JE, Monath TP (2008) Yellow fever: 100 years of discovery. JAMA 300:960–962. https://doi.org/10.1001/jama.300.8.960

Timmers AM, Newmark JA, Turunen HT et al (2020) Ocular inflammatory response to intravitreal injection of adeno-associated virus vector: relative contribution of genome and capsid. Hum Gene Ther 31:80–89. https://doi.org/10.1089/hum.2019.144

Wan Y, Simovic B, Walsh S (2015) Mechanistic insights into the oncolytic activity of vesicular stomatitis virus in cancer immunotherapy. Oncolytic Virother 4:157–167. https://doi.org/10.2147/ov.s66079

Wehmeier O, Loa A (2019) Animal cell biotechnology, methods and protocols. Methods Mol Biol 2095:17–25. https://doi.org/10.1007/978-1-0716-0191-4_2

Witkowski PT, Schuenadel L, Wiethaus J et al (2010) Cellular impedance measurement as a new tool for poxvirus titration, antibody neutralization testing and evaluation of antiviral substances. Biochem Biophys Res Commun 401:37–41. https://doi.org/10.1016/j.bbrc.2010.09.003

Zhao Z, Anselmo AC, Mitragotri S (2021) Viral vector-based gene therapies in the clinic. Bioeng Transl Med 7:e10258. https://doi.org/10.1002/btm2.10258

Zhen Z, Espinoza Y, Bleu T et al (2004) Infectious titer assay for adeno-associated virus vectors with sensitivity sufficient to detect single infectious events. Hum Gene Ther 15:709–715. https://doi.org/10.1089/1043034041361262

Chapter 12
Product-Related Impurities in Therapeutic Virus Bioprocessing

Judy King Man Ng

12.1 Introduction

There are more than 200 virus species that are relevant to human health to date (Woolhouse et al. 2012). Such virus particles come in different shapes and sizes: the shapes can be spherical, helical, complex, and enveloped, and their sizes can range between 17 nm and up to 2500 nm in diameter (Gelderblom 1996). The genome of a virus may be DNA or RNA, single-stranded or double-stranded, linear or circular (Gelderblom 1996).

Some viruses are exploited to be manufactured into products for vaccine (immunoprophylatic or immunotherapeutic) or virotherapeutic uses (e.g., viral vectors for gene therapy, oncolytic therapy, immunotherapy). In general, viral vectors are derived from viruses that have been isolated from nature and genetically modified for cell manipulation (Warnock et al. 2011). Unlike other gene delivery systems, they are useful vectors for their unique ability to infect cells, their diverse tropism, ability for gene transfer, and induction of a host immune response (Ura et al. 2014). The basic attributes used for selecting a specific virus vector are the following: (1) their natural or engineered tropism to specific cell types, (2) low toxicity to the host at specified dosages, (3) genomic stability (especially for self-replicating viruses), (4) safety against undesired pathogenicity in the target host, and (5) capacity to encode for transgenes (Lundstrom 2018).

By nature of their specific uses, the fundamental requirements for viral vaccine products compared to virus-based therapeutic products differ. Viral vaccines refer to preparations of virus used to induce an immune response against the virus itself, or against another virus where a key transgene is encoded in the viral vector. Viral

J. K. M. Ng (✉)
Viratherapeutics GmbH/Boehringer Ingelheim, Rum, Austria
e-mail: judy.ng@boehringer-ingelheim.com

S. Gautam et al. (eds.), *Bioprocess and Analytics Development for Virus-based Advanced Therapeutics and Medicinal Products (ATMPs)*,
https://doi.org/10.1007/978-3-031-28489-2_12

vectors used for gene therapy (e.g., adeno-associated virus AAV, adenoviruses, lentiviruses) carry a therapeutic cargo, a foreign transgene, and are ideally replication-defective (Li et al. 2012). Replication-competent oncolytic vectors (e.g., herpes simplex virus, vesicular stomatitis virus, measles virus, vaccinia virus, adenovirus) are used for cancer therapy and cancer gene therapy; they preferentially replicate in cancer cells and destroy them through their lytic virus life cycle, and foreign transgene(s) may be inserted for additional therapeutic functions (Fukuhara et al. 2016). While viral vaccines should elicit a strong immune response against itself, immunogenicity against the viral vector will likely reduce the efficacy of therapeutic virus products, especially if repeated applications are necessary (which might not be necessary for certain gene therapy and OV products). Mutations that arise during virus production for vaccine use, given that they do not change tropism for key host cells relevant for vaccine efficacy, may have negligible impact and may even be beneficial. However, this should be strategically minimized for therapeutic viral vectors used for gene therapy, as the fidelity of the transgene is often an essential component. Furthermore, the dosage typically required for vaccine use can vary by several orders of magnitude and are typically lower than for virotherapeutic use (refer to examples in Table 12.1), with direct implications for viral vector manufacturing (Kiesslich and Kamen 2020). An exception might be the inactivated or single-replication virus vaccine concepts (e.g., adenovirus vaccines), which require high amounts of virus due to lack of "in-vivo amplification," thereby inducing a more profound immunity.

The growing number of different viral vectors in development and in clinical trials justify efforts invested to ensure sufficient understanding of the composition of such products. Specifically, critical quality attributes should be well characterized to define tolerances in purity, potency, and safety for manufacturing (Wright 2014). As such, process-related impurities as well as product-related impurities are considered here. Process-related impurities are defined as contaminating components in a biopharmaceutical sample originating from the production process itself (e.g., host cell components, nutrient media, process additive enzymes, buffer components, leachables and extractables from consumable material). Product-related impurities refer to variants of the target product that differs in activity, efficacy, and potential safety found in the drug substance (e.g., molecular variants, variations in virus glycoprotein structures, post-translational modifications, and particle structure). This may also include any truncated or modified forms of the viral genome, virus particle aggregates, precursor, and degradation product. Unlike small molecule manufacturing, but similar to manufacturing of biologics (e.g., monoclonal antibody therapeutics), the manufacturing process of viral vectors is very closely tethered to the product-related quality attributes.

In this chapter, discussion is limited to the manufacturing of active (i.e., not inactivated) therapeutic virus products targeting mammalian cells of both replication-competent and replication-incompetent viral vectors used for gene transfer (in-vivo and ex-vivo), oncolytic therapy, and viral immunotherapy. The focus will be on product-related impurities produced in the upstream production process,

Table 12.1 Selected examples of viral vectors manufactured for vaccine and virotherapeutic uses

Virus (product name)	Use	Size	Shape	Enveloped	Dosage	Cell line	Status	Reference
Vesicular stomatitis virus (Ervebo)	Live vaccine	70 × 180 nm	Cylindrical/ bullet-shaped	Yes	10^7–10^8 pfu, im	Vero	Market	Suder et al. (2018)
Japanese encephalitis (JE-CV)	Live chimeric vaccine	50 nm	Spherical	Yes	10^4 pfu	Vero	Market	Appaiahgari and Vrati (2012)
Smallpox, vaccinia-based (ACAM2000)	live vaccine	250 × 270 × 360 nm	Brick-shaped	Yes	2.5–12.5 x10^5/ vial 0.3 mL	Vero	Market	Greenberg and Kennedy (2008)
Adenovirus (AZD1222)	Live non-replicating vaccine	70–100 nm	Icosahedral	No	2.5×10^8 IU/0.5 mL	HEK293	Market	EMA website
Herpesvirus (T-Vec: Imlygic)	Oncolytic	155–240 nm	Icosahedral	Yes	10^6–10^8 pfu, it	Vero, BHK	Market	Ungerechts et al. (2016)
Reovirus (Reolysin: Pelareorep)	Oncolytic	85 nm	Double capsid	No	3^{10} TCID50, iv	HEK293	Phase 2 trials	Ungerechts et al. (2016)
Vaccina Virus	Oncolytic	360 × 270 × 250 nm	Brick shaped	Yes	10^8–10^9 pfu, it, iv	HeLa, Vero	In development, various products in Phase 1/2	Ungerechts et al. (2016)
Adenovirus	Oncolytic	90 nm	Icosahedral	No	10^{10}–10^{12} vp, it, iv	HEK293, Per. C6, A549/ pTG6559, N52. E6	In development, various products in Phase 1/2	Ungerechts et al. (2016)

(continued)

Table 12.1 (continued)

Virus (product name)	Use	Size	Shape	Enveloped	Dosage	Cell line	Status	Reference
Measles	Oncolytic	100–300 nm	Helical nucleocapsid	Yes	10^9–10^{11} TCID50, it, ip, iv	Vero	In development, various products in Phase 1/2	Ungerechts et al. (2016)
Parvovirus: H-1	Oncolytic	26 nm	Icosahedral	No	10^6–10^9 pfu, it, iv	SV40 transformed human newborn kidney cells	In development, various products in Phase 1/2	Ungerechts et al. (2016)
Adeno-associated virus	Gene therapy	20–30 nm	Icosahedral	No	10^{12}–10^{13} gc/kg iv	Sf9, HEK293	Phase 1, 2, 3	Miesbach et al. (2018), www.Clincaltrials.gov
Lentivirus	Gene therapy, CAR-T cell production	80–100 nm	Spherical	Yes	n/a	HEK293T	Market	Martínez-Molina et al. (2020)

methods of characterization to define such impurities, and possible strategies to reduce these impurities in subsequent downstream processing.

12.2 Product-Related Impurities from Upstream Production

An important aim in viral vector bioprocessing, aside from achieving the target concentration, required production volume, and purity, is to minimize the ratio of total virus particles to infectious virus particles (particle to infectivity ratio (P:I ratio)) (McCormick and Mermel 2021). The particle-to-infectivity ratio provides a measure of the potency (and quality) of a virus preparation, in other words, the fraction of viral particles that can infect susceptible cells out of all the virus particles present. This ratio, which strongly varies between different viruses (McCormick and Mermel 2021) may be associated with impacting the therapeutic efficacy and drug toxicity or immunogenicity (Reece-Ford et al. 2008). The setup for upstream production is particularly important in determining the quality of virus particles produced, which are subsequently subjected to downstream processing to remove impurities, to potentially enrich for infectious particles, and is ideally devised to minimize loss of infectivity. Furthermore, the presence and quantity of similarly presented particles, namely extracellular vesicles and virus-like particles in the case of enveloped viruses, must also be considered and minimized.

In particular, the following aspects must be carefully considered: (1) choice of host cell, media, growth conditions, and culture system, (2) the quality of virus seed stock or plasmid transfection strategy, (3) infection/transfection step and subsequent culture parameters for virus replication, and (4) method of harvest to produce downstream processing feed.

12.2.1 Choice of Host Cell, Culture Medium, Growth Conditions, and Culture System

For viral vector production, cell-line-based production using chemically defined animal component-free culture media is preferred over previously established production in chicken eggs and primary cells due to ease of use and higher consistency (Kiesslich and Kamen 2020). Such cell lines are of human, mammalian, or insect origin. Some continuous cell lines that are commonly used for virus production include Vero cells (from kidney of African green monkey, HEK293 cells and its derivatives (human embryonic kidney cells), A549 cells (human alveolar basal epithelial cells), Per.C6 (human embryonic retinal cells), CHO (Chinese hamster ovary cells), and Sf9 (Spodoptera frugiperda insect cells) (Ungerechts et al. 2016).

The producer cell line chosen for a specific viral vector should be susceptible to virus infection to allow for replication, or plasmid transfection in the case of viral vectors produced by transfection. The host cells should also demonstrate genomic stability over a useful number of passages in culture for production (Ferreira et al. 2021). The production titer and P:I ratio can vary between different cell lines as well as growth conditions, due to specific properties which may influence the viral replication cycle that are not well-defined (Tan et al. 2021). The host cell determines not only the spectrum of soluble host cell impurities present, but also host cell components (proteins, nucleic acids, sugars, lipids) that are incorporated into the virus particle. In addition, the specific host cell and growth conditions influence the viral and host protein post-translational modifications (Wright 2014; Gale et al. 2019). This is especially critical for the production of enveloped viruses, where the virus takes on the host cell membrane including associated membrane proteins. Thus, the choice of cell line must be appropriate for its application. In recent years, utilization of cell line engineering has successfully increased virus yield; however, the impact of such efforts on the quality of virus particle produced remains to be determined in each case (Hoeksema et al. 2018). Virus particles produced are either secreted into the supernatant or found intracellularly, or both at various ratios dependent on culturing conditions. Extracellular vesicles (EVs) are typically released from producer cell cultures into the supernatant where secreted virus particles are collected (Minh et al. 2021). When the EVs released by cells share similar size and biophysical characteristics as enveloped viruses that are produced, separation of the two can be very challenging, and processing capacity for the target product is reduced (Minh et al. 2021).

Ideally, the producer cell line is adapted to a chemically defined, animal component-free growth medium free of inhibitory ingredients that also allow for productive virus infection. A high growth capacity is demanded from the culture medium because upon virus infection, the metabolic burden on the cells increase dependent on viral interference with the cell life cycle and viral load present in the culture while cell proliferation and viability tend to decline (Humphrey et al. 2004; Reece-Ford et al. 2008). With sufficient growth capacity from the medium, cell growth and viability is sustained for a longer period of time, which is beneficial to vector stability and yield of target viral product (Perry and Rayat 2021).

The choice of growth conditions (temperature, gas composition, feeding strategy, cell density, adherent or suspension culture, and in case of suspension cells shaker or stir conditions), as well as the choice of culture system (T-flask, cell factory, roller bottles, shake flasks, bioreactor) will directly influence the behavior and metabolism of the cells. These factors also affect the efficiency of virus spread from cell to cell. Alteration of culture parameters will result in various cell metabolic states, which lead to proteins produced with alteration in the glycosylation profile, charge variants resulting from chemical degradation, splice variants, and degree of aggregation or degradation (Torkashvand et al. 2017). Together, they directly affect the productivity of the cells as well as the incorporation of host cell proteins found in enveloped virus particles (Loo and Wright 2016; Gale et al. 2019). Furthermore, the culture system influences processing time and has a direct implication on scalability and risk (Reece-Ford et al. 2008).

12.2.2 Virus Production by Transfection or Infection, Quality of Virus Seed Stock

Cell culture-based production of viral vectors and vaccines are produced either by transfection of a packaging cell line, or infection of a host cell line for propagation.

A transient transfection production system involves one or several plasmids encoding the viral genome to be packed and to provide helper virus functions (Mátrai et al. 2010). For production of replication-incompetent viral vectors, such as lentiviruses, adenoviruses, and adeno-associated viruses, stable packaging cell lines can be generated to express viral genes specific to produce virus particles (Mátrai et al. 2010). In some cases, the viral proteins are provided by a helper virus (Zhou et al. 2002). To increase the safety of the viral vectors used, transcomplementation cell lines are also utilized to further attenuate viral vectors by additionally removing viral genes from the viral genome (Yeh et al. 1996). Efficiency of virus production by transient transfection is highly dependent on the constructs used and the characteristics of the packaging cell line. Scaling up transient transfection processes is often challenging and can lead to inconsistent production batches (Robert et al. 2017). Variations in plasmid quality as well as transfection efficiency of multiple plasmids may lead to variation in virus structural protein stoichiometry, ratio of empty to full capsids, and improperly packed nucleic acid content in the case of AAVs (Wright 2014). Undesired viral interactions with packaging cell lines or helper viruses can occur to produce helper virus instead, or even undesired replication-competent viruses (Miller and Buttimore 1986).

Virus production by infection requires first the production of a virus seed stock (master and working). The seed stock is typically produced by virus rescue (e.g., plasmid-based reverse genetics techniques, with or without helper viruses) to generate recombinant viruses followed by a defined small number of passages (Martínez-Sobrido and García-Sastre 2010). The quality of this initial seed stock must be carefully controlled through the design of initial virus generation, subsequent purification steps, early seed stock production, and afterward by deep genomic sequencing to minimize variation due to quasispecies and defective interfering particles when amplified during upstream production in host cells. For clinical production, a master virus seed stock (MVSS) must be produced in cells from a well-characterized master cell bank (MCB), and from the MVSS, a working virus seed stock (WVSS) is produced. The seed stocks are typically produced in a big batch, not necessarily purified, and frozen in many aliquots of a predetermined size useful for production and long-term storage. For every production, an aliquot is taken for infection of a culture; this allows for control of passage number and higher batch-to-batch consistency.

Viral quasispecies include variant genomes due to mutations occurring during virus production for both replication-competent and replication-incompetent viruses. For replication-competent viruses, additional changes in the viral quasispecies from the MVSS and WVSS may occur due to production conditions in cell culture (Domingo et al. 2021). A consensus sequence is typically defined for a viral vector product to define a limit to the amount and nature of mutations acceptable before further investigations to changes in biological properties are warranted. Some nucleotide changes are silent, if they produce no change in the amino acid sequence of the encoded protein and may have limited to no impact. For sequence changes that also change the amino acid sequence, it is necessary to control for changes in biological function of the mutated protein, such as effects on receptor binding or replication efficiency. In general DNA viruses show a lower mutation rate than RNA viruses depending on the polymerase used and proof-reading function. RNA viruses tend to keep a sustained level of genetic heterogeneity to be able to respond to constraints, whether they are kept in a static or a changing environment (Domingo et al. 2021). This is seen for clinically important viruses such as human immunodeficiency virus (HIV), hepatitis C virus, and influenza (Lauring and Andino 2010). During upstream production, the virus seed gets amplified according to defined culture parameters, and the resulting virus pool must be then sequenced again to monitor for possible population disequilibrium, resulting in deviation from the single consensus sequence (Domingo et al. 2021). Such genetic alterations can potentially change the safety and efficacy profile of a virotherapy.

Defective interfering (DI) virus particles are viral mutants with truncated genome due to defective replication or recombination. These are especially relevant for infection with RNA viruses, such as reoviruses, vesicular stomatitis virus, hepatitis viruses, and also HIV (Reta 2017). They are spontaneously derived from the parental virus preparation and are rendered non-infectious due to substantial loss or alteration of genome. However, they can still be replicated with the help of a fully functional virus. Importantly, the presence of DI particles can interfere with and decrease infectious virus yield in production because upon cell entry they compete for resources in the host cell during the replication and assembly phase (Strauss and Strauss 2008). Since defective interfering particles arising from RNA virus replication, for example, retain *cis*-acting sequences needed for replication and encapsidation into particles, sequencing will provide information on the identity of such DI particles, and northern blots can be made to monitor their presence (Strauss and Strauss 2008). Separating DI particles from fully functional virus particles in a scalable manner during downstream processing, however, is difficult to achieve due to the strong surface similarities between the two. Furthermore, these DI particles can stimulate the host immune system while simultaneously attenuating the efficacy of the virotherapy dose, which is a detrimental combination for viral therapeutics (Reta 2017).

When upstream virus production is achieved by infection of a cell culture, the three most critical parameters are multiplicity of infection (MOI), time of infection (TOI), and time of harvest (TOH). Multiplicity of infection refers to the ratio of infectious virus particles per cell at the time of infection. With the aím of managing manufacturing costs while achieving high production yields, it is usually chosen to balance a high final titer, short production time, and conservation of virus seed stock (Kiesslich and Kamen 2020). Typically, a low MOI is preferred to avoid the amplification of potential DI particles in a seed stock (Reta 2017; Bora et al. 2019; Linder et al. 2021). The time of infection parameter considers the optimal density and culture condition of the host cell to start virus production. If a low MOI is chosen, the cell culture can still grow until the virus replication phase completely takes over the culture for the production of virus particles and no further growth and viability decline takes place. At the time of infection, the cell culture should exhibit a high percentage of viable cells and be in exponential growth phase, while the culture medium still has ample capacity to sustain virus production. Lastly, the time (or times) of harvest should be chosen to balance high final titer and virus quality (e.g., P:I ratio), amount of cytopathic effect on host cells and host cell impurities released, and metabolic byproduct formation. During production, the duration between TOI and TOH, virus particle stability and cell growth in the specific production temperature, efficiency of virus spread from cell to cell, changes in pH and osmolarity, shear stress (in case of suspension cultures), and effects of host cell debris need to be considered (Perry and Rayat 2021). Furthermore, host cell metabolism and viability have a direct influence on post-translational modifications and virus protein stoichiometry as well as host cell protein incorporation (Beer et al. 2003) which in turn will affect the infectivity and immunogenicity of a particular virus particle (Moerdyk-Schauwecker et al. 2009).

12.2.3 *Method of Harvesting*

Harvesting of the virus particles produced in the cell culture concludes upstream virus production. The method of harvest is determined with the time of harvest, as virus particles may be differentially found inside the cells, still associated with cells, or in the supernatant. Viability of the infected cell culture is also considered: some viruses (e.g., adenoviruses) are harvested from viable cells while for other viruses, cells are lysed to release the virus into the supernatant. For example, recombinant adeno-associated virus (AAV) vectors are found distributed in both extracellular and intracellular compartments, the specific distribution influenced by culture and harvest conditions. Some enveloped viruses, such as influenza virus, may exclusively bud out of the cell plasma membrane into the supernatant (Samji 2009), while other viruses, such as poxviruses, may be found as a mature enveloped virion intracellularly and also extracellularly with distinct surface antigenic properties (Petersen et al. 2014). Depending on the specific virus produced and its preponderance at a specific location, harvesting is accomplished by host cell lysis to released intracellular virus particles, inducing release from cell surface association, by collection of the supernatant with a clarification step, or a combination thereof (Reece-Ford et al. 2008). If additives are used during harvesting (e.g., detergents, salts, protease inhibitors), their potential to modify virus particles, their stability in solution, and their association with impurities should be evaluated. In this step of harvesting, the pool of virus particles produced (infectious and non-infectious) as well as other impurities released during the upstream process (e.g., extracellular vesicles, host cell impurities, metabolic components) is then collected for subsequent downstream processing. Generating a deeper understanding of the target viral product and the life cycle of the viral vector will be beneficial to determine the best harvest strategy to facilitate successful downstream processing.

12.3 Characterization and Definition of Product-Related Impurities and Target Critical Quality Attributes

From upstream processing, a feed for downstream processing is produced. This feed, either taken from the cell fraction, the supernatant, or both, contains not only the product of interest, i.e., the therapeutic virus particles, but also many different components generated during the cultivation process: this includes host cell debris and soluble impurities, metabolic products and medium components, extracellular vesicles, and undesired modified forms of the virus (truncated, precursor, degraded, aggregated). The presence and amount of these non-target components can vary depending on the upstream process parameters as discussed in Sect. 12.2. Nonetheless, they should be characterized to evaluate the production strategy, and to understand their effect on the stability of the target virus particles and on the function and immunogenicity of the preparation.

Product-related impurities are primarily generated during culturing in the upstream production process, while during downstream processing and formulation undesired modifications of the target viral vector may manifest, such as degradation, aggregation, and in some cases, physiochemical modifications. The crude harvest is unnecessarily complex for direct analysis (Wright 2021). The aim here is to identify impurities that closely resemble the product itself and that cannot be easily removed by purification. Thus, an initial purification procedure is needed to concentrate and simplify the test sample, with an aim at concentrating the target species based on size and density separation, thereby mostly eliminating components such as host cell nucleic acid, as well as free proteins and other components soluble in the spent culture medium or process buffer. Harsh conditions and solubility changes should be avoided to minimize downstream modifications to the test sample. To test for effects of specific critical upstream parameters such as multiplicity of infection, incubation temperature, and time of harvest, these factors must be varied to produce separate harvests, purified, and then analyzed for differential ratios of target and target-similar impurities (Lodish and Porter 1980).

Viruses basically consist of protein and nucleic acid components, and for enveloped viruses also lipid components. Their intrinsic structural complexity allows for variations to be generated in the production process, as well as co-purification of structurally similar particles originating from the host cells

(Wright 2014). Key attributes for analysis fall into four main categories: physical properties, chemical properties, structural attributes, and biological efficacy and safety (Markova et al. 2021). Such attributes are typically measured for the entire virus particle population in the sample, which may contain differential degrees of heterogeneity in each test sample. Thus, the difficulty here is to derive meaning and assess impact from these measurements to describe the optimal target virus particle profile and related impurities.

12.3.1 Physical Attributes

Physical properties refer to particle shape and size, concentration, surface charge, homogeneity, and thermal stability (Markova et al. 2021). Light scattering measurement methods (e.g., dynamic light scattering (DLS), nanoparticle tracking analysis (NTA), and multi-angle light scattering (MALS)) are standard methods to monitor and infer size, shape, concentration, and physical homogeneity. They may also be used as a readout for thermal stability with temperature ramping experiments and changes in surface charge through pH changes. With light scattering methods, the entire population of virus particles in a sample is measured at once rather than individual particles. While critical information of the preparation can be obtained, these methods are of limited use to look at minority particle populations in a sample, for example, particles with subtle differences that render them more or less infectious. For viral particle quantification, high-performance liquid chromatography may be used with a column that allows for separation of viral particles and impurities (Sanyal et al. 2021). Examples of columns used for this purpose include ion exchange chromatography analytical columns as well as size exclusion chromatography columns, these have been shown suitable for both enveloped virus-like particles and non-enveloped viruses (Steppert et al. 2017; Gagnon et al. 2021). Additional to virus quantification, HPLC-based analysis can be used for in-depth characterization of a virus preparation through sample separation based on size and/or charge, combined with optical monitoring for intrinsic and extrinsic fluorescence, UV, and light scattering, as shown for AAV production (Gagnon et al. 2021). Viral genome quantification is achieved by either absolute quantification by droplet digital PCR (ddPCR) or by quantitative PCR (qPCR) using a standard curve (Dobnik et al. 2019). Analytical ultracentrifugation based on sedimentation velocity separates a sample by molecular weight and size of particles, distinct from density ultracentrifugation. This method is useful to distinguish and quantify particles with differential content and size, including empty and full capsids in the case of AAVs (Gimpel et al. 2021), and potentially virus particles from extracellular vesicles for enveloped viruses. Surface charge of the virus population in the sample as a whole may be monitored by change in zeta potential measurements by electrophoretic light scattering (ELS). Advanced imaging techniques such as electron microscopy and atomic force microscopy allow visualization of particle structure, allowing for monitoring of particle heterogeneity and presence of EVs (Markova et al. 2021).

12.3.2 Chemical Attributes

Chemical properties refer to identity, composition, impurities, degradation products, and chemical stability (Markova et al. 2021). Deep sequencing of viral genome will yield consensus genomic sequences that can reveal significant mutations, recombinations, and truncations, or even packaged nucleic acids from host cell origin (Loo and Wright 2016). Northern and Southern blots are useful to screen and monitor viral genomes (RNA and DNA respectively) for truncations and recombinations (dependent on the probe sequence). Virus composition (e.g., protein, lipid, post-translational modifications, viral protein stoichiometry) and carry-over impurities from purified fractions can be analyzed by liquid chromatography mass spectrometry (LC-MS) (Gale et al. 2019) as well as capillary electrophoresis, Western blot, ELISA and immunostaining methods (Baldwin et al. 2021). Degradation products can be monitored by HPLC analysis, based on differential pH, size, or surface charge. Chemical stability due to its composition may be inferred through stress studies using harsh chemical conditions that may lead to dissociation or conformational changes of the virus particles, as well as changes in receptor binding (Mateu 2011).

12.3.3 Structural Attributes

Structural attributes refer to the particle morphology, aggregation, and component structural complex organization (Markova et al. 2021). Morphology, including virus shape, integrity, and surface distribution of glycoproteins, and degree of particle aggregation in a sample can be directly assessed through high-resolution electron microscopy (EM; transmission EM or Cryo EM) (Ke et al. 2020). Immunostaining, by means of in vitro assays or by flow virometry, is useful to evaluate surface components. Conformational changes that expose or hide functionally relevant epitopes can be assessed by monoclonal antibody surface mapping (Baldwin et al. 2021).

12.3.4 Biological Attributes

Biological efficacy and safety refer to in vitro bioactivity in terms of potency (e.g., infectivity), and P:I ratio, immunogenicity, and cell targeting (Markova et al. 2021). To measure potency for oncolytic viruses, the concentration of replication-competent viral particles is determined through virus titration using end-point biological assays, such as 50% cell culture infectious dose ($TCID_{50}$) or plaque assay, which are set up to reflect infectivity in the target host. The potency of gene therapy vectors can be assessed by transduction efficiency and presence of the transgene in target host cells, and expression level and biological activity of gene of interest (Ramsey et al. 2021). Comparing this infectious titer result with total viral particle concentration or total genome copies measured will give the particle-to-infectivity ratio, which is a measure of the quality of the viral vector preparation. A low ratio is desired, to reach the desired effective dose while minimizing the amount of total viral particles applied. Replication kinetics as well as cell targeting assays can be performed in vitro to provide further functional information. Assessing viral vector-induced immune response is to look at both humoral and cellular immunity. This is commonly done in vivo in animal models, and to a limited extent in vitro using immunoassays, complement-activation assays, and antigen-stimulated immune cell assays, where the readout may be immune cell proliferation, recruitment, or activation of inflammatory signaling pathways and cytokine production (Baldwin et al. 2021).

12.3.5 Defining Target Product and Product-Related Impurities

A detailed multi-factorial profile of viral vectors produced using various critical upstream processing parameters can be created through analysis of the above-mentioned key attributes, which can be used to deduce product target profile (e.g., infectious virus particles) and to start defining less desirable product-related impurities (e.g., similarly presented non-infectious particles) (Table 12.2). An assessment as to the presence and proportion of various product-related impurities produced with specific upstream processing parameters should then be made, and a strategy to reduce these through data-driven changes in such parameters can be devised. As modified and defective viral particles in a clinical preparation provide unnecessary viral antigen that may also interfere with the effectiveness of infectious particles, likely increasing the risk of immunotoxicity, it is reasonable to presume a targeted reduction of such particles will potentiate a more efficacious lower dosage needed (Wright 2014). Finally, a risk assessment of the 'optimized' clinical preparation can be made with knowledge on particle stability for formulation and storage, and more importantly, the balance of efficacy and immunogenicity of the required therapeutic dose.

12.4 Implications for Downstream Processing

With a target profile as well as some definition of product-related impurities, process controls, and acceptance criteria for in-process controls as well as drug substance can be specified. By tweaking upstream production parameters to improve the quality of the feed utilized in downstream processing (as discussed in Sect. 12.2), the ratio of target virus particles to impurities may already be optimized, which also have

Table 12.2 Biophysical characterization of viral vector attributes and analytical methods available

Attribute category	Attributes to measure	Analytical methods available
Physical	Particle size, shape, concentration, physical homogeneity	Light Scattering methods (DLS, NTA, MALS)
	Thermal stability, pH stability	Ramping/titration experiments combined with light scattering method
	Viral particle quantification	HPLC-MALS (with, e.g., SEC, IEX, affinity columns)
	Viral genome quantification	ddPCR, qPCR
	Molecular weight and size	Analytical ultracentrifugation
	Surface charge, Zeta-potential	ELS
	Particle structure	Electron microscopy, atomic force microscopy
Chemical	Viral genome mutations, recombinations, truncations, presence of host cell genomic material	Deep sequencing, Northern blot, Southern blot
	viral composition and post-translational modifications	LC-MS, capillary electrophoresis, Western blot, ELISA, immunostaining
	Degradation products	HPLC analysis
	Chemical stability	Stress studies, followed by physical readout method
Structural	Particle morphology	Electron microscopy
	Surface components evaluation	Immunostaining, in-vitro assays, flow virometry
	Conformational changes	mAb surface mapping
Biological	Potency for replication-competent viruses	TCID50 or plaque assay
	Potency for replication-deficient viruses (e.g., for gene therapy)	Transduction and transgene presence in susceptible cells
	Viral vector-induced immune response	Treatment in in vivo animal model, in vitro cell-based assays

implications in the capacity needed for further processing (Wright 2014). Specifically, by reducing the amount of product-related impurities (e.g., non-infectious particles, aggregated or degraded particles, particles with undesired host cell components incorporated) produced in upstream culturing conditions, there is more capacity available to capture and process the target viral vector.

Downstream processing refers to the process steps taken to concentrate, polish and formulate the virus particles to a pre-defined concentration and purity level. With knowledge of the target profile, a strategy to exploit such physical characteristics (e.g., surface charge, size, stability) can be devised. For example, virus particles may be separated from a complex feed using ion exchange and also affinity chromatography columns, by size exclusion chromatography, and also by tangential flow filtration using hollow fibers or membranes that are designed for gentle handling for virus particles. Conditions that lead to modifications to the virus particles, degradation and aggregate formation should be largely avoided (Morenweiser 2005). Subjecting virus particles to shear forces by centrifugation or high pressure during filtration may lead to loss of infectivity, especially for enveloped viruses. Processes involving chemical precipitation to promote aggregation using polyethylene glycol or salts like ammonium sulfate and calcium phosphate can irreversibly damage virus particles due to osmotic pressure and also lead to irreversible aggregate formation. With the analytical methods used to establish a target profile, such undesired changes can be identified early during process development and steps can be taken to change the process parameters to minimize the impact of downstream processing steps on the target viral vector.

12.5 Conclusions

A good understanding of the target viral vector product profile is essential for determining the appropriate manufacturing process. Beyond producing high concentrations of infectious virus particles, it is equally important to assess product-related and process-related impurities. This has become clear as many new types of viral vectors and unique manufacturing processes have shown their implications on determining an effective dose and its associated risks. Critically looking at steps where such impurities are generated during manufacturing, and how to minimize them upstream to afford a more efficient downstream process is the way forward to devise manufacturing strategy based on quality by design. Reduction of both process and product-related impurities will free up downstream manufacturing capacities and importantly may lower the overall dose of virus particles applied for a desired efficacy. By profiling critical key attributes of viral vector preparations, correlations can be made between certain attributes and desired function, enabling the creation of a target profile of critical quality attributes for viral vectors.

References

Appaiahgari MB, Vrati S (2012) Clinical development of IMOJEV®—a recombinant Japanese encephalitis chimeric vaccine (JE-CV). Expert Opin Biol Th 12:1251–1263. https://doi.org/10.1517/14712598.2012.704908

Baldwin J, Piplani S, Sakala IG et al (2021) Rapid development of analytical methods for evaluating pandemic vaccines: a COVID-19 perspective. Bioanalysis 13:1805–1826. https://doi.org/10.4155/bio-2021-0096

Beer C, Meyer A, Müller K, Wirth M (2003) The temperature stability of mouse retroviruses depends on the cholesterol levels of viral lipid shell and cellular plasma membrane. Virology 308:137–146. https://doi.org/10.1016/s0042-6822(02)00087-9

Bora M, Yousuf RW, Dhar P et al (2019) Characterization of defective interfering (DI) particles of Pestedes petitsruminants vaccine virus Sungri/96 strain-implications in vaccine upscaling. Biologicals 62:57–64. https://doi.org/10.1016/j.biologicals.2019.09.008

Dobnik D, Kogovšek P, Jakomin T et al (2019) Accurate quantification and characterization of adeno-associated viral vectors. Front Microbiol 10:1570. https://doi.org/10.3389/fmicb.2019.01570

Domingo E, García-Crespo C, Perales C (2021) Historical perspective on the discovery of the quasispecies concept. Ann Rev Virol 8:51–72. https://doi.org/10.1146/annurev-virology-091919-105900

Ferreira MV, Cabral ET, Coroadinha AS (2021) Progress and perspectives in the development of lentiviral vector producer cells. Biotechnol J 16:2000017. https://doi.org/10.1002/biot.202000017

Fukuhara H, Ino Y, Todo T (2016) Oncolytic virus therapy: a new era of cancer treatment at dawn. Cancer Sci 107:1373–1379. https://doi.org/10.1111/cas.13027

Gagnon P, Goricar B, Mencin N et al (2021) Multiple-monitor HPLC assays for rapid process development, in-process monitoring, and validation of AAV production and purification. Pharmaceutics 13:113. https://doi.org/10.3390/pharmaceutics13010113

Gale TV, Horton TM, Hoffmann AR et al (2019) Host proteins identified in extracellular viral particles as targets for broad-spectrum antiviral inhibitors. J Proteome Res 18:7–17. https://doi.org/10.1021/acs.jproteome.8b00204

Gelderblom H (1996) Structure and classification of viruses. In: Baron S (ed) Medical microbiology, 4th edn. University of Texas Medical Branch at Galveston, Galveston (TX)

Gimpel AL, Katsikis G, Sha S et al (2021) Analytical methods for process and product characterization of recombinant adeno-associated virus-based gene therapies. Mol Ther Methods Clin Dev 20:740–754. https://doi.org/10.1016/j.omtm.2021.02.010

Greenberg RN, Kennedy JS (2008) ACAM2000: a newly licensed cell culture-based live vaccinia smallpox vaccine. Expert Opin Investig Drug 17:555–564. https://doi.org/10.1517/13543784.17.4.555

Hoeksema F, Karpilow J, Luitjens A et al (2018) Enhancing viral vaccine production using engineered knockout vero cell lines – a second look. Vaccine 36:2093–2103. https://doi.org/10.1016/j.vaccine.2018.03.010

Humphrey T, Brooks G, Emmett SR et al (2004) Cell cycle control, mechanisms and protocols. Methods Enzymol 296:197–218. https://doi.org/10.1385/1-59259-857-9:197

Ke Z, Oton J, Qu K et al (2020) Structures and distributions of SARS-CoV-2 spike proteins on intact virions. Nature 588:498–502. https://doi.org/10.1038/s41586-020-2665-2

Kiesslich S, Kamen AA (2020) Vero cell upstream bioprocess development for the production of viral vectors and vaccines. Biotechnol Adv 44:107608. https://doi.org/10.1016/j.biotechadv.2020.107608

Lauring AS, Andino R (2010) Quasispecies theory and the behavior of RNA viruses. PLoS Pathog 6:e1001005. https://doi.org/10.1371/journal.ppat.1001005

Li M, Husic N, Lin Y, Snider BJ (2012) Production of lentiviral vectors for transducing cells from the central nervous system. J Vis Exp e4031. https://doi.org/10.3791/4031

Linder A, Bothe V, Linder N et al (2021) Defective interfering genomes and the full-length viral genome trigger RIG-I after infection with vesicular stomatitis virus in a replication dependent manner. Front Immunol 12:595390. https://doi.org/10.3389/fimmu.2021.595390

Lodish HF, Porter M (1980) Heterogeneity of vesicular stomatitis virus particles: implications for virion assembly. J Virol 33:52–58. https://doi.org/10.1128/jvi.33.1.52-58.1980

van der Loo JCM, Wright JF (2016) Progress and challenges in viral vector manufacturing. Hum Mol Genet 25:R42–R52. https://doi.org/10.1093/hmg/ddv451

Lundstrom K (2018) Viral vectors in gene therapy. Diseases 6:42. https://doi.org/10.3390/diseases6020042

Markova N, Cairns S, Jankevics-Jones H et al (2021) Biophysical characterization of viral and lipid-based vectors for vaccines and therapeutics with light scattering and calorimetric techniques. Nato Adv Sci Inst Se 10:49. https://doi.org/10.3390/vaccines10010049

Martínez-Molina E, Chocarro-Wrona C, Martínez-Moreno D et al (2020) Large-scale production of lentiviral vectors: current perspectives and challenges. Pharmaceutics 12:1051. https://doi.org/10.3390/pharmaceutics12111051

Martínez-Sobrido L, García-Sastre A (2010) Generation of recombinant influenza virus from plasmid DNA. J Vis Exp. https://doi.org/10.3791/2057

Mateu MG (2011) Virus engineering: functionalization and stabilization. Protein Eng Des Sel 24: 53–63. https://doi.org/10.1093/protein/gzq069

Mátrai J, Chuah MK, VandenDriessche T (2010) Recent advances in lentiviral vector development and applications. Mol Ther 18:477–490. https://doi.org/10.1038/mt.2009.319

McCormick W, Mermel LA (2021) The basic reproductive number and particle-to-plaque ratio: comparison of these two parameters of viral infectivity. Virol J 18:92. https://doi.org/10.1186/s12985-021-01566-4

Miesbach W, Meijer K, Coppens M et al (2018) Gene therapy with adeno-associated virus vector 5–human factor IX in adults with hemophilia B. Blood 131:1022–1031. https://doi.org/10.1182/blood-2017-09-804419

Miller AD, Buttimore C (1986) Redesign of retrovirus packaging cell lines to avoid recombination leading to helper virus production. Mol Cell Biol 6:2895–2902. https://doi.org/10.1128/mcb.6.8.2895

Minh AD, Star AT, Stupak J et al (2021) Characterization of extracellular vesicles secreted in lentiviral producing HEK293SF cell cultures. Viruses 13:797. https://doi.org/10.3390/v13050797

Moerdyk-Schauwecker M, Hwang S-I, Grdzelishvili VZ (2009) Analysis of virion associated host proteins in vesicular stomatitis virus using a proteomics approach. Virol J 6:166. https://doi.org/10.1186/1743-422x-6-166

Morenweiser R (2005) Downstream processing of viral vectors and vaccines. Gene Ther 12:S103–S110. https://doi.org/10.1038/sj.gt.3302624

Perry C, Rayat ACME (2021) Lentiviral vector bioprocessing. Viruses 13:268. https://doi.org/10.3390/v13020268

Petersen BW, Karem KL, Damon IK (2014) Viral infections of humans, Epidemiology and Control 501–517. https://doi.org/10.1007/978-1-4899-7448-8_21

Ramsey JP, Khatwani SL, Lin M et al (2021) Overview of analytics needed to support a robust gene therapy manufacturing process. Curr Opin Biomed Eng 20:100339. https://doi.org/10.1016/j.cobme.2021.100339

Reece-Ford M, Hitchcock A, Lipinski K (2008) Aspects of process development for virus vector production to improve quality and quantity. Pharm Technol Eur

Reta A (2017) Defective interfering particles and their role in disease progression and persistence. Archives Microbiol Immunol 01:65–72. https://doi.org/10.26502/ami.9365009

Robert M, Chahal PS, Audy A et al (2017) Manufacturing of recombinant adeno-associated viruses using mammalian expression platforms. Biotechnol J 12:1600193. https://doi.org/10.1002/biot.201600193

Samji T (2009) Influenza A: understanding the viral life cycle. Yale J Biol Med 82:153–159

Sanyal G, Särnefält A, Kumar A (2021) Considerations for bioanalytical characterization and batch release of COVID-19 vaccines. Npj Vaccines 6:53. https://doi.org/10.1038/s41541-021-00317-4

Steppert P, Burgstaller D, Klausberger M et al (2017) Quantification and characterization of virus-like particles by size-exclusion chromatography and nanoparticle tracking analysis. J Chromatogr A 1487:89–99. https://doi.org/10.1016/j.chroma.2016.12.085

Strauss JH, Strauss EG (2008) Subviral Agents. Viruses Hum Dis 345–368. https://doi.org/10.1016/b978-0-12-373741-0.50012-x

Suder E, Furuyama W, Feldmann H et al (2018) The vesicular stomatitis virus-based Ebola virus vaccine: from concept to clinical trials. Hum Vacc Immunother 14:2107–2113. https://doi.org/10.1080/21645515.2018.1473698

Tan E, Chin CSH, Lim ZFS, Ng SK (2021) HEK293 cell line as a platform to produce recombinant proteins and viral vectors. Front Bioeng Biotechnol 9:796991. https://doi.org/10.3389/fbioe.2021.796991

Torkashvand F, Iran BRC Pasteur Institute of Iran, Tehran, Vaziri B (2017) Main quality attributes of monoclonal antibodies and effect of cell culture components. Iranian Biomed J 21:131–141. https://doi.org/10.18869/acadpub.ibj.21.3.131

Ungerechts G, Bossow S, Leuchs B et al (2016) Moving oncolytic viruses into the clinic: clinical-grade production, purification, and characterization of diverse oncolytic viruses. Mol Ther Methods Clin Dev 3:16018. https://doi.org/10.1038/mtm.2016.18

Ura T, Okuda K, Shimada M (2014) Developments in viral vector-based vaccines. Nato Adv Sci Inst Se 2:624–641. https://doi.org/10.3390/vaccines2030624

Warnock JN, Daigre C, Al-Rubeai M (2011) Viral vectors for gene therapy, methods and protocols. Methods Mol Biology 737:1–25. https://doi.org/10.1007/978-1-61779-095-9_1

Woolhouse M, Scott F, Hudson Z et al (2012) Human viruses: discovery and emergence. Philos Trans R Soc B Biol Sci 367:2864–2871. https://doi.org/10.1098/rstb.2011.0354

Wright JF (2014) Product-related impurities in clinical-grade recombinant AAV vectors: characterization and risk assessment. Biomedicines 2:80–97. https://doi.org/10.3390/biomedicines2010080

Wright JF (2021) Quality control testing, characterization and critical quality attributes of adeno-associated virus vectors used for human gene therapy. Biotechnol J 16:2000022. https://doi.org/10.1002/biot.202000022

Yeh P, Dedieu JF, Orsini C et al (1996) Efficient dual transcomplementation of adenovirus E1 and E4 regions from a 293-derived cell line expressing a minimal E4 functional unit. J Virol 70:559–565. https://doi.org/10.1128/jvi.70.1.559-565.1996

Zhou HS, Zhao T, Rao XM, Beaudet AL (2002) Production of helper-dependent adenovirus vector relies on helper virus structure and complementing. J Gene Med 4:498–509. https://doi.org/10.1002/jgm.301

Chapter 13
Process Analytical Technologies (PAT) and Quality by Design (QbD) for Bioprocessing of Virus-Based Therapeutics

Matthias Schad, Saurabh Gautam, Tanja A. Grein, and Friedrich Käß

Abbreviations

AAV	Adeno-Associated Virus
AEC	Anion Exchange Chromatography
AEX-HPLC	Anion Exchange High Performance Liquid Chromatography
API	Active Pharmaceutical Ingredient
ATMP	Advanced Therapy Medicinal Product
ATR-FTIR	Attenuated Total Reflection-Fourier Transform Infrared
AUC	Analytical Ultracentrifugation
BLI	Biolayer Interferometry
CAD	Charged Aerosol Detector
CAR-T	Chimeric Antigen Receptor T-Cell
CE	Capillary Electrophoresis
CE-SDS-LIF	Capillary Electrophoresis Sodium Dodecyl Sulfate Laser Induced Fluorescence
CMA	Critical Material Attribute
CPP	Critical Process Parameter

M. Schad (✉) · F. Käß
Boehringer Ingelheim Pharma GmbH & Co.KG, Biberach an der Riss, Germany
e-mail: matthias.schad@boehringer-ingelheim.com

S. Gautam
Boehringer Ingelheim International GmbH, Ingelheim, Germany

ViraTherapeutics GmbH, Rum, Austria

T. A. Grein
ProBioGen AG, Berlin, Germany

S. Gautam et al. (eds.), *Bioprocess and Analytics Development for Virus-based Advanced Therapeutics and Medicinal Products (ATMPs)*,
https://doi.org/10.1007/978-3-031-28489-2_13

CQA	Critical Quality Attribute
DAD	Diode Array Detector
ddPCR	digital droplet Polymerase Chain Reaction
DIP	Defective Interfering Particle
DLS	Dynamic Light Scattering
DO	Dissolved Oxygen
ELISA	Enzyme-linked Immunosorbent Assay
FDA	Food and Drug Administration
FLR	Fluorescence
FS	Fluorescence Spectrometry
FT-IR	Fourier-Transform Infrared
FT-MIR	Fourier-Transform-Mid Infrared
FV	Flow Virometry
GFP	Green Fluorescenting Protein
hcDNA	host cell DNA
HCP	Host Cell Protein
HPLC	High-Performance Liquid Chromatography
HTST	High Temperature Short Time
ICH	International Council for Harmonisation of Technical Requirements for Pharmaceuticals for Human Use
ICP-MS	Inductively Coupled Plasma-Mass Spectrometry
LAL	*Limulus* amebocyte lysate
LFC	Laser Force Cytometry
LPS	Lipopolysaccharides
mAb	monoclonal Antibody
MALDS	Siehe Multi-Angle Dynamic Light Scattering
MALS	Multi-Angle Light Scattering
MES	Manufacturing Execution System
MOA	Mode Of Action
MS	Mass Spectrometry
MVDA	Multivariate Data Analysis
NGS	Next-Generation Sequencing
NIR/MIR	Near/Mid Infrared
NOR	Normal Operating Range
NTA	Nanoparticle Tracking Analysis
OD	Optical Density
OFAT	One Factor At a Time
PAR	Proven Acceptable Range
PAT	Process Analytical Technology
pCQA	potential Critical Quality Attribute
PFU	Plaque Forming Units
PHA	Preliminary Hazard Analysis
PTM	Post Translational Modification
QbD	Quality by Design

qPCR	quantitative Polymerase Chain Reaction
QTPP	Quality Target Product Profile
RTCA	Real-Time Cell Analysis
SDS–PAGE	Sodium Dodecyl Sulfate–Polyacrylamide Gel Electrophoresis
SEC	Size Exclusion Chromatography
SEC-RP	Size Exclusion Chromatography-Reversed Phase Chromatography
SLS	Static Light Scattering
SPR	Surface Plasmon Resonance
$TCID_{50}$	Tissue Culture Infective Dose
TEM	Transmission Electon Microscopy
TRPS	Tunable Resistive Pulse Sensing
UV	Ultraviolet
UVRR	Ultraviolet Resonance Raman Spectroscopy
UV–vis	Ultraviolet–visual light
VLP	Virus-like particles

13.1 Introduction

Advanced therapy medicinal products (ATMPs) are a relatively new field in the (bio-)pharmaceutical industry. This category of pharmaceuticals—varying with source of definition—includes cell therapies, gene therapies, tissue-engineered medicines (EU 2007), and tumor vaccines (PEI 2019). Many of these therapies are either virus-based with respect to their active pharmaceutical ingredient or apply viral vectors for their generation. As the whole field is still in its early stages, clinical and commercial manufacturing has not yet reached the maturity grade of other more established biopharmaceuticals and will continue to evolve in the upcoming years.

It has previously been demonstrated that the biopharmaceutical industry can handle such high requests regarding quality and pureness in other product classes: there has been immense progress in protein production, especially for monoclonal antibodies (Farid 2007) and more recently in therapeutic mRNA vaccines as part of the COVID-19 pandemic response (Berg et al. 2021). Highly efficient manufacturing processes were developed with high titer, yield, high purity, and sound understanding of the critical process parameters (CPPs) that impact quality.

Back in 2004, the United States Food and Drug Administration (FDA) pushed for more systematical evaluation and understanding of CPP and critical quality attribute (CQA) relationships and the translation of this knowledge into process controls. This was done with the Quality by Design (QbD) and Process Analytical Technology (PAT) initiative (FDA 2004). Since then, PAT initiatives have evolved and are now major components in bioprocess development (Gillespie et al. 2021). The next logical step is now starting the implementation of QbD and PAT also in the manufacturing processes for ATMPs.

13.1.1 Oncolytic Viruses

At the beginning of the twentieth century, George Dock described a beneficial impact of an influenza infection in two leukemia patients. Both showed a temporary partial remission in tumor size or leucocyte number (Dock 1904). In 1971, Bluming and Ziegler published their observation about a patient with complete remission of Burkitt's lymphoma after measles infection (Bluming and Ziegler 1971).

Around a hundred years after the publication of George Dock, the first clinical trials with oncolytic viruses started and led to the first commercialized product, approved in 2015: talimogene laherparepvec (alias: T-VEC or Imlygic). Whereas T-VEC is based on a modified/attenuated Herpes Simplex Virus-1, there are a lot of other virus types, enveloped and not enveloped, in clinical studies for oncolytic therapy: Adeno-, Reo-, Pox-, Newcastle disease-, Coxsackie-, Polio-, Measles-, and Vesicular stomatitis viruses (Abd-Aziz and Poh 2021).

The common mode of action of all oncolytic virus types is the same (see Fig. 13.1). The patient's immune system and the cellular pathways defend against the infection with the therapeutic viruses, whereas in tumor cells, tumor tissue and its environment the viruses can replicate and trigger tumor cell lysis. The lysis attracts immune cells, which activates tumor recognition by the patient's immune system (Abd-Aziz and Poh 2021; Mueller et al. 2021).

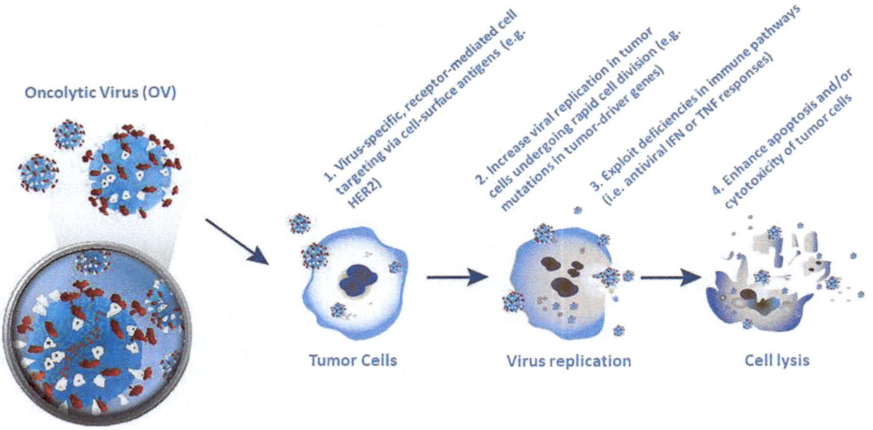

Fig. 13.1 Mode of action of an oncolytic virus. Healthy cells can be infected, but cell-internal defense mechanisms, e.g., interferon-based, prohibit virus replication and cell lysis. Tumor cells with compromised cell defense get infected, virus can replicate and tumor cells are lysed. Cell lysis leads to further oncolytic virus propagation, immune response, and inflammation to trigger tumor targeting. Figure reproduced under the terms of the Creative Commons Attribution License from (Carter et al. 2021)

13.1.2 Viral Vectors

Viral vectors are used to transfer genes into cells, both *in-vivo* and *ex-vivo*. For gene-therapy applications, mostly adeno-associated-(AAV), Retro-, Lenti-, and Adenoviruses are in use (Chen et al. 2018; Bulcha et al. 2021).

The mode of action of viral vectors in therapy is to transduce cells and tissues with a gene or genes of interest either integratively or non-integratively. These integrated genes can take over the role where a patient's gene is defective (e.g., lipoprotein lipase in lipoprotein lipase deficiency (Tremblay et al. 2018)). A second mode of action is to transduce cells (healthy or tumor cells) with a gene that encodes immune response activation or recruiting proteins, e.g., Chimeric antigen receptor T cells in CAR-T cell therapy (Sterner and Sterner 2021).

13.1.3 Vaccines

Traditional vaccination established in the 1920s has saved millions of lives over the last 100 years (Panhuis et al. 2013) and with the COVID-19 pandemic starting in 2019, the need and benefit of vaccines moved even more into the center of public attention.

In the case of viral infection diseases, traditionally attenuated, killed, split viruses or virus-like particles were manufactured to train the immune system for potential reaction to wild-type virus infection (Iwasaki and Omer 2020).

The relatively new field of therapeutic cancer vaccines uses the same principle as the oncolytic viruses, but the primary target is not to eliminate the tumor cells by infection and lysis. With cancer vaccines, the mode of action is more focused on lysing tumor cells, combined with expression of chemokines, cytokines, and antigens to attract and activate immune cells (Saxena et al. 2021).

Whereas for cancer vaccines, the used viruses are usually the same as for oncolytic viruses (Saxena et al. 2021), the number of different virus types in traditional virus-based vaccines is broad and diverse (Iwasaki and Omer 2020), including virus-like particles (VLPs). As suggested by the name, VLPs are not fully infectious virus particles with genome and supportive proteins, but consist of self-assembling viral modules like capsids, envelopes, or core proteins that form nanoscale structures (Nooraei et al. 2021).

13.1.4 Defective Interfering Particles

Another class of virus-based therapeutics are defective interfering particles (DIPs). DIPs are viruses with an incomplete genome that lacks sequences for viral replication. So, they can infect cells but not replicate and spread without wild-type helper virus infecting the same cell (Frensing et al. 2014).

DIPs interfere with the replication cycle of pathogenic viruses and by stimulation and activation of the patient's immune response, DIPs showed the capability of protecting mice from lethal virus infections (Scott et al. 2011).

According to Yang and colleagues, in 2019 there were no DIPs in clinical trials, but they listed a broad summary of tested virus types as Influenza-, Mumps-, Ebola-, Measles-, Rota-, Hepatitis-, Adeno-, and other viruses (Yang et al. 2019).

13.1.5 Virus Manufacturing

Virus propagation in eukaryotic cells—e.g., for influenza vaccines—is often based on the infection of embryonated hen eggs. This is a well-established, low-cost, and fast process to supply the high demands of these vaccines (Rajaram et al. 2020).

In recent years, more and more cell culture-based techniques have begun to emerge (Milián and Kamen 2015). With the possibility of virus production under fully defined and controlled process parameters, bioreactor processes have quickly become the standard upstream platform for oncolytic viruses (Ungerechts et al. 2016; Mueller et al. 2021) and viral vectors (Loo and Wright 2016). DIP manufacturing also tends in the direction of mammalian cell culture processes (Wasik et al. 2018).

In traditional vaccines, relatively low amounts of ranging from 10^4 to 10^8 plaque-forming units (PFU) per dose are needed (Frey et al. 2002; McCormack and Keam 2009; Patel et al. 2009), so there is minor demand for an extended purification process, as some impurities can simply be diluted below specification limits with the adjustment of the final drug product concentration. Some data even indicate that impurities like aggregates, host cell components, or virus proteins can enhance potency in a comparable manner to adjuvants (Jones et al. 2011; Blazevic et al. 2015).

With new modalities came the need for higher virus doses. As oncolytic virus doses of around 10^9 to 10^{10} PFU are common (Mondal et al. 2020; Porosnicu et al. 2022) for gene therapy modalities, concentrations up to 10^{14} vector genomes/kg patient's weight are needed (Mendell et al. 2020). Such high virus particle titers are either reachable by increasing the specific and/or the volumetric productivity. For both improvements, it is crucial to know the starting materials and the production process. Whereas the cell-specific productivity (higher virus yield per host cell) does not necessarily impact the impurity burden negatively, a higher volumetric productivity goes along with high-density cell cultures, which in turn brings more cell culture-dependent impurities. The relatively higher titer of newer viral vector processes makes means that there is no chance to dilute impurities below limit. This makes multi-step purification necessary, using primarily a series of chromatography and filtration steps. Downstream purification of these modalities, therefore, has

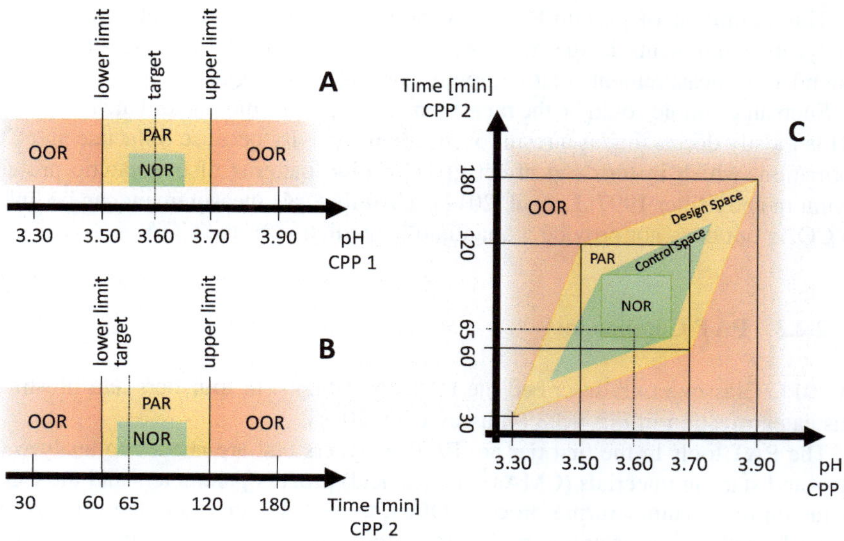

Fig. 13.2 Visualization of the hypothetical example 1 of a virus inactivation process step. (**a** and **b**) Show parameter ranges from the minimal approach including normal operating range (NOR), proven acceptable range (PAR), and out-of-range (OOR) for pH (**a**) and incubation time (**b**). (**c**) Illustrates parameters in a two-dimensional graph for the minimal approach PAR (black square) and NOR (green square) and for a PAT/ObD approach including design space (yellow diamond) and control space (green diamond)

upper limit 3.65 (see Fig. 13.2a). For incubation time the following NORs were defined: target 65 min, lower limit 62 min, and upper limit 100 min (see Fig. 13.2b).

The same process step can be defined with a two-factorial design space. In this case, it was proven in a Design of Experiment (DoE) approach that the combination of the CPPs pH and time directly correlates with the reduction rate of adventitious viruses. This correlation and the resulting design and control spaces are shown in Fig. 13.2c as yellow (design space) and green (control space) diamonds. Considering operational practice and the benefit of deepened process understanding, it can be concluded that the concept of scientifically sound spaces gives more control and flexibility compared to the ranges (NOR/PAR).

13.2.2.2 Pseudo-PAT

The term pseudo-PAT was distinguished from PAT by the teams of Mendhe and Rathore as follows: “These approaches do not involve direct measurement of product quality attributes, but are operationally simpler to implement” (Mendhe et al. 2015; Rathore et al. 2018).

This definition of pseudo-PAT somehow overlaps with the definition of PAT analyzers/components in the next section (PAT tools in the first tier) but can be extended to measurements that indirectly correlate with CQAs.

Such an example could be the measurement of pH in a bioreactor batch stage. The pH is mostly decreasing as a result of metabolic activity because of lactate and CO_2 formation, which in turn can affect glycosylation patterns of therapeutic proteins (Kimura and Miller 1997; Li et al. 2012). Therefore, pH measurement can be linked to CQAs but does not provide a quantitative prediction of the CQA.

13.2.2.3 PAT Components

In 2011, Glassey et al. described the PAT components in four tiers and illustrated this categorization in Fig. 13.3 (Glassey et al. 2011):

The PAT tools in the first tier are PAT analyzers that are needed to analyze the raw- and starting materials (CMAs), the (critical) process parameters and the CQAs of the current manufacturing process. Often, these PAT analyzers are used synonymously with inline or online probes to generate data instantly. Depending on the

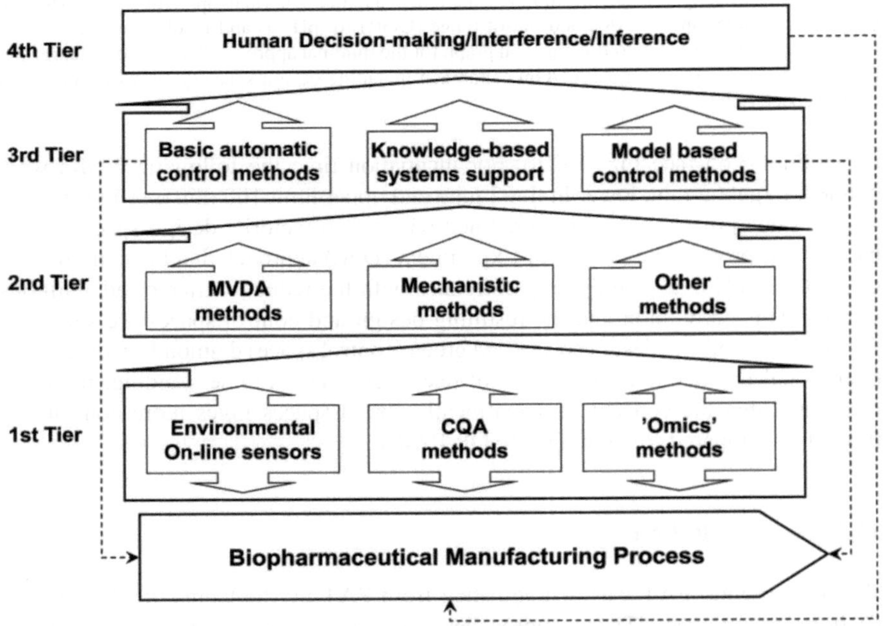

Fig. 13.3 Categorization of PAT tools into levels/tiers, beginning with data acquisition (first tier), followed by data processing (second tier) and tools for process control and data-based support (third tier). Level 3 tools can directly control processes or support human decision-making and control (tier 4); MVDA = Multivariate Data Analysis; CQA = Critical Quality Attribute. Figure adapted with permission from Glassey et al. (2011)

QbD design and the control strategy, this is not always necessary. If for example the manganese concentration in a cell culture process is a CPP which impacts a glycosylation pattern of a mAb, it can be sufficient to analyze the manganese concentration in the media powder batches and either mix powder batches or dose manganese in the cell culture process based on the CMA data. In this case, an offline mass-spectrometry method is a valuable PAT analyzer without delivering online data. So, PAT analyzers include all sensors, probes, and analytical instruments that generate process data in a timely manner to react on the gathered knowledge within the process (process reaction time).

In the second tier, Glassey and colleagues categorized data processing and data analysis systems and techniques. The tools in this tier are especially relevant for PAT analyzers that have no output that can be directly correlated to an action. In the example above, the measured manganese concentration can be directly translated to a need for a manganese dosage in cell culture media, and therefore no tier two data processing is needed. In case of a more complex data output, for example, a Raman spectrum or, if multiple datasets have to be processed to come to a read out (e.g., cell culture pH, oxygen demand and temperature → product quality) further data analysis is essential. These tools can include multivariate data analysis (MVDA), mechanistic methods, neuronal networks, machine learning and so on. What is essential here is to understand and validate this data processing to assure a sophisticated, robust, and reproducible outcome over the process lifecycle. Especially for black-box models like machine learning, this can be challenging.

The third tier includes tools and techniques to control the process based on the gathered (tier 1) and processed (tier 2) information. The simplest tool is a basic automatic control method, like an automated base addition based on a pH probe (tier 1) in a bioreactor. Such an automated addition, e.g., of a substrate can also be triggered by data model (tier 2) which is based on a Raman spectrum (tier 1). So this is a model-based control method. These tier 3 tools can also feed their trigger for process control, ideally combined with additional knowledge from process development studies or gathered process knowledge data, into a system or interface to support human decision-making.

Such a human decision-making system was categorized by Glassey et al. into tier 4. For example, a feedback loop to feed results and information into a manufacturing execution system (MES), or an operational procedure to include control information into a paper-based batch record system, or the supply of all relevant information into a data collection interface to supply the quality control officer with the necessary information for an online batch release.

13.2.2.4 PAT Integration

There are typically four principles to integrate analyzers (tier 1 PAT tools) into a manufacturing process: Inline, Online, Atline, and Offline. The classical description of these categories depends on the location of the analyzer/probe and the need of

Table 13.1 List of criteria to be fulfilled for the purpose-based analyzer nomenclature

Term	Real-time?	Automated/connected?	In situ?	Sample conditioning?	Typical example
Inline	Yes	Yes	Yes	No	Spectroscopic immersion probe, temperature probe, etc.
Online	Yes	Yes	No	Possible	Spectroscopic flow cell in by-pass; connected and automated GC fast enough for the process requirement
Atline	Possible	No	No	Possible	Remote titrator with manual sample transport
Offline	No	No	No	Yes	Color measurement in process lab; remote GC with manual feeding

Table reproduced with permission from (Minnich et al. 2016)

sampling. The US FDA 2004 Guidance for industry (FDA 2004) described it in this classical way:

- Inline: No sample needed. Invasive or noninvasive measurement directly in process stream.
- Online: Sample in form of a split stream needed for measurement. Sample could be returned to process.
- Atline: Sample to be drawn from process stream. Analyzer in close proximity to process stream.
- Offline: Sample needed. Analyzer is not close to the process stream.

In 2016, Minnich and colleagues re-defined this classification away from a sample and location-based classification toward a "purpose-based nomenclature" (Minnich et al. 2016) depicted in Table 13.1.

13.3 PAT Applications in Virus-Based Products

13.3.1 PAT for CMAs

As critical materials in virus-based product manufacturing are comparable to materials used in other biologics such as mAbs and other therapeutic proteins, the critical material attributes in virus manufacturing are also comparable.

A very special focus of the health authorities regarding raw materials is on materials with animal origin or potential contamination with animal-derived components, as described by the European Commission guidance note (EC 2011).

As described previously, raw materials can usually be ordered and tested prior to the manufacturing process with adequate lead-time. Therefore, there is no need for inline, online, or atline analyzers to measure attributes in real time. So, systems with relatively long analysis time up to days or weeks can be suitable PAT analyzers.

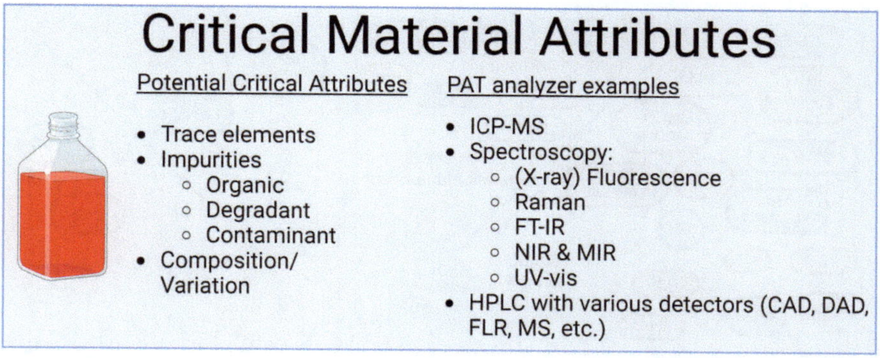

Fig. 13.4 Potential critical material attributes and PAT analyzers in virus manufacturing. *ICP-MS* Inductively Coupled Plasma-Mass Spectrometry, *FT-IR* Fourier-Transform Infrared, *NIR/MIR* Near/Mid Infrared, *UV–vis* Ultraviolet–visual light, *HPLC* High Performance Liquid Chromatography, *CAD* Charged Aerosol Detector, *DAD* Diode Array Detector, *FLR* Fluorescence, *MS* Mass Spectrometry. Created with BioRender.com

Dickens et al. and Ryder described in their reviews the potential material attributes and analytical methods to address them (Dickens et al. 2018; Ryder 2018). Figure 13.4 lists an overview of these attributes and potential PAT methods.

13.3.2 PAT for CQAs/CPPs

As already described, PAT should be fast enough (real-time) to be able to control the manufacturing process based on feedback from the analytical data. Depending on the structural integrity and intermediate stability of a virus-based API, the feasible maximum process response time from analytical result to controller output for the PAT can differ. This means that some of the following methods are restricted in their application and may only be suitable for a subset of virus products.

Snyder et al. categorized the quality attributes for AAV into six classes, namely titer, purity, identity, potency, safety, and stability (Snyder et al. 2011). In the following subsection we merged titer and potency and do not follow up on stability, as stability testing is independent from the manufacturing process and therefore from PAT testing.

Figure 13.5 summarizes the CQA classes, potential CQAs and methods that are more or less applicable for their use in PAT for the specific attributes.

13.3.2.1 Identity

The focus of identity testing is to confirm and validate the active ingredient in drug substance and drug product.

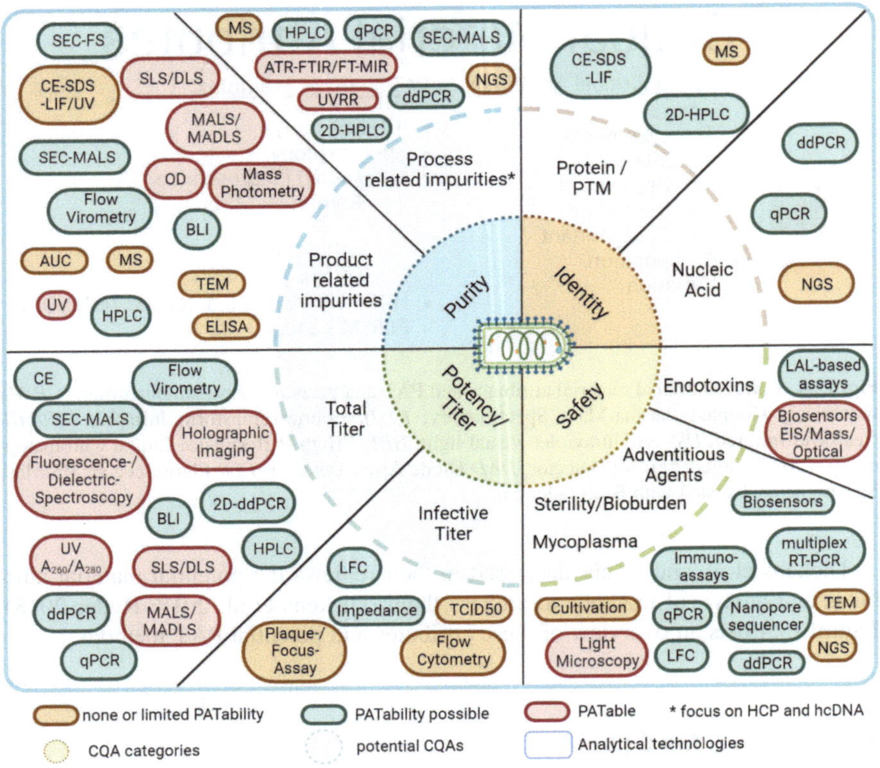

Fig. 13.5 CQAs of virus-based products and potential PAT methods that could be implemented in a PAT-controlled process. Analysis methods are categorized into: PATable = fast/real-time methods usable in PAT, PATability possible = not real-time methods with longer turnaround time and sample processing, none or limited PATability = long turnaround time for assay or extensive sample handling and processing which makes use in a PAT process unlikely. *CQA* Critical Quality Attribute, *PAT* Process Analytical Technology, *SEC-FS* Size Exclusion Chromatography-Fluorescence Spectrometry, *CE-SDS-LIF/UV* Capillary Electrophoresis-Sodium Dodecyl Sulfate-Laser Induced Fluorescence/Ultraviolet, *SEC-MALS* Size Exclusion Chromatography-Multi-Angle Light Scattering, *SLS/DLS* Static/Dynamic Light Scattering, *MALS/MADLS* Multi-Angle (Dynamic) Light Scattering, *OD* Optical Density, *BLI* Biolayer Interferometry, *AUC* Analytical Ultracentrifugation, *MS* Mass Spectrometry, *TEM* Transmission Electron Microscopy, *UV* Ultraviolet, *ELISA* Enzyme-Linked Immunosorbent Assay, *HPLC* High Performance Liquid Chromatography, *CE* Capillary Electrophoresis, *2D-ddPCR* two Dimensional digital droplet Polymerase Chain Reaction, *UV A260/A280* Ultraviolet Absorbance ratio at 260 nm and 280 nm, *qPCR* quantitative Polymerase Chain Reaction, *ddPCR* digital droplet Polymerase Chain Reaction, *LFC* Laser Force Cytometry, *NGS* Next Generation Sequencing, *LAL* Limulus Amebocyte Lysate, *RT-PCR* Real-Time-Polymerase Chain Reaction, *EIS* Electrochemical Impedance Spectroscopy, *2D-HPLC* two Dimensional-Hight Performance Liquid Chromatography, *UVRR* Ultraviolet Resonance Raman, *ATR-FTIR* Attenuated Total Reflection-Fourier Transformation Infrared, *FT-MIR* Fourier Transformation-Mid Infrared. Created with BioRender.com

To real-time analyze identity attributes and make control decisions in the manufacturing process, the controllability is limited. In most cases a confirmed, failed identity specification is a go/no-go decision for a batch release, not a trigger to change process parameters on results.

Identity testing of recombinant AAV by next-generation sequencing (NGS) was described by Guerin and colleagues in 2020. They developed a NGS method to identify the DNA genomes and to detect presence of DNA of contaminating AAV. They also stated that the method can "[. . .] detect other sequences or events, such as reverse packaging of plasmid backbone or DNA from the packaging cell line" (Guerin et al. 2020), which can besides identity testing be an application for purity testing.

There is a certain overlap of identity and purity testing, as e.g., the identification of a distinct post-translational modification (PTM) pattern can be crucial for identifying a virus (Giles et al. 2018). However, there are always low percentage variants detectable with slight pattern variations which might be classified as impurities, but which does not mean a failed identity specification of the manufacturing batch.

A protein-based identity test was described by García-Canas et al., who developed a 2D-HPLC size exclusion chromatography combined with reversed-phase (SEC-RP) method. This method allows to distinguish between different influenza types or even when the same types are produced with different manufacturing processes (García-Cañas et al. 2007).

In general, identity testing in virus-based therapeutics is based on protein and nucleic acid (except for VLP) identification. So, advanced methods for protein identification, like mass spectrometry (MS) or capillary electrophoresis sodium dodecyl sulfate laser induced fluorescence (CE-SDS-LIF) could be used (Pierson et al. 2016; Zhang et al. 2021). The advanced method for nucleic acid identification is the quantitative or digital droplet polymerase chain reaction (qPCR/ddPCR) analysis (Escandell et al. 2022).

13.3.2.2 Potency/Viral Titer

The potency of a virus-based product can correlate either with the total number of virus or virus-like particles, e.g., in the case of vaccines. In modalities where the infection of patient cells is part of the mode of action, the number of infective particles directly correlates with the product's potency. Depending on the mode of action, additional potency analytics are necessary. For example, the activation of an immune response for vaccines or the gene transfer and their expression in the target cell in gene therapy or CAR-T applications.

Infective Titer

For infective particles, the infective titer is the most important quality attribute as it is the main attribute that defines potency. To determine the infective particle titer, cell-based assays are still the gold standard, although they come with a high workload for cell cultivation, sample preparation, dilutions, assay execution, incubation and evaluation.

Recently Lothert et al. compared three cell-based assays to quantify infective viral particles (Lothert et al. 2022):

1. The classical tissue culture infective dose assay ($TCID_{50}$) where reporter cells are infected with dilutions of samples. The concentration at which half of the reporter cells showed a cytopathic effect is used to calculate the infective concentration (Spearman 1908; Kärber 1931; Reed and Muench 1938). A recent publication showed that for a TCID50 with assay optimization and automation, the hands-on-time could be shortened from 8 to 2.2 h and an improved variation coefficient of ~20% is possible (Hochdorfer et al. 2022). But there is still a need for 3 days incubation (cells with virus) before readout and the variation is still high compared to assays which are not cell-based.
2. For the plaque titration and focus-forming assay, adherent reporter cells are infected with dilutions of the sample to be analyzed and covered by a layer of medium. To determine the number of infective particles, the number of plaques = areas with lysed cells, are counted and used to calculate the initial number of infective particles. In these assays also a long incubation time (2–14 days) is needed for a readout and variation is high (Baer and Kehn-Hall 2014).
3. As a third option, Lothert et al. listed the flow cytometric titration. Therefore, the infection of reporter cells with a virus must come with the integration of a fluorescent marker to detect infected cells with a flow cytometer or impaired sensitivity plate reader. As the integration of a marker is necessary for this method, flow cytometric titration is not an option for a clinical or commercial manufacturing process. To overcome this problem, Guerreiro and colleagues published a label-free method to analyze infective titer by flow cytometry titration. They used reporter cells, modified with cyclized green fluorescent protein (GFP) with a cleavage site for a viral protease. An infection with adenoviruses or human immunodeficiency 1 virus led to the activation of the GFP. The GFP signal could then be quantified by a flow cytometer or plate reader (Guerreiro et al. 2019). But still, a 42 h incubation time is needed and the variability coefficient is still relatively high at 9%.

A faster but still cell-based analytic of infective particles was published by Herbert and colleagues. Their technique uses Laser Force Cytometry (LFC), which showed a 4.5-fold faster titer determination (3 days →16 h) for vesicular stomatitis virus infection in Vero cells (Hebert et al. 2018). Charretier and colleagues also improved and shortened their cell-based assay by using their so-called real-time cell analysis (RTCA) which is based on impedance measurement of the infected indicator cells. They stated a "[. . .] 5 times less labor-intense and 3.5 times cheaper [. . .]" assay compared to their standard cell culture infective dose assay (Charretier et al. 2018).

As described, there is a high demand for analyzers or assays with shorter assay time and better precision, especially for PAT applications. One PAT-like application where these cell-based assays could be used to control a process is if, e.g., a process intermediate or drug substance could be stored frozen until the assay data is available. Then the intermediate could be processed based on the result to dilute or concentrate to target.

In some virus-based processes, the intermediate process hold at room temperature could be a viable option for waiting on assay results, with regard to product stability. But in terms of process flow, this is often not a preferred option. For some viruses and processes, neither frozen nor liquid intermediate holds are a viable option. For example, lentiviruses tend to degrade quickly at room temperature (Valkama et al. 2020). Therefore, an actual real-time analyzer for infective titer monitoring would be a game changer in process control without additional storage steps in between.

Another possibility to control a process with infective titer could be the use of a PATable surrogate parameter or attribute or a combination. Let us assume it was proven that in a cell culture process the infective/total titer ratio stays consistent if pH and dissolved oxygen concentration (DO) are stable within the controlled range. So, with a total titer PAT analyzer like fluorescence spectroscopy (see next section) in combination with inline probes for pH and DO, infective titer could be monitored or controlled in a (pseudo-) PAT manner. Or the three sensor outputs (tier 1) could feed a software model with validated correlation to create a PAT soft-sensor (tier 2) to directly control the upstream process (tier 3) or serve as a support for human decision-making (tier 4) (see Fig. 13.3).

We assume that such a soft sensor PAT analysis is the most promising approach for implementing a real-time infective titer PAT analyzer as the other options are still too slow for a sophisticated monitoring and control in process.

Non-infective/Total Titer

For the determination of the total number of viral particles there are lots of methods published, which must be chosen based on the product to be analyzed and the monitored unit operation.

Pais et al., Grein et al., and Käßer et al. published some work for monitoring AAV and measles virus particles in cell culture. They focused on inline probes and described the use of fluorescence spectroscopy (Pais et al. 2019), holographic imaging (Pais et al. 2020b) and dielectric spectroscopy (Grein et al. 2018; Käßer et al. 2019; Pais et al. 2020a), which are all promising PAT or pseudo-PAT analyzers for real-time monitoring of titers in cell culture. These methods are based on the viruses and their interactions with the host cells and are therefore only applicable in upstream, but not in purification processes where no host cells are present. For example, with the monitoring of the actual titer in real-time the optimized timing of the harvest or controls of a continuous processing could be triggered.

For nucleic acid-containing particles, there are some established titer methods available based on qPCR, ddPCR (Dobnik et al. 2019), and two-dimensional ddPCR

(Furuta-Hanawa et al. 2019). Sommer and colleagues used the ratio for ultraviolet (UV) absorption at 260 nm/280 nm (Sommer et al. 2003) to determine total viral particle count. It is known that absorption at 260 nm correlates with the nucleic acid content, whereas absorption at 280 nm measures the protein content. So, whereas the qPCR and ddPCR give specific results for the reporter genes of interest and can therefore be used in most of the cell culture and buffer matrices in upstream and downstream processing, the UV method is not specific and needs purified intermediates/matrices without foreign, interfering DNA or proteins. As both principles (PCR and UV) have limitations as labor-intensive sample preparation and long analytic times (PCR) and potential matrix effects/interference (UV), the capability as PAT analyzer must be assessed specifically for process and unit operation.

Another principle of measuring the total viral particle count is the determination of virus-specific proteins and/or their binding affinity. Lothert and colleagues listed in this category the surface plasmon resonance (SPR), protein identification methods like capillary electrophoresis (CE) and finally the broad field of analytical chromatography (Lothert et al. 2022).

The surface plasmon resonance and the related biolayer interferometry (BLI) technology are promising technologies to monitor viral particles based on binding affinities. It was demonstrated that SPR can be used for the detection and quantification of human enterovirus 71 by Prabowo et al. in 2017, where they immobilized anti-VP1 (virus capsid) antibodies to the sensor surface. They could detect virus particle concentrations down to approximately 67 virus particles per milliliter (Prabowo et al. 2017). For the BLI technology, Carvalho and team showed capability to quantify influenza virus-like particles (Carvalho et al. 2017).

Capillary electrophoresis could be validated to quantify reliably adenovirus concentration in a manufacturing process. The method showed good precision and standard deviation and sample preparation was only necessary for high DNA concentration (Tricht et al. 2017).

A well-established analytical method for AAV characterization (see purity) and titer determination is the anion exchange high-performance liquid chromatography (AEX-HPLC) with a detector for UV absorption at 260 nm. Whitfield et al. also showed applicability for the method for adenoviruses and the method was capable of diverse matrices from crude cell lysate to pure virus (Whitfield et al. 2009). Transfiguracion and colleagues successfully tested their AEX-HPLC method for influenza particles and stated that "No significant sample preparation except for filtration prior to injection on the HPLC column is needed." Such a simple sample application paves the way for this technology to be used as a PAT analyzer (Transfiguracion et al. 2015). A quite extensive review on chromatography-based virus and VLP analytics including anion exchange-, size exclusion- and reversed phase chromatography was done in 2015 by Kramberger and colleagues (Kramberger et al. 2015). Depending on the specific method protocol some techniques could be used for real-time monitoring, but for some labor-extensive sample preparation, the analysis time can disqualify the methods for real-time PAT applications.

Another promising technique for determining the total number of nucleic acid viruses could be the flow virometry (FV), e.g., described by Rossi and team. In their work they showed the applicability for counting total numbers of filovirus particles and made a comparison to real-time qPCR and transmission electron microscopy (TEM) (Rossi et al. 2015).

Quantification methods are already established for viruses or virus-like particles based on (image-based) particle analytics without assessing them for integrability in PAT monitoring and control (Lothert et al. 2022). They listed light scattering methods, centrifugal sedimentation, Nanoparticle tracking analysis (NTA), and Tunable resistive pulse sensing (TRPS). Centrifugal sedimentation and TRPS were rated as difficult, complex, and currently not (yet) capable as standard analytics for virus process analytics (Lothert et al. 2022).

Light scattering methods like dynamic light scattering (DLS), multi-angle dynamic light scattering (MALDS), static light scattering (SLS), multi-angle (static) light scattering (MALS) and NTA—which is also a type of light scattering method—are powerful methods for counting and physically characterizing particles such as viruses and virus-like particles. The methods are relatively fast and easy to implement even in a process stream. However, a major drawback of these assays is the variability in the matrix compositions of the samples. Bigger and smaller particles and high particle concentrations interfere with the optical methods and lead to poor resolution (Lothert et al. 2022). There are publications to overcome these limitations to be able to also analyze more complex samples that are not fully purified. But these come with the cost of additional sample treatment, like nuclease digestion (Hohl et al. 2017), labeling (Driskell et al. 2011), or combination with separation techniques like size exclusion chromatography (SEC) (McIntosh et al. 2021). This in turn would counteract the advantages like fast and easy monitoring of attributes in process stream.

Altogether, there are methodologies available for a total titer monitoring in upstream and downstream. All have advantages and drawbacks that should be balanced for the actual API and the process to be implemented. Some of the methods could also deliver results regarding product purity, which should also be considered for the right method choice.

13.3.2.3 Purity

The purity of a product confirms that no undesired components are contained in drug substance/drug product or that the unwanted components are below defined specifications.

There are two categories of undesired components or impurities, defined by the origin: process-related impurities and product-related impurities.

Product-Related Impurities

Product-related impurities are undesired components that are related to the product. For example, if the desired active ingredient is an infective virus particle for an oncolytic mode of action, non-infective particles could be defined as product-related impurity. In contrast to that, for a vaccine application the desired active, stimulatory ingredient could be a non-infective virus particle. Therefore, active viruses must be defined as product-related impurity. Other product-related impurities could be virus fragments, virus aggregates, virus particles without or partial genome, VLPs with genome or active viruses in a DIP process. As we already addressed the analytics of infective particles and nucleic acids in the sections above, we will now focus on the other types of product-related impurities.

There are a lot of publications available for AAV manufacturing. So, there is a lot of knowledge about AAV-specific product-related impurities and the corresponding analytical methods. Gimpel and colleagues defined two product-related impurities as AAV CQAs: non-full capsids and aggregates (Gimpel et al. 2021).

Non-full capsids can be either empty capsids without genetic load or capsids with parts of the intended or foreign genetic load. Both types are primarily unwanted types as these non-functional particles can trigger and activate the patient's immune response to also fight against the API. Secondly, infective non-functional particles can compete with the API and therefore lower the transfection efficiency in target cells. However, there is data that empty capsids can have beneficial effects as they could intercept patient's anti-drug antibodies (Wright 2014).

Gimpel et al. listed analyses methods for the analysis of capsid numbers and content ratio (full vs. empty capsid): anion-exchange chromatography (AEC) e.g., with a HPLC (Joshi et al. 2021), analytical ultracentrifugation (AUC) (Burnham et al. 2015), BLI (Wheatley et al. 2015), MS (Pierson et al. 2016), enzyme-linked immunosorbent assay (ELISA) (Kuck et al. 2007), flow virometry (Rossi et al. 2015; Niu et al. 2021), mass photometry (Refeyn 2021), optical density (OD) (Sommer et al. 2003), SEC coupled with MALS (McIntosh et al. 2021), SLS/DLS (Zhang et al. n.d.) and TEM (Dobnik et al. 2019).

Some of the mentioned methods and assays are limited for PAT use as—comparable to the mentioned cell-based assays—the time and workload from sampling to result is too high to directly react in a (purification) process. Methods like TEM, AUC, MS, and ELISA need several hours of preparation and analysis to deliver results. Other methods like HPLC, BLI, FV, and SEC-MALS have turnaround times of less than one hour, which leads to a chance to enable a PAT-based process. Methods with short turnaround times of a few minutes like mass photometry, OD and SLS/DLS do come with disadvantages such as limited repeatability or range. But depending on the process requirements these methods could bring the highest benefit in a PAT-based process (Gimpel et al. 2021).

Another method principle, known in protein manufacturing (such as monoclonal antibodies), which is promising for monitoring virus capsids is capillary electrophoresis. This method was used not just to monitor total capsids or content ratio, but also to characterize and monitor the different subtypes and variants of the capsid proteins.

An AAV virion is typically built of the capsid proteins named VP1, VP2, and VP3. These should be present in a ratio of around 1:1:10. Such a subtype analysis could be done with traditional sodium dodecyl sulfate–polyacrylamide gel electrophoresis (SDS–PAGE) with silver staining for analysis. The more automated method with CE was published with the use of CE-SDS-laser-induced fluorescence detection (CE-SDS-LIF) or CE-SDS-UV to differentiate virus protein subtypes (Zhang and Meagher 2017; Zhang et al. 2021).

As already mentioned in the identity section, the A_{260}/A_{280} measurement of non-enveloped viruses such as AAV could be a promising method for content ratio (empty to full particles) monitoring as it is a fast and simple measurement (Sommer et al. 2003). However, it is not the most reliable or specific method because the absorbance at both these wavelengths (260 or 280 nm) could be affected by a number of other factors, such as any host cell DNA (hcDNA) or host cell protein (HCP) present within the feedstream or sample.

For product-related impurities, such as aggregates and fragments, Gimpel et al. listed the methods AUC, SEC-fluorescence spectrometry (SEC-FS), SEC-MALS and SLS/DLS for AAVs. As stated for the capsid impurities, AUC with a turnaround time of about 6 h is hard to integrate in a PAT-based process, whereas SEC-FS and SEC-MALS with turnaround times of 30 min are PATable, depending on the process and its requirements. The fastest method, SLS/DLS, with a few minutes might be the most promising PAT method for monitoring aggregates or fragments (Gimpel et al. 2021). Patel and team already used inline-MALS as cutting detector in a mAb process (Patel et al. 2018). But the transferability to virus-based processes was not yet shown.

There is one PAT-analyzer for monitoring product and even process-related impurities (indirectly) we did not mention so far, which is established in most of the purification processes: UV detectors in chromatography. In many virus or VLP purification processes an ion exchange chromatography bind-elute step is implemented for capture, purification or polishing. For some particles like AAV it is known that a pH or salt gradient leads to a separate elution of impurities like capsids or aggregates (Qu et al. 2007). It is also standard that the API collection is controlled by UV (e.g., 300 nm or 280 nm) absorbance. In conclusion, a lot of processes have implemented a real-time (pseudo-) PAT analyzer system, consisting of ion exchange chromatography and UV detector that directly controls (start/stop collection) the purity-CQA.

Besides the already established PAT methods the light scattering techniques seem to be the most promising analyzers to find their way into processes, at least in process units where the API is already in a relatively purified form. The methods are fast and relatively easy to implement and can—besides product-related impurities—deliver data about total titer. And as all virus-based products are also particle-based, the techniques should be applicable for most entities.

Process-related Impurities

Every manufacturing process is run with raw materials, starting materials, equipment and purification supports/materials. For every component used in a process, there is a chance of the introduction of impurities, e.g., for every filter or chromatography resin there is a likelihood that substances like ligands or preservative agents can be carried over into the process stream.

The FDA already listed sources and possible sources or categories for process-related impurities in their guideline (FDA 1999). Based on this, we created a summary table (Table 13.2).

With the diversity of viruses and virus-like products, there is a multiplied diversity of possible manufacturing processes and their combination of unit operations and raw- and starting materials. Just considering the variability in commercially available media, media components and serum sources, it must be concluded that there is no simple list of analysis methods for addressing all process-related impurities, which could possibly be found in a manufacturing process or even in the final drug substance or drug product. Such an assessment must be done and is only possible in a product- and process-specific manner.

In general, there are two impurities that are highly variable but are immanent in all cell culture processes and must be addressed in every process development and commercialization: host cell proteins and host cell DNA (hcDNA) (FDA 1999). As every cell culture process is affected by these impurities, there are some techniques available from mAb manufacturing that might be used in virus-based therapeutics as well.

Table 13.2 Process-related impurities and possible sources/categories

Category	Components
Cell culture-derived	Media components
	Serum components
	Antibiotics
	Transfection reagents
	Expression Enhancers/Inducing Agents Extractables/Leachables
Cell substrate-derived	Host cell proteins
	Host cell or helper cell DNA
	Vesicular particles
	Other host cell-derived substances (e.g., lipids)
Downstream-derived	Buffer components and salts
	Enzymes (e.g., nucleases)
	Chemical and biochemical processing reagents
	Solvents
	Carriers
	Ligands Extractables/Leachables

For analyzing hcDNA the same analytical principles are true as described for the identity testing in the identity section. Therefore, also methods like NGS, qPCR, and ddPCR can be methods for analyzing and quantifying hcDNA (Wang et al. 2019). Again, these methods are relatively slow and labor-intensive, especially the NGS method, and maybe not be applicable in a PAT approach.

Gagnon and colleagues published nucleic acid analysis based on HPLC methods in the context of AAV analysis and characterization. In their work, they used Picogreen fluorescence dye to label free DNA. Therefore, this technique could also be used to monitor hcDNA (Gagnon et al. 2021).

Another approach for monitoring DNA and RNA in a cell culture broth was described by Ashton et al. and based on a mAb process. They used ultraviolet resonance Raman spectroscopy (UVRR) and successfully monitored Raman spectra over the harvest process steps as centrifugation and depth filtration and challenged the method against potential interfering agents (Ashton et al. 2015). Whether this technique could be used for nucleic acid containing particles must be challenged, but for empty/virus-like particles this method could be promising.

For the monitoring of HCPs in a virus-based processing there is little published experience. Bracewell and his team covered in their opinion paper about "Analytics of host cell proteins (HCPs): lessons from biopharmaceutical mAb analysis for Gene therapy products" in 2021 the question of whether MS methods could be beneficial (PAT) analytics for HCP monitoring in mAb processes and also for virus-based processes. In the context of virus products, they stated that "[...] MS based monitoring, detection and characterization of HCPs in gene therapy based products has been explored, although is comparatively in its infancy" (Bracewell et al. 2021). Why virus-based products are hard to analyze regarding HCPs was described in the publication of Sviben et al. In viruses, there are multiple proteins coded in the virus genome that are part of the virus particle or are co-expressed by the host cells (Sviben et al. 2018). To distinguish virus-immanent proteins from free or extracellular particle-associated proteins is challenging.

Capito et al. described in two publications in 2013 and 2015 how to use Fourier transform-mid infrared analysis (FT-MIR) at-line in a recombinant antibody production process to monitor product titer, aggregate amount and HCP concentration. The authors limited the applicability in their setup to higher HCP concentrations down to 700 ng/mL and stated that "[...] HCP monitoring following intermediate and polishing chromatography steps is currently not a well-suited application" (Capito et al. 2013, 2015). As described above, the applicability and transferability to virus processes is difficult and must be shown.

A completely different challenge regarding process- and product-related impurities on column resins was described by Boulet-Audet and colleagues in their publication in 2016. For a monoclonal antibody process, they successfully monitored protein (product and HCP) accumulation (resin fouling) on their capture column by using attenuated total reflection-Fourier transform infrared (ATR-FTIR) (Boulet-Audet et al. 2016). This could be a reasonable method to develop a reuse process of bind-elute columns like Anion-Exchange (AEX), Cation- Exchange (CEX) or affinity columns like AAV affinity resins.

There are different publications where variations of HPLC-coupled techniques are described to analyze and characterize HCPs in e.g. influenza manufacturing processes like two-dimensional HPLC (García-Cañas et al. 2007; Lorbetskie et al. 2011) or reversed phase HPLC (Creskey et al. 2012). But as Kramberger et al. stated: "However, due to quite extensive sample preparation they are mostly not most suitable for in-process monitoring for the DSP"(Kramberger et al. 2015). But again, this must be assessed specifically for the product and process combination for which a PAT approach is to be planned.

So, at least for HCP and hcDNA there are some analytical methods available (FT-IR, UVRR) that have potential for a fast real-time application. But, as stated, these methods were tested for use in mAb processes and must be verified in a process for a virus-based product.

13.3.2.4 Safety

The quality attributes for safety are mostly product- and process-unspecific and are mandatory attributes in most manufacturing processes. In the category of safety attributes contaminations are listed that could pose risks to patients if applied with the drug (Snyder et al. 2011). For virus-based products the monitoring and control of contaminants and especially biological contaminants is even more crucial compared to e.g., therapeutic proteins as virus reduction unit operations like pH inactivation and virus filtration in the process stream might not be applicable. Established bioburden reduction filtrations with 0.2 μm filter materials or mycoplasma reduction with 0.1 μm filters could have negative impact on titer, as virus or virus-like particles can be filtered out because of size separation or interaction with the membrane material.

There are strategies to overcome the limitations. Size-based separation like SEC or 0.2 or 0.1 μm filtration might be used for large- or small-sized virus reduction. To validate such a unit operation and state the reduction performance is difficult.

As reduction of adventitious agents is still challenging, it is even more important to target for a fully closed process operation without open handling. Additionally, the starting materials and raw materials should be monitored and to some extent treated for virus reduction methods like virus filtration, high-temperature short time (HTST) treatment or UV treatment.

But as a completely closed processing and raw material treatment is not yet fully established, the testing for biological contaminants is still essential for patient safety. Biological contaminant testing for

- Sterility
- Mycoplasma
- Adventitious agents
- Endotoxin

are well mandatory for intermediates and final product (Snyder et al. 2011).

Sterility, Mycoplasma and Adventitious Viruses

Morris et al. recently reviewed the techniques for adventitious agent detection with including viruses, bacteria, and mycoplasma. They described four categories of methods for adventitious agent detection (Morris et al. 2021):

1. Cultivation-, cell-based, and in vivo assays are the traditional methods for detecting adventitious agents. The assays are well established, simple, and non-specific and are capable of detecting most of the agents. On the other hand, these assays are labor- and time-consuming and cell-based and *in vivo* assays are known to be unprecise and error-prone. These disadvantages make them suboptimal or unsuitable for a PAT approach.
2. Well established and widespread are light microscopic examinations for micro-organisms such as bacteria and mycoplasma (Nims and Price 2017) and electron microscopic methods for adventitious viruses. Light microscopic methods are fast and simple with some need for sample preparation but for small organisms there is a limit of detection. Especially mycoplasmas that adhere onto host cells, spores and non-mobile organisms are hard to detect and is highly operator dependent. Process- and product-related impurities like host cell debris lowers sensitivity. The same is true for the detection of virus by electron microscopy like TEM (Richert-Pöggeler et al. 2019). For that technique the turnover time and the costs are clear drawbacks for use as a PAT application. Both microscopic methods are non-specific and therefore there is no need to specify the organisms or viruses to be detected.
3. The third category of detection methods for adventitious agents listed by Morris and team is the nucleic acid-based detection. The already mentioned that methods such as qPCR, ddPCR (Quan et al. 2018), NGS (Khan et al. 2017), real-time PCR (Pallás et al. 2018) or relatively new techniques like nanopore sequencer (Tyler et al. 2018) can be used to detect nucleic acid patterns of potential adventitious agents. The also discussed pros and cons of these techniques hold still true: sample preparation, assay time, interference with other nucleic acids. Additionally, to detect organisms with these methods there must be prior knowledge available of the genome sequences of the organisms and viruses to detect. If that is the case, a specific identification and quantification is possible.
4. In the fourth category of immunoassays there is a wide field of techniques summarized. It begins with the basics like ELISA and cell-based activation immunoassays, and spans to more advanced assays like cellulase-linked immunomagnetic-(Pankratov et al. 2020), immuno-chromatographic-(Tominaga 2018), magnetophoretic (Kim et al. 2017) immuno-assays or biosensors (Bueno and Cerezo 2014) and immunoelectron microscopy (Richert-Pöggeler et al. 2019). For all the subtypes specific binders are needed for the agents to be detected. So, unknown organisms or viruses are not detectable and need to be addressed by other methods.

In addition to the described methods and categories of Morris and team, Hayes and colleagues described the use of LFC. This technique we already mentioned in the analytics for infective virus titers and the same principle is used here for adventitious viruses. A reporter cell culture is spiked with the sample to be analyzed, potential viruses infect the reporter cells and changes in the cellular features due to the infection are detected by the laser force cytometer (Hayes et al. 2022).

In summary, the implementation of fast and reliable techniques for the detection sterility, mycoplasma and virus is difficult. The PCR methods are promising for getting a sensitive, specific, and relatively fast result regarding nucleic acid-containing organisms and particles. For monitoring a virus or virus-based purification process in real-time or even to manage a drug substance real-time release, there is still a way to go.

Endotoxins

Endotoxins or lipopolysaccharides (LPS) are a class of pyrogens, which can induce unfavorable, strong immune response in patients. These LPS molecules consists of "[. . .] a heteropolysaccharide (O-antigen), a core oligosaccharide, and a nonpolar lipid A tail" (Schneier et al. 2020).

Some traditional detection methods were based on *in vivo* tests with rabbits or blood assays with human or bovine blood. The most established traditional methods are based on *Limulus* amebocyte lysate assay (LAL) in which the clotting mechanism of blood from horseshoe crabs with LPS is used. Meanwhile, there is an alternative to the need of crab-blood as the relevant Factor C can be produced recombinantly. These LAL-based tests, sub-categorized in clot, turbidity and chromogenic methods are well established and widespread in the biopharmaceutical industry. These tests can be applicable for use in a PAT process as the assay turnaround time is around one hour or less. There is sample handling and preparation needed but is restricted within a reasonable range. However, there are issues with interfering agents that could mask LPS and lead to false-negative results (Schwarz et al. 2017) and there are factors that can create false-positive results (Elin and Wolff 1973).

Nonetheless, there are ambitions to expedite the monitoring of endotoxin tests and to increase robustness using biosensors. Three types of systems are currently in development: electrochemical-, optical- and mass-based sensors. All showed promising results with lower limits of detection and quantification, less false-positive and -negative results and faster turnaround time, even real-time. The current drawbacks are that some sensors are single-use and expensive in manufacturing.

Altogether, fast endotoxin detection and monitoring assays and techniques are available for an implementation in PAT. For real-time endotoxin PAT analyzers there is still basic research to be completed.

13.4 Discussion

In this chapter, we describe the basic principles of QbD, PAT, and pseudo-PAT, including CQAs and CPPs in the context of virus-based products. We list potential CQAs for these products and give an overview about analytical methods that are or could be used to monitor, characterize and control these attributes including an evaluation about the applicability of the methods and assays for a PAT-based manufacturing process. Figure 13.5 illustrates this overview.

For now, there is basis and research work to be done and there is still a lack of real-time analyzers. But the existing methodologies can already be used not as real-time controlling within a running process, but between the runs. The implementation of the methodologies could shorten the long analytics and evaluation phase for development runs. Therefore, the readout and planning of the next development package could be minimized and an optimized laboratory utilization could be reached.

For other, more established biologicals like mAbs, the PAT technologies are already an inherent part of manufacturing processes. Most of the technologies implemented are not directly monitoring CQAs, but control CPPs or in a pseudo-PAT manner: pH, conductivity, dissolved oxygen, UV, weighing, and turbidity. We did not elaborate on these parameters extensively in this work. Nonetheless, direct CQA monitoring and control are widely established as well, like Raman, Liquid Chromatography coupled with MS, at-line titer and reversed phase-HPLC (Gillespie et al. 2021).

To our opinion, if we as developers of virus-based products take this progress of the other biologicals in PAT seriously, extract the gain and transfer the applicable learnings to our virus-based products, there is a chance to bring the virus-based therapeutics to the same level within the next 5–10 years.

There is also a chance that the virus process development can even outpace traditional biologics regarding implementing the next level of PAT—the monitoring and control of CQAs regularly in a process. The increased complexity and lower stability of virus-based therapeutics may be just the incentive needed for the establishment of more sophisticated analytical tools, and the trust in real-time evaluation instead of less flexible approaches.

References

Abd-Aziz N, Poh CL (2021) Development of oncolytic viruses for cancer therapy. Transl Res. https://doi.org/10.1016/j.trsl.2021.04.008

Ashton L, Hogwood CEM, Tait AS et al (2015) UV resonance Raman spectroscopy: a process analytical tool for host cell DNA and RNA dynamics in mammalian cell lines. J Chem Technol Biotechnol 90:237–243. https://doi.org/10.1002/jctb.4420

Baer A, Kehn-Hall K (2014) Viral concentration determination through plaque assays: using traditional and novel overlay systems. J Vis Exp e52065. https://doi.org/10.3791/52065

van de Berg D, Kis Z, Behmer CF et al (2021) Quality by design modelling to support rapid RNA vaccine production against emerging infectious diseases. Npj Vaccines 6:65. https://doi.org/10.1038/s41541-021-00322-7

Blazevic V, Malm M, Arinobu D et al (2015) Rotavirus capsid VP6 protein acts as an adjuvant in vivo for norovirus virus-like particles in a combination vaccine. Human Vacc Immunother 12:740–748. https://doi.org/10.1080/21645515.2015.1099772

Bluming Avrum Z, Ziegler John L (1971) Regression of Burkitt's lymphoma in association with measles infection. Lancet 298:105–106. https://doi.org/10.1016/s0140-6736(71)92086-1

Boulet-Audet M, Kazarian SG, Byrne B (2016) In-column ATR-FTIR spectroscopy to monitor affinity chromatography purification of monoclonal antibodies. Sci Rep UK 6:30526. https://doi.org/10.1038/srep30526

Bracewell DG, Smith V, Delahaye M, Smales CM (2021) Analytics of host cell proteins (HCPs): lessons from biopharmaceutical mAb analysis for Gene therapy products. Curr Opin Biotech 71: 98–104. https://doi.org/10.1016/j.copbio.2021.06.026

Bueno J-L, Cerezo B (2014) P05 A sterile procedure to obtain and store high concentrated platelet rich plasma (HC-PRP) as a source of growth factors to be used in the non transfusional setting. Transfus Apher Sci 50:S10–S11. https://doi.org/10.1016/s1473-0502(14)50022-5

Bulcha JT, Wang Y, Ma H et al (2021) Viral vector platforms within the gene therapy landscape. Signal Transduct Target Ther 6:53. https://doi.org/10.1038/s41392-021-00487-6

Burnham B, Nass S, Kong E et al (2015) 556. Analytical ultracentrifugation as an approach to characterize recombinant AAV vectors. Mol Ther 23:S223. https://doi.org/10.1016/s1525-0016(16)34165-x

Capito F, Skudas R, Kolmar H, Stanislawski B (2013) Host cell protein quantification by Fourier transform mid infrared spectroscopy (FT-MIR). Biotechnol Bioeng 110:252–259. https://doi.org/10.1002/bit.24611

Capito F, Skudas R, Kolmar H, Hunzinger C (2015) At-line mid infrared spectroscopy for monitoring downstream processing unit operations. Process Biochem 50:997–1005. https://doi.org/10.1016/j.procbio.2015.03.005

Carter ME, Koch A, Lauer UM, Hartkopf AD (2021) Clinical trials of oncolytic viruses in breast cancer. Front Oncol 11:803050. https://doi.org/10.3389/fonc.2021.803050

Carvalho SB, Moleirinho MG, Wheatley D et al (2017) Universal label-free in-process quantification of influenza virus-like particles. Biotechnol J 12:1700031. https://doi.org/10.1002/biot.201700031

Charretier C, Saulnier A, Benair L et al (2018) Robust real-time cell analysis method for determining viral infectious titers during development of a viral vaccine production process. J Virol Methods 252:57–64. https://doi.org/10.1016/j.jviromet.2017.11.002

Chen YH, Keiser MS, Davidson BL (2018) Viral vectors for gene transfer. Curr Protoc Mouse Biol 8:e58. https://doi.org/10.1002/cpmo.58

Creskey MC, Li C, Wang J et al (2012) Simultaneous quantification of the viral antigens hemagglutinin and neuraminidase in influenza vaccines by LC–MSE. Vaccine 30:4762–4770. https://doi.org/10.1016/j.vaccine.2012.05.036

Dobnik D, Kogovšek P, Jakomin T et al (2019) Accurate quantification and characterization of adeno-associated viral vectors. Front Microbiol 10:1570. https://doi.org/10.3389/fmicb.2019.01570

Dock G (1904) The influence of complicating diseases upon Leukaemia. Am J Med Sci 1827-1924: 563–592

Dickens J, Khattak S, Matthews TE et al (2018) Biopharmaceutical raw material variation and control. Curr Opin Chem Eng 22:236–243. https://doi.org/10.1016/j.coche.2018.10.007

Driskell JD, Jones CA, Tompkins SM, Tripp RA (2011) One-step assay for detecting influenza virus using dynamic light scattering and gold nanoparticles. Analyst 136:3083–3090. https://doi.org/10.1039/c1an15303j

EC (2011) Note for guidance on minimising the risk of transmitting animal spongiform encephalopathy agents via human and veterinary medicinal products (EMA/410/01 rev.3). https://www.

ema.europa.eu/en/documents/scientific-guideline/minimising-risk-transmitting-animal-spongiform-encephalopathy-agents-human-veterinary-medicinal_en.pdf. Accessed 29 Sep 2022

Elin RJ, Wolff SM (1973) Nonspecificity of the limulus amebocyte lysate test: positive reactions with polynucleotides and proteins. J Infect Dis 128:349–352. https://doi.org/10.1093/infdis/128.3.349

Escandell JM, Pais DA, Carvalho SB et al (2022) Leveraging rAAV bioprocess understanding and next generation bioanalytics development. Curr Opin Biotech 74:271–277. https://doi.org/10.1016/j.copbio.2021.12.009

EU (2007) Regulation (EC) No 1394/2007 of the European parliament and of the council of 13 November 2007 on advanced therapy medicinal products and amending Directive 2001/83/EC and Regulation (EC) No 726/2004. Official Journal of the European Union

Farid SS (2007) Process economics of industrial monoclonal antibody manufacture. J Chromatogr B 848:8–18. https://doi.org/10.1016/j.jchromb.2006.07.037

FDA (1999) Guidance for Industry Q6B Specifications: Test Procedures and Acceptance Criteria for Biotechnological/Biological Products

FDA (2004) PAT — A Framework for Innovative Pharmaceutical Development, Manufacturing, and Quality Assurance

Frensing T, Pflugmacher A, Bachmann M et al (2014) Impact of defective interfering particles on virus replication and antiviral host response in cell culture-based influenza vaccine production. Appl Microbiol Biotechnol 98:8999–9008. https://doi.org/10.1007/s00253-014-5933-y

Frey SE, Newman FK, Cruz J et al (2002) Dose-related effects of smallpox vaccine. New Engl J Med 346:1275–1280. https://doi.org/10.1056/nejmoa013431

Furuta-Hanawa B, Yamaguchi T, Uchida E (2019) Two-dimensional droplet digital PCR as a tool for titration and integrity evaluation of recombinant adeno-associated viral vectors. Hum Gene Ther Methods 30:127–136. https://doi.org/10.1089/hgtb.2019.031

Gagnon P, Goricar B, Mencin N et al (2021) Multiple-monitor HPLC assays for rapid process development, in-process monitoring, and validation of AAV production and purification. Pharmaceutics 13:113. https://doi.org/10.3390/pharmaceutics13010113

García-Cañas V, Lorbetskie B, Bertrand D et al (2007) Selective and quantitative detection of influenza virus proteins in commercial vaccines using two-dimensional high-performance liquid chromatography and fluorescence detection. Anal Chem 79:3164–3172. https://doi.org/10.1021/ac0621120

Giles AR, Sims JJ, Turner KB et al (2018) Deamidation of amino acids on the surface of adeno-associated virus capsids leads to charge heterogeneity and altered vector function. Mol Ther 26:2848–2862. https://doi.org/10.1016/j.ymthe.2018.09.013

Gillespie C, Wasalathanthri DP, Ritz DB et al (2021) Systematic assessment of process analytical technologies for biologics. Biotechnol Bioeng. https://doi.org/10.1002/bit.27990

Gimpel AL, Katsikis G, Sha S et al (2021) Analytical methods for process and product characterization of recombinant adeno-associated virus-based gene therapies. Mol Ther Methods Clin Dev 20:740–754. https://doi.org/10.1016/j.omtm.2021.02.010

Glassey J, Gernaey KV, Clemens C et al (2011) Process analytical technology (PAT) for biopharmaceuticals. Biotechnol J 6:369–377. https://doi.org/10.1002/biot.201000356

Grein TA, Loewe D, Dieken H et al (2018) High titer oncolytic measles virus production process by integration of dielectric spectroscopy as online monitoring system. Biotechnol Bioeng 115:1186–1194. https://doi.org/10.1002/bit.26538

Guerin K, Rego M, Bourges D et al (2020) A novel next-generation sequencing and analysis platform to assess the identity of recombinant adeno-associated viral preparations from viral DNA extracts. Hum Gene Ther 31:664–678. https://doi.org/10.1089/hum.2019.277

Guerreiro MR, Freitas DF, Alves PM, Coroadinha AS (2019) Detection and quantification of label-free infectious adenovirus using a switch-on cell-based fluorescent biosensor. Acs Sensors 4:1654–1661. https://doi.org/10.1021/acssensors.9b00489

Hayes PR, Przybycien TM, Schneider JW (2022) Viral adventitious agent detection using laser force cytology: intrinsic cell property changes with infection and comparison to in vitro testing. Biotechnol Bioeng 119:134–144. https://doi.org/10.1002/bit.27957

Hebert CG, DiNardo N, Evans ZL et al (2018) Rapid quantification of vesicular stomatitis virus in Vero cells using Laser Force Cytology. Vaccine 36:6061–6069. https://doi.org/10.1016/j.vaccine.2018.09.002

Hochdorfer D, Businger R, Hotter D et al (2022) Automated, label-free TCID50 assay to determine the infectious titer of virus-based therapeutics. J Virol Methods 299:114318. https://doi.org/10.1016/j.jviromet.2021.114318

Hohl A, Ramms AS, Dohmen C et al (2017) Adenovirus particle quantification in cell lysates using light scattering. Hum Gene Ther Methods 28:268–276. https://doi.org/10.1089/hgtb.2017.052

ICH (2009) Pharmaceutical development Q8(R2). https://database.ich.org/sites/default/files/Q8_R2_Guideline.pdf

Iwasaki A, Omer SB (2020) Why and how vaccines work. Cell 183:290–295. https://doi.org/10.1016/j.cell.2020.09.040

Jones JC, Settles EW, Brandt CR, Schultz-Cherry S (2011) Virus aggregating peptide enhances the cell-mediated response to influenza virus vaccine. Vaccine 29:7696–7703. https://doi.org/10.1016/j.vaccine.2011.07.133

Joshi PRH, Bernier A, Chahal PS, Kamen A (2021) Development and validation of an anion exchange high-performance liquid chromatography method for analysis of empty capsids and capsids encapsidating genetic material in a purified preparation of recombinant adeno-associated virus serotype 5. Hum Gene Ther 32:1390–1402. https://doi.org/10.1089/hum.2020.317

Kärber G (1931) Beitrag zur kollektiven Behandlung pharmakologischer Reihenversuche. Naunyn-schmiedebergs Archiv Für Exp Pathologie Und Pharmakologie 162:480–483. https://doi.org/10.1007/bf01863914

Käßer L, Zitzmann J, Grein T, et al (2019) Turbidimetry and dielectric spectroscopy as process analytical technologies for mammalian and insect cell cultures. In: Animal cell biotechnology, methods and protocols, pp. 335–364

Khan AS, Ng SHS, Vandeputte O et al (2017) A multicenter study to evaluate the performance of high-throughput sequencing for virus detection. Msphere 2:e00307–e00317. https://doi.org/10.1128/msphere.00307-17

Kim J, Lee K-S, Kim EB et al (2017) Early detection of the growth of Mycobacterium tuberculosis using magnetophoretic immunoassay in liquid culture. Biosens Bioelectron 96:68–76. https://doi.org/10.1016/j.bios.2017.04.025

Kimura R, Miller WM (1997) Glycosylation of CHO-derived recombinant tPA produced under elevated pCO2. Biotechnol Prog 13:311–317. https://doi.org/10.1021/bp9700162

Kramberger P, Urbas L, Štrancar A (2015) Downstream processing and chromatography based analytical methods for production of vaccines, gene therapy vectors, and bacteriophages. Hum Vacc Immunother 11:1010–1021. https://doi.org/10.1080/21645515.2015.1009817

Kuck D, Kern A, Kleinschmidt JA (2007) Development of AAV serotype-specific ELISAs using novel monoclonal antibodies. J Virol Methods 140:17–24. https://doi.org/10.1016/j.jviromet.2006.10.005

van der Loo JCM, Wright JF (2016) Progress and challenges in viral vector manufacturing. Hum Mol Genet 25:R42–R52. https://doi.org/10.1093/hmg/ddv451

Li J, Wong CL, Vijayasankaran N et al (2012) Feeding lactate for CHO cell culture processes: impact on culture metabolism and performance. Biotechnol Bioeng 109:1173–1186. https://doi.org/10.1002/bit.24389

Lorbetskie B, Wang J, Gravel C et al (2011) Optimization and qualification of a quantitative reversed-phase HPLC method for hemagglutinin in influenza preparations and its comparative evaluation with biochemical assays. Vaccine 29:3377–3389. https://doi.org/10.1016/j.vaccine.2011.02.090

Lothert K, Eilts F, Wolff MW (2022) Quantification methods for viruses and virus-like particles applied in biopharmaceutical production processes. Expert Rev Vaccines 1–16. https://doi.org/10.1080/14760584.2022.2072302

McCormack PL, Keam SJ (2009) Rotavirus Vaccine RIX4414 (Rotarix™). Pediatr Drugs 11:75–88. https://doi.org/10.2165/0148581-200911010-00025

McIntosh NL, Berguig GY, Karim OA et al (2021) Comprehensive characterization and quantification of adeno associated vectors by size exclusion chromatography and multi angle light scattering. Sci Rep UK 11:3012. https://doi.org/10.1038/s41598-021-82599-1

Mendell JR, Al-Zaidy SA, Rodino-Klapac LR et al (2020) Current clinical applications of in vivo gene therapy with AAVs. Mol Ther 29:464–488. https://doi.org/10.1016/j.ymthe.2020.12.007

Mendhe R, Thukkaram M, Patil N, Rathore AS (2015) Comparison of PAT based approaches for making real-time pooling decisions for process chromatography – use of feed forward control. J Chem Technol Biotechnol 90:341–348. https://doi.org/10.1002/jctb.4448

Milián E, Kamen AA (2015) Current and emerging cell culture manufacturing technologies for influenza vaccines. Biomed Res Int 2015:504831. https://doi.org/10.1155/2015/504831

Minnich C, Hardy S, Krämer S (2016) Stopping the Babylonian confusion: an updated nomenclature for process analyzers in PAT applications. Chem Ing Tech 88:694–697. https://doi.org/10.1002/cite.201500188

Mondal M, Guo J, He P, Zhou D (2020) Recent advances of oncolytic virus in cancer therapy. Hum Vacc Immunother 16:1–14. https://doi.org/10.1080/21645515.2020.1723363

Morris C, Lee YS, Yoon S (2021) Adventitious agent detection methods in bio-pharmaceutical applications with a focus on viruses, bacteria, and mycoplasma. Curr Opin Biotechnol 71:105–114. https://doi.org/10.1016/j.copbio.2021.06.027

Mueller D, Garcia AP, Grein TA, et al (2021) Process for producing a purified rhabdovirus from cell culture

Nims RW, Price PJ (2017) Best practices for detecting and mitigating the risk of cell culture contaminants. Vitro Cell Dev Biol Anim 53:872–879. https://doi.org/10.1007/s11626-017-0203-9

Niu Q, Ma L, Zhu S et al (2021) Quantitative assessment of the physical virus titer and purity by ultrasensitive flow virometry. Angew Chem Ger Edit 133:9437–9442. https://doi.org/10.1002/ange.202100872

Nooraei S, Bahrulolum H, Hoseini ZS et al (2021) Virus-like particles: preparation, immunogenicity and their roles as nanovaccines and drug nanocarriers. J Nanobiotechnol 19:59. https://doi.org/10.1186/s12951-021-00806-7

Pais DAM, Portela RMC, Carrondo MJT et al (2019) Enabling PAT in insect cell bioprocesses: in situ monitoring of recombinant adeno-associated virus production by fluorescence spectroscopy. Biotechnol Bioeng 116:2803–2814. https://doi.org/10.1002/bit.27117

Pais DAM, Brown C, Neuman A et al (2020a) Dielectric spectroscopy to improve the production of rAAV used in gene therapy. PRO 8:1456. https://doi.org/10.3390/pr8111456

Pais DAM, Galrão PRS, Kryzhanska A et al (2020b) Holographic imaging of insect cell cultures: online non-invasive monitoring of adeno-associated virus production and cell concentration. PRO 8:487. https://doi.org/10.3390/pr8040487

Pallás V, Sánchez-Navarro JA, James D (2018) Recent advances on the multiplex molecular detection of plant viruses and viroids. Front Microbiol 09:2087. https://doi.org/10.3389/fmicb.2018.02087

van Panhuis WG, Grefenstette J, Jung SY et al (2013) Contagious diseases in the United States from 1888 to the Present. New Engl J Med 369:2152–2158. https://doi.org/10.1056/nejmms1215400

Pankratov D, Bendixen M, Shipovskov S et al (2020) Cellulase-linked immunomagnetic microbial assay on electrodes: specific and sensitive detection of a single bacterial cell. Anal Chem 92:12451–12459. https://doi.org/10.1021/acs.analchem.0c02262

Patel M, Shane AL, Parashar UD et al (2009) Oral rotavirus vaccines: how well will they work where they are needed most? J Infect Dis 200:S39–S48. https://doi.org/10.1086/605035

Patel BA, Gospodarek A, Larkin M et al (2018) Multi-angle light scattering as a process analytical technology measuring real-time molecular weight for downstream process control. MAbs 10: 945–950. https://doi.org/10.1080/19420862.2018.1505178

PEI (2019) ATMP. https://www.pei.de/DE/arzneimittel/atmp/atmp-node.html. Accessed 10 Jul 2022

Pierson EE, Keifer David Z, Asokan A, Jarrold MF (2016) Resolving adeno-associated viral particle diversity with charge detection mass spectrometry. Anal Chem 88:6718–6725. https://doi.org/10.1021/acs.analchem.6b00883

Porosnicu M, Quinson A-M, Crossley K et al (2022) Phase I study of VSV-GP (BI 1831169) as monotherapy or combined with ezabenlimab in advanced and refractory solid tumors. Future Oncol 18:2627–2638. https://doi.org/10.2217/fon-2022-0439

Prabowo BA, Wang RYL, Secario MK et al (2017) Rapid detection and quantification of Enterovirus 71 by a portable surface plasmon resonance biosensor. Biosens Bioelectron 92:186–191. https://doi.org/10.1016/j.bios.2017.01.043

Qu G, Bahr-Davidson J, Prado J et al (2007) Separation of adeno-associated virus type 2 empty particles from genome containing vectors by anion-exchange column chromatography. J Virol Methods 140:183–192. https://doi.org/10.1016/j.jviromet.2006.11.019

Quan P-L, Sauzade M, Brouzes E (2018) dPCR: a technology review. Sensors 18:1271. https://doi.org/10.3390/s18041271

Rajaram S, Wojcik R, Moore C et al (2020) The impact of candidate influenza virus and egg-based manufacture on vaccine effectiveness: literature review and expert consensus. Vaccine 38: 6047–6056. https://doi.org/10.1016/j.vaccine.2020.06.021

Rathore AS, Singh SK, Kumar J, Kapoor G (2018) Chapter 48 - Implementation of QbD for manufacturing of biologics—has it met the expectations. In: Biopharmaceutical processing - development, design, and implementation of manufacturing processes, pp 1051–1073

Reed LJ, Muench H (1938) A simple method of estimating fifty per cent endpoints12. Am J Epidemiol 27:493–497. https://doi.org/10.1093/oxfordjournals.aje.a118408

Refeyn (2021) Mass photometry of AAVs

Richert-Pöggeler KR, Franzke K, Hipp K, Kleespies RG (2019) Electron microscopy methods for virus diagnosis and high resolution analysis of viruses. Front Microbiol 9:3255. https://doi.org/10.3389/fmicb.2018.03255

Rossi CA, Kearney BJ, Olschner SP et al (2015) Evaluation of ViroCyt® virus counter for rapid filovirus quantitation. Viruses 7:857–872. https://doi.org/10.3390/v7030857

Ryder AG (2018) Cell culture media analysis using rapid spectroscopic methods. Curr Opin Chem Eng 22:11–17. https://doi.org/10.1016/j.coche.2018.08.008

Saxena M, van der Burg SH, Melief CJM, Bhardwaj N (2021) Therapeutic cancer vaccines. Nat Rev Cancer 21:360–378. https://doi.org/10.1038/s41568-021-00346-0

Schneier M, Razdan S, Miller AM et al (2020) Current technologies to endotoxin detection and removal for biopharmaceutical purification. Biotechnol Bioeng 117:2588–2609. https://doi.org/10.1002/bit.27362

Schwarz H, Gornicec J, Neuper T et al (2017) Biological activity of masked endotoxin. Sci Rep UK 7:44750. https://doi.org/10.1038/srep44750

Scott PD, Meng B, Marriott AC et al (2011) Defective interfering influenza virus confers only short-lived protection against influenza virus disease: evidence for a role for adaptive immunity in DI virus-mediated protection in vivo. Vaccine 29:6584–6591. https://doi.org/10.1016/j.vaccine.2011.06.114

Snyder RO, Audit M, Francis JD (2011) Adeno-associated virus, methods and protocols. Methods Mol Biol 807:405–428. https://doi.org/10.1007/978-1-61779-370-7_17

Sommer JM, Smith PH, Parthasarathy S et al (2003) Quantification of adeno-associated virus particles and empty capsids by optical density measurement. Mol Ther 7:122–128. https://doi.org/10.1016/s1525-0016(02)00019-9

Spearman C (1908) The method of 'right and wrong cases' ('constant stimuli') without Gauss's formulaE. Br J Psychol 1904–1920(2):227–242. https://doi.org/10.1111/j.2044-8295.1908.tb00176.x

Sterner RC, Sterner RM (2021) CAR-T cell therapy: current limitations and potential strategies. Blood Cancer J 11:69. https://doi.org/10.1038/s41408-021-00459-7

Sviben D, Forcic D, Halassy B et al (2018) Mass spectrometry-based investigation of measles and mumps virus proteome. Virol J 15:160. https://doi.org/10.1186/s12985-018-1073-9

Tanaka T, Hanaoka H, Sakurai S (2020) Optimization of the quality by design approach for gene therapy products: a case study for adeno-associated viral vectors. Eur J Pharm Biopharm 155: 88–102. https://doi.org/10.1016/j.ejpb.2020.08.002

Tominaga T (2018) Rapid detection of Klebsiella pneumoniae, Klebsiella oxytoca, Raoultella ornithinolytica and other related bacteria in food by lateral-flow test strip immunoassays. J Microbiol Methods 147:43–49. https://doi.org/10.1016/j.mimet.2018.02.015

Transfiguracion J, Manceur AP, Petiot E et al (2015) Particle quantification of influenza viruses by high performance liquid chromatography. Vaccine 33:78–84. https://doi.org/10.1016/j.vaccine.2014.11.027

Tremblay K, Brisson D, Gaudet D (2018) Lipoprotein lipase gene replacement therapy long-term effect on familial hyperchylomicronemia gene expression profile. Atherosclerosis 275:e34. https://doi.org/10.1016/j.atherosclerosis.2018.06.086

van Tricht E, Geurink L, Backus H et al (2017) One single, fast and robust capillary electrophoresis method for the direct quantification of intact adenovirus particles in upstream and downstream processing samples. Talanta 166:8–14. https://doi.org/10.1016/j.talanta.2017.01.013

Tyler AD, Mataseje L, Urfano CJ et al (2018) Evaluation of Oxford Nanopore's MinION Sequencing Device for Microbial Whole Genome Sequencing Applications. Sci Rep UK 8:10931. https://doi.org/10.1038/s41598-018-29334-5

Ungerechts G, Bossow S, Leuchs B et al (2016) Moving oncolytic viruses into the clinic: clinical-grade production, purification, and characterization of diverse oncolytic viruses. Mol Ther Methods Clin Dev 3:16018. https://doi.org/10.1038/mtm.2016.18

Valkama AJ, Oruetxebarria I, Lipponen EM et al (2020) Development of large-scale downstream processing for lentiviral vectors. Mol Ther Methods Clin Dev 17:717–730. https://doi.org/10.1016/j.omtm.2020.03.025

Wang Y, Cooper R, Kiladjian A et al (2019) A digestion-free method for quantification of residual host cell DNA in rAAV gene therapy products. Mol Ther Methods Clin Dev 13:526–531. https://doi.org/10.1016/j.omtm.2019.05.005

Wasik MA, Eichwald L, Genzel Y, Reichl U (2018) Cell culture-based production of defective interfering particles for influenza antiviral therapy. Appl Microbiol Biotechnol 102:1167–1177. https://doi.org/10.1007/s00253-017-8660-3

Wheatley DW, Saunders D, Welsh J et al (2015) Influenza vaccine titer determination using biolayer interferometry (BLI). BMC Proc 9:P75. https://doi.org/10.1186/1753-6561-9-s9-p75

Whitfield RJ, Battom SE, Barut M et al (2009) Rapid high-performance liquid chromatographic analysis of adenovirus type 5 particles with a prototype anion-exchange analytical monolith column. J Chromatogr A 1216:2725–2729. https://doi.org/10.1016/j.chroma.2008.11.010

Wright JF (2014) Product-related impurities in clinical-grade recombinant AAV vectors: characterization and risk assessment. Biomedicines 2:80–97. https://doi.org/10.3390/biomedicines2010080

Yang Y, Lyu T, Zhou R et al (2019) The antiviral and antitumor effects of defective interfering particles/genomes and their mechanisms. Front Microbiol 10:1852. https://doi.org/10.3389/fmicb.2019.01852

Zhang C-X, Meagher MM (2017) Sample stacking provides three orders of magnitude sensitivity enhancement in SDS capillary gel electrophoresis of adeno-associated virus capsid proteins. Anal Chem 89:3285–3292. https://doi.org/10.1021/acs.analchem.6b02933

Zhang Z, Park J, Barrett H et al (2021) Capillary electrophoresis-sodium dodecyl sulfate with laser-induced fluorescence detection as a highly sensitive and quality control-friendly method for monitoring adeno-associated virus capsid protein purity. Hum Gene Ther 32:628–637. https://doi.org/10.1089/hum.2020.233

Zhang X, Wang W, Kenrick S n.d. AN5007: characterization of AAV-based viral vectors by DynaPro DLS/SLS instruments

Chapter 14
Methods and Practical Considerations in Imaging Viral Therapeutics

Martin Dass and Dongyue Xin

14.1 Introduction

Vectors for gene therapy can be divided into two groups: non-viral vectors and viral vectors. One big advantage of viral vectors is the defined size and morphology, which can be used for classification and characterization (Parvez 2020). Most common viruses in gene therapy are the Adeno-Associated Virus (AAV) and its recombinant Adeno-Associated viral vector (Dobnik et al. 2019) or Lenti-viral vectors which are already in clinical trials (Bulcha et al. 2021).

To successfully develop viral therapeutics, a broad range of analytical methods ranging from microbiological assays and particle analysis to chromatography and imaging should be combined to understand the morphology, antigenicity, infectivity, cell–receptor interaction, and behavior of viruses in cells or tissue. For example, it is important to discriminate between full/empty AAV capsids and between active/non-active viral particles for other viruses used in gene therapy like lentivirus or vesicular stomatitis viruses (Zhao et al. 2008) since only full and active vectors can deliver genomic material. This morphological information can be readily identified via electron microscopy, and therefore electron microscopy often supports the development of other analytical methods like size exclusion chromatography (SEC) or analytical ultracentrifugation (AUC) (Meng et al. 2022; Werle et al. 2021).

Electron microscopy techniques can provide high-resolution images of viruses and viral vectors. For many applications, 2D imaging either by negative staining or cryogenic transmission electron microscopy (TEM) (Subramanian et al. 2019) is

M. Dass (✉)
Boehringer Ingelheim Pharma GmbH & Co KG, Biberach an der Riß, Germany
e-mail: martin.dass@boehringer-ingelheim.com

D. Xin
Boehringer Ingelheim Pharmaceutical, Inc., Ridgefield, CT, USA

S. Gautam et al. (eds.), *Bioprocess and Analytics Development for Virus-based Advanced Therapeutics and Medicinal Products (ATMPs)*,
https://doi.org/10.1007/978-3-031-28489-2_14

sufficient. For 3D investigations using electron tomography, special equipment and powerful workstations for data processing are recommended. For instance, to investigate the virus structure, cryoTEM and cryo-electron tomography (cryoET) can be used as an alternative to X-ray crystallography especially when samples do not crystallize (Sevvana et al. 2021). All together cryoTEM and tomography are useful tools in biology, virology (Lee and Gui 2016; Luque and Castón 2020), drug discovery and lead optimization (Ceska et al. 2019). In this chapter we focus on imaging methods to visualize the virus particle morphology and interaction with cells or tissue. The state-of-the-art preparation methods as well as future perspectives of high-resolution imaging via electron microscopy are described.

14.2 Electron Microscopy: Theory and Instrumentation

When electron beam interacts with the specimen, elastic and inelastic scattering occur. Elastic scattering is essential for contrast in TEM images without energy loss while inelastic scattering occurs with energy loss which corelates to specimen material. High energy losses or scattering occurs with increasing atomic numbers and is used in techniques like EELS or EDX. An overview of interactions of the primary electron beam with matter is illustrated in Fig. 14.1. Image contrast is generated by the interaction of electrons with investigated samples and higher scattering and better image contrast can be achieved when high atomic number elements are present.

Electron beam is generated by different types of electron sources including tungsten filament, LaB6 and CeB6 monocrystals or field emission gun (FEG) (Schmidt 1994). Tungsten filaments are cheap and easy to change but has the shortest lifetime of around 50–100 h and generate the lowest resolution in electron microscopy imaging compared with the other electron sources. LaB6 and CeB6 monocrystals generate higher resolutions and lifetime is longer but a higher vacuum is required which ends up with additional ion getter pumps on the electron source. Highest resolution can be achieved with FEGs in which an external electric field was used to induce electron emission with thin beam geometry and better signal/noise ratio. Cold FEGs are running at room temperature whereas thermal FEGs utilize high electric field to reduce the potential barrier of electrons for thermionic emission. The emission current density increases and generate brighter images with increasing resolution in the order of tungsten filament, LaB6/CeB6 monocrystals and FEGs.

Generated electrons are controlled by an illumination system to transmit or interact with the sample. There are two common illuminations: spot/probe illumination and parallel broad illumination of the specimen (Williams and Carter 2009). For conventional bright field imaging parallel broad illumination is used whereas spot illumination is used in imaging modes like scanning transmission electron microscopy (STEM), high-angle annular dark-field imaging (HAADF) or electron energy loss spectroscopy (EELS), while the latter two techniques are more common in material science. Electron microscopy is performed in high vacuum to generate a

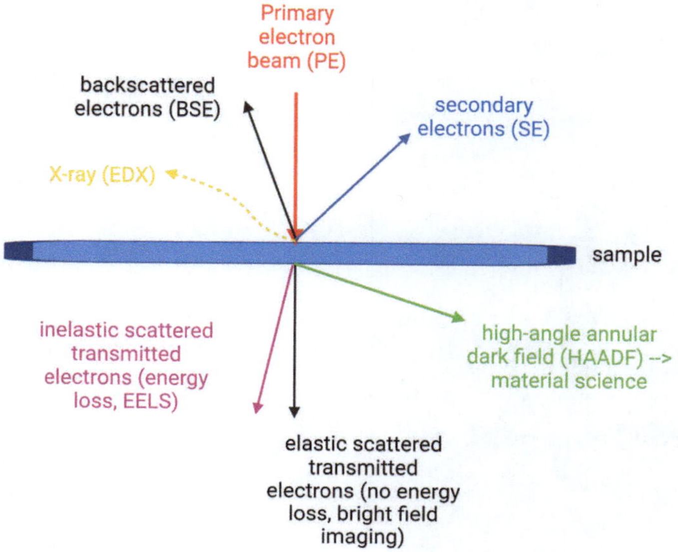

Fig. 14.1 Interaction of primary electron beam with matter/sample, generating different types of electrons for detection: secondary electrons (SE) for surface information in SEM, Backscattered electrons (BSE for material contrast information in SEM and X-rays for energy dispersive X-ray detection for elemental composition information. Transmitted electrons are elastic and inelastic scattered and generate different information Elastic scattered electrons are the main information for conventional bright field imaging, whereas inelastic scattered electrons contain information about the sample by defined loss of energy during scattering on the sample. If a high-angle scattering occurs special HAADF detectors are used in the microscopes

path with minimal deflection for electrons. There are two main types of electron microscope, transmission electron microscopes (TEM) (Fig. 14.2) and scanning electron microscopes (SEM) (Fig. 14.3).

In TEM, after electrons have transmitted the specimen the electron image/signal is collected by detectors (STEM or HAADF) or controlled by apertures and electromagnetic lenses to a fluorescence screen or CCD camera for brightfield imaging.

SEM is mainly developed for imaging of bulky samples (Fig. 14.3). The electron beam path is shorter and ends below the objective lens to a chamber with a stage for a sample holder. Around the sample different detectors are placed. For surface analysis mainly SE-, BSE-, EDX-detectors are placed on top of the sample and for detection of transmitted electrons a STEM detector are used below the sample. In a focused ion beam-scanning electron microscope (FIB-SEM) an additional column is attached to generate an ion beam to mill the sample to very thin TEM lamellas for further TEM investigations or for block face imaging as mentioned in Sect. 14.5.5.

Figure 14.4 summarizes options to run an electron microscope for specific applications. Microscopes are modular designed so depending on the scientific questions different detectors as well as cooling options for cryogenic applications can be attached. For morphological investigations of viruses or viral vectors

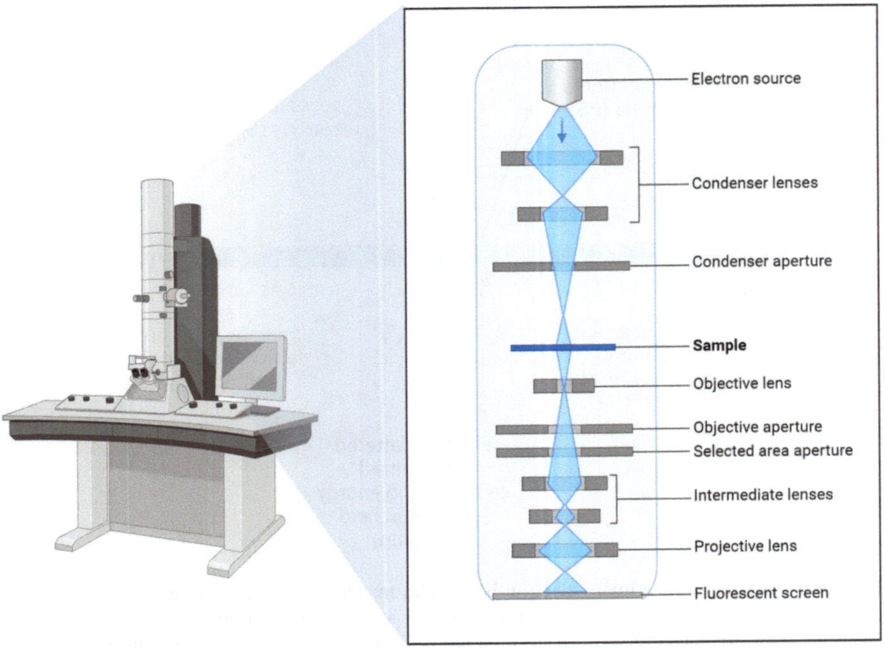

Fig. 14.2 Schematic illustration of the electron beam path in a transmission electron microscope (Biorender n.d.)

conventional bright field imaging on a TEM is a routine and the first-choice method for 2D imaging. Alternatively, a SEM with a TEM grid holder and STEM detector can be used. 3D imaging of viruses or viral vectors requires either a TEM equipped with a tilting stage for tomography or serial block face imaging via FIB-SEM or SEM with an integrated microtome or serial sectioning (see Sect. 14.5.5). Besides a well-equipped electron microscope, a good preparation of the sample is essential for high-quality images. Therefore, the next section will provide an insight into preparation techniques for electron microscopy.

14.3 Overview of Sample Preparation Techniques

To investigate viruses or viral vectors by electron microscopy a dedicated preparation procedure is required which stabilizes the sample for high vacuum condition and makes visualization by electron interaction with sample possible. Depending on the aim of investigation and condition of sample material, different workflows can be followed according to Fig. 14.5. To stabilize the sample for high vacuum they are dried or embedded in solid state of vitrified water (plunge freezing, high-pressure

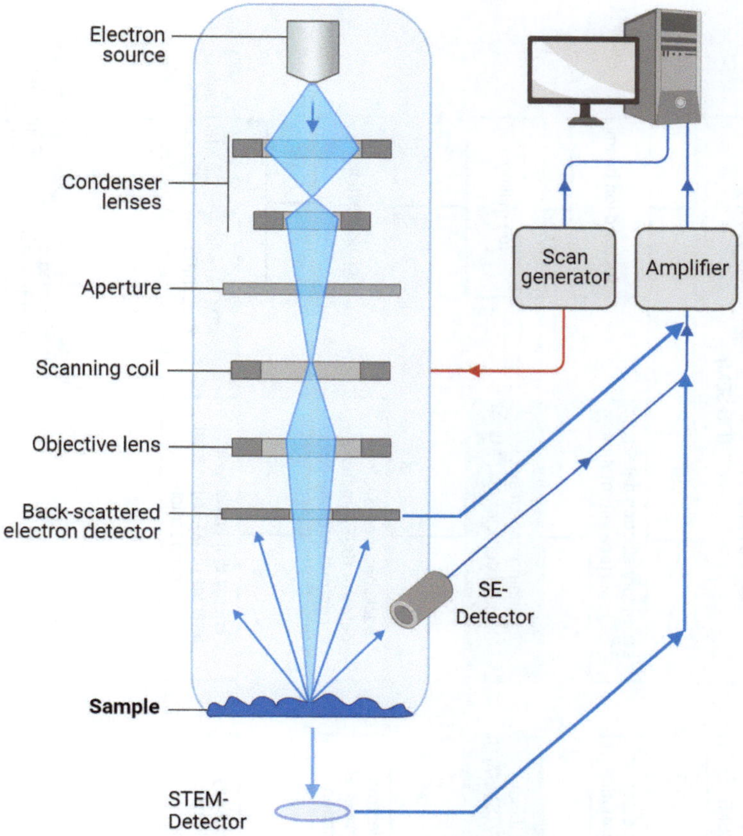

Fig. 14.3 Schematic illustration of the electron beam path in a scanning electron microscope (Biorender n.d.)

freezing and cryoEM of vitreous sections) or resin (embedding after chemical fixation or freeze substitution).

Negative staining with heavy metal salts performed at room temperature is historically and commonly used in the investigation of virus particles whereas chemical fixation, staining and embedding is used for biological samples like cells and tissues. Thin and small samples like viruses can be investigated directly after negative staining and drying whereas thicker samples like cells and tissue have to be thinned by sectioning. Contrary to tissue staining where it is important to stain specific structures of interest like membranes, in negative staining the staining solution is used as dark background surrounding structure of interest (Fig. 14.6).

During the last years cryo-electron microscopy (cryoEM) became popular as one big advantage of cryoEM is material can be investigated in native state without artifacts caused by drying and chemical fixation. This achievement of cryoEM was made possible by more sensitive detectors which allow the imaging of samples with

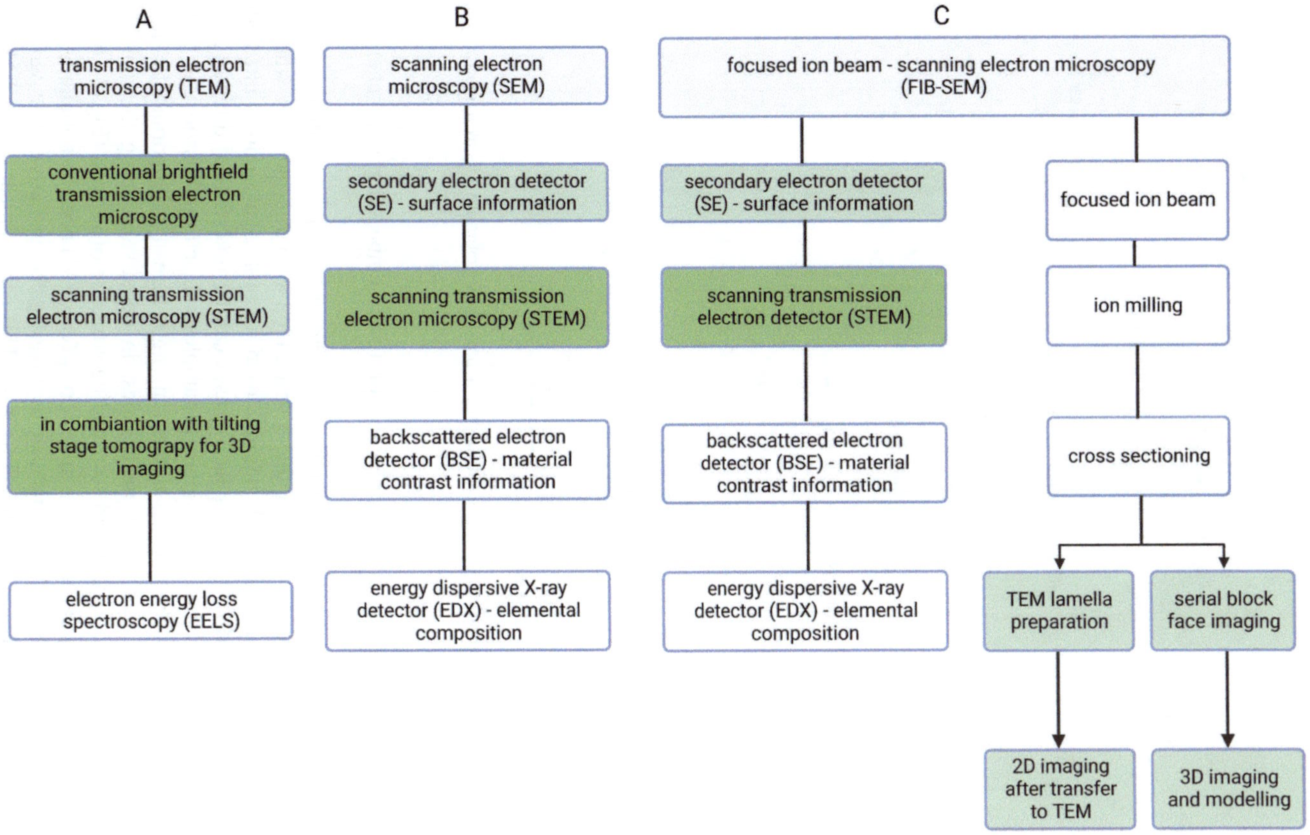

Fig. 14.4 Overview of the most common electron microscope types and corresponding detector options. (**a**) Transmission electron microscope, (**b**) scanning electron microscope and (**c**) focused ion beam-scanning electron microscope. The main applications for virus and viral vector diagnostics are highlighted in green. The gradient of light green and no color are options for investigations on viruses and viral vectors for specific questions

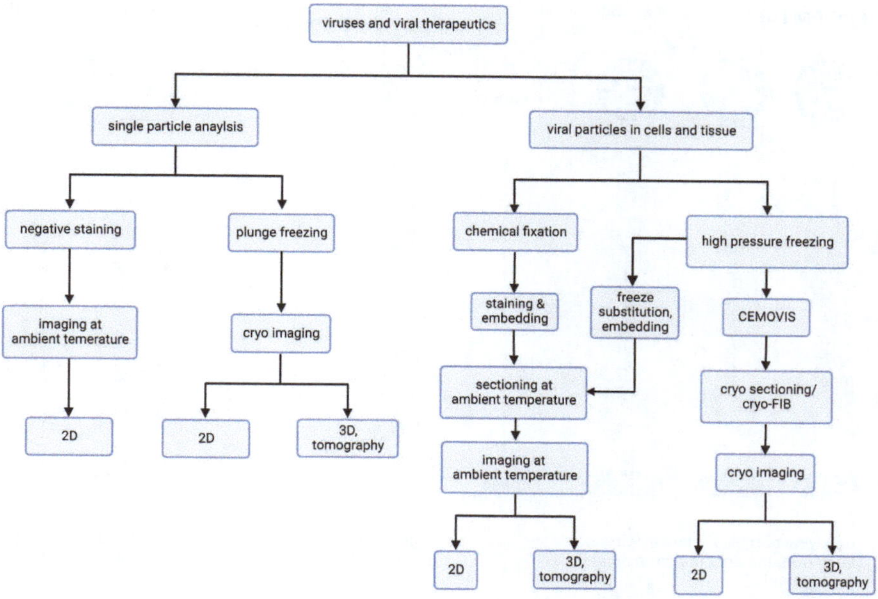

Fig. 14.5 Preparation procedures and workflows for electron microscopy investigation. Workflows are divided in workflows for room temperature conditions and deep frozen cryo conditions

low atomic numbers, especially biological samples which are composed of low atomic number elements like carbon, nitrogen, sulfur, and phosphorus. Unlike negative staining TEM, material is embedded in vitrified water, so the opposite contrast is observed in cryoTEM imaging (Fig. 14.6).

Next sections are focused on essential preparation techniques starting with negative staining as standard routine method in virus analysis followed by cryoTEM preparation for investigation of viral particles and essentials for the preparation of tissue or cells samples.

14.4 Analyzing Morphology of Viral Vectors

Viruses used as vectors are in a size range of around 20–300 nm which requires high-resolution imaging based on fluorescence (Robb 2022) or electron microscopy. Furthermore, the morphology and especially surface proteins may influence biodistribution and infectivity of viruses and viral vector (Vincent et al. 2021). Therefore, TEM which can reveal much of the needed information regarding virus structure is one of the main techniques for virus characterization (Roingeard et al. 2019; Richert-Pöggeler et al. 2019).

A minimum virus concentration of around 10^6 virions or higher is recommended. Lower concentrations of virus particles around 5×10^4 particles/mL require

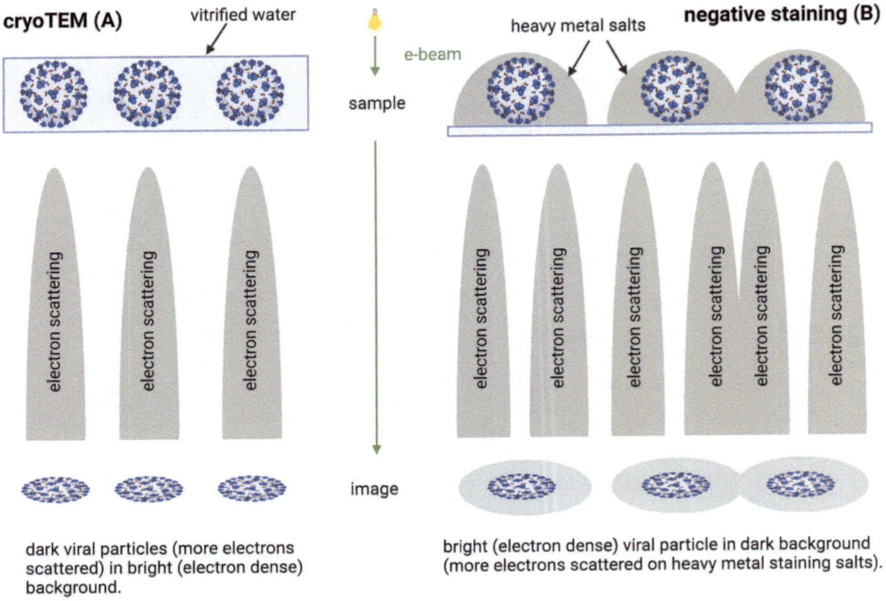

Fig. 14.6 Image generation by scattering of electrons induced by applied preparation method. Material which is embedded in vitrified water (cryoTEM) induces more scattering of electrons and appears dark in a bright background (crystalline ice induces also scattering of electrons and appears as homogenous dark). Biological material negatively stained with heavy metals appears bright in a dark background (Biorender n.d.)

additional ultracentrifugation step to concentrate virus material (Laue and Bannert 2010). Viral vectors need to be purified to remove cell debris or other contaminants via production process at pharmaceutical scale or using techniques such as sucrose cushion centrifugation at small scale (Kirchdoerfer et al. 2019). A sample volume of around 3 μL per TEM-grid is commonly used, and for several preparations and a general TEM investigation, a total sample volume of 25–50 μL is sufficient.

14.4.1 Negative Staining

For many years negative staining was and is still a standard technique for virus characterization. This method is fast and imaging can be performed with a conventional transmission electron microscope.

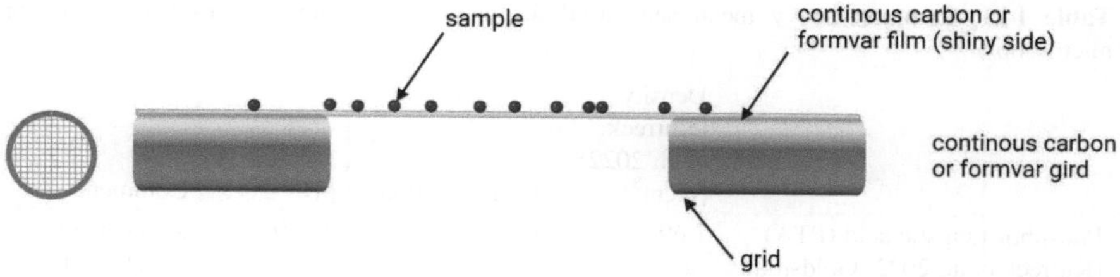

Fig. 14.7 TEM grid for negative staining with a continuous carbon or formvar film (Biorender n. d.)

14.4.1.1 Sample Preparation

For negative staining a carbon coated or formvar coated TEM-grid (Fig. 14.7) and staining solutions based on heavy metals are used. In general, stains with a very low granularity are preferred. Most common are uranyl acetate and phosphor tungsten acid.

The contrast by negative staining depends on the incubation time with the staining solution like uranyl acetate and longer incubation can stain inner structures as shown in the case of an orthopoxvirus (Laue 2010). Aqueous solution of uranyl acetate is acidic with a pH range of 2.0–5.0 depending on the concentration (0.5% to 3%). Phosphotungstic acid can be used as an alternative to uranyl acetate at a concentration range from 0.5% to 2%, while in many cases 0.5% is sufficient. For both uranyl acetate and phosphotungstic acid the granularity is very fine and pH range can be adjusted as needed for the structure to image, commonly pH 7.2–pH 7.6 for biological samples. Common heavy metal salts for negative staining are summarized in Table 14.1.

The preparation protocol for negative staining is quick and easy to perform. It contains the minimum following steps:

(1) Select the closed continuous carbon or formvar side (shiny side) of a continuous carbon/formvar-coated TEM grid.
(2) Glow discharge the coated TEM grid on the shiny side for a few seconds. Do not perform glow discharge too long as harsh conditions will also thin the carbon or formvar layer. Glow discharge makes the grid hydrophilic, and the sample drop will spread easily and material will attach more easily to the grid.
(3) Add around 3 μL sample suspension to the shiny side of the grid either by placing the drop on the grid (drop on grid) or putting the shiny side of the grid on the top of virus suspension (grid on drop).
(4) Remove the rest of suspension with a filter paper.
(5) Place the grid with side of attached particles on top of a drop of staining solution or add the staining solution to the grid for a few seconds.
(6) Remove the staining solution with a filter paper carefully from the side of the grid.

Table 14.1 Common heavy metal salts used for negative staining in transmission electron microscopy

	Density (Kurreck et al. 2022) [g/cm^3]	Concentrations	pH	Comments
Phosphor tungstic acid (PTA) (Kurreck et al. 2022; Goldsmith and Miller 2009; Mulisch and Welsch 2015)	1.69	0.5%–3%	4.0–9.0	Adjustment with 0.1 M KOH
Methylamin tungstic acid (MAT) (Kurreck et al. 2022; Fera et al. 2012)	3.88	2% aq.	3.0–10.0 Most common 6.0–7.5	Nonionic, no positive contrast
Uranyl acetate (UA) (Kurreck et al. 2022; Goldsmith and Miller 2009; Mulisch and Welsch 2015)	2.89	0.05%–3%	2.0–5.0	a. Dest or ethanol as solvent, fixative
Ammonium molybdate (Kurreck et al. 2022; Goldsmith and Miller 2009; Harris et al. 1999)	2.28	0.05%–5%	5.0–8.0	Good labeling of membranes and fibrillose proteins

(7) Dry the grid and store it in a TEM-grid box until imaging.
(8) Wash with water between steps 5 and 7 if needed.

Drop-on-grid and grid-on-drop methods should be compared during method development as surface tension on the top of sample drop may lead to higher particle concentration on the grid (Fig. 14.8). Additionally, staining, washing and different staining agents need to be optimized and alcian blue can be used to increase the adhesion of virus material to the grid (Laue and Bannert 2010; Goldsmith and Miller 2009). Inactivation of viruses may be necessary for safety reasons. Aldehydes like paraformaldehyde (PFA) or glutaraldehyde (GA) with concentrations typically between 1% and 4% are well-established fixatives in electron microscopy (Möller et al. 2015). The concentration and inactivation depend on the virus and should be tested by plaque assay or TCID50 assay. For example, vesicular stomatitis virus (VSV) can be inactivated with 0.05% glutaraldehyde or 0.5% paraformaldehyde (Zimmer et al. 2013).

A collection of negative stained viruses was summarized in the VirusExplorer DEM which is a database for diagnostic electron microscopy of viruses generated by the Robert-Koch-Institute (RKI) The VirusExplorer DEM Image Catalogue | Zenodo. AAV, VSV and Orf viruses are shown as examples in Fig. 14.9.

Negative staining is a fast and easy way to increase the contrast and can be further optimized for high throughput (Rames et al. 2014) or performed in combination with tomography (Fera et al. 2012) to optimize the visualization of structures of interest. For example, in investigating lipid-rich nanoparticles or liposomes the embedding in methylcellulose films combined with negative staining provides an alternative

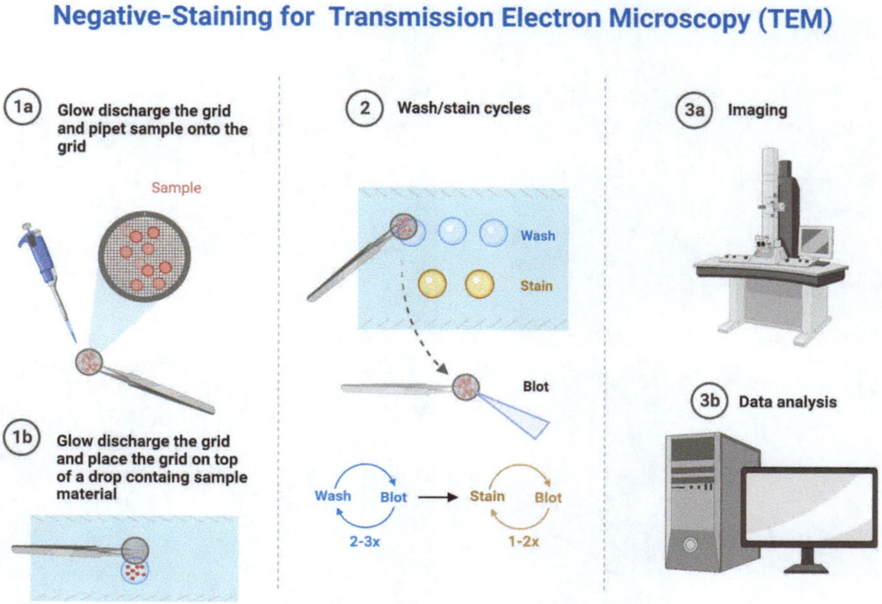

Fig. 14.8 Schematic procedure of negative staining for transmission electron microscopy (Biorender n.d.)

approach to cryoTEM (Asadi et al. 2017). Also, the combination of negative staining and cryoTEM can be performed (Harris et al. 1999). So negative staining has an essential role in the analysis of viruses or viral vectors.

14.4.2 cryoTEM Imaging: Plunge Freezing and High-Pressure Freezing

Since its inception, cryoEM has immensely influenced molecular imaging and this technique has won Jacques Dubochet, Joachim Frank, and Richard Henderson the Nobel Prize in chemistry in 2017 (Cressey and Callaway 2017; Shen 2018). CryoTEM requires dedicated equipment, preparation techniques, and workflows to achieve high image quality.

Vitrified ice is essential for high-quality imaging in cryoTEM (Adrian et al. 1984). Vitrification can be achieved with a fast-freezing process in which ice crystal formation is slower than the freezing process (Dubochet 2007). Therefore, two approaches are common. For thin samples of 50–300 nm like viruses plunge freezing is well established (Fig. 14.10). For samples in the micrometer range (1–500 μm) such as cell cultures high-pressure freezing is recommended (Studer et al. 1989). A high pressure of around 2000 bar prevents ice crystal formation during freezing with

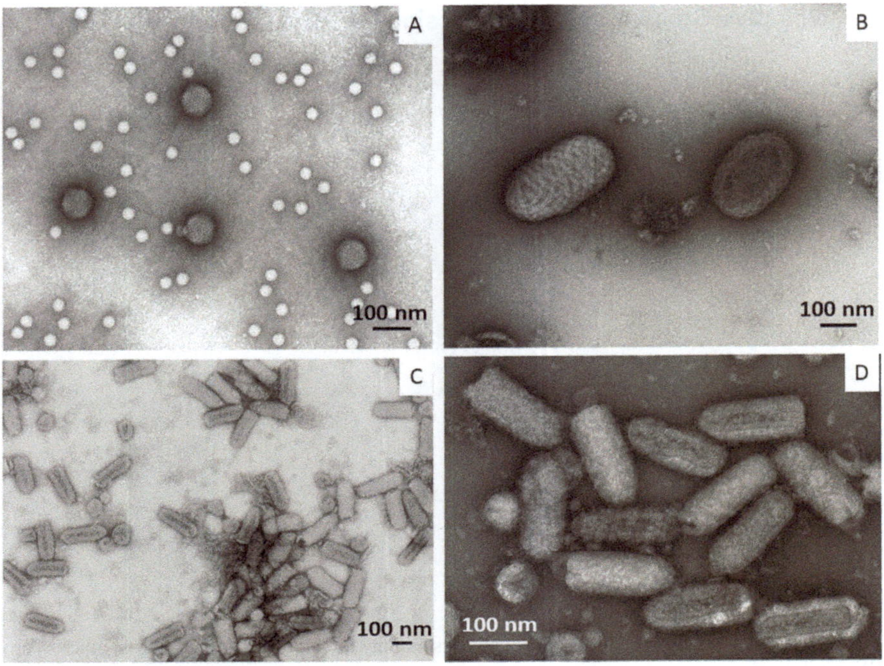

Fig. 14.9 Negative staining of viruses. (**a**) Human adenovirus 2 (larger icosahedrical adenovirus particle) and adeno-associated virus (smaller round AAV particles), staining: UA, (**b**) Orf virus, left = M-type, right = C-type, Staining MAT, PTA, UA, (**c**) vesicular stomatitis virus (VSV), Staining: UA, (**d**) vesicular stomatitis virus (VSV), staining: PTA

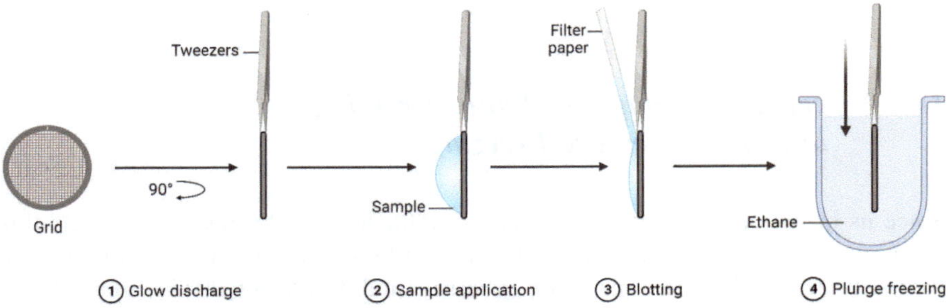

Fig. 14.10 Schematic plunge freezing procedure for cryo-electron microscopy of small samples in water (Biorender n.d.)

liquid nitrogen (Moor 1987; Studer et al. 2001). Both techniques require dedicated equipment and procedures.

A plunge freezer contains a humidity chamber to control the humidity around the specimen during preparation. Additionally, a temperature control in this chamber is recommended. The cryogen is placed in a liquid-nitrogen precooled container below

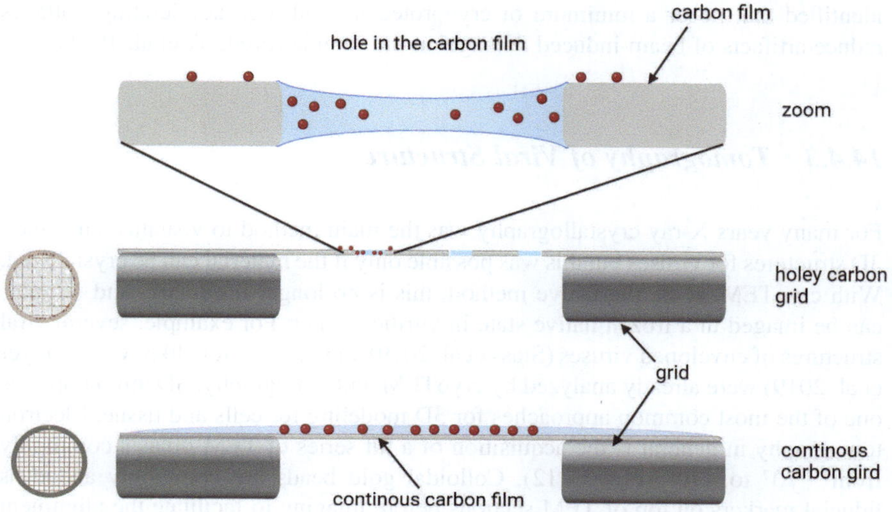

Fig. 14.11 Preparation options of grids for cryo-electron microscopy (schematic) (Biorender n.d.)

the humidity chamber and contacts the specimen for milliseconds. Liquid propane and ethane are most common instead of liquid nitrogen as they eliminate the Leidenfrost effect that a gas phase of the evaporating gas is shielding the specimen and prevent further cooling and freezing of the specimen (Stewart 2022).

In plunge freezing 3–6 μL of a suspension is added on a holey carbon grid, a continuous carbon grid, or graphene oxide substrate (Palovcak et al. 2018) (Fig. 14.11). After blotting to remove surplus liquid the grid is injected into the cryogen for plunge freezing (Fig. 14.10). To increase the number of viral particles per image, the operator can increase the sample volume, application cycles, incubation times and blotting procedure. For better contact of water-based suspensions with the grid a glow discharge procedure for around 30–60 s prior to grid preparation is recommended. Thickness of the vitrified film depends on the application side of the sample to the grid. Thinner vitrified films are generated by applying the sample on the carbon film (shiny side) in combination with glow discharge.

A self-contained cooling procedure from freezing of the sample until imaging in the cryoTEM is important for all cryoEM techniques. This means the frozen samples/grids need to be stored in liquid nitrogen and transferred in cold conditions to the precooled cryoTEM holder with specially designed containers and adaptors. Additionally, a closed container is recommended especially for longer preparation times to avoid ice formation from humidity condensed on cooled equipment and the liquid nitrogen.

Cryoprotectants are often not needed (Tuijtel et al. 2019; Vanhecke et al. 2010). Usage depends on application and can be considered in combination with other methods (Tyree et al. 2018). Cryoprotectants like glycerol, propylene glycol, or methanol, i.e., were tested in combination with chemical fixation of tissue and it was identified that no or a minimum of cryoprotectant and high accelerating voltages reduce artifacts of beam-induced damage on the sample (Frederik et al. 1991).

14.4.3 Tomography of Viral Structure

For many years X-ray crystallography was the main method to visualize and study 3D structures for viruses but this was possible only if the material can be crystallized. With cryoTEM as an alternative method, this is no longer necessary, and samples can be imaged in a frozen native state in vitrified water. For example, several viral structures of enveloped viruses (Stass et al. 2019) and viral vectors like AAV (Meyer et al. 2019) were already analyzed by cryoTEM and tomography. 3D tomography is one of the most common approaches for 3D modeling for cells and tissue. Electron tomography in general is the acquisition of a tilt series of TEM images commonly from −70° to +70° (Fig. 14.12). Colloidal gold beads are commonly added as fiducial markers on top of TEM sections before imaging to facilitate the alignment of image series to a 3D model.

It should be considered that the electron beam needs high power to penetrate a bigger volume related to increased tilting of the sample so a powerful TEM with 200 or 300 kV is recommended (Walther et al. 2018). Common slice thickness for tomography is around 500 nm (Vijayakrishnan et al. 2020) and in the last years slice thickness between 700 nm and 1 µm is also possible to image, ideally with 300 kV

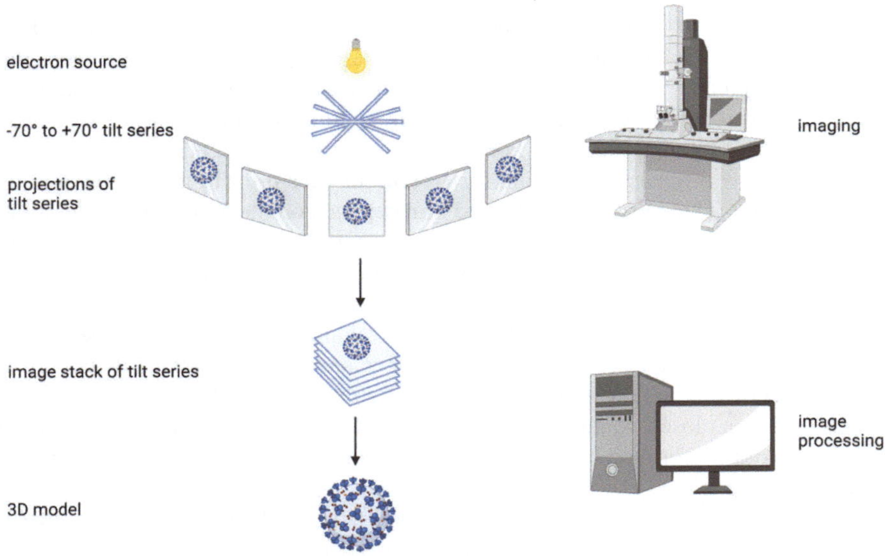

Fig. 14.12 Schematic procedure of 3D tomography in transmission electron microscopy (Biorender n.d.)

and chromatic aberration corrected microscopes. For cryoTEM tomography embedding medium is vitrified ice and fiducial markers have to be identified within the sample or should be added before freezing.

During the last years serial block face imaging and FIB-SEM was established as an alternative to tomography. This approach can be performed in combination with fluorescence microscopy following the correlative light and electron microscopy (CLEM) setup (Lucas et al. 2012). More details about CLEM are summarized in Sect. 14.5.6.

In the context of single particle analysis of viruses and viral vectors 3D EM tomography is the most instrumentation-demanding approach compared with cryoTEM and negative staining TEM and can generate a detailed understanding of the morphology of the virus and viral therapeutics.

14.5 Viral Vectors and Virus Particles in Cells and Tissue

Beside the single viral particle analysis and single particle reconstruction (SPR) (Lee and Gui 2016; Sigworth 2016), the interaction of virus and viral therapeutics with host cells and tissue is quite important to understand mechanisms of infections (Romero-Brey and Bartenschlager 2015) and biodistribution. Histopathological investigations on tissue or cells are common for highly virulent and fatal pathogens (Martines et al. 2015; McFadden 2005). The same approaches can be used in the development of targeted viral-based therapeutics to understand tissue and cellular tropism of viruses like AAV vector serotypes in adult retina (Hellström et al. 2009) or for treatment of Parkinson disease (Lewis et al. 2010). Both tissue and cells can be prepared either via classical chemical fixation or via physical fixation like the cryo workflow and high-pressure freezing.

14.5.1 Chemical Fixation and Embedding

Chemical fixation preparation can be performed with common equipment in the lab and imaging can be performed on a conventional transmission electron microscope. Fixation and staining can be performed at room temperature or at 4 °C using fixatives like 1–2% glutaraldehyde or 2–4% paraformaldehyde and staining solutions such as 0.5–2% osmium tetroxide and/or 0.5–2% uranyl acetate. Uranyl acetate binds to lipids and proteins and is recommended for staining of nucleic acids DNA and RNA because it binds to phosphate groups. This is the reason why it is commonly used together with osmium tetroxide which stains the membranes of tissue or cells. To further increase the contrast, the OTO method is worth mentioning. The OTO method increases the contrast of membranes by double staining with osmium tetroxide in combination with thiocarbohydrazide (TCH) (Seligman et al. 1966).

Table 14.2 Common embedding media for electron microscopy

Resin	Polymerization	Usage, Comments
Araldit (epoxy resin)	at 60 °C	– semithin and ultrathin sectioning, – investing of ultrastructure,
Epon™ (epoxy resin)	at 60 °C	– semithin and ultrathin sectioning, – investing of ultrastructure,
Spurr (epoxy resin)	at 60 °C	– semithin and ultrathin sectioning, – investing of ultrastructure.
LR White (acryl resin)	With accelerator at RT, without accelerator at 60 ° C or at 4 °C with UV, avoid oxygen during polymerization	– brittle, – infiltration of cells and tissue is challenging, – immunolabeling,
LR Gold (acryl resin)	With catalyst below 1 °C with light or UV, avoid oxygen during polymerization	– brittle, – immunolabeling,
Lowicryl® K4M (acryl resin)	With UV until −40 °C	– sectioning more difficult, – immunolabeling, – freeze substitution,
Lowicryl® HM20 (acryl resin)	With UV until −40 °C	– easy sectioning, – immunolabeling, – freeze substitution,
Unicryl™ (acryl resin)	With UV at low temperature or alternatively at 60 ° C	– brittle, – semithin and ultrathin sectioning, – investing of ultrastructure, – immunolabeling.

Alternatively, a post-embedding staining of sections with lead citrate can be performed to enhance the contrast (Reynolds 1963).

After staining the sample has to be dehydrated using alcohols like isopropanol or ethanol or propylene oxide because many of the embedding media are hydrophobic. Available embedding media are summarized in Table 14.2 (Mulisch and Welsch 2015). Most common resins for room temperature sectioning are epoxy resins (Araldit, Epon™, Spurr (Spurr 1969)) and acryl resins (LR-White, LR-Gold, Lowicryl® K4M, HM20, Unicryl) (Mulisch and Welsch 2015) and should be selected based on the aim of the analysis. For the study of viruses, the common standard is embedding in epoxy resins like Epon™ or Spurr (1969). These epoxy resins are easy to cut and sections of around 50 nm are common. In case of immune labeling lowicryls are more common, i.e., HM20 or K4M (Schwarz and Humbel 1989). Lowicryls can be hardened at low temperatures at around −20 °C with UV light but they are more brittle than epoxy resins during sectioning. This low-temperature embedding could be beneficial for immune labeling. Sections of resin-embedded cells or tissue can be used for conventional transmission electron microscopy and stored at room temperature for many years.

14.5.2 Cryo-embedding and Cryo-sectioning

Alternatively, samples can be embedded in vitrified water and cryo-sectioning can be performed. The complete workflow from freezing to cryo-sectioning and cryoTEM is called CEMOVIS and requires cooled condition during the complete process (Al-Amoudi et al. 2004). For immune labeling cryo workflows are recommended (Möbius 2009). In electron microscopy the Tokuyasu preparation protocol based on sucrose is common and used in several applications (Möbius and Posthuma 2018). The low contrast in cryo-sections can be adjusted by using staining solutions known from conventional electron microscopy (Tokuyasu 1978). Alternatively new detectors and cameras became more sensitive within the last years and imaging with low contrast became possible. In the cryogenic preparation workflow, the sample volume has to be considered for fast freezing process. Highest freezing rates for larger samples are achieved with high-pressure freezing.

14.5.3 High-Pressure Freezing

For investigation of fast biological processes chemical fixation should be replaced by a physical fixation method like high-pressure freezing. A detailed and well-preserved ultrastructure can be achieved. This has an advantage for the investigation of interactions of viral therapeutics with cells or tissue. Due to the larger sample volume of cells and tissue in the cryogenic preparation, plunge freezing has limited success and high-pressure freezing should be the method of choice (McDonald and Auer 2006). High-pressure freezing preserves the state of biological process within milliseconds, significantly faster compared to well-established chemical fixation where samples are fixed within 15–60 min. High-pressure freezing allows fast freezing rates and freezing of sample volumes within the μm range and it can be performed with biopsies, adherent cells, or with suspension cells.

Adherent cells can be cultivated on carbon-coated sapphire discs. The carbon layer between sapphire disc and adherent cells acts as separating layer after embedding. The cell layer on the carbon-coated sapphire disc will be frozen in a dedicated sample holder in the high-pressure freezer within milliseconds at around 2000 bar. To keep the sample safe, a sandwich of two aluminum plates around the sapphire disc covers the cells. Alternatively, new assembly of two sapphire discs with a spacer is used as well (Walther et al. 2018; Monaghan et al. 2003). 1-Hexadecene is commonly used as pressure mediator (Studer et al. 1989). The frozen sample can be stored in liquid nitrogen until further preparation.

Suspension cells can be absorbed into cellulose capillaries and high-pressure frozen (Hohenberg et al. 1994). Short segments of the capillary can be placed in the aluminum plates or dedicated sample holders can be used. Cellulose capillaries are not sealed and are difficult to handle due to their transparency in liquid. Alternatively, sealable copper capillaries are available for freezing and sectioning (Yakovlev and Downing 2011).

14.5.4 Freeze Substitution and Embedding

Freeze substitution is a preparation method that combines the advantage of good preservation of ultrastructure achieved by high-pressure freezing with advantage of room temperature sectioning and storage of plastic-embedded sections. It is important to keep temperatures constantly at controlled cool conditions until samples are transferred in the freeze substitution equipment and further processed. Commonly samples are stored in liquid nitrogen and transferred rapidly into a container in the freeze substitution chamber filled with precooled freeze substitution media. Commonly used substitution media are alcohols and acetone which are liquid at a low temperature of around −90 °C. After the transfer of the frozen samples into −90 °C freeze substitution media the temperature will be increased following a defined temperature ramp. A common freeze substitution protocol takes overnight or 24 h where 2/3 of the freeze substitution is performed below −40 °C and then the samples warm up until room temperature within a few hours. The water content of around 5% in the freeze substation media can have a positive impact and increase the contrast of membranes (Walther and Ziegler 2002).

Fixatives or heavy metals like osmium tetroxide or uranyl acetate that can react at low temperatures can be added to fix and/or stain the ultrastructure. Without additional fixatives and staining agents, the preservation of the ultrastructure is poor. On the other hand, fixatives and staining agents may influence antigenicity. So staining salts should be selected by the aim of the experiment and reduced or avoided in experiments for immunolabeling. Samples can be further embedded in resins following embedding procedures described previously in Sect. 14.5.1. For example, Giddings et al. described a protocol to increase the contrast of membranes and generate samples suitable for immunolabeling if fixatives like 0.25–3% glutaraldehyde and staining agents like 2% osmium tetroxide in combination with 0.1% tannic acid are used (Giddings 2003). For immunolabeling lowicryls like HM20 or K4M are recommended as an embedding medium (Gang et al. 2017; Schwarz and Humbel 1989). Lowicryls can be polymerized at −20 °C with UV light which can preserve the antigenicity. A common freeze substitution protocol for the investigation of ultrastructure is listed in Table 14.3 and one for immunolabeling (Mulisch and Welsch 2015) in Table 14.4.

Table 14.3 Freeze substitution procedure with final embedding in Epon™

solvent	Process step	Fixative/Staining	Duration [h]	Temperature
Acetone	Freeze substitution, staining, fixation	0.1% Uranyl acetate +0.2% Osmium tetroxide +5% water	Overnight, around 12–24 h	−90 °C to +20 °C
Epon™: Acetone 1:2	Substitution of solvent to embedding media		1 h	+20 °C
Epon™: Acetone 1:1			1–2 h	+20 °C
Epon™: Acetone 2:1			1–2 h	+20 °C
Epon™			2 h	+20 °C
Epon™			5 h	+ 20°C
Epon™			5 h	+20 °C
Epon™	Embedding, Polymerization		Up to 72 h	+ 60 °C

Table 14.4 Freeze substitution procedure in HM20 for immunolabeling

Solvent	Process step	Fixative/ Staining	Duration [h]	Temperature
Acetone	Freeze substitution, staining, fixation	0.5% Uranyl acetate	24 h	−90 °C to −50 °C
HM20:Acetone 1:2	Substitution of solvent to embedding media		1–2 h	−50 °C
HM20:Acetone 1:1			1–2 h	
HM20:Acetone 2:1			1–2 h	
HM20			2 h	
HM20			12 h	
HM20			2 h	
HM20	UV-Polymerization		24 h	

14.5.5 Sectioning

After embedding of cells or tissue the samples are sectioned for electron microscopy investigations. The aim of sectioning is to thin the sample so that electrons can pass the specimen. Therefore, samples need to be embedded in hard resins or vitrified ice.

In electron microscopy, ultramicrotomes are used for sectioning at room temperature (Noda 2017) or with a cryo chamber for cryo-sectioning (Chlanda and Sachse 2013). In a standard process of sectioning resin embedded tissue or cells at room temperature, a two-step approach is performed. First semithin sections with maximum 1 × 1 mm and around 300–1000 nm thick are cut for preselection of specific regions of interest like inflammation or infections. They can be collected single or as ribbon (Blumer et al. 2002) on glass slides, wafers or similar and can be stained by light microscopy stainings used in histology like hematoxylin-eosin, toluidine blue or alcian blue (Litwin and Kasprzyk 1980; Craighead 1979; Eyden 2012; Burns 1978; Berezovskiĭ 1978). After the selection of a region in the semithin section on a conventional light microscope only ¼ of the semithin section area is used for ultrathin sectioning with around 50 nm thickness in TEM. To minimize cutting artifacts (Al-Amoudi et al. 2005) by compression of the section, oscillating diamond knives were developed and can be used instead of conventional diamond knives (Al-Amoudi et al. 2003). Sections are collected on TEM grids with different mesh numbers. Also, TEM grids without a further support film, i.e. 300 mesh grids or smaller field of view can be used to achieve a brighter image with more details. Sections on the grid can be stored over years.

In comparison, cryo-sectioning is performed dry with cryo diamond knives. For an easier transfer of sections, micromanipulators and ionizators are useful optional equipment (Ladinsky et al. 2006; Studer et al. 2014). The correct temperature and equilibrating time for the sample to reach the optimal temperature for cryo-sectioning in the cryo chamber are important factors. To keep the vitreous state of water, temperature should be below −130 °C which makes handling more challenging (Al-Amoudi et al. 2005).

For investigations on interactions of viral therapeutics with tissue or cells, serial sectioning, automated tape collecting ultra-microtome scanning electron microscopy (ATUM-SEM) or block face imaging could be an option. Serial sectioning was developed to collect serial sections on a tape or wafer and image them in scanning electron microscope (SEM) known as array tomography (Peddie et al. 2022). To generate a continuous ribbon of sections trapezoid trimmed blocks are essential. Alternatively, block face imaging can be performed to serial sectioning either by an integrated ultra-microtome in the SEM or by using a focused ion beam-scanning electron microscope (FIB-SEM). The integrated microtome uses a moving diamond knife and achieves a similar slice thickness as on conventional microtomes for TEM sectioning. The FIB-SEM is able to alter the slice thickness by adjusting the power and current of the focused ion beam. So thin and precise sections can be generated but a well-trained operator is required. Such a workflow is also possible under cryo conditions by combining cryo-microtome and a cryo-SEM to do block face imaging on frozen tissue or cells (Walther and Müller 1999). 3D models can be produced by image processing of the image series acquired with the sections.

14.5.6 Correlative Microscopy, CLEM

A technique combining the advantage of fluorescence microscopy and electron microscopy is called correlative microscopy or correlative light and electron microscopy (CLEM). The advantages of CLEM when used for investigations of virus–host interactions are summarized in a previous review by Bykov et al. (2016).

In CLEM, images can be acquired on light microscope (LM) first, followed by embedding procedure for TEM. One advantage of starting at light microscope before embedding is a broad range of fluorescence labels which are commercially available. The number of immunogold labels for EM has increased and can be custom designed on request but the number is still low compared with LM labels. On the other hand, there could be an offset between the LM and EM images, caused by the EM preparation after LM imaging. In cases that stable labels and dyes can be identified for both imaging techniques, EM preparation can be performed first, followed by LM and then EM imaging. A conjugated label containing one fluorophore for fluorescence microscopy and one nanogold label for EM analysis could be an option but the conjugate should be able to penetrate to region of interest as larger fluorophores or gold nanoparticles (commonly 3–10 nm) can limit this penetration.

One challenge of correlative microscopy is the accurate identification of the same region in both microscopes, especially on the EM where structures have to be identified at higher magnifications. Prelabelled sample holders were developed to recognize the same region precisely. This can be achieved by either direct imaging of cells grown on TEM finder grids or on prelabeled sapphire discs with a self-made or commercially available imprinted coordinate system.

Gold grids need to be used instead of copper grids due to the toxicity of copper in cell culture. Cells can grow directly on the gold grids with formvar film coated with carbon as carbon film alone can be easily destroyed in the liquid cell culture media. Additionally, the orientation of the grid should be tested. Kommnick et al. proposed to place the grid with carbon-formvar film facing downwards so that cells can grow on different heights and leading to an easier focusing during imaging (Kommnick et al. 2019). Additionally, this approach is recommended for thin cells and in combination with a powerful cryoTEM where the electron beam is able to penetrate the cell volume.

Correlative microscopy on resin sections at room temperature enable sample storage but this is limited by stable correlative labels and dyes. So several similar labels have to be tested to identify one label which is suitable for the complete workflow. Additionally, bleaching can occur during imaging or after storage. For storage, it is recommended to store samples in the same condition as labels and dyes, quite often dark and at 4–8 °C. The combination of correlative microscopy and immunolabeling on sections is possible and was summarized by H. Schwarz and B. H. Humbel (2013).

A post-embedding labeling on the TEM sections could be performed but the preparation protocol for electron microscopy should be tested for individual labels carefully. The fixatives and staining agents may hinder the binding of antibodies for immunolabeling due to their chemical interactions with proteins and potential inactivation of the epitopes. Different concentrations for each added chemical should be tested to ensure the binding of labeled antibodies. CryoTEM has the advantage of not requiring additional fixatives or staining agents.

CLEM technique can also be performed under cryogenic conditions. As one example, Hampton et al. described a protocol to perform correlative fluorescence microscopy and cryo-electron tomography of virus-infected or transfected mammalian cells (Hampton et al. 2017). Gold NH2 London finder grids with Quantifoil holey carbon film grids were used in cell culture. The cryo preparation was performed via plunge freezing followed by cryo fluorescence light microscopy and cryoTEM imaging. For cryo fluorescence microscopy a low laser power is recommended to image vitrified sections, like the cryogenic super-resolution optical fluctuation imaging (cryo-SOFI) (Moser et al. 2019).

A correlative approach combining light microscopy and scanning electron microscopy is also possible for the investigation of cellular surface. Larger sample holders are available to combine light and scanning electron microscopy because of larger sample stages in the SEM by using glass slides or cover slips. SEM providers offer shuttle systems and software tools for recognition of the positions on each microscope.

14.5.7 In Vivo Imaging

TEM is used to image the viral vectors or ex vivo in cell-based experiments. For non-invasive in vivo imaging radio labels can be used to image the gene expression (Nichol and Kim 2001; Vassaux and Groot-Wassink 2003). Most common techniques are magnetic resonance imaging (MRI), positron emission tomography (PET), and single-photon emission computed tomography (SPECT). This field of imaging is linked to nuclear medicine using radiotracer (PET) or gamma rays (SPECT). Common radiotracers or radioactive labels are radioactive isotopes of hydrogen, carbon, phosphorus, sulfur, and iodine.

Overall, the investigation of viruses and viral therapeutics interacting with cells and tissue requires a different preparation workflow compared with single particle analysis because of the larger volume. Imaging technologies can provide a deeper insight into the biology of viruses and viral therapeutics in interaction with cells and tissue and their biodistribution.

14.6 Imaging, Data Acquisition, and Data Processing

Electrons are collected with different detectors as mentioned in Sect. 14.2. Images are acquired with a charged coupled device (CCD) camera in which a single YAG crystal or phosphor powder scintillator is illuminated by the electron beam and coupled with fiber or lens optics with the CCD chip (Meyer and Kirkland 2000). Cameras with more square pixels have limitations when imaging is performed with accelerating voltages of more than 100 kV (Tietz et al. 2008). To overcome this limit at high accelerating voltages, a decelerating of electrons in the camera is an option

(Downing and Mooney 2008). When EM images are acquired in 2D mode, there is no issue with the dataset size but when large data sets are generated for 3D modeling, gigabyte or terabyte of data has to be handled and stored (Peddie et al. 2022; Guerin and Lippens 2021), especially in the field of volume electron microscopy or correlative microscopy (Peddie et al. 2022; Guerin and Lippens 2021).

Data acquisition is performed with a series of microscope parameters like beam current and voltage, field of view (FOV), scalebars, pixels size, etc., and these parameters are typically stored in the header of the image. For tomography or array tomography additional software is needed to control the movement of the sample holder (TEM) or stage (SEM) accurately. Microscope suppliers offer software packages for their tomography systems or freeware can be used (Peddie et al. 2022). Acquired image series are aligned to generate an image volume as mentioned in the previous tomography section. Plugins like TrakEM2 in ImageJ/Fiji can also be used for image analysis. Segmentation to highlight structures of interest can be performed manually or semiautomated by software solutions such as ilastik, IMOD, Amira, Imaris or Fiji-Plugins like reconstruct or TrakEM2 (Peddie et al. 2022). Manual segmentation is more time-consuming and therefore semiautomated or automated approaches are preferred. Automated segmentation is mainly based on machine learning and tools like DeepImageJ, ilastik neuronal network, deepMIB, CDeep3EM or Apeer, and others are available (Peddie et al. 2022). After segmentation, data can be visualized, and 3D models can be generated. Because of the large data volume, powerful working stations are recommended and software tools like IMOD and MIB or Fiji plugins like Bigdataviewer or MoBIE can be used (Peddie et al. 2022). In summary, 2D images are easy to acquire in high image quality while 3D data sets require more software solutions to handle large data sets. Both approaches can support the understanding of viruses and viral vectors either for single viral particles investigation or by generating 3D models of viruses or viral vectors in interaction with cells and tissue.

14.7 Future of Imaging Methods: Automatization and High Throughput

Imaging methods are essential for the understanding of morphology and interaction with cells, tissue, and hosts in the development of viral therapeutics. In the last 10 years, imaging and preparation methods have been significantly improved. With high accelerating voltages and correctors on electron microscopes and new software, higher resolutions up to 2 Å were achieved (Namba and Makino 2022; Wang 2015).

Besides resolution enhancement, another focus in development is the automatization of processes. Automatization allows higher throughput and standardization during preparation. Higher throughput can be achieved with a rapid transfer system for CLEM (Verkade 2008), by optimizing established preparation protocols for cryo lamella preparation with FIB-SEM (Buckley et al. 2020) or via automated tape collection microtome (ATUM) for serial sectioning (Baena et al. 2019). In addition, advances have been made in the automation of imaging and data collection and this will be an active area for future development (Kim et al. 2021). New microscopes and software have enabled automated collection of large data sets of images in high quality.

Data processing for large data sets is complicated with several software tools and powerful hardware involved (Baldwin et al. 2018). For example, when studying the post-fusion structure of the SARS-CoV-2 spike glycoprotein (Fan et al. 2020), data processing starts with motion correction (i.e., motionCor2) contrast adjustment (CTFFind4), particle picking and 2D classification, generation of initial density map (i.e., Relion), 3D classification of particles, auto-refinement of the final density map and finally calculation of local resolution of the final density map (ResMap). One emerging area in automating data processing is the application of machine learning algorithms especially convolutional neuronal network (Li et al. 2020; Maruthi et al. 2020) in data analysis.

Due to the high cost, high-end microscopes are often placed in EM clusters or EM centers and support many researchers and groups in parallel. On the other hand, benchtop systems make it possible to perform electron microscopy investigations directly in laboratories as a complementary technique to other analytical methods. In the development of viruses or viral vectors as therapeutics, characterization with a combination of techniques such as mass spectrometry (Banerjee et al. 2021), Raman spectroscopy and microscopy and light and fluorescence microscopy is desirable. This demand leads to the development of specialty instruments that can produce information from multiple techniques simultaneously, such as integrated microscopes cryo-FLM in SEM, Raman at SEM or integrated preparation (i.e., sectioning) into the microscope (integrated microtomes in SEM) or FIB. A range of equipment is now available and will be more diversified in the coming years.

References

Adrian M, Dubochet J, Lepault J, McDowall AW (1984) Cryo-electron microscopy of viruses. Nature 308:32–36

Al-Amoudi A, Dubochet J, Gnaegi H, Lüthi W, Studer D (2003) An oscillating cryo-knife reduces cutting-induced deformation of vitreous ultrathin sections. J Microsc 212:26–33

Al-Amoudi A et al (2004) Cryo-electron microscopy of vitreous sections. EMBO J 23:3583–3588

Al-Amoudi A, Studer D, Dubochet J (2005) Cutting artefacts and cutting process in vitreous sections for cryo-electron microscopy. J Struct Biol 150:109–121

Asadi J et al (2017) Enhanced imaging of lipid rich nanoparticles embedded in methylcellulose films for transmission electron microscopy using mixtures of heavy metals. Micron 99:40–48

Baena V, Schalek RL, Lichtman JW, Terasaki M (2019) Chapter 3 Serial-section electron microscopy using automated tape-collecting ultramicrotome (ATUM). Methods Cell Biol 152:41–67

Baldwin PR et al (2018) Big data in cryoEM: automated collection, processing and accessibility of EM data. Curr Opin Microbiol 43:1–8

Banerjee A, Bhakta S, Sengupta J (2021) Integrative approaches in cryogenic electron microscopy: recent advances in structural biology and future perspectives. Iscience 24:102044

Berezovskiĭ ME (1978) Method of staining of semi-thin sections with hematoxylin-eosin. Ark Patol 40:69–70
Biorender (n.d.) Imaging tool
Blumer MJF, Gahleitner P, Narzt T, Handl C, Ruthensteiner B (2002) Ribbons of semithin sections: an advanced method with a new type of diamond knife. J Neurosci Methods 120:11–16
Buckley G et al (2020) Automated cryo-lamella preparation for high-throughput in-situ structural biology. J Struct Biol 210:107488
Bulcha JT, Wang Y, Ma H, Tai PWL, Gao G (2021) Viral vector platforms within the gene therapy landscape. Signal Transduct Target Ther 6:53
Burns. (1978) Thick sections: technique and applications. Wiley, New York. https://doi.org/10.1002/micr.1920010115
Bykov YS, Cortese M, Briggs JAG, Bartenschlager R (2016) Correlative light and electron microscopy methods for the study of virus–cell interactions. FEBS Lett 590:1877–1895
Ceska T, Chung C-W, Cooke R, Phillips C, Williams PA (2019) Cryo-EM in drug discovery. Biochem Soc Trans 47:281–293
Chlanda P, Sachse M (2013) Electron microscopy. Methods Mol Biol 1117:193–214
Craighead JE (1979) Diagnostic electron microscopy, Volume I, Edited by Benjjamin F. Trump and Rymond T. Jones, 346 pp, Illus, Wiley, Somerset, NJ, 1978. $29.50. Microsurgery 1, 87–88
Cressey D, Callaway E (2017) Cryo-electron microscopy wins chemistry Nobel. Nature 550:167–167
Dobnik D et al (2019) Accurate quantification and characterization of adeno-associated viral vectors. Front Microbiol 10:1570
Downing KH, Mooney PE (2008) A charge coupled device camera with electron decelerator for intermediate voltage electron microscopy. Rev Sci Instrum 79:043702
Dubochet J (2007) The physics of rapid cooling and its implications for cryoimmobilization of cells. Methods Cell Biol 79:7–21
Eyden B (2012) Diagnostic electron microscopy – a practical guide to interpretation and technique. 153–180. https://doi.org/10.1002/9781118452813.ch5
Fan X, Cao D, Kong L, Zhang X (2020) Cryo-EM analysis of the post-fusion structure of the SARS-CoV spike glycoprotein. Nat Commun 11:3618
Fera A, Farrington JE, Zimmerberg J, Reese TS (2012) A negative stain for electron microscopic tomography. Microsc Microanal 18:331–335
Frederik PM, Bomas PHH, Stuart MCA (1991) The ultrastructure of cryo-sections and intact vitrified cells—the effects of cryoprotectants and acceleration voltage on beam induced bubbling. Scanning Microsc Suppl 5:S43–S52
Gang Y et al (2017) Plastic embedding immunolabeled large-volume samples for three-dimensional high-resolution imaging. Biomed Opt Express 8:3583
Giddings TH (2003) Freeze-substitution protocols for improved visualization of membranes in high-pressure frozen samples. J Microsc 212:53–61
Goldsmith CS, Miller SE (2009) Modern uses of electron microscopy for detection of viruses. Clin Microbiol Rev 22:552–563
Guerin CJ, Lippens S (2021) Correlative light and volume electron microscopy (vCLEM): How community participation can advance developing technologies. J Microsc 284:97–102
Hampton CM et al (2017) Correlated fluorescence microscopy and cryo-electron tomography of virus-infected or transfected mammalian cells. Nat Protoc 12:150–167
Harris JR et al (1999) Application of the negative staining technique to both aqueous and organic solvent solutions of polymer particles. Micron 30:289–298
Hellström M et al (2009) Cellular tropism and transduction properties of seven adeno-associated viral vector serotypes in adult retina after intravitreal injection. Gene Ther 16:521–532
Hohenberg H, Mannweiler K, Müller M (1994) High-pressure freezing of cell suspensions in cellulose capillary tubes. J Microsc 175:34–43
Kim PT, Noble AJ, Cheng A, Bepler T (2021) Learning to automate cryo-electron microscopy data collection with Ptolemy. Arxiv

Kirchdoerfer RN, Saphire EO, Ward AB (2019) Cryo-EM structure of the Ebola virus nucleoprotein–RNA complex. Acta Crystallogr Sect F 75:340–347

Kommnick C, Lepper A, Hensel M (2019) Correlative light and scanning electron microscopy (CLSEM) for analysis of bacterial infection of polarized epithelial cells. Sci Rep UK 9:17079

Kurreck J, Engels JW, Lottspeich F (2022) Bioanalytik. https://doi.org/10.1007/978-3-662-61707-6

Ladinsky MS, Pierson JM, McIntosh JR (2006) Vitreous cryo-sectioning of cells facilitated by a micromanipulator. J Microsc 224:129–134

Laue M (2010) Chapter 1 electron microscopy of viruses. Methods Cell Biol 96:1–20

Laue M, Bannert N (2010) Detection limit of negative staining electron microscopy for the diagnosis of bioterrorism-related micro-organisms. J Appl Microbiol 109:1159–1168

Lee KK, Gui L (2016) Dissecting virus infectious cycles by cryo-electron microscopy. PLoS Pathog 12:e1005625

Lewis TB, Glasgow JN, Glandon AM, Curiel DT, Standaert DG (2010) Transduction of brain dopamine neurons by adenoviral vectors is modulated by CAR expression: rationale for tropism modified vectors in PD gene therapy. PLoS One 5:e12672

Li Y, Cash JN, Tesmer JJG, Cianfrocco MA (2020) High-throughput cryo-EM enabled by user-free preprocessing routines. Structure 28:858–869.e3

Litwin JA, Kasprzyk JM (1980) Alcian blue staining of semithin Epon sections for light microscopy. Acta Histochem 67:265–271

Lucas MS, Günthert M, Gasser P, Lucas F, Wepf R (2012) Chapter 17 Bridging microscopes 3D correlative light and scanning electron microscopy of complex biological structures. Methods Cell Biol 111:325–356

Luque D, Castón JR (2020) Cryo-electron microscopy for the study of virus assembly. Nat Chem Biol 16:231–239

Martines RB, Ng DL, Greer PW, Rollin PE, Zaki SR (2015) Tissue and cellular tropism, pathology and pathogenesis of Ebola and Marburg viruses. J Pathol 235:153–174

Maruthi K, Kopylov M, Carragher B (2020) Automating decision making in the cryo-EM pre-processing pipeline. Structure 28:727–729

McDonald KL, Auer M (2006) High-pressure freezing, cellular tomography, and structural cell biology. BioTechniques 41:137–143

McFadden G (2005) Poxvirus tropism. Nat Rev Microbiol 3:201–213

Meng H et al (2022) Size exclusion chromatography with dual wavelength detection as a sensitive and accurate method for determining the empty and full capsids of recombinant adeno-associated viral vectors. Hum Gene Ther 33:202–212

Meyer RR, Kirkland AI (2000) Characterisation of the signal and noise transfer of CCD cameras for electron detection. Microsc Res Tech 49:269–280

Meyer NL et al (2019) Structure of the gene therapy vector, adeno-associated virus with its cell receptor. AAVR Elife 8:e44707

Möbius W (2009) Cryopreparation of biological specimens for immunoelectron microscopy. Ann Anat Anatomischer Anzeiger 191:231–247

Möbius W, Posthuma G (2018) Sugar and ice: immunoelectron microscopy using cryosections according to the Tokuyasu Method. Tissue Cell 57:90–102

Möller L et al (2015) Evaluation of Virus inactivation by formaldehyde to enhance biosafety of diagnostic electron microscopy. Viruses 7:666–679

Monaghan P, Cook H, Hawes P, Simpson J, Tomley F (2003) High-pressure freezing in the study of animal pathogens. J Microsc 212:62–70

Moor H (1987) Cryotechniques in biological electron. Microscopy:175–191. https://doi.org/10.1007/978-3-642-72815-0_8

Moser F et al (2019) Cryo-SOFI enabling low-dose super-resolution correlative light and electron cryo-microscopy. Proc Natl Acad Sci 116:4804–4809

Mulisch M, Welsch U (2015) Romeis -Mikroskopische Technik. https://doi.org/10.1007/978-3-642-55190-1.

Namba K, Makino F (2022) Recent progress and future perspective of electron cryomicroscopy for structural life sciences. Microscopy 71:i3–i14
Nichol C, Kim EE (2001) Molecular imaging and gene therapy. J Nucl Med 42:1368–1374
Noda T (2017) Ebolaviruses, methods and protocols. Methods Mol Biol 1628:243–250
Palovcak E et al (2018) A simple and robust procedure for preparing graphene-oxide cryo-EM grids. J Struct Biol 204:80–84
Parvez MK (2020) Geometric architecture of viruses. World J Virol 9:5–18
Peddie CJ et al (2022) Volume electron microscopy. Nat Rev Methods Primers 2:51
Rames M, Yu Y, Ren G (2014) Optimized negative staining: a high-throughput protocol for examining small and asymmetric protein structure by electron microscopy. J Vis Exp Jove 51087. https://doi.org/10.3791/51087
Reynolds ES (1963) The use of lead citrate at high pH as an electron-opaque stain in electron microscopy. J Cell Biol 17:208–212
Richert-Pöggeler KR, Franzke K, Hipp K, Kleespies RG (2019) Electron microscopy methods for virus diagnosis and high resolution analysis of viruses. Front Microbiol 9:3255
Robb NC (2022) Virus morphology: insights from super-resolution fluorescence microscopy. Biochim Biophys Acta B Mol Basis Dis 1868:166347
Roingeard P, Raynal P, Eymieux S, Blanchard E (2019) Virus detection by transmission electron microscopy: still useful for diagnosis and a plus for biosafety. Rev Med Virol 29:e2019
Romero-Brey I, Bartenschlager R (2015) Viral infection at high magnification: 3D electron microscopy methods to analyze the architecture of infected cells. Viruses 7:6316–6345
Schmidt PF (1994) Praxis der Rasterelektronenmikroskopie und Mikrobereichsanalyse
Schwarz H, Humbel BM (1989) Influence of fixatives and embedding media on immunolabelling of freeze-substituted cells. Scanning Microsc Suppl 3, 57–63; discussion 63–4
Schwarz H, Humbel BM (2013) Electron microscopy, methods and protocols. Methods Mol Biol 1117:559–592
Seligman AM, Wasserkrug HL, Hanker JS (1966) A new staining method (OTO) for enhancing contrast of lipid-containing membranes and droplets in osmium tetroxide-fixed tissue with osmiophilic thiocarbohydrazide (TCH). J Cell Biol 30:424–432
Sevvana M, Klose T, Rossmann MG (2021) Encyclopedia of virology. Encycl Virol 257–277. https://doi.org/10.1016/b978-0-12-814515-9.00033-3
Shen PS (2018) The 2017 nobel prize in chemistry: cryo-EM comes of age. Anal Bioanal Chem 410:2053–2057
Sigworth FJ (2016) Principles of cryo-EM single-particle image processing. Microscopy 65:57–67
Spurr AR (1969) A low-viscosity epoxy resin embedding medium for electron microscopy. J Ultra Mol Struct Res 26:31–43
Stass R, Ng WM, Kim YC, Huiskonen JT (2019) Structures of enveloped virions determined by cryogenic electron microscopy and tomography. Adv Virus Res 105:35–71
Stewart SM (2022) Leidenfrost drop dynamics: a forgotten past and modern day rediscoveries. Eur J Phys 43
Studer D, Michel M, Müller M (1989) High pressure freezing comes of age. Scanning Microsc Suppl 3, 253–68; discussion 268–9
Studer D, Graber W, Al-Amoudi A, Eggli P (2001) A new approach for cryofixation by high-pressure freezing. J Microsc 203:285–294
Studer D, Klein A, Iacovache I, Gnaegi H, Zuber B (2014) A new tool based on two micromanipulators facilitates the handling of ultrathin cryosection ribbons. J Struct Biol 185:125–128
Subramanian S et al (2019) Filling adeno-associated virus capsids: estimating success by cryo-electron microscopy. Hum Gene Ther 30:1449–1460
Tietz D, Tietz H, Nickell S, Baumeister W, Plitzko JM (2008) EMC 2008 14th European Microscopy Congress 1–5 September 2008, Aachen, Germany, Volume 1: Instrumentation and Methods. 89–90. https://doi.org/10.1007/978-3-540-85156-1_45
Tokuyasu KT (1978) A study of positive staining of ultrathin frozen sections. J Ultra Mol Struct Res 63:287–307

Tuijtel MW, Koster AJ, Jakobs S, Faas FGA, Sharp TH (2019) Correlative cryo super-resolution light and electron microscopy on mammalian cells using fluorescent proteins. Sci Rep UK 9: 1369

Tyree TJ, Dan R, Thorne RE (2018) Density and electron density of aqueous cryoprotectant solutions at cryogenic temperatures for optimized cryoprotection and diffraction contrast. Acta Crystallogr Sect D 74:471–479

Vanhecke D et al (2010) Lamellar body ultrastructure revisited: high-pressure freezing and cryo-electron microscopy of vitreous sections. Histochem Cell Biol 134:319–326

Vassaux G, Groot-Wassink T (2003) In vivo noninvasive imaging for gene therapy. J Biomed Biotechnol 2003:92–101

Verkade P (2008) Moving EM: the Rapid Transfer System as a new tool for correlative light and electron microscopy and high throughput for high-pressure freezing. J Microsc 230:317–328

Vijayakrishnan S, McElwee M, Loney C, Rixon F, Bhella D (2020) In situ structure of virus capsids within cell nuclei by correlative light and cryo-electron tomography. Sci Rep UK 10:17596

Vincent MP et al (2021) The combination of morphology and surface chemistry defines the immunological identity of nanocarriers in human blood. Adv Ther 4:2100062

Walther P, Müller M (1999) Biological ultrastructure as revealed by high resolution cryo-SEM of block faces after cryo-sectioning. J Microsc 196:279–287

Walther P, Ziegler A (2002) Freeze substitution of high-pressure frozen samples: the visibility of biological membranes is improved when the substitution medium contains water. J Microsc 208:3–10

Walther P et al (2018) STEM tomography of high-pressure frozen and freeze-substituted cells: a comparison of image stacks obtained at 200 kV or 300 kV. Histochem Cell Biol 150:545–556

Wang H (2015) Cryo-electron microscopy for structural biology: current status and future perspectives. Sci China Life Sci 58:750–756

Werle AK et al (2021) Comparison of analytical techniques to quantitate the capsid content of adeno-associated viral vectors. Mol Ther Methods Clin Dev 23:254–262

Williams DB, Carter CB (2009) Transmission electron microscopy – a textbook for material science

Yakovlev S, Downing KH (2011) Freezing in sealed capillaries for preparation of frozen hydratedsections. J Microsc 244:332–332

Zhao Y, Keating K, Dolman C, Thorpe R (2008) Characterization of Complete Particles (VSV-GSIN-GFP) and Empty Particles (VSV-GEMPTY) in human immunodeficiency virus type 1-based lentiviral products for gene therapy potential applications for improvement of product quality and safety. Hum Gene Ther 19:475–486

Zimmer B, Summermatter K, Zimmer G (2013) Stability and inactivation of vesicular stomatitis virus, a prototype rhabdovirus. Vet Microbiol 162:78–84

Chapter 15
Application of NMR Spectroscopy in Viral Assembly Characterization

Dongyue Xin and Michael Hawley

15.1 Introduction

NMR spectroscopy is a technique that observes the local magnetic fields around atomic nuclei (Günther 2013). In the presence of a high external magnetic field (B_0), different spin-states of NMR active nuclei will differ in energy. Irradiation of a nucleus with radio frequency corresponding to the exact energy difference between different spin states will cause excitation of the nucleus from low to high spin energy state. Local electron density around the nucleus creates a magnetic field that opposes the applied field and reduces the field experienced at the nucleus (Fig. 15.1). The exact resonance frequency is also influenced by the existence of neighboring NMR active nuclei. In a complex molecule, a given type of nuclei will resonate at different frequencies depending on their chemical environment which can be characterized by the position (chemical shift, δ) and pattern (splitting) of the NMR signal. Due to the sensitivity of NMR signals to changes in chemical environment, atomic-level understanding of structure and dynamics is attainable in many NMR spectroscopy studies.

NMR spectroscopy can be performed in solution or solid state with specialized hardware and experimental setups. Solution NMR spectroscopy is suitable for analyzing soluble species with an upper size limit of about 100 kDa when the analyte is directly investigated in NMR experiments (Puthenveetil and Vinogradova 2019). Studies of large macromolecules with solution NMR spectroscopy are complicated by the increased linewidth caused by slower tumbling and spectral overlap. In solid-state NMR spectroscopy, materials are studied as crystalline or non-crystalline solids. There is no size limit and solubility requirement for solid-

D. Xin (✉) · M. Hawley
Material & Analytical Sciences, Boehringer Ingelheim Pharmaceuticals, Inc., Ridgefield, CT, USA
e-mail: dongyue.xin@boehringer-ingelheim.com

S. Gautam et al. (eds.), *Bioprocess and Analytics Development for Virus-based Advanced Therapeutics and Medicinal Products (ATMPs)*,
https://doi.org/10.1007/978-3-031-28489-2_15

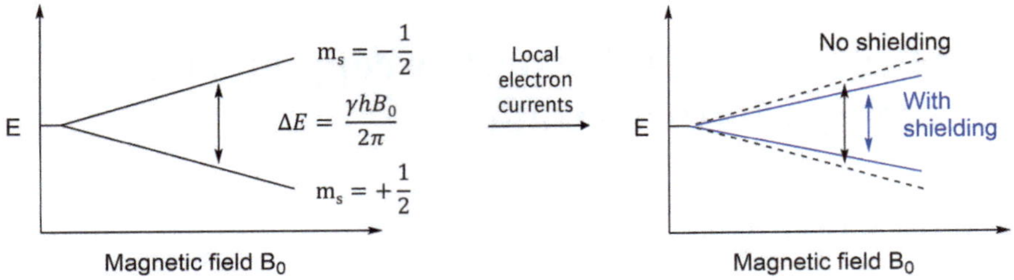

Fig. 15.1 In the presence of an external magnetic field, nuclei at different spin energy states will differ in energy. Shielding by local electron current and interaction with neighboring NMR active nuclei lead to unique resonance for a given type of nuclei in different chemical environments

state NMR but due to the lack of solution-state rapid random tumbling, anisotropic or orientation-dependent spin interactions lead to broader signals and less sensitivity, so special techniques have been developed to improve line shape and sensitivity (Müller and Geppi 2021).

15.2 Solid-State NMR Applications

Solid-state NMR has been successfully employed in investigating structure features for native viruses or VLPs (Quinn and Polenova 2017; Porat-Dahlerbruch et al. 2021). Although viruses are quite large, viral proteins typically assemble into the capsid with high symmetry, thus each virus particle contains many copies of viral proteins in a similar chemical environment and can therefore be detected. With solid-state NMR experiments, resonances of nuclei being studied can be potentially fully resolved and high dimensional experiments that reveal the distance between nuclei in space enable the understanding of 3D structures and functions of biomolecules in viruses or VLPs at the atomic level.

Similar to NMR studies on other biological systems, the most routinely investigated NMR active nuclei for viral assemblies are ^{1}H, ^{13}C, and ^{15}N. Due to the low natural abundance of ^{13}C (1.11%) and ^{15}N (0.37%), target virus or VLPs need universal or site-selected isotope labeling to enable resonance assignments and in-depth structural characterization via 2D or 3D NMR experiments (Renault et al. 2010). For large biological assemblies, partial deuterium (^{2}H) labeling is also frequently used to reduce or eliminate ^{1}H–^{1}H dipolar interactions, improve proton spectral resolution, reduce spectral complexity, or isolate the site of interest. Residue-selective or universal isotope labels can be introduced to virus material by utilizing isotope-labeled cell culture medium components in the production step (Lacabanne et al. 2019). After sufficient purification, the virus or VLPs material needs to be packed into the rotor by sedimentation. Within the rotor, virus and VLPs are fully hydrated to maintain their native states during the course of the experiment.

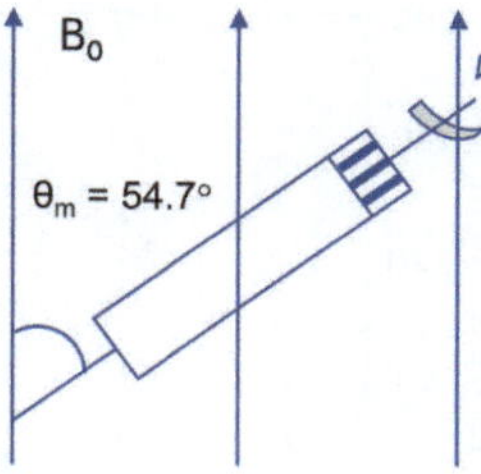

Fig. 15.2 In MAS NMR, the rotor containing the sample is rotated with high frequency at the magic angle $\theta_{\mathrm{m}} = 54.7°$ with regard to the orientation of the external magnetic field

In solid state, orientation-dependent anisotropic interactions and strong dipolar coupling contribute to line broadening and reduced resolution and sensitivity. One technique to address this issue is mechanical rotation of the sample rotor at high frequency. The effect of improved spectral resolution is most prominent at a magic angle of 54.74° against the orientation of external magnetic field and therefore the technique is called magic-angle spinning NMR (MAS NMR) (Fig. 15.2). In a typical MAS NMR experiment, the rotor containing viruses or VLPs is spun in the NMR probe at frequencies on the order of thousands of revolutions per second (kHz).

One advantage of ssNMR over many other atomic-level resolution structural analysis techniques such as cryoEM or X-ray crystallography is that studies can be performed over a wide range of experimental conditions including pH, ionic strength, concentration, and temperature. Virus or VLPs can be investigated in both their native states or under conditions that will induce desired changes in their structure or dynamics. This feature is especially important when studying membrane proteins in enveloped virus or VLPs. Another advantage is that chemical shifts are very sensitive to chemical environments, which makes ssNMR a suitable method to understand fine details of interactions, dynamics, or structure changes in complex viral molecular assemblies. Applications of ssNMR in viral systems have been recently reviewed (Quinn and Polenova 2017; Porat-Dahlerbruch et al. 2021), so only a general description of technology advances in this area will be discussed here.

Two major limitations of MAS ssNMR in studying complex biomolecular systems are the low sensitivity due to the necessity of observing nuclei ^{13}C and ^{15}N with low gyromagnetic ratio γ and limited peak resolution for large systems. High and ultra-high magnetic fields will improve both sensitivity and resolution. The signal-to-noise ratio scales approximately with $B_0^{3/2}$ while data acquisition time to achieve the same S/N ratio is proportional to B_0^{3} under the same experimental condition. Resolution improvement is achieved by both reduced linewidth and increased chemical shift dispersion. This is even more pronounced in high-dimensional experiments as the resolution of n-dimensional experiments is proportional to B_0^{n}. The introduction of the latest 1.2 GHz ultra-high field spectrometers creates new breakthroughs in the field as every new generation of high-field NMR spectrometers provides a major step forward (Quinn et al. 2017; Schwalbe 2017). The use of a high-field or ultra-high-field NMR spectrometer is required for the structural study of virus or VLPs due to their structural complexity.

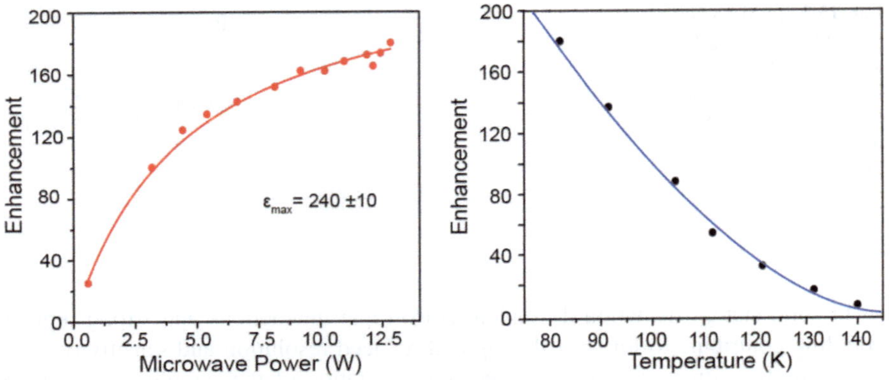

Fig. 15.3 Left: ^{13}C signal enhanced by a maximum of 240 times at high microwave power in a sample of 1 M U-[^{13}C, ^{15}N] urea with 10 mM TOTAPOL at 80 K. Right: temperature dependence of ^{13}C DNP enhancement at 12.5 W μw with 7 kHz spinning rate. Reprinted with permission from Ni et al. (2013). Copyright 2013 American Chemical Society

Another powerful technique to drastically improve sensitivity is dynamic nuclear polarization (DNP). In DNP experiments, polarization is transferred from unpaired electrons of microwave-irradiated paramagnetic polarizing agents to the surrounding nuclei of interest. With currently available hardware, sensitivity can be improved by >100 fold under ideal conditions (Fig. 15.3). The theoretical basis for DNP enhancement has been reviewed (Ni et al. 2013; Thankamony et al. 2017). However, the presence of paramagnetic radicals and the necessity to perform experiments at cryogenic temperatures (~100 K) lead to line broadening of DNP NMR spectra of biomolecules, so either the full advantage of DNP sensitivity enhancement has to be compromised by using higher experimental temperature (Geiger et al. 2016) or additional techniques are needed to improve the resolution (Jaudzems et al. 2018, 2019).

In MAS ssNMR, one major parameter is the spinning rate of rotors in the probe. Ultra-fast spinning probes have provided a significant improvement in spectral resolution, especially for ^{1}H by averaging the strong proton–proton dipolar interactions. With the latest probe design, experiments performed at an ultra-fast spinning rate (>100 kHz) allow for the acquisition of ^{1}H-detected high-resolution solid-state NMR spectra (Marchetti et al. 2012; Fricke et al. 2017). In addition, since ^{1}H has a higher gyromagnetic ratio ($\gamma/(2\pi) = 42.5$ MHz/T) than ^{13}C($\gamma/(2\pi) = 10.7$ MHz/T) and ^{15}N ($\gamma/(2\pi) = -4.3$ MHz/T), a significant mass sensitivity improvement can be achieved using proton detection experiments, reducing the requested sample amount by nearly two orders of magnitudes. For the analysis of viral particles or VLPs, ultra-fast spinning with a 0.7 mm rotor can reduce the sample requirement to about 0.6 mg (Yan et al. 2013; Lecoq et al. 2019).

With the advancement of hardware and commercialization of some key components, the new techniques described above are becoming more accessible. State-of-the-art ssNMR studies typically utilize some or all of these new techniques together

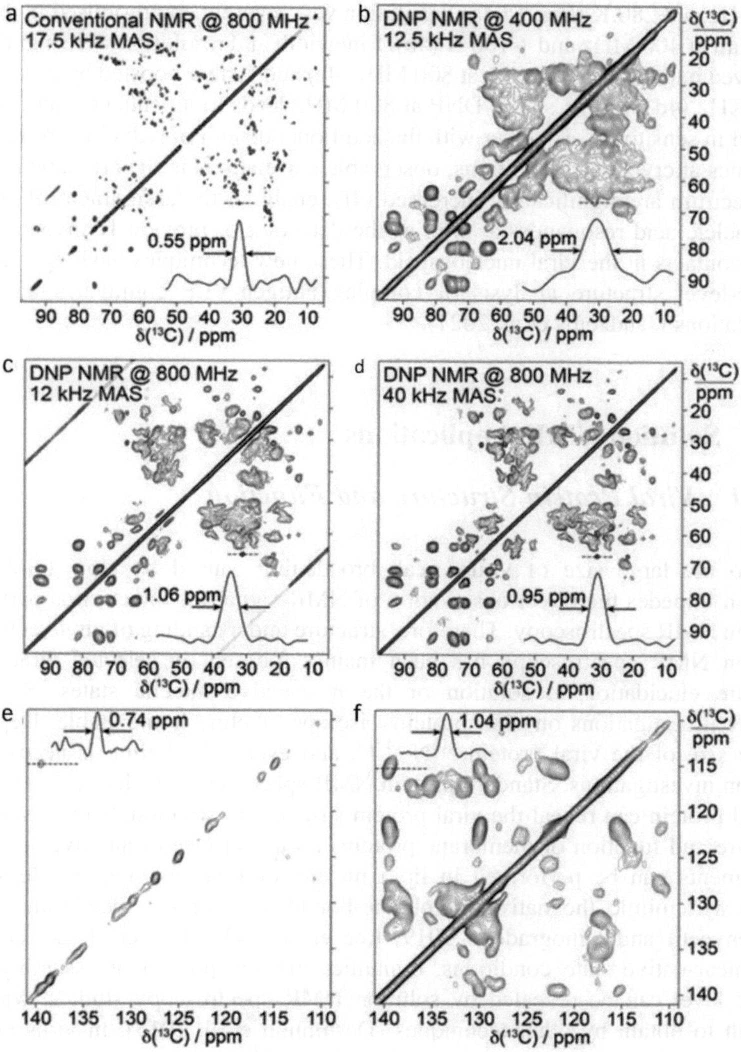

Fig. 15.4 The aliphatic-to-aliphatic DARR (Takegoshi et al. 2001) spectrum at (**a**) 800 Hz, 280 K with 17.5 kHz MAS, (**b**) 400 MHz and 107 K using DNP with 12.5 kHz MAS, (**c**) 800 MHz and 103 K using DNP with 12 kHz MAS and (**e**) the aromatic-to-aromatic regions of a DARR spectrum at 800 Hz, 280 K with 17.5 kHz MAS. The aliphatic-to-aliphatic (**d**) and aromatic-to-aromatic (**f**) regions of a CORD (Hou et al. 2013) spectrum at 800 MHz and 115 K using DNP with 40 kHz MAS. Peak widths at half height in ppm were extracted from 1D spectrum for the Cα/Cβ and Cε2/ζ2 correlations of Val104 and Trp97. Adapted with permission from Jaudzems et al. (2018)

to study complex virus or VLPs. The benefit of combining these techniques was well demonstrated in a recent study on ^{13}C, ^{15}N-labeled AP205 VLPs (Fig. 15.4) (Jaudzems et al. 2018). Compared with conventional ssNMR spectrum acquired at

800 MHz and 280 K (Fig. 15.4a), resolution was severely compromised in the DNP spectrum at 400 MHz and ~ 100 K (4b). Linewidth and resolution were significantly improved in the DNP spectrum at 800 MHz (4c) and further boosted by fast spinning at 40 KHz (4d). In this study, DNP at 800 MHz leads to an enhancement of about 17-fold in sensitivity. Together with the additional benefit of reducing aromatic ring dynamics at cryogenic conditions, observable correlations in the aromatic region of the spectrum are significantly increased (4f), enabling the assignment of aromatic and nucleic acid resonances as well as the detection of previously invisible, long-range contacts in the viral nucleocapsid. These new techniques have even allowed atomic-level structure analysis of complex antigen-VLP conjugates in vaccine formulations (Jaudzems et al. 2021).

15.3 Solution NMR Applications

15.3.1 Viral Protein Structure and Function

Due to the large size of virus, peak broadening caused by slow tumbling in solution impedes the direct observation of NMR signal for intact virus particles in solution NMR spectroscopy. Therefore, structure understanding of viral vectors with solution NMR spectroscopy has been mainly focused on selected viral protein structure elucidation in solution or the near-native micelle states. Similar to ssNMR investigations on viral proteins, isotope labeling is inevitable. Depending on the size of the viral protein, ^{15}N, ^{13}C, and even ^{2}H labeling is necessary for solution investigations. Standard protein NMR spectroscopy techniques on isotope-labeled protein can reveal the viral protein structure in solution. To investigate the structure and function of membrane proteins on the surface of an enveloped virus, experiments can be performed in lipid micelle or lipid-polymer bicelle systems which will mimic the native membrane-bound state of the membrane proteins (Puthenveetil and Vinogradova 2019; Lee et al. 2021; Piai et al. 2021). Under these near-native state conditions, dynamics of viral protein or assembly at the atomic level can be revealed by solution NMR spectroscopy studies, which are difficult to obtain by other techniques (Deshmukh et al. 2015). In virus or VLPs studies, protein structure investigation by solution NMR spectroscopy is sometimes combined with other techniques such as cryoEM to gain atomic-level understanding of the whole virus structure. For instance, fitting of solution NMR structures into low-resolution cryoEM electron densities can lead to atomic resolution of full structural model of complex viral assemblies (Byeon et al. 2009; Cuniasse et al. 2017).

15.3.2 Molecular Interactions

Besides structural studies, certain NMR techniques can provide insight into molecular interactions in solution. These techniques have been routinely applied in drug discovery to select active ligands that can bind with a receptor. Some early applications of NMR spectroscopy in virus molecular interaction understanding rely on NMR chemical shift titrations. In an NMR titration experiment, binding of a ligand with a slow-tumbling, large receptor will lead to changes in chemical shift of the ligand as well as line broadening. This method requires stoichiometric amounts of ligand and receptor. Even if multiple binding sites exist on the virus surface, large quantities of the virus at mg/mL concentration are required in a single experiment (Hanson et al. 1992). Therefore, direct observation of chemical shift changes or line shape of ligand upon binding with virus is impractical for routine investigation of interactions (Benie et al. 2003). Several new NMR techniques have been developed to investigate molecular interactions involving large receptors at low receptor concentrations. Specifically, two ligand-based techniques, STD NMR and waterLOGSY (WL), showed promising results for investigating molecular interactions involving virus or VLPs. Compared with titration experiments, the amount of virus was reduced by approximately two orders of magnitude, and these techniques allowed for atomic-level understanding of intact virus–ligand interactions.

15.3.2.1 Saturation Transfer Difference Spectroscopy (STD NMR)

In NMR spectroscopy, if the target nuclei are pre-irradiated by radio frequency to equalize the populations at higher and lower spin energy states, the target nuclei will no longer be able to further absorb radiation and result in null signal in the subsequent pulse sequence. This is called saturation of the selected nuclei. STD NMR relies on the transfer of saturation between receptors and ligands. In STD NMR, a specific proton on the receptor is irradiated and saturation can be transferred within the receptor by spin diffusion and to a bound small molecule ligand via through-space nuclear Overhauser effect (NOE) and lead to reduced ligand signal intensity. The efficiency of saturation transfer increases with larger receptor size, so viral particles are ideal candidates for STD NMR. After the initial transfer of the saturation to a transiently bound ligand, saturation can be propagated to the large excess of ligands in the bulk solution via rapid exchange, amplifying the final signal. In a typical STD experiment (Fig. 15.5), an on-resonance spectrum is acquired when the receptor resonance is irradiated. On the same sample, an off-resonance spectrum can be acquired using the same experimental setup except that the irradiation frequency is way out of the region of receptor signal. Then, the two spectra are subtracted to yield the difference spectrum. Only the ligands that can bind with the receptor provide a signal in the difference spectrum. Since saturation transfer relies on the observation of solution ligand signal from exchange between bound and

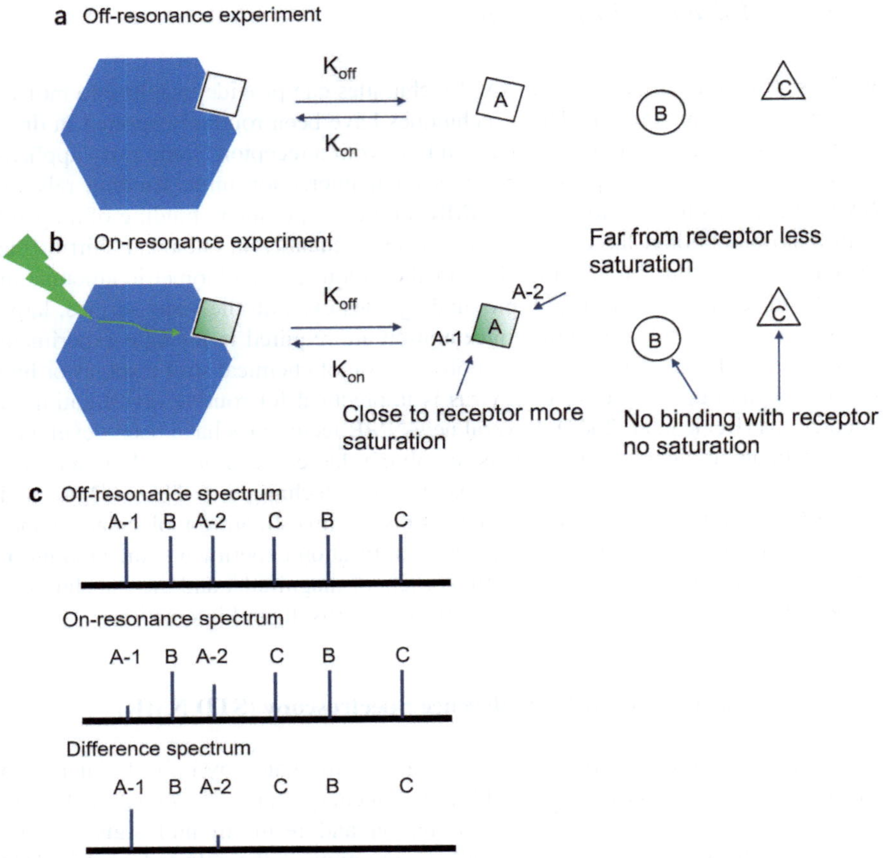

Fig. 15.5 (**a**) Off-resonance experiment without irradiating the receptor proton. (**b**) Saturation transfers from irradiated receptor proton to bound ligand. Exchange between bound and free ligands leads to observable changes in ligand signals for only binders. More saturation can be observed for ligand fragment that is closer to receptor surface. (**c**) Final STD difference spectrum is obtained by subtracting the on-resonance spectrum from the off-resonance spectrum

unbound ligands, STD NMR is suitable for the detection of low and medium binding ligands with a fast off-rate.

Saturation transfer efficiency is associated with the distance of the proton from the receptor surface (Mayer and Meyer 2001). Therefore, under controlled conditions, epitope mapping is possible to determine the part of the ligand that can interact more closely with the receptor. In the example shown in Fig. 15.5b, A-1 proton of ligand A is closer to receptor surface in the receptor bound state and therefore receives more saturation transfer from the receptor in STD experiment. This will lead to further reduced signal in the on-resonance spectrum and a higher peak intensity in the STD difference spectrum than A-2, another proton that is further away from the receptor surface in the receptor bound state. A systematic analysis of all resonances in the ligand can reveal the protons that are closer to the surface of receptor and thus reveal the orientation of the bound ligand.

In STD NMR, exchange between bound and unbound ligands amplifies the signal of receptor information if the residence time of a ligand is significantly shorter than the saturation time and if the ligand concentration is higher than receptor concentration. The effect can be quantified with STD amplification factor (STD-AF) using the following equation (Mayer and Meyer 2001):

$$\mathrm{STD{-}AF} = \frac{I_0 - I_{\mathrm{sat}}}{I_0} \times \frac{c_{\mathrm{ligand}}}{c_{\mathrm{receptor}}}$$

where I_O is the off-resonance signal integral, I_{sat} is the signal intensity from on-resonance experiment, and c_{ligand} and c_{receptor} are the concentrations of ligand and receptor.

STD experiments can be performed on a series of ligand concentrations to determine the apparent dissociation constant $K_D{}^{\mathrm{app}}$. Since the overall STD amplification factor is dependent on the ratio of bound versus free ligands, $K_D{}^{\mathrm{app}}$ can be calculated by curve fitting to a Langmuir isotherm equation at different ligand concentrations (Angulo and Nieto 2011; Hall et al. 2020)

$$\mathrm{STD{-}AF} = \frac{\alpha_{\mathrm{STD}}[L]}{K_D^{\mathrm{app}} + [L]}$$

where α_{STD} is the maximum STD amplification factor, $[L]$ is the ligand concentration, and STD-AF is the experimental amplification factor in a ligand concentration titration series.

The apparent $K_D{}^{\mathrm{app}}$ measured with this method is dependent on experimental conditions such as saturation time, relative concentration of species, monitored proton, etc. The artifacts are most prominent at low ligand/receptor ratio and may lead to an underestimate of the binding affinity ($K_D{}^{\mathrm{app}} \geq K_D{}^{\mathrm{real}}$). The discrepancies were found to be caused by the fast rebinding of saturated ligand to the receptor and accumulation of saturated ligand in solution (Angulo and Nieto 2011; Hall et al. 2020). To obtain a more accurate K_D, a method has been proposed to use the initial STD-AF build-up rate STD-AF$_0$ for K_D calculation (Angulo et al. 2010). In this approach, a STD-AF build-up curve with various saturation time t_{sat} was measured for each ligand concentration instead of one STD experiment at a single saturation time. Each STD-AF build-up curve can be fitted to the equation:

$$\mathrm{STD{-}AF}(t_{\mathrm{sat}}) = \alpha_{\mathrm{STD}} \times \left(1 - e^{-k_{\mathrm{sat}} t_{\mathrm{sat}}}\right)$$

in which k_{sat} is the saturation rate. The initial slopes STD-AF$_0$ are calculated by the equation

$$\mathrm{STD{-}AF_0} = \left[\frac{d(\mathrm{STD{-}AF}(t_{\mathrm{sat}}))}{dt}\right]_{t=0} = \alpha_{\mathrm{STD}} \times k_{\mathrm{sat}}$$

Then the initial STD-AF build-up rates can be used to obtain K_D^{app}.

$$\mathrm{STD{-}AF_0} = \frac{\alpha_{\mathrm{STD}}[L]}{K_D^{\mathrm{app}} + [L]}$$

The calculated K_D^{app} from this approach is comparable to thermodynamic K_D^{real} measured by other techniques.

Direct investigation of binding by STD NMR is limited to medium- to low-affinity ligands (mM to μM). For high-affinity ligands, the STD effect will be very small due to the slow off-rate. It has been demonstrated that competitive STD titration experiments can be employed to reveal the presence of a strong binding ligand. The addition of a strong binding ligand to a mixture of receptor and a weak ligand decreases the STD signal of the weak ligand when the two ligands compete for the same receptor binding site. This approach allows the detection of high-affinity ligands and an estimate of K_D in NMR screening experiments (Wang et al. 2004).

15.3.2.2 WaterLOGSY (WL)

An alternative method in ligand–receptor interaction investigation is waterLOGSY (WL) (Huang and Leung 2018). In WL, bulk water is irradiated and magnetization was transferred from water to the receptor and ligands in both bound and free states (Fig. 15.6). In the presence of specific interactions between the ligand and receptor, magnetization is transferred from receptor and bound water to the ligand, leading to negative NOE due to the slow tumbling rate of the large receptor complex. NOE is also transferred from bulk water to all the soluble ligands, resulting in positive NOE signal. The final WL spectrum is an average of the NOE signals from two opposite phases for ligands that bind with receptor (Huang and Leung 2018). Under optimal experimental conditions, signal from binders is opposite to signals of non-binders.

K_D can also be measured by WL experiment (Dalvit et al. 2001). Since free ligand also contributes to the final averaged signal, two sets of titration experiments, with and without receptor, are typically performed so that the signal of free ligand can be subtracted from WL signal to yield the binding isothermal. Low protein concentration was found to increase the accuracy of K_D measurement, likely by reducing the influence from rebinding of ligands to receptor (Huang et al. 2017). As with STD NMR, strong binding ligands are detectable by competitive experiments with a reference compound having a similar binding mode to the receptor. Solvent accessibility of the bound ligand can be mapped by WL using a short mixing time. Buried ligand protons mainly receive magnetization from the receptor and therefore develop negative NOE. If ligands are partially exposed to solvent in bound state,

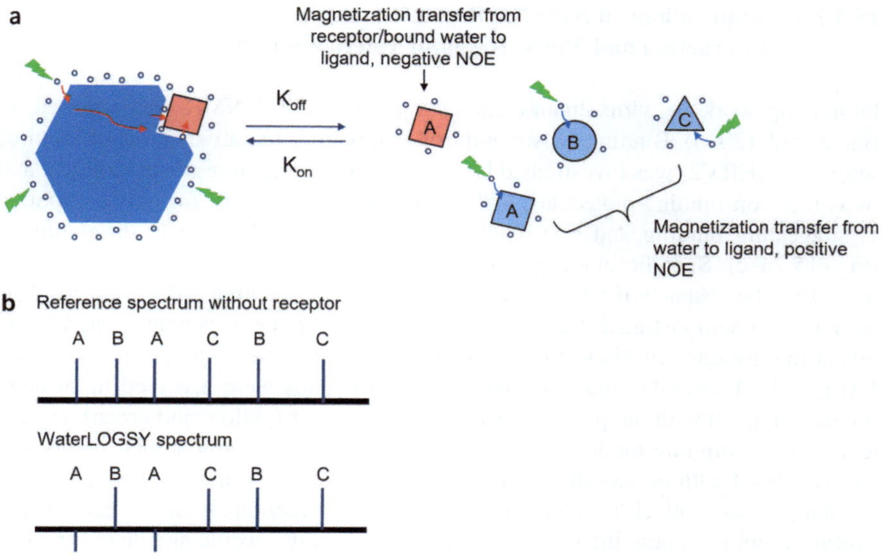

Fig. 15.6 (**a**) Magnetization transfer between bulk water with receptor and ligands after initial excitation of bulk water resonance. (**b**) Binding and non-binding ligands are differentiated by relative peak phase and intensity in WL spectrum

solvent-exposed protons may retain the positive NOE signal developed in their free state when there is not sufficient time for magnetization transfer (Ludwig et al. 2008).

Comparing the two methods, WL was shown to be >10× more sensitive than STD NMR and it does not require much experimental optimization on saturation power and the selection of the optimal "on" resonance saturation frequency for the receptor (Antanasijevic et al. 2014). However, the STD experiment is more established to provide unique insight into the proximity of the ligand to the receptor surface and orientation of the ligand. Both STD NMR and WL are observing the ligand signal, experiments are typically performed with low uM concentration of receptor in the presence of mM or sub-mM concentration of ligands. This is especially advantageous for virus or VLP due to challenges in producing these materials in large quantities. In addition, both techniques detect ^{1}H signals at natural abundance and no isotope labeling is required on the ligand or the receptor, which can save significant efforts and resources in preparing isotope-labeled material. Moreover, STD NMR and WL do not require immobilization of either the ligand or the receptor and experiments can be performed under a wide range of conditions. Since these techniques are compatible with large molecular assemblies, they have been successfully applied in the characterization of virus–ligand interaction or virus–receptor interactions using native virus particles or virus-like particles (VLPs).

15.3.2.3 Applications of STD NMR and WL in Virus–Ligand Interaction and Virus–Receptor Interaction Studies

Pioneering work on virus–ligand interactions using STD NMR was reported by Benie et al. (2003). Binding of two antiviral compounds to native human rhinovirus serotype 2 (HRV2) was investigated by STD NMR. In a mixture of compound 1 and two other non-binding molecules, STD signals were observable for only compound 1, suggesting binding and non-binding compounds can be readily differentiated (Fig. 15.7a–c). Specific binding of compound 1 to intact HRV2 was further confirmed by the absence of STD signals in a thermally denatured HRV2 sample. The authors also demonstrated that competitive STD titration experiments were able to reveal the presence of high-affinity ligands. Group epitope mapping of compound 1 (Fig. 15.7d) revealed that the two aromatic moieties were involved in stronger interaction (red) with the protein surface than the linker (yellow and green), consistent with the binding mode in X-ray structures of a structurally similar compound cocrystallized with human rhinovirus serotype 14 mutant (Hadfield et al. 1995).

Many viruses attach to host cells via specific interactions with the cell surface glycan receptors. These interactions are typically highly specific and determined by glycan-binding pockets on the virus surface. NMR experiments have been applied in understanding these interactions. As one example, screening a library of mono-, di-, and trisaccharides against RHDV (rabbit hemorrhagic disease virus) VLPs by STD NMR revealed the structural requirements for receptor recognition (Table 15.1) (Rademacher et al. 2008). Epitope mapping allowed the determination of binding modes to RHDV for several oligosaccharides at atomic level resolution. In this study, STD NMR was shown to readily differentiate the binding affinity of structurally similar oligosaccharides.

NMR studies on the interaction of intact virus particles or VLPs with ligands or glycans have been summarized in Table 15.2. These investigations were performed on various types of virus/VLPs with sizes ranging from 30 to 100 nm. As long as the virus concentration is sufficient, there is no theoretical limit on the size of virus particles to be studied by STD NMR or waterLOGSY. In fact, interaction of ligand with cellular membrane-bound receptors on living cells has been successfully investigated by a variation of the STD NMR technique (Claasen et al. 2005).

15.4 Other Applications

Besides structural analysis and molecular interaction studies, successful examples have been reported in virus particle detection or characterization using sensor or molecular constructs specially designed for NMR experiments.

In a proof-of-concept study, ^{19}F NMR spectroscopy has been demonstrated to reveal the assembly state of ^{19}F-labeled VLPs (Leung et al. 2017). ^{19}F signal is broad in large intact VLPs and disassembly of VLPs into individual proteins leads to

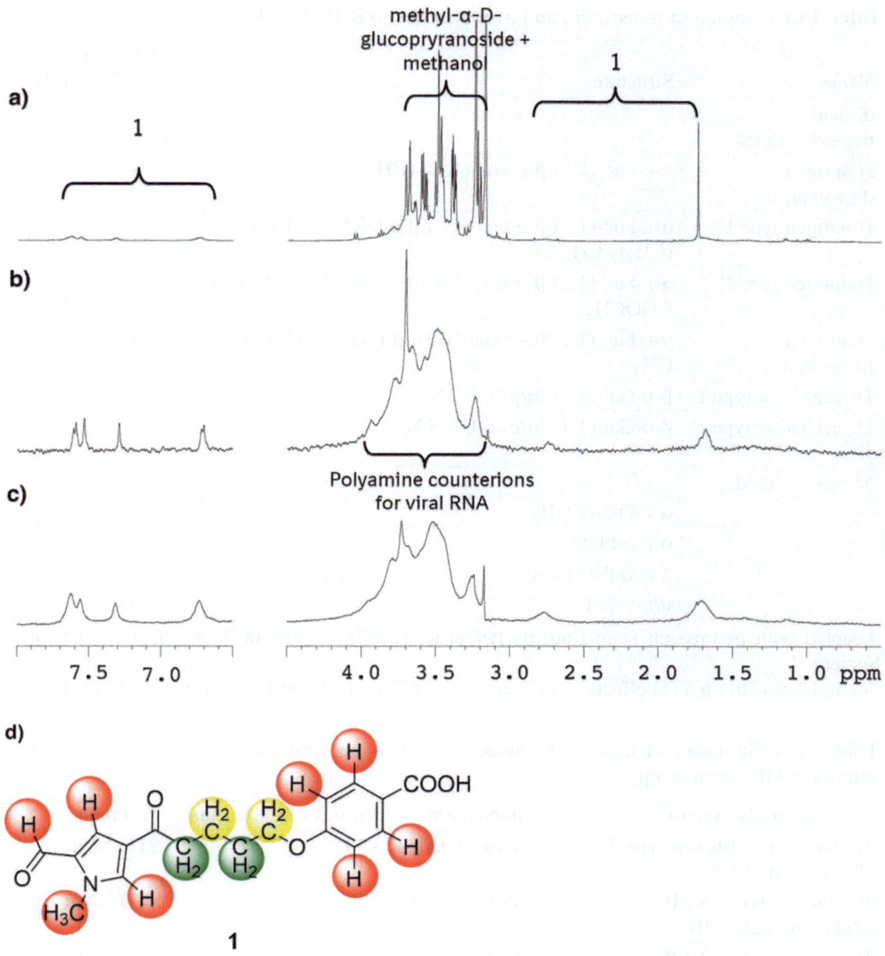

Fig. 15.7 (**a**) 1D reference spectrum of compound 1 (500 μM) with HRV2 (40 nM) in the presence of methyl-a-D-glucopryranoside (5 mM) and methanol (6 mM). (**b**) STD spectrum for compound 1 and HRV2. (**c**) STD spectrum showing the signal of compound 1 but no signal from methyl-a-D-glucopryranoside (5 mM) and methanol (6 mM). (**d**) Structure of compound 1 and epitope mapping results. Percentage of STD signal. Red ≥80%, yellow: 80%–40%, green: ≤ 40%. Adapted with permission from Benie et al. (2003). Copyright 2003 American Chemical Society

sharper peak in ^{19}F spectra. Significant efforts are necessary to introduce ^{19}F-containing non-natural amino acid into VLPs, but the high sensitivity of ^{19}F NMR and low background ^{19}F signal in biological samples allow for direct studies on VLPs status in complex cellular systems.

In several studies (Perez et al. 2003, 2004; Koh et al. 2008), virus particles or the interaction of virus particles with antibodies were detected with specially designed NMR probes in biological medium. These studies were based on the change of

Table 15.1 Binding of mono-, di-, and trisaccharides to RHDV VLPs

Name	Structure	Binding to RHDV VLPs[a]
di- and trisaccharides:		
H-antigen disaccharide	α-L-Fuc-(1,2)-β-D-Gal-(1,*O*)-$(CH_2)_7$-CH_3	++
H-antigen type I	α-L-Fuc-(1,2)-β-D-Gal-(1,3)-β-D-GlcNAc-(1,*O*)-$(CH_2)_7$-CH_3	–
H-antigen type II	α-L-Fuc-(1,2)-β-D-Gal-(1,4)-β-D-GlcNAc-(1,9)-$(CH_2)_8$-$COOCH_3$	++
B-antigen trisaccharide	α-L-Fuc-(1,2)-(α-D-Gal-(1,3))-β-D-Gal-(1,*O*)-$(CH_2)_7$-CH_3	++
Disaccharide type I	β-D-Gal-(1,3)-(α/β)-D-GlcNAc	–
Disaccharide type II	β-D-Gal-(1,4)-(α/β)-D-GlcNAc	+
Monosaccharides:		
	α-L-Fuc-*O*-CH_3	++
	α/β-D-Fuc	–
	β-D-Gal-*O*-CH_3	–
	α/β-L-Gal	++

[a]++ indicates strong STD effects, + indicates weak STD effects, and – indicates no STD observed

Table 15.2 Summary of molecular interactions involving viral assembly systems explored by solution NMR spectroscopy

Viral assembly system	Interaction	Virus/VLPs size (nm)	Technique
Human rhinovirus serotype 2 (Benie et al. 2003)	Virus–ligand	~30	STD NMR
Bovine norovirus VLP (Zakhour et al. 2009)	VLP–glycan	~30	STD NMR
Human norovirus VLP (Fiege et al. 2012, 2017)	VLP–glycan	~30	STD NMR
Human norovirus VLP (Rademacher et al. 2011)	VLP–ligand	~30	STD NMR
Rabbit hemorrhagic disease virus (RHDV) VLP (Rademacher et al. 2008)	VLP–glycan	35–40	STD NMR
Rhesus Rotavirus (Haselhorst et al. 2011)	Virus–glycan	~70	STD NMR
Avian influenza H5-containing VLP (Haselhorst et al. 2008)	VLP–glycan	~100	STD NMR
VLP based on HIV virus core (Antanasijevic et al. 2016)	VLP–ligand	~100	STD NMR + WL

relaxation rate of bulk water in the presence of magnetic nanoparticles. The presence of well-dispersed magnetic nanoparticles significantly shortened the spin-spin relaxation time T_2 of surrounding water molecules. In developing virus-specific probes, the surface of magnetic nanoparticles was modified to conjugate with mAbs targeting the virus particles. The presence of viral particles induced aggregation of magnetic nanoparticle probes, resulting in less paramagnetic relaxation of the bulk water and longer T_2 relaxation time. In one study, as few as 50 HSV-1 virus particles per 100 μL were detected in serum (Perez et al. 2003). The magnetic nanoprobes need to be carefully designed and prepared but this approach provides an alternative method for viral particle detection with high sensitivity and selectivity.

15.5 Summary

NMR spectroscopy has recently emerged as a powerful technique in the characterization of viruses or VLPs in native or near-native states. Investigation of virus structure, dynamics, and molecular interactions by solution or solid-state NMR provides atomic-level information not accessible by other characterization techniques. Advances in hardware and the development of new experiments have resulted in a large sensitivity improvement that has reduced the required amount of material. An additional benefit has been a significant enhancement in spectral resolution which has enabled the study of larger viruses or VLPs. When NMR spectroscopy is combined with other state-of-the-art techniques, its application can be expanded to more challenging viral assembly systems.

References

Angulo J, Enríquez-Navas PM, Nieto PM (2010) Ligand–receptor binding affinities from Saturation Transfer Difference (STD) NMR spectroscopy: the binding isotherm of STD initial growth rates. Chem Eur J 16:7803–7812. https://doi.org/10.1002/chem.200903528

Angulo J, Nieto PM (2011) STD-NMR: application to transient interactions between biomolecules—a quantitative approach. Eur Biophys J 40:1357–1369. https://doi.org/10.1007/s00249-011-0749-5

Antanasijevic A, Kingsley C, Basu A et al (2016) Application of virus-like particles (VLP) to NMR characterization of viral membrane protein interactions. J Biomol NMR 64:255–265. https://doi.org/10.1007/s10858-016-0025-1

Antanasijevic A, Ramirez B, Caffrey M (2014) Comparison of the sensitivities of WaterLOGSY and saturation transfer difference NMR experiments. J Biomol NMR 60:37–44. https://doi.org/10.1007/s10858-014-9848-9

Benie AJ, Moser R, Bäuml E et al (2003) Virus–ligand interactions: identification and characterization of ligand binding by NMR spectroscopy. J Am Chem Soc 125:14–15. https://doi.org/10.1021/ja027691e

Byeon I-JL, Meng X, Jung J et al (2009) Structural convergence between cryo-EM and NMR reveals intersubunit interactions critical for HIV-1 capsid function. Cell 139:780–790. https://doi.org/10.1016/j.cell.2009.10.010

Claasen B, Axmann M, Meinecke R, Meyer B (2005) Direct observation of ligand binding to membrane proteins in living cells by a saturation transfer double difference (STDD) NMR spectroscopy method shows a significantly higher affinity of integrin αIIbβ3 in native platelets than in liposomes. J Am Chem Soc 127:916–919. https://doi.org/10.1021/ja044434w

Cuniasse P, Tavares P, Orlova EV, Zinn-Justin S (2017) Structures of biomolecular complexes by combination of NMR and cryoEM methods. Curr Opin Struc Biol 43:104–113. https://doi.org/10.1016/j.sbi.2016.12.008

Dalvit C, Fogliatto G, Stewart A et al (2001) WaterLOGSY as a method for primary NMR screening: practical aspects and range of applicability. J Biomol NMR 21:349–359. https://doi.org/10.1023/a:1013302231549

Deshmukh L, Ghirlando R, Clore GM (2015) Conformation and dynamics of the Gag polyprotein of the human immunodeficiency virus 1 studied by NMR spectroscopy. Proc Natl Acad Sci 112:3374–3379. https://doi.org/10.1073/pnas.1501985112

Fiege B, Leuthold M, Parra F et al (2017) Epitope mapping of histo blood group antigens bound to norovirus VLPs using STD NMR experiments reveals fine details of molecular recognition. Glycoconj J 34:679–689. https://doi.org/10.1007/s10719-017-9792-5

Fiege B, Rademacher C, Cartmell J et al (2012) Molecular details of the recognition of blood group antigens by a human norovirus as determined by STD NMR spectroscopy. Angew Chem Int Ed 51:928–932. https://doi.org/10.1002/anie.201105719

Fricke P, Chevelkov V, Zinke M et al (2017) Backbone assignment of perdeuterated proteins by solid-state NMR using proton detection and ultrafast magic-angle spinning. Nat Protoc 12:764–782. https://doi.org/10.1038/nprot.2016.190

Geiger M-A, Orwick-Rydmark M, Märker K et al (2016) Temperature dependence of cross-effect dynamic nuclear polarization in rotating solids: advantages of elevated temperatures. Phys Chem Chem Phys 18:30696–30704. https://doi.org/10.1039/c6cp06154k

Günther H (2013) NMR spectroscopy: basic principles, concepts and applications in chemistry, 3rd edn

Hadfield AT, Oliveira MA, Kim KH et al (1995) Structural studies on human rhinovirus 14 drug-resistant compensation mutants. J Mol Biol 253:61–73. https://doi.org/10.1006/jmbi.1995.0536

Hall JL, Sohail A, Cabrita EJ et al (2020) Saturation transfer difference NMR on the integral trimeric membrane transport protein GltPh determines cooperative substrate binding. Sci Rep UK 10:16483. https://doi.org/10.1038/s41598-020-73443-z

Hanson JE, Sauter NK, Skehel JJ, Wiley DC (1992) Proton nuclear magnetic resonance studies of the binding of sialosides to intact influenza virus. Virology 189:525–533. https://doi.org/10.1016/0042-6822(92)90576-b

Haselhorst T, Fiebig T, Dyason JC et al (2011) Recognition of the GM3 ganglioside glycan by rhesus rotavirus particles. Angew Chem Int Ed 50:1055–1058. https://doi.org/10.1002/anie.201004116

Haselhorst T, Garcia J, Islam T et al (2008) Avian influenza H5-containing virus-like particles (VLPs): host-cell receptor specificity by STD NMR spectroscopy. Angew Chem Int Ed 47:1910–1912. https://doi.org/10.1002/anie.200704872

Hou G, Yan S, Trébosc J et al (2013) Broadband homonuclear correlation spectroscopy driven by combined R2nv sequences under fast magic angle spinning for NMR structural analysis of organic and biological solids. J Magn Reson 232:18–30

Huang R, Bonnichon A, Claridge TDW, Leung IKH (2017) Protein-ligand binding affinity determination by the waterLOGSY method: an optimised approach considering ligand rebinding. Sci Rep UK 7:43727. https://doi.org/10.1038/srep43727

Huang R, Leung IKH (2018) Protein–small molecule interactions by WaterLOGSY. Methods Enzymol 615:477–500. https://doi.org/10.1016/bs.mie.2018.08.020

Jaudzems K, Bertarello A, Chaudhari SR et al (2018) Dynamic nuclear polarization-enhanced biomolecular NMR spectroscopy at high magnetic field with fast magic-angle spinning. Angew Chem Int Ed 57:7458–7462. https://doi.org/10.1002/anie.201801016

Jaudzems K, Kirsteina A, Schubeis T et al (2021) Structural analysis of an antigen chemically coupled on virus-like particles in vaccine formulation. Angew Chem Int Ed 60:12847–12851. https://doi.org/10.1002/anie.202013189

Jaudzems K, Polenova T, Pintacuda G et al (2019) DNP NMR of biomolecular assemblies. J Struct Biol 206:90–98. https://doi.org/10.1016/j.jsb.2018.09.011

Koh I, Hong R, Weissleder R, Josephson L (2008) Sensitive NMR sensors detect antibodies to influenza. Angew Chem Int Ed 47:4119–4121. https://doi.org/10.1002/anie.200800069

Lacabanne D, Fogeron M-L, Wiegand T et al (2019) Protein sample preparation for solid-state NMR investigations. Prog Nucl Mag Res Spectrosc 110:20–33. https://doi.org/10.1016/j.pnmrs.2019.01.001

Lecoq L, Schledorn M, Wang S et al (2019) 100 kHz MAS proton-detected NMR spectroscopy of hepatitis B virus capsids. Front Mol Biosci 6:58. https://doi.org/10.3389/fmolb.2019.00058

Lee J, Kreutzberger AJB, Odongo L et al (2021) Ebola virus glycoprotein interacts with cholesterol to enhance membrane fusion and cell entry. Nat Struct Mol Biol 28:181–189. https://doi.org/10.1038/s41594-020-00548-4

Leung RLC, Robinson MDM, Ajabali AAA et al (2017) Monitoring the disassembly of virus-like particles by 19F-NMR. J Am Chem Soc 139:5277–5280. https://doi.org/10.1021/jacs.6b11040

Ludwig C, Michiels PJA, Wu X et al (2008) SALMON: solvent accessibility, ligand binding, and mapping of ligand orientation by NMR spectroscopy. J Med Chem 51:1–3. https://doi.org/10.1021/jm701020f

Marchetti A, Jehle S, Felletti M et al (2012) Backbone assignment of fully protonated solid proteins by 1H detection and ultrafast magic-angle-spinning NMR spectroscopy. Angew Chem Int Ed 51:10756–10759. https://doi.org/10.1002/anie.201203124

Mayer M, Meyer B (2001) Group epitope mapping by saturation transfer difference NMR to identify segments of a ligand in direct contact with a protein receptor. J Am Chem Soc 123:6108–6117. https://doi.org/10.1021/ja0100120

Müller K, Geppi M (2021) Solid state NMR: principles, methods, and applications

Ni QZ, Daviso E, Can TV et al (2013) High frequency dynamic nuclear polarization. Acc Chem Res 46:1933–1941. https://doi.org/10.1021/ar300348n

Perez JM, Josephson L, Weissleder R (2004) Use of magnetic nanoparticles as nanosensors to probe for molecular interactions. Chembiochem 5:261–264. https://doi.org/10.1002/cbic.200300730

Perez JM, Simeone FJ, Saeki Y et al (2003) Viral-induced self-assembly of magnetic nanoparticles allows the detection of viral particles in biological media. J Am Chem Soc 125:10192–10193. https://doi.org/10.1021/ja036409g

Piai A, Fu Q, Sharp AK et al (2021) NMR model of the entire membrane-interacting region of the HIV-1 fusion protein and its perturbation of membrane morphology. J Am Chem Soc 143:6609–6615. https://doi.org/10.1021/jacs.1c01762

Porat-Dahlerbruch G, Goldbourt A, Polenova T (2021) Virus structures and dynamics by magic-angle spinning NMR. Ann Rev Virol 8:219–237. https://doi.org/10.1146/annurev-virology-011921-064653

Puthenveetil R, Vinogradova O (2019) Solution NMR: a powerful tool for structural and functional studies of membrane proteins in reconstituted environments. J Biol Chem 294:15914–15931. https://doi.org/10.1074/jbc.rev119.009178

Quinn CM, Polenova T (2017) Structural biology of supramolecular assemblies by magic-angle spinning NMR spectroscopy. Q Rev Biophys 50:e1. https://doi.org/10.1017/s0033583516000159

Quinn CM, Wang M, Polenova T (2017) Protein NMR, methods and protocols. Methods Mol Biol 1688:1–35. https://doi.org/10.1007/978-1-4939-7386-6_1

Rademacher C, Guiard J, Kitov PI et al (2011) Targeting norovirus infection—multivalent entry inhibitor design based on NMR experiments. Chem Eur J 17:7442–7453. https://doi.org/10.1002/chem.201003432

Rademacher C, Krishna NR, Palcic M et al (2008) NMR experiments reveal the molecular basis of receptor recognition by a calicivirus. J Am Chem Soc 130:3669–3675. https://doi.org/10.1021/ja710854r

Renault M, Cukkemane A, Baldus M (2010) Solid-state NMR spectroscopy on complex biomolecules. Angew Chem Int Ed 49:8346–8357. https://doi.org/10.1002/anie.201002823

Schwalbe H (2017) Editorial: New 1.2 GHz NMR spectrometers—New Horizons? Angew Chem Int Ed 56:10252–10253. https://doi.org/10.1002/anie.201705936

Takegoshi K, Nakamura S, Terao T (2001) 13C–1H dipolar-assisted rotational resonance in magic-angle spinning NMR. Chem Phys Lett 344:631–637. https://doi.org/10.1016/s0009-2614(01)00791-6

Thankamony ASL, Wittmann JJ, Kaushik M, Corzilius B (2017) Dynamic nuclear polarization for sensitivity enhancement in modern solid-state NMR. Prog Nucl Mag Res Spectrosc 102:120–195. https://doi.org/10.1016/j.pnmrs.2017.06.002

Wang Y, Liu D, Wyss DF (2004) Competition STD NMR for the detection of high-affinity ligands and NMR-based screening. Magn Reson Chem 42:485–489. https://doi.org/10.1002/mrc.1381

Yan S, Suiter CL, Hou G et al (2013) Probing structure and dynamics of protein assemblies by magic angle spinning NMR spectroscopy. Acc Chem Res 46:2047–2058. https://doi.org/10.1021/ar300309s

Zakhour M, Ruvoën-Clouet N, Charpilienne A et al (2009) The αGal epitope of the Histo-blood group antigen family is a ligand for bovine norovirus Newbury2 expected to prevent cross-species transmission. PLoS Pathog 5:e1000504. https://doi.org/10.1371/journal.ppat.1000504

Chapter 16
Improved Production Strategies for Oncolytic Measles Viruses as a Therapeutic Cancer Treatment

Dustin Eckhardt, Sascha Bossow, Jan-Philip Klee, Björn Boshof, Guy Ungerechts, Peter Czermak, and Denise Salzig

16.1 Introduction

Oncolytic viruses (OVs) used for cancer immunotherapies offer an emerging treatment option for solid and hematologic malignancies. Therefore, a considerable variety of viruses, ranging from small picornaviruses to large poxviruses, are currently investigated as potential candidates. Several virus families are currently evaluated as cancer treatments in clinical trials (Fu et al. 2019; Nettelbeck et al. 2021).

D. Eckhardt · J.-P. Klee · B. Boshof
Institute of Bioprocess Engineering and Pharmaceutical Technology (IBPT), University of Applied Sciences Mittelhessen (THM), Gießen, Germany

S. Bossow
Clinical Cooperation Unit Virotherapy, German Cancer Research Center (DKFZ), Heidelberg, Germany

G. Ungerechts
Clinical Cooperation Unit Virotherapy, German Cancer Research Center (DKFZ), Heidelberg, Germany

Department of Medical Oncology, National Center for Tumor Diseases (NCT) and Heidelberg University Hospital, Heidelberg, Germany

Cancer Therapeutics Program, Ottawa Hospital Research Institute, Ottawa, ON, Canada

P. Czermak · D. Salzig (✉)
Institute of Bioprocess Engineering and Pharmaceutical Technology (IBPT), University of Applied Sciences Mittelhessen (THM), Gießen, Germany

Faculty of Biology and Chemistry, Justus-Liebig University of Giessen, Giessen, Germany
e-mail: denise.salzig@lse.thm.de

S. Gautam et al. (eds.), *Bioprocess and Analytics Development for Virus-based Advanced Therapeutics and Medicinal Products (ATMPs)*,
https://doi.org/10.1007/978-3-031-28489-2_16

The only commercially approved virus construct in the Western hemisphere to date is T-VEC (Imlygic®, Amgen) for the treatment of advanced melanoma. It is an attenuated and genetically engineered herpes simplex type 1 virus (HSV-1), characterized by an additional integration of the gene encoding human granulocyte-macrophage colony-stimulating factor (hGM-CSF), intended to trigger an enhancement of the virus-mediated anti-tumor immune response (Rehman et al. 2016).

OVs represent a panel of therapeutically active viruses that differ considerably with respect to their structure, size, genome, replication mechanisms, and host interactions including triggered host immune responses (Ungerechts et al. 2016). This new class of anti-cancer drugs could prospectively complement surgery, chemotherapy, irradiation, targeted small molecules, and antibodies in routine clinical oncology. Despite the diversity of OVs and their pleiotropic modes of action, they have at least two modes in common: (i) a direct tumor cell killing by productive infection and viral spread (Bell and McFadden 2014; Russell et al. 2012), and (ii) an induction of a local and systemic anti-tumor immunity via the release of tumor-associated antigens, triggering an immunogenic tumor microenvironment (Russell et al. 2012, 2019; Breitbach et al. 2016; Cattaneo et al. 2008; Fisher et al. 2021; Harrington et al. 2019; Pikor et al. 2015; Russell and Barber 2018; Woller et al. 2014; Kaufman et al. 2015). With the increase of immunotherapies in clinical oncology, the induction of anti-tumor immunity by viral oncolysis has received growing attention, with OVs being explored as the components of combination regimens that induce primary anti-tumor responses and/or trigger immune cell infiltration of tumors.

OVs must be safe for clinical applications. A defining feature of OVs in terms of safety is a tumor-selective infection or replication. This oncotropism can be achieved by attenuated virus vaccine strains, or by an exploitation of defective anti-viral host responses in cancer cells, allowing for an application of interferon-sensitive viruses (Pikor et al. 2015).

One promising candidate among the OVs is the oncolytic Measles virus. In the development of oncolytic measles viruses (MeV) derived from an attenuated vaccine strain, the safety aspect is a beneficial starting point, as the Measles vaccine (nowadays available in the triple measles-mumps-rubella vaccine M-M-R®II) provides an excellent safety profile (Measles, Mumps, Rubella (MMR) 2022). The attenuated Measles vaccine strains only trigger an immune response, without any clinical symptoms of the disease (WHO 2017; Di Pietrantonj et al. 2020).

Measles viruses are enveloped particles of pleomorphic size and belong to the *morbillivirus* genus in the *Paramyxoviridae* family (Aref et al. 2016; Knipe and Howley 2021). The MeV genome is a helical nucleocapsid and consists of a non-segmented, single-stranded RNA of negative polarity with a length of approximately 16,000 nucleotides. The RNA genome of *Paramyxoviridae* cannot integrate into the DNA genome of infected host cells, thus preventing genotoxicity. The envelope is lined by the matrix (M) protein and carries two transmembrane glycoproteins—the hemagglutinin (H) and the fusion (F) protein—which define the tropism and mediate the entry into the host cell. Wild-type MeV uses CD150/

SLAM as an attachment receptor for entering different immune cells (macrophages, dendritic cells, activated or memory B and T cells), and Nectin-4/PVRL4 on cells of the airway epithelium to exit the host via the respiratory route (Tatsuo et al. 2000; Mühlebach et al. 2011). Attenuated replication-competent vaccine strains of MeV exhibit a natural oncotropism by preferentially entering tumor cells through the additional receptor CD46 (Dörig et al. 1993), a membrane protein that is typically overexpressed in malignant cells (Anderson et al. 2004).

16.2 MeV Engineering

In addition to their excellent safety record and natural oncotropism, the possibilities of multi-level genetic engineering make MeV a promising candidate for (immune) virotherapy (Leber et al. 2020b; Engeland and Ungerechts 2021). Notably, recent preclinical and clinical research has revealed a pleiotropic therapeutic activity of MeV by genetically engineering MeV, following a panel of different strategies (Fig. 16.1), aiming at (a) a restricted biodistribution and high tumor specificity ("Targeting"), (b) an extension of therapeutic applications via an insertion of transgenes that can trigger anti-tumor immune responses or bystander killing ("Arming"), and (c) the modification of the viral envelope to prevent an inactivation of oncolytic MeV by neutralizing antibodies in vaccinated patients after an intravenous application ("Shielding").

16.2.1 Targeting

When considering the use of replication-competent viruses as therapeutic agents for the treatment of cancer, tumor specificity is of critical importance to ensure both patient safety and therapeutic efficacy. By genetic engineering, not only off-target effects of the oncolytic agent shall be prevented, but also efficient tumor-homing is intended. OVs that may bind to many different cell types, or that get sequestered in, e.g., the liver, might not reach the tumor in sufficient numbers, thus limiting their efficacy, especially when administered systemically.

The natural oncotropism of MeV was first described in a well-known case report of a young boy whose Burkitt's lymphoma regressed, following an infection with wild-type MeV (Bluming and Ziegler 1971).

Currently, most of the preclinical and clinical studies use oncolytic MeV constructs derived from vaccine strains. The molecular basis for a natural tumor selectivity of MeV is primarily based on its receptor usage and its sensitivity toward the anti-viral interferon (IFN) response, which is often compromised in cancer cells. Vaccine strains preferentially use CD46 (a negative regulator of complement activity) for cell entry, which is constitutively overexpressed in a wide variety of tumor cells (Anderson et al. 2004). Interestingly, some of the vaccine strains (including

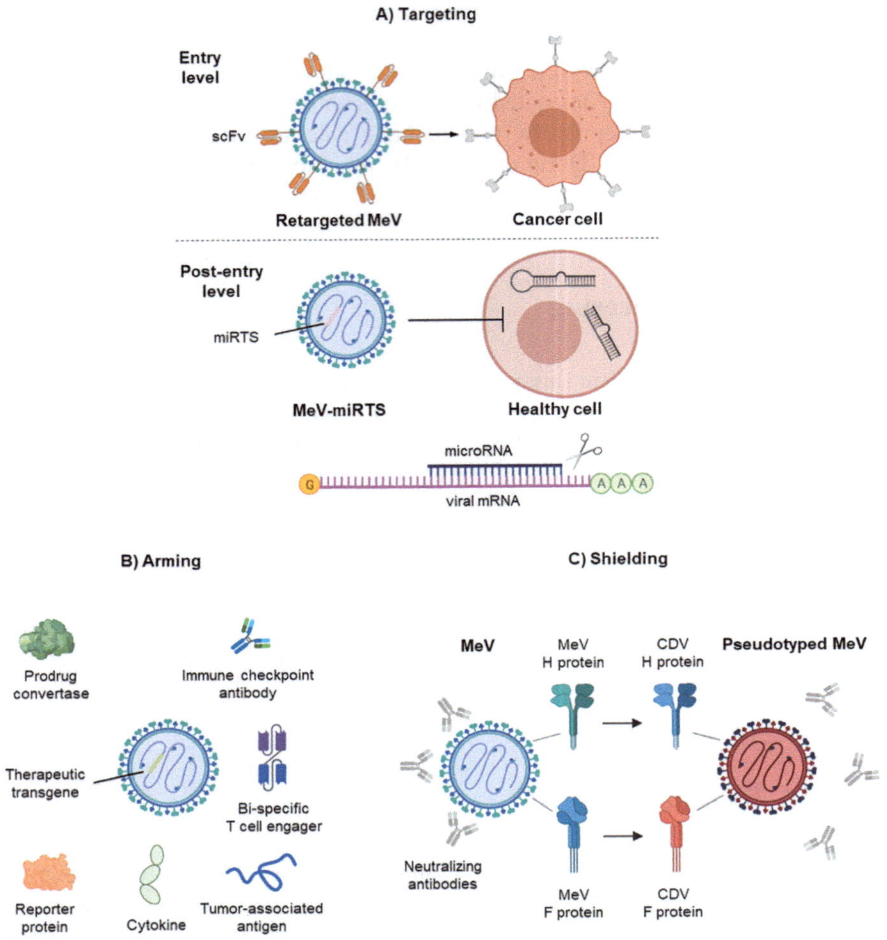

Fig. 16.1 Strategies for oncolytic MeV engineering—Targeting, Arming, and Shielding. (**a**) Targeting and tumor-specificity of oncolytic MeV can be achieved on multiple levels. Entry targeting: A fully retargeted MeV is exemplarily shown. It displays a distinct single-chain antibody (scFv) fused to the MeV-"blinded" H protein to recognize a corresponding tumor surface antigen. Post-entry targeting: The oncolytic MeV carries target sites (miRTS) against distinct microRNA species, which are present in healthy cells, but lost in malignant cells. This microRNA-controlled MeV is strongly attenuated in healthy cells expressing the cognate microRNAs, but remains fully effective against tumor cells. (**b**) Oncolytic MeV are engineered to encode therapeutic transgenes, e.g., prodrug-converting enzymes and immune-modulating proteins. (**c**) Shielding of oncolytic MeV. The unmodified oncolytic MeV can be neutralized by pre-existing anti-MeV antibodies. To avoid neutralization, the regular MeV surface proteins F and H are replaced by envelope glycoproteins from related viruses, e.g., the canine distemper virus (CDV). This pseudotyped MeV cannot be recognized by neutralizing anti-MeV antibodies anymore but retains its oncolytic activity

Edmonston B, from which oncolytic MeV, currently used in clinical trials, are derived) carry mutations in the phosphoprotein (P) gene, which dampens their ability to control the host's anti-viral IFN response (Devaux et al. 2008, 2011; Haralambieva et al. 2010; Naniche et al. 2000; Naniche 2009; Ohno et al. 2004). Since the IFN response is frequently dysregulated in malignant cells, this accounts for, at least in part, the preferential replication of MeV vaccine strains in transformed cells.

Engineered tumor specificity, also referred to as "targeting," has been achieved on the entry and post-entry levels. In the case of entry targeting, the recognition of the natural receptor is disabled by altering the amino acids responsible for receptor-binding within the hemagglutinin (H) gene. In a second step, a novel tropism is conveyed by genetic fusion of targeting ligands to the "blinded" H gene. A variety of targeting molecules were successfully tested, including different kinds of (single chain) antibodies recognizing tumor-related surface molecules (Nakamura et al. 2005; Bossow et al. 2011; Ungerechts et al. 2007a, b; Zaoui et al. 2012; Allen et al. 2006; Bach et al. 2013), DARPins (Friedrich et al. 2013; Hanauer et al. 2019), and ligands for cytokine or growth factor receptors (Allen et al. 2008; Jing et al. 2014).

In the case of post-entry targeting, microRNA target sites have been explored. For this concept, reverse-complementary target sequences were inserted into the MeV genome against selected microRNAs, which are lost in tumor cells, but expressed in healthy tissues. We showed that this approach suppresses the spread of oncolytic MeV in microRNA-expressing healthy tissues, while the replication kinetics and the subsequent destruction of cancer cells remain unchanged (Leber et al. 2011, 2018; Baertsch et al. 2014).

16.2.2 Arming

Three different attenuated MeV constructs were administered to cancer patients in clinical trials, one unmodified (Heinzerling et al. 2005), and two with encoding reporter transgenes mainly for biotracing, i.e., the sodium iodide symporter (NIS) and carcinoembryonic antigen (CEA) (Galanis et al. 2010; Dispenzieri et al. 2017). A strong argument for the introduction of therapeutic transgenes is that future clinical trials will yield even better patient outcomes. This approach utilizes the capacity of MeV to locally express a given protein in infected cancer cells. A tumor-restricted activation of chemotherapeutics was explored, using prodrug convertases. Early experiments with virally encoded prodrug convertases used *Escherichia coli* purine nucleoside phosphorylase (PNP) in conjunction with fludarabine to concentrate toxic ATP analogs in cancer cells (Ungerechts et al. 2007a, b, 2010). Due to toxicity concerns related to a systemic fludarabine administration, current chemo-virotherapeutic approaches switched to a fusion protein of cytosine deaminase and uracil phosphoribosyltransferase (CD-UPRT) from yeast (Kaufmann et al. 2013; Lampe et al. 2013; Lange et al. 2013; Maurer et al. 2019). CD-UPRT converts the prodrug 5-fluorocystosine into 5-fluorouracil, and then into 5-fluorouridine monophosphate (5-FUMP). This process occurs exclusively in infected cancer cells, due to the tumor-restricted MeV-mediated expression of CD-UPRT. Cellular enzymes in cancer cells then process 5-FUMP into toxic metabolites that interfere with DNA repair and other cellular mechanisms, leading to enhanced tumoricidal bystander effects (Lampe et al. 2013).

The recent field of preclinical transgene research has an emphasis on immunomodulation. Specifically, transgenes that activate or enhance the patient's immune response against the tumor are desirable. A previous study suspected that the host inflammatory response, particularly neutrophils, contributed to the therapeutic effect of MeV in SCID mice, which was supported by encoding the cytokine granulocyte-macrophage colony-stimulating factor (GM-CSF) (Grote et al. 2003). To address the influence of the adaptive immune response toward the therapy, the MeV-GM-CSF approach was revisited, using an immunocompetent murine colon adenocarcinoma model with MC38cea cells permissive to retargeted MeV-antiCEA. This study demonstrated that MeV-antiCEA, armed with GM-CSF, not only enhanced the median overall survival but also led to a durable complete remission in one-third of the mice. Furthermore, mice treated with MeV-GM-CSF-antiCEA rejected tumors upon re-challenge, demonstrating an adaptive memory immune response with a lasting protection (Grossardt et al. 2013).

Immune checkpoint inhibitors are a breakthrough in cancer immunotherapy that enable persistent activation of tumor-infiltrating lymphocytes (TILs); for this and other reasons, OVs are an ideal partner for a combination therapy. OVs attract cytotoxic T lymphocytes to the tumor and upregulate the expression of PD-L1 on both cancer and the immune cells (Hardcastle et al. 2017; Liu et al. 2017). In an attempt to combine MeV with immune checkpoint inhibition (ICI), we virally encoded PD-L1 and CTLA-4 antibodies and found that the MeV virotherapy was greatly enhanced by ICI in an immune-competent mouse melanoma model (Engeland et al. 2014). This demonstrated a synergy between the two treatment modalities and showed how they can be engineered into a single functional monotherapy. Bi-specific T-cell engagers (BiTEs) are another strategy to engage TILs toward cancer cells. BiTEs are single-chain variable fragments with a dual specificity for T cells and a selected tumor antigen, thereby bringing them in proximity to form an artificial immunological synapse (Dietz and Engeland 2020). We have shown in a murine melanoma model using transgenic B16-CD20-CD46 cells that MeV encoding CD3xCD20 BiTE causes a significant increase in CD8+ TILs, and that it induces a protective anti-tumor immunity (Speck et al. 2018). In another study, MeV-encoded IL-12 likewise enhanced an abundance of CD8+ TILs. Depletion experiments showed that this T-cell subtype was crucial for its efficacy (Veinalde et al. 2017). MeV encoding IL-12 also showed a superior anti-tumor efficacy in MC38cea and B16-hCD46 tumor models, when compared to an IL-15 encoding variant (Backhaus et al. 2019). Finally, MeV-encoding tumor-associated antigens were engineered as an oncolytic cancer vaccine platform, and showed to successfully prime and activate CD8+ T cells (Busch et al. 2020).

16.2.3 Shielding

While oncolytic MeV offers many advantages as a therapeutic agent (excellent safety profile; possibility for multi-level genetic engineering), a major hurdle for a systemic application is the pre-existing neutralizing antibodies. This is the case, as most people in the Western population were previously vaccinated to confer an immunity against pathogenic MeV strains (Bankamp et al. 2011; Polack et al. 2000). Thus, the neutralizing antibodies may inactivate the virus before it reaches the tumor location (Russell and Whye Peng 2009) and pose a challenge for a systemic administration of the therapeutic virus. A clinical trial of an intraperitoneal MeV treatment in ovarian cancer patients with neutralizing antibodies toward MeV was encouraging. However, it was hypothesized that the results could be improved by removing the pre-existing neutralizing antibodies (Galanis et al. 2015). This was probably due to the success of a clinical trial for disseminated myeloma that utilized a systemic administration of MeV in patients without neutralizing antibodies toward the virus (Russell et al. 2014).

While these results were promising, novel strategies emerged to circumvent the potential obstacle of pre-existing neutralizing antibodies. One approach is pseudotyping, i.e., the substitution of envelope glycoproteins of one virus species with another. Since MeV-neutralizing antibodies recognize and achieve neutralization via binding to the viral F and H glycoproteins (Polack et al. 2000), exchanging these surface proteins with those of a closely related virus, which the general population has not yet been exposed to, would confer a humoral immune evasion to the administered pseudotyped MeV. The envelope exchange was achieved with several viruses with varying degrees of success. MeV was successfully pseudotyped with the closely related canine distemper virus (CDV). This "shielded" MeV-CDV vector was able to evade anti-MeV-neutralizing antibodies, but still had similar oncolytic properties compared to the original MeV (Miest et al. 2011; Muñoz-Alía and Russell 2019). On the other hand, pseudotyping MeV with Tupaia paramyxovirus (TPMV; also from the *Paramyxoviridae* family, but more distantly related) generated a virus that was too attenuated for a viable oncolytic therapy option (Hudacek et al. 2013). In two other attempts, MeV surface proteins were exchanged with the glycoprotein G of vesicular stomatitis virus (VSV; Rhabdovirus) and the surface proteins of the Newcastle disease virus (NDV; Paramyxovirus), which are both not endemic in the general human population. Unfortunately, both pseudotyped constructs did not yield functional viral particles (Neault et al. 2022), showing that an envelope exchange might need more fine-tuning by designing sophisticated chimeric surface proteins.

16.3 Manufacturing as the Key to Bring MeV into Clinics

Early clinical trials of oncolytic MeV indicated that the treatment is well tolerated with only mild unwanted effects (Galanis et al. 2010). A standard vaccine dose is a 1000 tissue culture infective dose of 50% ($TCID_{50}$) administered subcutaneously, compared to the highest intravenous oncolytic MeV dose of 10^{11} $TCID_{50}$ given intravenously within a clinical trial to treat disseminated myeloma (Russell et al. 2014). Although these patients were predominantly older, had many courses of previous chemotherapy, and were therefore more frail and immune-compromised to some degree, they tolerated the treatment well.

Galanis et al. (2010) recruited 21 patients with recurrent or progressive ovarian or primary peritoneal cancer, all of whom had been previously treated with chemotherapy. As an inclusion criterion, patients had to have a proven immunity to MeV. In this trial, MeV immunity was determined by anti-measles IgG levels >20 ELISA units mL^{-1}. In order to monitor infectivity, the MeV Edmonston vaccine strain was engineered to express CEA, an oncofetal antigen that is produced in trace amounts by normal adult cells. MeV-CEA was injected into the intraperitoneal cavity with escalating doses ranging from 10^3 to 10^9 $TCID_{50}$, but no dose-limiting toxicity was observed. The median overall survival of the patients was 12.15 months.

Another clinical trial in ovarian cancer used MeV to encode the sodium iodine symporter (NIS) for the monitoring of biodistribution (Galanis et al. 2015). This time, only the higher doses of MeV, at 10^8 and 10^9 $TCID_{50}$, were administered intraperitoneally.

There are several ongoing MeV clinical trials, including patients with either glioblastoma multiforme (NCT00390299), mesothelioma (NCT01503177), squamous cell carcinoma of the head and neck (NCT01846091), peripheral nerve sheath tumor (NCT02700230), multiple myeloma, (NCT02192775, NCT00450814), ovarian, fallopian, or peritoneal cancer (NCT00408590, NCT02364713). An interim report showed an improvement in two heavily pre-treated patients with myeloma (Russell et al. 2014). This was the first report about using MeV via an intravenous injection, which supports the safety of a systemic administration at a dose of 10^{11} $TCID_{50}$. The dramatic improvement seen in these patients can be explained, at least in part, by their low levels of neutralizing measles antibodies.

As a consequence of the observed dose–response relationships in early clinical trials (Galanis et al. 2010, 2015; Russell et al. 2014), the highest feasible doses are currently administered in trials, even though the maximally tolerated dose (MTD) for applied oncolytics apparently was not reached. Currently, the feasible dosage is limited by the manufacturing process. It is without a doubt that the huge doses needed for a clinical application of oncolytic MeV require a sophisticated production concept. A large-scale manufacturing of the required high-titer, highly purified, good manufacturing practice (GMP)-grade recombinant MeV remains challenging (Ungerechts et al. 2016), despite the development of upstream and downstream processes including the production in serum-free cell cultures, tangential flow filtration, and diafiltration (Langfield et al. 2011; Eckhardt et al. 2021; Loewe

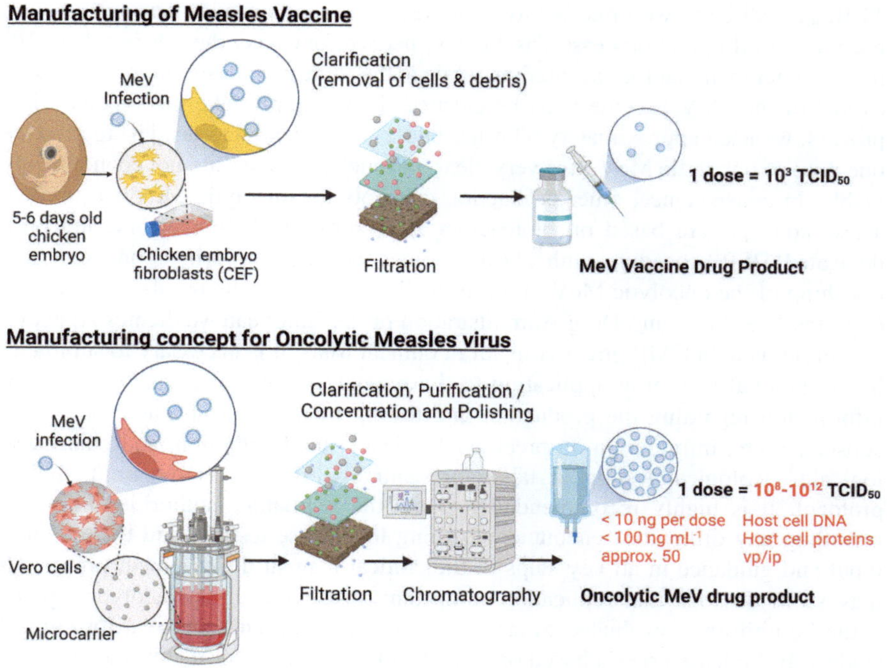

Fig. 16.2 Schematic overview of the manufacturing effort for a MeV vaccine dose, compared to a MeV dose for cancer treatment (Figure has been created with BioRender.com)

et al. 2020, 2022). The data situation for detailed information on production concepts for oncolytic MeV has, so far, been rather sparse.

As all oncolytic MeV are based on MeV vaccine strains (Buijs et al. 2015), the question might arise as to why the production process for MeV vaccine is not simply adapted to the need of an oncolytic MeV. To understand why this is not as simple, the MeV vaccine production is depicted in Fig. 16.2. A production process is generally divided into two phases: the upstream processing (here, the set-up of the host cells/producer cells for the MeV production process, and their cultivation and propagation up to the respective production scale in the bioreactor), and the downstream processing (here, the MeV purification and concentration steps). The upstream processing (USP) of the MeV vaccine is done with chicken embryo fibroblasts as host cells. These cells are derived from 5- to 6-day-old embryonated eggs (Betáková et al. 2013). The host cells are grown in a production system, such as roller bottles or cell factories, and infected with the MeV. The MeV infects the cells, which then form syncytia, get lysed, and release new MeV particles. The DSP for the MeV vaccine is then very simple, as the whole supernatant is just clarified once, to remove the cell debris and any larger particles. Then the final vaccine is more or less ready, and for the formulation, just lactose, mannitol, and sorbitol are added to stabilize the MeV (Betáková et al. 2013). One MeV vaccine dose contains 10^3

$TCID_{50}$. And here, we directly have one key consideration with the regard to the necessity for different processes, as the dose needed for cancer therapy is at least 10^6 times higher than that for vaccination, while at the same time, a similar purity profile as that of the MeV vaccine must be ensured. If we simply take the vaccine MeV process, we need approximately 63 roller bottles or 40 cell factories, just to produce one dose of oncolytic MeV. It is very clear that such processes are not economically viable. In consequence, other production concepts for oncolytic MeV are needed. These concepts are based on bioreactors, e.g., stirred-tank bioreactors, and on a dedicated DSP procedure with clarification, concentration/purification, and final polishing of the oncolytic MeV to reach the limits given by the regulatory agencies (e.g., the U.S. Food and Drug Administration or the European Medicines Agency).

Before using a GMP-grade virus lot in clinical trials, it is necessary to submit an investigational new drug application to the regulatory authorities that contains all information regarding the production and testing of the clinical-grade MeV drug substance, i.e., information on preclinical efficacy, biodistribution, and pharmacological/toxicological testing in laboratory animals, as well as the draft clinical protocol. It is highly recommended to keep the regulatory authorities informed regarding any difficulties encountered during lot release testing, and to seek their input and guidance in all key steps in the clinical reagent development process. It may seem obvious that replication-competent viruses, such as oncolytic agents, could be uniformly evaluated in terms of manufacturing and approval for clinical studies. But release criteria based on validated test methods for the characterization of the critical quality attributes—related to the identity, purity, potency, and safety of the OV products (Ungerechts et al. 2016)—can vary, taking the nature of the respective virus type into account due to its inherent biologically specific properties such as size, host species, tropism, etc. Thus, to find common ground, most release criteria of clinical OV platforms are derived from those defined in GMP specifications of viral vaccine drugs, most of them consisting of formulations with low amounts of total viral particles (vp) versus infectious particles (ip), sufficient for establishing the desired vaccination effect.

Current MeV platforms require large doses of the viral product, well in excess of vaccine doses, in order to see an effective delivery to the tumor sites, and the related therapeutic effects. Current batch preparations meet the requirements for clinical trials, but purity may be further improved. MeV products are tested for identity, purity, potency, adventitious agents, contaminating host cell DNA (HCD) and host cell proteins (HCP) (Ungerechts et al. 2016). The ratio between total vp and ip is a critical parameter for the quality of virus preparations. Although a certain vp/ip ratio is not specified by the regulatory agencies, a value of ≈ 50 in MeV preparations can be assumed on average. Higher virus concentrations often come along with an accumulation of impurities, such as HCD and HCP. Eliminating these by more extensive purification procedures, the integrity of infectious viral particles may suffer, resulting in a higher ratio of vp:ip, often depending on the sensitivity of the virus type.

As some host cell lines can harbor tumorigenic or retroviral sequences that may be capable of transmitting an infection, the U.S. Food and Drug Administration

(FDA) guidelines recommend to take steps to minimize the biological activity of any residual DNA associated with the manufactured viral preparation. This can be accomplished by reducing the size of the DNA to below the size of a functional gene, and by decreasing the amount of residual DNA, e.g., through a nuclease process step. It is recommended to limit the amount of residual DNA for continuous non-tumorigenic cells to less than 10 ng/dose, and the DNA size to below approximately 200 base pairs (FDA n.d.; FDA Chemistry, Manufacturing, and Control (CMC) n.d.; U.S. Food and Drug Administration n.d.). The limit of HCP is given as 100 ng mL^{-1}. As these are general recommendations, based on the production of viral vaccines, the limit of 10 ng HCD per dose MeV vaccine drug with only 1000 $TCID_{50}$ can easily be fulfilled. However, for a large-scale production of high-titer oncolytic drug products with at least 10^8 $TCID_{50}$ per dose, the same limit of residual 10 ng HCD per dose poses a challenge—even after a Benzonase treatment and purification plus concentration steps.

In contrast, specific HCD and HCP limits per dose are not given in the European Medical Agency's (EMA) current guideline on quality, non-clinical, and clinical requirements for investigational advanced therapy medicinal products (ATMP) in clinical trials (Guideline on quality n.d.). EMA recommends that process-related impurities (e.g., media residues, growth factors, HCP, HCD, cell fragments or non-viable cells, precursors, degradation products, aggregates) should be kept to a minimum, or a risk assessment must be provided. Based on the identified risks, some consideration should be given to the maximum amount of HCP and HCD for the highest clinical dose, and also an estimation of the clearance of impurities should be provided.

This shows that the release criteria (e.g., limit of impurities) of the final GMP-grade drug product must be coordinated with the local regulating authorities responsible for the site where the clinical study is to be conducted.

16.4 Upstream Process Considerations for MeV

Oncolytic MeV, based on genetically engineered attenuated vaccine strains, is currently under investigation in clinical trials as a promising approach for the treatment of cancer, with the potential to induce or enhance an immune-mediated tumor rejection. In contrast to using MeV as a vaccine agent, the oncolytic activity as an ATMP depends on a high concentration of infectious particles. While the size range of pleomorphic MeV particles is often quoted as 100–300 nm (Knipe and Howley 2021), in practice, MeV must be treated as 1 μm particles that are extremely shear-sensitive, in order to maximize recoveries and to retain infectivity. Therefore, the entire production and purification process must take place under gentle and aseptic GMP conditions (Langfield et al. 2011). Clinical batches of MeV are produced in Vero cells adapted to a serum-free growth in cell factory multilayer vessels, resulting in 50% of the virus in the supernatant, and 50% staying associated to the cells. The supernatant is clarified by filtration and treated with Benzonase to digest the nucleic acids of the host cell. MeV particles are then concentrated and purified using tangential flow filtration and diafiltration, followed by a final passing through a clarifying filter prior to vialing and storage at <−65 °C.

Previously, various compounds and excipient formulations were tested as stabilizers to protect the integrity of the viral envelope (i.e., viral infectivity) and to inhibit an aggregation of virus particles. MeV formulated with a buffered sucrose solution, additionally containing magnesium chloride, showed an enhanced physical stability, providing a virus longevity of more than 6 years, with a stable infectivity of the sensitive agent.

16.4.1 MeV Stability

The challenge to build up an efficient MeV production process lies in the fact that the MeV itself is a sensitive virus. As mentioned above, MeV is an enveloped virus, which is prone to any physiochemical stress. At temperatures of 37 °C, which are ideal for the growth of most host cells, the MeV half-life time is less than 1 h (Weiss et al. 2013). In USP, it is necessary to find a compromise between a sufficient growth of the host cells and the thermal degradation of the MeV. To solve this conflict, USP can be divided into two phases: a first growth phase of the host cells at 37 °C, in which the host cells proliferate and build up biomass, and a second phase at 30–32 °C, during which the MeV infection and production occurs. Further, the MeV is degraded in an environment with an acidic pH. For USP, this means that the pH value must be kept in the range of 7 or higher. As the host cells also prefer pH values of 7.2–7.4, this is in agreement with the MeV stability. However, due to the metabolism of the host cells, usually the pH value drops due to the formation of acidic metabolites (e.g., lactate). This must be avoided, or the pH must be regulated to keep it above 7.0. Unfortunately, the MeV has its isoelectric point at pH 6.8 (Loewe et al. 2019a), which, in consequence, means that if the high MeV titers, which are needed for cancer therapy, are reached, the virus tends to aggregate even more around its isoelectric point. This increases the need to ensure a pH value above 7.0. Further, the cell culture medium might be supplemented with additives to prevent the MeV from an unwanted aggregation.

In most cases, USP will take place in bioreactors. Most of the systems are able to apply shear stress to the host cells, simply by getting motion into the system. Although the applied shear stress is not damaging for the host cells, it does have an effect on the MeV titer (Grein et al. 2019). Whether this effect really means a damage of the MeV, or is rather more related to a hindered budding process, must be investigated further. A second aspect in a larger bioreactor system is that an active aeration is needed to satisfy the oxygen demand of the host cells. Unfortunately, a bubble aeration decreases the MeV titer in production runs compared to runs with head space aeration. Before setting up a USP for oncolytic MeV, a profound knowledge of the viral properties must be present, which then has to be included into process design considerations.

16.4.2 Host Cells

Anchorage-dependent Vero cells are the most used continuous cell lines to produce viruses for viral vaccines, vectors, and disease treatments. This is due to their many advantages, such as their broad susceptibility to many viruses, the long-term experience in cell culture, and the regulatory portfolio associated with vaccine manufacturing organizations and health authorities worldwide (Kiesslich and Kamen 2020; Kiesslich et al. 2021).

Also, for the production of MeV, the Vero cell line is currently the optimal host cell line. In 2018, Grein et al. screened several cell lines for their suitability to produce oncolytic MeV. In this screening, several cancer cell lines, beside Vero cells, were analyzed, given the fact that MeV has a natural preference to infect cancer cells. Interestingly, a high MeV infection did not automatically lead to high MeV titers in the supernatant. For example, the cell lines HeLa, RPMI8226, and SCLC-21H had an infection efficiency of 100%, but the measured MeV titers in the supernatant were below 10^3 $TCID_{50}$ mL^{-1}. Among all tested cancer cell lines, the BJAB (a Burkitt lymphoma cell line) cells showed high MeV titers in preliminary experiments in T-flasks, but also in shaking flasks. However, in the stirred-tank bioreactor, the MeV titer decreased dramatically. Efficient replication kinetics in a bioreactor system are essential for a large-scale manufacturing of GMP-grade oncolytic MeV. Depending on the genetic engineering of MeV constructs, titers are typically found in the range of 10^6–10^8 $TCID_{50}$ mL^{-1} in the supernatant of infected Vero cells, seeded in 2D systems (e.g., T-flasks and cell factory stacks) (Backhaus et al. 2019; Grein et al. 2017c). In a stirred-tank bioreactor, however, the Vero cells show a robust MeV replication, reaching high MeV titers of 10^9 $TCID_{50}$ mL^{-1}.

However, anchorage-dependent host cells, just as Vero cells, show limitations in the scale-up in manufacturing, due to long seed-trains, leakages in scale-up considerations, and higher costs due to surface-providing materials such as microcarriers, or the usage of enzymes for cell detachment. Therefore, a suspension Vero cell line in a serum-free medium (SFM) is desirable for a sustainable large-scale production (Kiesslich and Kamen 2020; Fulber et al. 2021).

Accordingly, the adaptation process for the anchorage-dependent Vero cells in a suspension culture has been of interest for several years. First efforts were achieved in 1992, when Litwin et al. adapted Vero cells to grow in aggregates in a serum-free suspension culture (Litwin 1992). Finally, in 2009, Paillet et al. were able to establish a single-cell Vero suspension culture to enable a high titer production for the vesicular stomatitis virus, the herpes simplex virus 1, and the polio virus 1, using perfusion bioreactors on a 4.5 L scale. However, the adaptation process was lengthy and lasted about 17 weeks to achieve an anchorage-independent growth of Vero cells (Paillet et al. 2009).

This adaptation process was shortened to 5–8 weeks by Shen et al. (2019), and Rourou et al. (2019). Both groups developed their own SFM and adaptation protocols for the Vero cells, as no commercial media enabled the growth of Vero suspension cells (Shen et al. 2019; Rourou et al. 2019).

For the overall viral production, high cell densities are favorable to enable larger manufacturing scales for vaccines. Shen et al. showed that Vero cells in suspension showed growth limitations when no media was exchanged in batch cultures (Shen et al. 2019). Therefore, the perfusion-mode seems to be the most successful cultivation mode for a high cell-density virus production (Kiesslich and Kamen 2020; Kiesslich et al. 2021; Fulber et al. 2021; Shen et al. 2019; Rourou et al. 2019; Lee et al. 2020). The group of Amine A. Kamen (Kiesslich and Kamen 2020; Kiesslich et al. 2021; Fulber et al. 2021; Shen et al. 2019) has provided the latest data about the production of viruses, such as the recombinant vesicular stomatitis virus for vaccine candidates against COVID-19 or HIV in suspended Vero cells, and also includes the work from Shen, Kiesslich, and Fulber et al. (Kiesslich and Kamen 2020; Kiesslich et al. 2021; Fulber et al. 2021; Shen et al. 2019).

We adapted Vero cells to a suspension using the media of Shen et al. (2019), but also succeeded with a new SFM, developed by one of our cooperation partners. We needed 10 weeks to adapt the cells, and used different additives and strategies to achieve the single cell suspension. We carried out MeV production runs, using our own adapted Vero suspension cells, and measured high MeV titers in the supernatant. Therefore, suspension cell cultures are a viable option for MeV production and could help to overcome some bottlenecks related to anchorage-dependent Vero cells in larger-scale manufacturing.

16.4.3 Cell Culture Medium

The cell culture medium is an essential part of USP, as it supplies nutrients (e.g., glucose or other carbon sources, glutamine, and amino acids, or other nitrogen sources), salts, vitamins, and a variety of growth factors, hormones, cytokines, and chemokines, which are essential for a host cell growth and an MeV production. First investigations on an MeV production were all made with a serum-containing medium (SCM) including 10% fetal calf serum (Scott and Choppin 1982). As the use of fetal calf-serum is outdated due to various reasons discussed elsewhere (Gstraunthaler and Lindl 2017), SFM, or ideally chemically defined media (CDM), are established for MeV production. For the USP of MeV, it is necessary to consider that, depending on the process set-up, two different media may be needed: a first medium, optimized for the growth of the Vero host cells, and a second medium, optimized for the MeV production (including additives, stabilizing the MeV against thermal degradation and to avoid aggregation). Grein et al. showed that Vero cell growth and MeV production are possible in several SFM, e.g., VP-SFM or OptiPro (Grein et al. 2019). Nonetheless, the serum has a protective effect on the MeV, as the shear sensitivity of the MeV in the SFM was much higher than in the serum-containing reference medium. We further investigated the MeV production in CDM (Fig. 16.3). Although the Vero cells grew slower in CDM

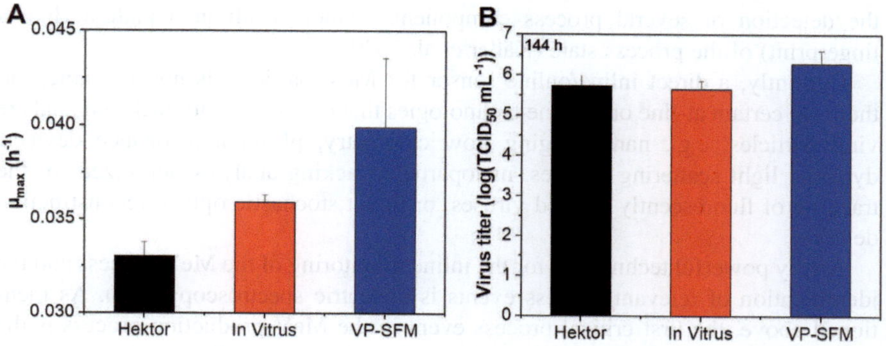

Fig. 16.3 (**a**) Maximum growth rate during the cultivation of Vero cells in the CDM Hektor (black) and InVitrus (red), and the hydrolysate-containing reference medium VP-SFM (blue). (**b**) Final MeV titers at TOH (144 h p.i)

(Hektor and InVitrus, Cell Culture Technologies), the reached MeV titers were as high as in the serum-free VP-SFM. Nonetheless, it is important to keep in mind that each change in USP directly impacts the outcome in DSP. We showed that the production of MeV in SFM reduces the DSP yield of different unit operations (e.g., TFF or IEX) by at least one log-unit (Loewe et al. 2022).

16.4.4 Process Monitoring

The USP for MeV is complex and requires the knowledge and the identification of critical process events. Using anchorage-dependent Vero cells as host cells, the MeV-USP has four important process events.

The first event is the attachment of the Vero cells to the growth surface (e.g., in stirred-tank bioreactors, to microcarriers). Depending on the seeding density, this phase can be followed by a growth phase of the host cells. If the cell density is high enough, the cells are infected by the MeV. At a high multiplicity of infection (MOI), almost all cells are infected in the following 45 h at 37 °C, and so-called syncytia (fused giant cells) are formed. This is the second critical process event. After syncytia formation, the MeV budding and release begins, until finally all host cells are lysed. Between the third critical process event (the start of MeV release) and the fourth (complete lysis of the host cells), the optimal time of harvest (TOH) must be found (Grein et al. 2017b). In agreement with the PAT strategy of the regulatory agencies, those critical process events must be determined at-line or on-line (Manser and Glenz 2022) during the MeV production. Certain inline technologies are available, allowing the monitoring and control of standard parameters such as temperature, pH, dissolved oxygen, and carbon dioxide, but also of cellular parameters such as cell growth or the consumption of glucose. Inline spectroscopy technologies allow the detection of several process components, which result in a pattern (like a fingerprint) of the process state (Käßer et al. 2020).

Currently, a direct inline/online sensor for MeV particles is not available, but there are certain at-line or off-line technologies that can be used to track and analyze viral particles, e.g., nano-imaging flow cytometry, plasmon resonance devices, dynamic light scattering devices, nanoparticle tracking analysis optimized for the tracking of fluorescently labeled viruses, or direct stochastic optical reconstruction devices.

A very powerful technology for the inline monitoring of the MeV process and the identification of relevant process events is dielectric spectroscopy (DS). As mentioned above, the first critical process event in the MeV production process is the proper attachment of the Vero host cells to the microcarrier. This can be monitored by the DS very well, as the attachment correlates with a decrease of the permittivity signal, given the fact that, due to the cell attachment, a strong morphological change in the cell population occurs. A similar correlation was also suggested by Downey et al. (2014). Additionally, with a decrease in permittivity, the conductivity changes. Due to the cellular attachment, the charge density of the microcarrier surface changes and leads to an increase in the degree of freedom of the previously bound ions. This, in turn, results in an increase in conductivity, as observed during DS. Further, DS is an ideal technology to monitor cell growth on microcarriers. Shown for several processes, the growth determined inline correlated very well with any data determined offline. As DS is not only a tool for the measurement of biomass but also offers much information about cellular morphology, it also allows the direct measurement of the second important process event of MeV production, the syncytia formation. After an infection with the virus, the syncytia formation can be identified by an increasing permittivity signal, due to the fusion of the cells to the giant cells. In this way, the global maximum of permittivity could be determined by the maximum of syncytial formation. A simultaneous analysis of the Cole-Cole correction factor α and f_c confirmed that this was actually the detection of a morphological change. The optimal TOH of the MeV USP occurred approximately 40 h after the detection of the global permittivity maximum. Using DS, it could be determined that the optimal harvest time for measles viruses can vary by up to 60 hours, even though the exact same bioreactor system, the same working cell bank, the same virus stock, and the same process parameters were used. However, only the TOH varied, but not the maximum achievable virus titer (Grein et al. 2017a). This shows that the MeV-USP is much more complex and sensitive than previously thought.

16.5 Downstream Process Considerations for MeV

To reach the required critical quality attributes of the final oncolytic MeV drug product, a DSP concept is needed, which consists of several unit operations for the clarification, purification, concentration, and polishing of the MeV (Fig. 16.4).

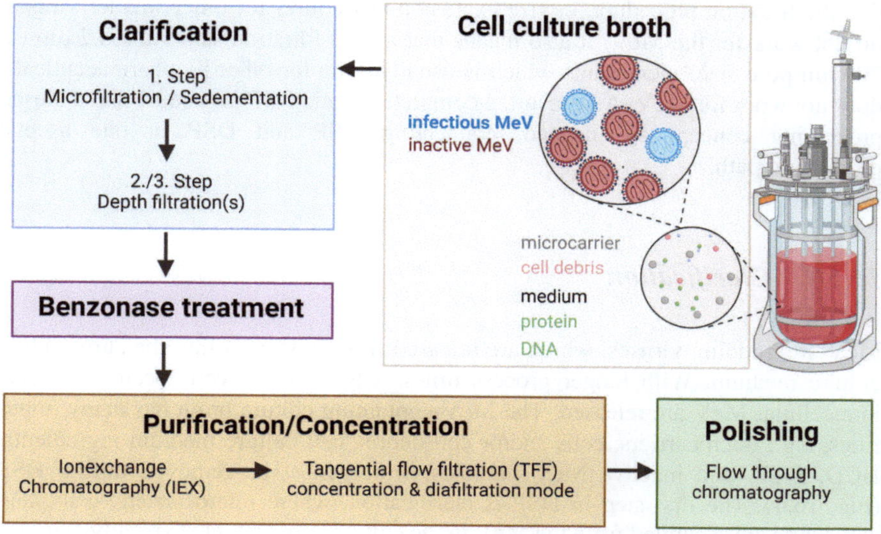

Fig. 16.4 Schematic overview of the impurities of the cell culture broth after USP and the necessary DSP operation units to produce a final MeV drug product. (Figure has been created with BioRender.com)

The described properties regarding the sensitivity of MeV must also be considered in the development of DSP. Due to the MeV temperature sensitivity (Weiss et al. 2013; Loewe et al. 2019a), any process steps should take place at room temperature (RT) or below, and should be kept as short as possible. This does not mean that high fluxes (L m^{-2} h^{-1}) in filtrations are useful to shorten process times, as it increases the shear rates of MeV and can lead to the destruction of infectious MeV particles. Too low shear rates in filtrations, on the other hand, can promote the adsorption of MeV to the filter material (Loewe et al. 2019a) and thus reduce their yield. During diafiltration, chromatography, and as well as in final formulations of MeV therapeutics, a special attention must be paid to the pH value of the virus suspension. In slightly alkaline pH ranges (pH 7–9), the MeV infectivity remains stable, whereas in slightly acidic pH ranges (pH 6–7), it decreases significantly. With all buffers used, a pH value <6 should be avoided at all costs, as the infectivity falls massively in this range. With regard to the ionic strength (0–60 mS cm^{-1}) and the osmolality (0–1000 mOsm kg^{-1}), the MeV has proven to be stable under these conditions in the presence of NaCl, KCl, glucose, or sucrose. Even the use of common buffers such as PBS, Tris, HEPES, phosphate, and citrate-phosphate does not negatively affect MeV infectivity (Loewe et al. 2019a). The addition of magnesium chloride to stabilize MeV suspensions was first described by Rapp et al. (Rapp et al. 1965). Also, 5% [w/v] sucrose turned out to be a stabilizer in potential formulation buffers (Loewe et al. 2019a).

The main challenge for the MeV-DSP is the viral size. A MeV is very large (300–800 nm, often considered as a 1 μm particle), which, in consequence, means that the filtration procedure, or size exclusion procedures for other smaller viruses, do not work for the MeV. It also means that a final filtration step via a 0.2 μm (= 200 nm pore size) membrane, which is usually done for other biopharmaceuticals, does not work for MeV. As a result, a complete aseptic DSP procedure and a closed processing concept are needed, connecting USP and DSP in one aseptic production path.

16.5.1 Clarification

MeV are budding viruses, which are released from host cells into the surrounding culture medium. With longer process times, a lysis of the cells occurs and also intracellular MeV are released. The MeV-containing culture broth has many impurities, e.g., microcarriers, cells and/or cell debris, cell culture medium ingredients, HCD, HCP, and inactive MeV particles, which must be removed during DSP (Fig. 16.4). The first step in DSP is clarification, where microcarriers, cells, and cell debris are separated from the MeV in the culture supernatant. Typically, a multi-stage clarification process is used.

In the first clarification step, larger particles, such as microcarriers and cells, were removed. On a small scale, this can be done by centrifugation (Sviben et al. 2016; Hall and Martin 1973). However, centrifugation processes are limited in their scalability, and due to the present shear forces, high losses of infectious MeV in the culture broth can occur (Sviben et al. 2016; Loewe 2021). In addition, during centrifugation, cells and cell fragments may be sheared into smaller particles, which, in turn, complicate further DSP (Hutchinson et al. 2006).

Filtrations, such as membrane filtration or depth filtration, are widely used unit operations, which are scalable more easily, and are also suitable for continuous processing (Loewe et al. 2020). Here, solid particles are separated from MeV feeds, depending on the selected pore size (cutoff or nominal rating).

Membrane filtration in a dead-end mode can quickly lead to the formation of a filter cake on the retentate side of the membrane filter. Under certain circumstances, the compressed filter cake acts like another filtration layer with a smaller cutoff, which may also retain MeV in the retentate. For example, the direct dead-end microfiltration of a MeV-containing cell culture supernatant with a 0.45 μm PVDF filter membrane resulted in a loss of 42% of the infectious MeV. Depending on the USP and the produced culture broth, a multi-stage membrane filtration with a decreasing cutoff is recommended.

In contrast to membrane filters, depth filters offer several advantages that justify their use in the clarification of MeV-containing culture broths. Their porous, more three-dimensional filter structure can both retain a large amount of variously sized insoluble particles without forming a filter cake and, depending on the filter matrix used, bind soluble impurities, such as HCD and HCP, via intermolecular interactions (Thakur et al. 2020; Nejatishahidein and Zydney 2021). Depth filters are therefore the preferred filters in comparison to membrane filters—especially, if only a two-stage clarification process is used.

It is worth considering, especially for processes with adherent cells on microcarriers and suspension cells with a high cell density at harvest time, to perform a clarification in a three-stage process. Here, the first clarification step serves the separation of comparatively very large particles (>10 μm), such as microcarriers and intact cells. Removing these large particles in advance lowers the burden for following DSP steps, e.g., a depth filtering with one-fifth of the filter capacity can be used afterward (Eckhardt et al., unpublished data). Especially in large-scale processes, such as those required for the clinical application of oncolytic MeV, the use of depth filters with a much smaller capacity would be very advantageous from both an economical and an ecological point of view.

In laboratory-scale STR processes with microcarriers, the sedimentation of the carriers can be considered. For example, the Cytodex 1 microcarriers used in a process for a high-titer MeV production (Grein et al. 2017a) have a sedimentation velocity of 90 cm h^{-1} (Cytiva life sciences 2022). Depending on the STR geometry and the working volume, an almost complete sedimentation of the microcarriers is possible <30 min after switching off the agitation. The MeV-containing culture broth can then be harvested via a higher-positioned port on the STR, and processed further.

If sedimentation is not an option in a larger-scale process, a membrane filtration/sieving with pore sizes of 30–100 μm should be considered in order to separate microcarriers in a first clarification step. This procedure was also used, for example, in the large-scale production (2× 350 L) of a polio vaccine with Vero cells on microcarriers. Here, the microcarriers were separated by filtration through a sieve with a 75-μm mesh (Thomassen et al. 2013).

For further clarification, a two-stage process using depth filters is recommended. Cells and larger debris can be separated by filtration with pore sizes of 2.5–5 μm, and smaller particles by microfiltration (pore size >0.1 μm). Since non-aggregated MeV have a diameter of 100–300 nm (Ungerechts et al. 2016), the microfiltration membranes should have a pore size of at least 0.4 μm to ensure that the MeV can pass through the filter membrane and do not remain in the retentate. Depending on the characteristics of the surrounding medium, especially the pH value, the MeV aggregates with a particle size of up to 600 nm may also be present (Loewe et al. 2019a). Accordingly, an experimental testing of cutoffs in the range of 0.4–1.0 μm is recommended for a microfiltration.

Depth filters made of polypropylene (Eckhardt et al. 2021; Loewe et al. 2019b; Steppert et al. 2022) and cellulose (Loewe 2021) have proven to be effective for clarifying MeV-containing culture broths. However, especially depending on the medium used in the USP (SCM, SFM, CDM), the optimal process parameters with respect to filter loading and flux must be determined experimentally. The results so far are contradictory, indicating that MeV produced in SFM either adsorb to a higher degree on the depth filter material, resulting in a lower yield of infectious MeV compared to SCM (Loewe 2021), or see no adsorptive effect in SFM (Fig. 16.5). While the content of soluble protein impurities does not change significantly after a

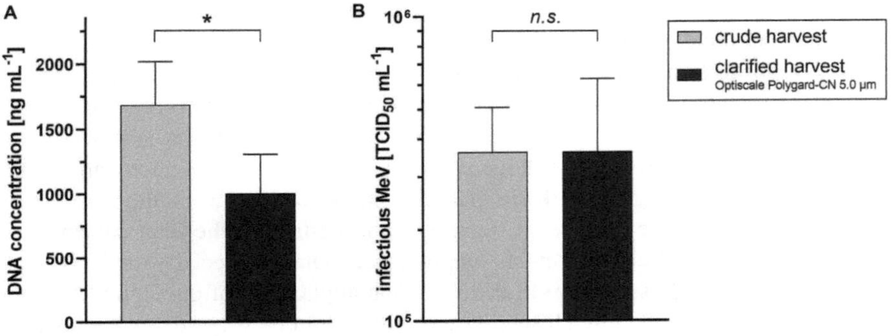

Fig. 16.5 Crude/clarified (Polygard-CN 5.0 μm depth filter) harvest of MeV production in T-flasks with Vero cells in VP-SFM, $n = 3$; (**a**) DNA concentration, (**b**) infectious MeV; group comparison with ratio paired t-test, * = significant ($P = 0.0202$), n.s. = not significant

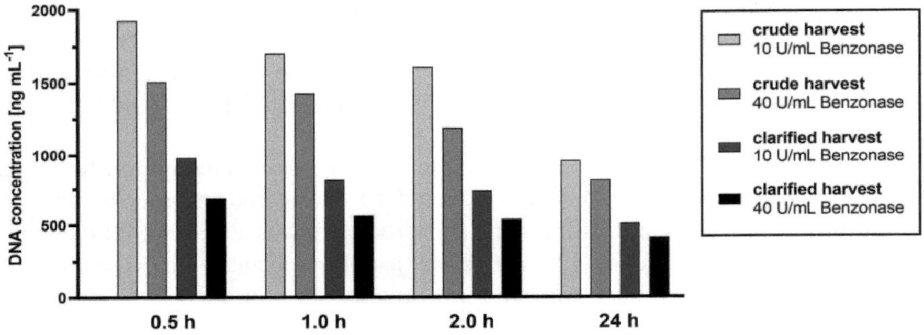

Fig. 16.6 Time- and concentration-dependent Benzonase treatment of a crude/clarified (centrifugation: 300× g, 10 min, RT) DNA model solution (confluent Vero cell culture in T-flasks in VP-SFM, lysed by three freeze/thaw cycles and vortexing) at RT

depth filtration, HCD can already be reduced by 33% (SCM, data not shown) and 45% (SFM, Fig. 16.5), respectively. Since clarifications by centrifugation provide similar results (Fig. 16.6), it can be assumed that the HCD separated here is probably associated with insoluble particles in the culture broth.

16.5.2 Benzonase Treatment

A potential therapeutic end product with infectious MeV must contain less than 10 ng per dose of HCD (U.S. Food and Drug Administration n.d.). To achieve this goal, a Benzonase treatment should be integrated in DSP (Langfield et al. 2011; Steppert et al. 2022). Depending on the incubation time and temperature, 10–50 U/mL of Benzonase in the presence of 2 mM $MgCl_2$ is sufficient to already digest a large fraction of dissolved HCD. Regarding incubation time and temperature, 1–4 h at RT is recommended, as MeV have a comparatively short half-life at temperatures above RT. The pH value of the MeV suspension, on the other hand, can be set to pH = 8.0 (optimum of the Benzonase) (Loewe et al. 2019a).

As mentioned previously, the portion of HCD, associated with insoluble particles of the culture broth, can already be separated via clarification. If a Benzonase treatment is performed after clarification, there is less HCD left to digest. Thus, less Benzonase is required, and/or shorter incubation times can be applied. For example, in a clarified MeV suspension, 10 U/mL and 1 h incubation at RT achieves similar final HCD concentrations as 40 U/mL and 24 h incubation at RT in an unclarified culture broth (Fig. 16.6).

Therefore, if Benzonase treatment is considered within DSP to reduce HCD, it is recommended to be performed after clarification.

16.5.3 Purification/Concentration

Chromatographic and tangential flow filtration methods are suitable for the purification and concentration of MeV. The applied modes and the sequence of methods vary depending on the research group.

We investigated the suitability of ion exchange chromatography for the concentration and purification of oncolytic MeV, using resin-based stationary phases. In screening experiments, anion exchange resins, cation exchange resins, and hybrid materials were investigated. Here, the Eshmuno ® CPX (Merck) material proved to be the most promising resin for MeV purification by cation exchange chromatography (CEX) in a bind-and-elute mode (Eckhardt et al. 2021). CEX was optimized by adjusting the pH value of the MeV feed and buffer to pH = 7.0, which is closer to the pI of MeV (~6.7) (Loewe et al. 2019a) and should, thus, increase the fraction of positive charges on the MeV surface. In addition, the flow rate during sample loading was reduced from 120 cm h^{-1} to 60 cm h^{-1}. With these optimizations, 80.7% of infectious MeV particles were eluted. The MeV eluate was rather pure, as the total protein and the total DNA content were reduced by 98.3% and 80.5%, respectively (Eckhardt et al. 2021).

Steppert et al. (2022) used anion-exchange chromatography and obtained only a moderate MeV yield of 11% in screening experiments, using the anion-exchange resin Eshmuno Q. Interestingly, they also investigated the flow-through chromatography with a CaptoTM Core 700 (which is often used at a later stage in the DSP for a polishing unit operation) at an early stage in the DSP of oncolytic MeV. The principle of flow-through chromatography is that the large MeV cannot enter the pores and flows through the material, whereas impurities, such as proteins and small DNA fragments, can be trapped inside the pores of the CaptoCore resins. Steppert et al. directly applied a clarified MeV suspension to the CaptoCore700 and obtained 61% MeV while 75% of dsDNA and 73% of the total protein impurities were separated. Interestingly, a previously performed Benzonase treatment led to a better DNA depletion of up to 95%, but also to significantly lower MeV yields of maximum 28%. Subsequently, unspecific MeV binding to the filter-based frit of the chromatography column were shown. This was avoided by changing to a net-based frit. Overall, they recovered 95% of the MeV yield and strongly reduced the dsDNA and the total protein content to <15%. As there is no interaction between MeV and CaptoCore700, this flow-through chromatography process can serve as a platform for (i) MeV with engineered envelopes, or (ii) other oncolytic viruses (Steppert et al. 2022). However, this still must be proven, and the performance of the process on a larger scale is still unknown. Further, in this process, no concentration of the MeV is reached, so that a second unit operation (ideally filtration-based) is needed to concentrate the MeV and reach the doses of MeV per mL needed for an oncolytic therapy.

Tangential flow filtration (TFF) is such a unit operation and is often used in DSP in a concentration mode and for the rebuffering of pharmaceutical products into the final formulation by diafiltration. Also, when developing an appropriate TFF strategy for a MeV suspension, the virus sensitivity must be considered. Processing times at room temperature should be kept to a minimum, but a too high flux during TFF should be avoided to not increase the shear rates beyond a detrimental value (Loewe et al. 2019a).

We tested TFF cassettes of different materials (polyethersulfone, (PES) and regenerated cellulose (xRC)) and various molecular weight cutoffs (molecular weight cutoffs (MWCOs); 100 kDa; 300 kDa, 100 nm) with respect to the yield of infectious MeV and the separation of impurities (HCP and HCD) from MeV suspensions in SCM and SFM in a TFF concentration mode (Loewe et al. 2019b). TFF cassettes using PES as well as xRC with MWCO ≤300 kDa are suitable for the lossless concentration of MeV, independent of SCM or SFM. However, regarding the efficiency of the HCP and HCD clearance in the concentration mode of MeV suspensions, there are significant differences depending on the MWCO of the membrane and the surrounding medium. While TFF cassettes with 100 and 300 MWCO result in a similarly good protein separation in SFM (79–87%), in SCM the proteins are partially concentrated in TFF with a 100 MWCO membrane (only 29–37% separation). Also, the HCD clearance was more effective with a 300 kDa membrane (permeability: 0.75 in SFM, 0.51 in SCM) than with 100 kDa membranes (permeability: 0.36–0.53 in SFM, 0.24–0.36 in SCM) (Loewe et al. 2019b).

Based on the results above, Loewe et al. (2022) used the 300 kDa PES cassette for further investigations regarding TFF in the concentration and the diafiltration (DF) mode. It was confirmed that a lossless concentration of the MeV suspension is possible in both the SCM and the SFM medium. Also, during DF with a formulation buffer (50 mM Tris/HCl, 5% [w/v] sucrose, pH 7.4), no significant reductions in infectivity of the MeV suspension were observed, regardless of the medium. Regarding the depletion of impurities, however, significant differences were found, depending on the surrounding medium. Thus, the residual HCD content after concentration and diafiltration with 5 diavolumes was 58% for MeV in SCM, compared to the initial MeV suspension, whereas it could be reduced to 6% for MeV in SFM. Regarding HCP, the differences were less striking, yet still a clear trend was evident (remaining HCP after concentration and DF: 24% in SCM, 5% in SFM) (Loewe et al. 2022).

16.6 Conclusions

Altogether, engineered MeVs as multitasking agents offer promising possibilities for the treatment of heterogeneous tumors, e.g., an avoidance of resistance development, and the implementation of combined therapies. The efforts to develop oncolytic MeV for clinical applications seem worthwhile, given the versatility of the MeV as an oncolytic vector platform (Leber et al. 2020b), the excellent safety record of MeV vaccines (WHO 2017; Di Pietrantonj et al. 2020), as well as its biosafety profile (Baldo et al. 2016) and the genetic stability (Leber et al. 2020a) of recombinant MeV. As the applicable dosages are currently limited by manufacturing, this is a critical step for a clinical realization of the oncolytic MeV therapy concepts.

In general, huge MeV doses (10^{11} $TCID_{50}$ per patient or even more) are needed for oncolytic therapies. The manufacturing of Measles vaccines cannot be adapted easily to the production of oncolytic MeV. Therefore, sophisticated production concepts must be established for oncolytic MeV. During the entire manufacturing process, the unique properties of the MeV must be considered which include its sensitivity toward physiochemical parameters, but also its large size, which requires a complete aseptic manufacturing process (also for DSP). For the USP, the host cells are essential. For this purpose, Vero cells in suspension seem to be the most viable option today. The stirred-tank bioreactor seems to be a good choice for USP, because it is well characterized and allows a tight control of relevant process events. The stirred-tank bioreactor can be equipped with in−/online sensors to control the process. Here, dielectric spectroscopy is useful, as many process-relevant steps during the MeV USP can be monitored. Further, the choice of the cell culture medium is important, as media changes can impact the DSP and decrease the MeV yield during DSP. Serum-free, chemically defined media should be used, and a better understanding of the impact of the media on the MeV and on its behavior during USP and DSP is needed. The DSP of oncolytic MeV is complex and contains several unit operations for clarification and purification/concentration. Different DSP concepts have been published, and currently, it is not clear which one is the optimal solution. For clarification, depth filters have revealed good results for MeV. For purification, several unit operations, such as ion exchange chromatography, flow-through chromatography, tangential flow filtration, and diafiltration, were successfully applied for different MeV strains. From a process point of view, it might be advantageous to build the DSP on unit operations with different separation principles, e.g., combining IEX (in bind-and-elute mode) and filtration (TFF or diafiltration), or flow-through chromatography and filtration (TFF or diafiltration).

As only one oncolytic product (herpes simplex virus Imlygic®, Amgen) has reached the market yet, the release criteria of a final GMP-grade MeV drug product must be coordinated with the local regulating authorities responsible for the site, where the clinical study is planned to be conducted. As the authorities often define country-specific acceptance criteria, this might pose a problem for clinical trials intended as multicenter studies, especially when recruiting a large pool of patients in several continents for phase III trials. Therefore, a global harmonization of the approval criteria and processes between the local regulatory authorities (e.g., FDA, Health Canada, EMA, and other agencies) could be a helpful development for future attempts to bring oncolytic virus platforms into the clinic at a faster pace and to a higher extent.

References

Allen C, Vongpunsawad S, Nakamura T, James CD, Schroeder M, Cattaneo R, Giannini C, Krempski J, Peng K-W, Goble JM et al (2006) Retargeted oncolytic measles strains entering via the EGFRvIII receptor maintain significant antitumor activity against gliomas with increased tumor specificity. Cancer Res 66:11840–11850. https://doi.org/10.1158/0008-5472.CAN-06-1200

Allen C, Paraskevakou G, Iankov I, Giannini C, Schroeder M, Sarkaria J, Puri RK, Russell SJ, Galanis E (2008) Interleukin-13 displaying retargeted oncolytic measles virus strains have significant activity against gliomas with improved specificity. Mol Ther 16:1556–1564. https://doi.org/10.1038/mt.2008.152

Anderson BD, Nakamura T, Russell SJ, Peng K-W (2004) High CD46 receptor density determines preferential killing of tumor cells by oncolytic measles virus. Cancer Res 64:4919–4926. https://doi.org/10.1158/0008-5472.CAN-04-0884

Aref S, Bailey K, Fielding A (2016) Measles to the rescue: a review of oncolytic measles virus. Viruses 8. https://doi.org/10.3390/v8100294

Bach P, Abel T, Hoffmann C, Gal Z, Braun G, Voelker I, Ball CR, Johnston ICD, Lauer UM, Herold-Mende C et al (2013) Specific elimination of CD133+ tumor cells with targeted oncolytic measles virus. Cancer Res 73:865–874. https://doi.org/10.1158/0008-5472.CAN-12-2221

Backhaus PS, Veinalde R, Hartmann L, Dunder JE, Jeworowski LM, Albert J, Hoyler B, Poth T, Jäger D, Ungerechts G et al (2019) Immunological effects and viral gene expression determine the efficacy of oncolytic measles vaccines encoding IL-12 or IL-15 agonists. Viruses 11. https://doi.org/10.3390/v11100914

Baertsch MA, Leber MF, Bossow S, Singh M, Engeland CE, Albert J, Grossardt C, Jäger D, Kalle CV, Ungerechts G (2014) MicroRNA-mediated multi-tissue detargeting of oncolytic measles virus. Cancer Gene Ther 21:373–380. https://doi.org/10.1038/cgt.2014.40

Baldo A, Galanis E, Tangy F, Herman P (2016) Biosafety considerations for attenuated measles virus vectors used in virotherapy and vaccination. Hum Vaccin Immunother 12:1102–1116. https://doi.org/10.1080/21645515.2015.1122146

Bankamp B, Takeda M, Zhang Y, Xu W, Rota PA (2011) Genetic characterization of measles vaccine strains. J Infect Dis 204(Suppl 1):S533–S548. https://doi.org/10.1093/infdis/jir097

Bell J, McFadden G (2014) Viruses for tumor therapy. Cell Host Microbe 15:260–265. https://doi.org/10.1016/j.chom.2014.01.002

Betáková T, Svetlíková D, Gocník M (2013) Overview of measles and mumps vaccine: origin, present, and future of vaccine production. Acta Virol 57:91–96. https://doi.org/10.4149/av_2013_02_91

Bluming A, Ziegler J (1971) Regression of Burkitt's Lymphoma in association with Measles infection. Lancet 298:105–106. https://doi.org/10.1016/S0140-6736(71)92086-1

Bossow S, Grossardt C, Temme A, Leber MF, Sawall S, Rieber EP, Cattaneo R, Kalle CV, Ungerechts G (2011) Armed and targeted measles virus for chemovirotherapy of pancreatic cancer. Cancer Gene Ther 18:598–608. https://doi.org/10.1038/cgt.2011.30

Breitbach CJ, Lichty BD, Bell JC (2016) Oncolytic viruses: therapeutics with an identity crisis. EBioMedicine 9:31–36. https://doi.org/10.1016/j.ebiom.2016.06.046

Buijs PRA, Verhagen JHE, van Eijck CHJ, van den Hoogen BG (2015) Oncolytic viruses: from bench to bedside with a focus on safety. Hum Vaccin Immunother 11:1573–1584. https://doi.org/10.1080/21645515.2015.1037058

Busch E, Kubon KD, Mayer JKM, Pidelaserra-Martí G, Albert J, Hoyler B, Heidbuechel JPW, Stephenson KB, Lichty BD, Osen W et al (2020) Measles vaccines designed for enhanced CD8+ T cell activation. Viruses 12. https://doi.org/10.3390/v12020242

Cattaneo R, Miest T, Shashkova EV, Barry MA (2008) Reprogrammed viruses as cancer therapeutics: targeted, armed and shielded. Nat Rev Microbiol 6:529–540. https://doi.org/10.1038/nrmicro1927

Cytiva life sciences. Cytodex 1, Cytodex 3 – Instructions for Use. Available online: https://cytiva-delivery.sitecorecontenthub.cloud/api/public/content/digi-11574-pdf (accessed on 24 June 2022)

Devaux P, Hodge G, McChesney MB, Cattaneo R (2008) Attenuation of V- or C-defective measles viruses: infection control by the inflammatory and interferon responses of rhesus monkeys. J Virol 82:5359–5367. https://doi.org/10.1128/JVI.00169-08

Devaux P, Hudacek AW, Hodge G, Reyes-Del Valle J, McChesney MB, Cattaneo R (2011) A recombinant measles virus unable to antagonize STAT1 function cannot control inflammation and is attenuated in rhesus monkeys. J Virol 85:348–356. https://doi.org/10.1128/JVI.00802-10

Di Pietrantonj C, Rivetti A, Marchione P, Debalini MG, Demicheli V (2020) Vaccines for measles, mumps, rubella, and varicella in children. Cochrane Database Syst Rev 4:CD004407. https://doi.org/10.1002/14651858.CD004407.pub4

Dietz L, Engeland CE (2020) Immunomodulation in oncolytic measles virotherapy. Methods Mol Biol 2058:111–126. https://doi.org/10.1007/978-1-4939-9794-7_7

Dispenzieri A, Tong C, LaPlant B, Lacy MQ, Laumann K, Dingli D, Zhou Y, Federspiel MJ, Gertz MA, Hayman S et al (2017) Phase I trial of systemic administration of Edmonston strain of measles virus genetically engineered to express the sodium iodide symporter in patients with recurrent or refractory multiple myeloma. Leukemia 31:2791–2798. https://doi.org/10.1038/leu.2017.120

Dörig RE, Marcil A, Chopra A, Richardson CD (1993) The human CD46 molecule is a receptor for measles virus (Edmonston strain). Cell 75:295–305. https://doi.org/10.1016/0092-8674(93)80071-L

Downey BJ, Graham LJ, Breit JF, Glutting NK (2014) A novel approach for using dielectric spectroscopy to predict viable cell volume (VCV) in early process development. Biotechnol Prog 30:479–487. https://doi.org/10.1002/btpr.1845

Eckhardt D, Dieken H, Loewe D, Grein TA, Salzig D, Czermak P (2021) Purification of oncolytic measles virus by cation-exchange chromatography using resin-based stationary phases. Sep Sci Technol:1–11. https://doi.org/10.1080/01496395.2021.1955267

Engeland CE, Ungerechts G (2021) Measles virus as an oncolytic immunotherapy. Cancers (Basel) 13. https://doi.org/10.3390/cancers13030544

Engeland CE, Grossardt C, Veinalde R, Bossow S, Lutz D, Kaufmann JK, Shevchenko I, Umansky V, Nettelbeck DM, Weichert W et al (2014) CTLA-4 and PD-L1 checkpoint blockade enhances oncolytic measles virus therapy. Mol Ther 22:1949–1959. https://doi.org/10.1038/mt.2014.160

FDA (n.d.) Guidance for industry – characterization and qualification of cell substrates and other biological materials used in the production of viral vaccines for infectious disease indications. Available online: https://www.fda.gov/media/78428/download

FDA Chemistry, Manufacturing, and Control (CMC) (n.d.) Information for Human Gene Therapy Investigational New Drug Applications (INDs); Guidance for Industry. Available online: https://www.fda.gov/media/113760/download

Fisher K, Hazini A, Seymour LW (2021) Tackling HLA deficiencies head on with oncolytic viruses. Cancers (Basel) 13. https://doi.org/10.3390/cancers13040719

Friedrich K, Hanauer JR, Prüfer S, Münch RC, Völker I, Filippis C, Jost C, Hanschmann K-M, Cattaneo R, Peng K-W et al (2013) DARPin-targeting of measles virus: unique bispecificity, effective oncolysis, and enhanced safety. Mol Ther 21:849–859. https://doi.org/10.1038/mt.2013.16

Fu L-Q, Wang S-B, Cai M-H, Wang X-J, Chen J-Y, Tong X-M, Chen X-Y, Mou X-Z (2019) Recent advances in oncolytic virus-based cancer therapy. Virus Res 270:197675. https://doi.org/10.1016/j.virusres.2019.197675

Fulber JPC, Farnós O, Kiesslich S, Yang Z, Dash S, Susta L, Wootton SK, Kamen AA (2021) Process development for newcastle disease virus-vectored vaccines in serum-free vero cell suspension cultures. Vaccine 9:1335. https://doi.org/10.3390/vaccines9111335

Galanis E, Hartmann LC, Cliby WA, Long HJ, Peethambaram PP, Barrette BA, Kaur JS, Haluska PJ, Aderca I, Zollman PJ et al (2010) Phase I trial of intraperitoneal administration of an oncolytic measles virus strain engineered to express carcinoembryonic antigen for recurrent ovarian cancer. Cancer Res 70:875–882. https://doi.org/10.1158/0008-5472.CAN-09-2762

Galanis E, Atherton PJ, Maurer MJ, Knutson KL, Dowdy SC, Cliby WA, Haluska P, Long HJ, Oberg A, Aderca I et al (2015) Oncolytic measles virus expressing the sodium iodide symporter to treat drug-resistant ovarian cancer. Cancer Res 75:22–30. https://doi.org/10.1158/0008-5472.CAN-14-2533

Grein TA, Loewe D, Dieken H, Salzig D, Czermak P (2017a) High titer oncolytic measles virus production process by integration of dielectric spectroscopy as online monitoring system. Biotechnol Bioeng. https://doi.org/10.1002/bit.26538

Grein TA, Weidner T, Czermak P (2017b) Concepts for the production of viruses and viral vectors in cell cultures. In Gowder SJT (Ed) New insights into cell culture technology. InTech, ISBN 978-953-51-3133-5

Grein TA, Schwebel F, Kress M, Loewe D, Dieken H, Salzig D, Weidner T, Czermak P (2017c) Screening different host cell lines for the dynamic production of measles virus. Biotechnol Prog 33:989–997. https://doi.org/10.1002/btpr.2432

Grein TA, Loewe D, Dieken H, Weidner T, Salzig D, Czermak P (2019) Aeration and shear stress are critical process parameters for the production of oncolytic measles virus. Front Bioeng Biotechnol 7:78. https://doi.org/10.3389/fbioe.2019.00078

Grossardt C, Engeland CE, Bossow S, Halama N, Zaoui K, Leber MF, Springfeld C, Jaeger D, von Kalle C, Ungerechts G (2013) Granulocyte-macrophage colony-stimulating factor-armed oncolytic measles virus is an effective therapeutic cancer vaccine. Hum Gene Ther 24:644–654. https://doi.org/10.1089/hum.2012.205

Grote D, Cattaneo R, Fielding AK (2003) Neutrophils contribute to the measles virus-induced antitumor effect: enhancement by granulocyte macrophage colony-stimulating factor expression. Cancer Res 63:6463–6468

Gstraunthaler G, Lindl T (2017) Auf der Suche nach brauchbaren Serumalternativen. BIOspektrum 23:724–725. https://doi.org/10.1007/s12268-017-0843-z

Guideline on quality, non-clinical and clinical requirements for investigational advanced therapy medicinal products in clinical trials | European Medicines Agency (n.d.). Available online: https://www.ema.europa.eu/en/guideline-quality-non-clinical-clinical-requirements-investigational-advanced-therapy-medicinal#current-version%2D%2Dsection (accessed on 12 August 2022)

Hall WW, Martin SJ (1973) Purification and characterization of measles virus. J Gen Virol 19:175–188. https://doi.org/10.1099/0022-1317-19-2-175

Hanauer JRH, Koch V, Lauer UM, Mühlebach MD (2019) High-affinity DARPin allows targeting of MeV to glioblastoma multiforme in combination with protease targeting without loss of potency. Mol Ther Oncolytics 15:186–200. https://doi.org/10.1016/j.omto.2019.10.004

Haralambieva IH, Ovsyannikova IG, Dhiman N, Vierkant RA, Jacobson RM, Poland GA (2010) Differential cellular immune responses to wild-type and attenuated edmonston tag measles virus strains are primarily defined by the viral phosphoprotein gene. J Med Virol 82:1966–1975. https://doi.org/10.1002/jmv.21899

Hardcastle J, Mills L, Malo CS, Jin F, Kurokawa C, Geekiyanage H, Schroeder M, Sarkaria J, Johnson AJ, Galanis E (2017) Immunovirotherapy with measles virus strains in combination with anti-PD-1 antibody blockade enhances antitumor activity in glioblastoma treatment. Neuro-Oncology 19:493–502. https://doi.org/10.1093/neuonc/now179

Harrington K, Freeman DJ, Kelly B, Harper J, Soria J-C (2019) Optimizing oncolytic virotherapy in cancer treatment. Nat Rev Drug Discov 18:689–706. https://doi.org/10.1038/s41573-019-0029-0

Heinzerling L, Künzi V, Oberholzer PA, Kündig T, Naim H, Dummer R (2005) Oncolytic measles virus in cutaneous T-cell lymphomas mounts antitumor immune responses in vivo and targets interferon-resistant tumor cells. Blood 106:2287–2294. https://doi.org/10.1182/blood-2004-11-4558

Hudacek AW, Navaratnarajah CK, Cattaneo R (2013) Development of measles virus-based shielded oncolytic vectors: suitability of other paramyxovirus glycoproteins. Cancer Gene Ther 20:109–116. https://doi.org/10.1038/cgt.2012.92

Hutchinson N, Bingham N, Murrell N, Farid S, Hoare M (2006) Shear stress analysis of mammalian cell suspensions for prediction of industrial centrifugation and its verification. Biotechnol Bioeng 95:483–491. https://doi.org/10.1002/bit.21029

Jing Y, Zaias J, Duncan R, Russell SJ, Merchan JR (2014) In vivo safety, biodistribution and antitumor effects of uPAR retargeted oncolytic measles virus in syngeneic cancer models. Gene Ther 21:289–297. https://doi.org/10.1038/gt.2013.84

Käßer L, Zitzmann J, Grein T, Weidner T, Salzig D, Czermak P (2020) Turbidimetry and dielectric spectroscopy as process analytical technologies for mammalian and insect cell cultures. Methods Mol Biol 2095:335–364. https://doi.org/10.1007/978-1-0716-0191-4_20

Kaufman HL, Kohlhapp FJ, Zloza A (2015) Oncolytic viruses: a new class of immunotherapy drugs. Nat Rev Drug Discov 14:642–662. https://doi.org/10.1038/nrd4663

Kaufmann JK, Bossow S, Grossardt C, Sawall S, Kupsch J, Erbs P, Hassel JC, von Kalle C, Enk AH, Nettelbeck DM et al (2013) Chemovirotherapy of malignant melanoma with a targeted and armed oncolytic measles virus. J Invest Dermatol 133:1034–1042. https://doi.org/10.1038/jid.2012.459

Kiesslich S, Kamen AA (2020) Vero cell upstream bioprocess development for the production of viral vectors and vaccines. Biotechnol Adv 44:107608. https://doi.org/10.1016/j.biotechadv.2020.107608

Kiesslich S, Kim GN, Shen CF, Kang CY, Kamen AA (2021) Bioreactor production of rVSV-based vectors in Vero cell suspension cultures. Biotechnol Bioeng 118:2649–2659. https://doi.org/10.1002/bit.27785

Knipe DM, Howley PM (eds) (2021) Fields virology. Wolters Kluwer, Philadelphia, PA

Lampe J, Bossow S, Weiland T, Smirnow I, Lehmann R, Neubert W, Bitzer M, Lauer UM (2013) An armed oncolytic measles vaccine virus eliminates human hepatoma cells independently of apoptosis. Gene Ther 20:1033–1041. https://doi.org/10.1038/gt.2013.28

Lange S, Lampe J, Bossow S, Zimmermann M, Neubert W, Bitzer M, Lauer UM (2013) A novel armed oncolytic measles vaccine virus for the treatment of cholangiocarcinoma. Hum Gene Ther 24:554–564. https://doi.org/10.1089/hum.2012.136

Langfield KK, Walker HJ, Gregory LC, Federspiel MJ (2011) Manufacture of measles viruses. Methods Mol Biol 737:345–366. https://doi.org/10.1007/978-1-61779-095-9_14

Leber MF, Bossow S, Leonard VHJ, Zaoui K, Grossardt C, Frenzke M, Miest T, Sawall S, Cattaneo R, von Kalle C et al (2011) MicroRNA-sensitive oncolytic measles viruses for cancer-specific vector tropism. Mol Ther 19:1097–1106. https://doi.org/10.1038/mt.2011.55

Leber MF, Baertsch M-A, Anker SC, Henkel L, Singh HM, Bossow S, Engeland CE, Barkley R, Hoyler B, Albert J et al (2018) Enhanced control of oncolytic measles virus using microRNA target sites. Mol Ther Oncolytics 9:30–40. https://doi.org/10.1016/j.omto.2018.04.002

Leber MF, Hoyler B, Prien S, Neault S, Engeland CE, Förster JM, Bossow S, Springfeld C, von Kalle C, Jäger D et al (2020a) Sequencing of serially passaged measles virus affirms its genomic stability and reveals a nonrandom distribution of consensus mutations. J Gen Virol 101:399–409. https://doi.org/10.1099/jgv.0.001395

Leber MF, Neault S, Jirovec E, Barkley R, Said A, Bell JC, Ungerechts G (2020b) Engineering and combining oncolytic measles virus for cancer therapy. Cytokine Growth Factor Rev 56:39–48. https://doi.org/10.1016/j.cytogfr.2020.07.005

Lee D-K, Park J, Seo D-W (2020) Suspension culture of Vero cells for the production of adenovirus type 5. Clin Exp Vaccine Res 9:48–55. https://doi.org/10.7774/cevr.2020.9.1.48

Litwin J (1992) The growth of Vero cells in suspension as cell-aggregates in serum-free media. Cytotechnology 10:169–174. https://doi.org/10.1007/bf00570893

Liu Z, Ravindranathan R, Kalinski P, Guo ZS, Bartlett DL (2017) Rational combination of oncolytic vaccinia virus and PD-L1 blockade works synergistically to enhance therapeutic efficacy. Nat Commun 8:14754. https://doi.org/10.1038/ncomms14754

Loewe DM (2021) Untersuchung eines filtrationsbasierten Aufreinigungsprozesses für die Applikation onkolytischer Masernviren in der Krebstherapie: Charakterisierung des Masernvirus sowie der Auswirkung des Zellkulturmediums auf den Aufreinigungsprozess, 1. Auflage; Shaker: Düren. ISBN 9783844079722

Loewe D, Häussler J, Grein TA, Dieken H, Weidner T, Salzig D, Czermak P (2019a) Forced degradation studies to identify critical process parameters for the purification of infectious measles virus. Viruses 11. https://doi.org/10.3390/v11080725

Loewe D, Grein TA, Dieken H, Weidner T, Salzig D, Czermak P (2019b) Tangential flow filtration for the concentration of oncolytic measles virus: the influence of filter properties and the cell culture medium. Membranes (Basel):9. https://doi.org/10.3390/membranes9120160

Loewe D, Dieken H, Grein TA, Weidner T, Salzig D, Czermak P (2020) Opportunities to debottleneck the downstream processing of the oncolytic measles virus. Crit Rev Biotechnol 40:247–264. https://doi.org/10.1080/07388551.2019.1709794

Loewe D, Dieken H, Grein TA, Salzig D, Czermak P (2022) A combined ultrafiltration/diafiltration process for the purification of oncolytic measles virus. Membranes (Basel) 12:105. https://doi.org/10.3390/membranes12020105

Manser B, Glenz M (2022) Regulatory and quality considerations of continuous bioprocessing. In Subramanian G (Ed) Process control, intensification, and digitalisation in continuous biomanufacturing. Wiley-VCH, Weinheim, pp 351–375. ISBN 9783527347698

Maurer S, Salih HR, Smirnow I, Lauer UM, Berchtold S (2019) Suicide gene-armed measles vaccine virus for the treatment of AML. Int J Oncol 55:347–358. https://doi.org/10.3892/ijo.2019.4835

Measles, Mumps, Rubella (MMR) Vaccine | CDC. Available online: https://www.cdc.gov/vaccinesafety/vaccines/mmr-vaccine.html (accessed on 12 August 2022)

Miest TS, Yaiw K-C, Frenzke M, Lampe J, Hudacek AW, Springfeld C, Messling VV, Ungerechts G, Cattaneo R (2011) Envelope-chimeric entry-targeted measles virus escapes neutralization and achieves oncolysis. Mol Ther 19:1813–1820. https://doi.org/10.1038/mt.2011.92

Mühlebach MD, Mateo M, Sinn PL, Prüfer S, Uhlig KM, Leonard VHJ, Navaratnarajah CK, Frenzke M, Wong XX, Sawatsky B et al (2011) Adherens junction protein nectin-4 is the epithelial receptor for measles virus. Nature 480:530–533. https://doi.org/10.1038/nature10639

Muñoz-Alía MA, Russell SJ (2019) Probing morbillivirus antisera neutralization using functional chimerism between measles virus and canine distemper virus envelope glycoproteins. Viruses 11. https://doi.org/10.3390/v11080688

Nakamura T, Peng K-W, Harvey M, Greiner S, Lorimer IAJ, James CD, Russell SJ (2005) Rescue and propagation of fully retargeted oncolytic measles viruses. Nat Biotechnol 23:209–214. https://doi.org/10.1038/nbt1060

Naniche D (2009) Human immunology of measles virus infection. Curr Top Microbiol Immunol 330:151–171. https://doi.org/10.1007/978-3-540-70617-5_8

Naniche D, Yeh A, Eto D, Manchester M, Friedman RM, Oldstone MB (2000) Evasion of host defenses by measles virus: wild-type measles virus infection interferes with induction of Alpha/Beta interferon production. J Virol 74:7478–7484. https://doi.org/10.1128/jvi.74.16.7478-7484.2000

Neault S, Bossow S, Achard C, Bell JC, Diallo JS, Leber MF, Ungerechts G (2022) Robust envelope exchange platform for oncolytic measles virus. J Virol Methods 302:114487. https://doi.org/10.1016/j.jviromet.2022.114487

Nejatishahidein N, Zydney AL (2021) Depth filtration in bioprocessing—new opportunities for an old technology. Curr Opin Chem Eng 34:100746. https://doi.org/10.1016/j.coche.2021.100746

Nettelbeck DM, Leber MF, Altomonte J, Angelova A, Beil J, Berchtold S, Delic M, Eberle J, Ehrhardt A, Engeland CE et al (2021) Virotherapy in Germany-recent activities in virus engineering, preclinical development, and clinical studies. Viruses 13. https://doi.org/10.3390/v13081420

Ohno S, Ono N, Takeda M, Takeuchi K, Yanagi Y (2004) Dissection of measles virus V protein in relation to its ability to block alpha/beta interferon signal transduction. J Gen Virol 85:2991–2999. https://doi.org/10.1099/vir.0.80308-0

Paillet C, Forno G, Kratje R, Etcheverrigaray M (2009) Suspension-Vero cell cultures as a platform for viral vaccine production. Vaccine 27:6464–6467. https://doi.org/10.1016/j.vaccine.2009.06.020

Pikor LA, Bell JC, Diallo J-S (2015) Oncolytic viruses: exploiting cancer's deal with the devil. Trends Cancer 1:266–277. https://doi.org/10.1016/j.trecan.2015.10.004

Polack FP, Lee SH, Permar S, Manyara E, Nousari HG, Jeng Y, Mustafa F, Valsamakis A, Adams RJ, Robinson HL et al (2000) Successful DNA immunization against measles: neutralizing antibody against either the hemagglutinin or fusion glycoprotein protects rhesus macaques without evidence of atypical measles. Nat Med 6:776–781. https://doi.org/10.1038/77506

Rapp F, Butel JS, Wallis C (1965) Protection of measles virus by sulfate ions against thermal inactivation. J Bacteriol 90:132–135

Rehman H, Silk AW, Kane MP, Kaufman HL (2016) Into the clinic: Talimogene laherparepvec (T-VEC), a first-in-class intratumoral oncolytic viral therapy. J Immunother Cancer 4:53. https://doi.org/10.1186/s40425-016-0158-5

Rourou S, Ben Zakkour M, Kallel H (2019) Adaptation of Vero cells to suspension growth for rabies virus production in different serum free media. Vaccine 37:6987–6995. https://doi.org/10.1016/j.vaccine.2019.05.092

Russell SJ, Barber GN (2018) Oncolytic viruses as antigen-agnostic cancer vaccines. Cancer Cell 33:599–605. https://doi.org/10.1016/j.ccell.2018.03.011

Russell SJ, Whye Peng K (2009) Measles virus for cancer therapy. Curr Top Microbiol Immunol 330:213–241

Russell SJ, Peng K-W, Bell JC (2012) Oncolytic virotherapy. Nat Biotechnol 30:658–670. https://doi.org/10.1038/nbt.2287

Russell SJ, Federspiel MJ, Peng K-W, Tong C, Dingli D, Morice WG, Lowe V, O'Connor MK, Kyle RA, Leung N et al (2014) Remission of disseminated cancer after systemic oncolytic virotherapy. Mayo Clin Proc 89:926–933. https://doi.org/10.1016/j.mayocp.2014.04.003

Russell L, Peng KW, Russell SJ, Diaz RM (2019) Oncolytic viruses: priming time for cancer immunotherapy. BioDrugs 33:485–501. https://doi.org/10.1007/s40259-019-00367-0

Scott JV, Choppin PW (1982) Enhanced yields of measles virus from cultured cells. J Virol Methods 5:173–179. https://doi.org/10.1016/0166-0934(82)90007-6

Shen CF, Guilbault C, Li X, Elahi SM, Ansorge S, Kamen A, Gilbert R (2019) Development of suspension adapted Vero cell culture process technology for production of viral vaccines. Vaccine 37:6996–7002. https://doi.org/10.1016/j.vaccine.2019.07.003

Speck T, Heidbuechel JPW, Veinalde R, Jaeger D, von Kalle C, Ball CR, Ungerechts G, Engeland CE (2018) Targeted BiTE expression by an oncolytic vector augments therapeutic efficacy against solid tumors. Clin Cancer Res 24:2128–2137. https://doi.org/10.1158/1078-0432.CCR-17-2651

Steppert P, Mosor M, Stanek L, Burgstaller D, Palmberger D, Preinsperger S, Pereira Aguilar P, Müllner M, Csar P, Jungbauer A (2022) A scalable, integrated downstream process for production of a recombinant measles virus-vectored vaccine. Vaccine 40:1323–1333. https://doi.org/10.1016/j.vaccine.2022.01.004

Sviben D, Forčić D, Kurtović T, Halassy B, Brgles M (2016) Stability, biophysical properties and effect of ultracentrifugation and diafiltration on measles virus and mumps virus. Arch Virol 161: 1455–1467. https://doi.org/10.1007/s00705-016-2801-3

Tatsuo H, Ono N, Tanaka K, Yanagi Y (2000) SLAM (CDw150) is a cellular receptor for measles virus. Nature 406:893–897. https://doi.org/10.1038/35022579

Thakur G, Hebbi V, Parida S, Rathore AS (2020) Automation of dead end filtration: an enabler for continuous processing of biotherapeutics. Front Bioeng Biotechnol 8:758. https://doi.org/10.3389/fbioe.2020.00758

Thomassen YE, van t'Oever AG, van Oijen MGCT, Wijffels RH, van der Pol LA, Bakker WAM (2013) Next generation inactivated polio vaccine manufacturing to support post polio-eradication biosafety goals. PLoS One 8:e83374. https://doi.org/10.1371/journal.pone.0083374

U.S. Food and Drug Administration (n.d.) Guidance for Industry – Chemistry, Manufacturing, and Control (CMC) Information for Human Gene Therapy Investigational New Drug Applications (INDs) [Online], 2021. Available online: https://www.fda.gov/regulatory-information/search-fda-guidance-documents/chemistry-manufacturing-and-control-cmc-information-human-gene-therapy-investigational-new-drug (accessed on 10 March 2021)

Ungerechts G, Springfeld C, Frenzke ME, Lampe J, Johnston PB, Parker WB, Sorscher EJ, Cattaneo R (2007a) Lymphoma chemovirotherapy: CD20-targeted and convertase-armed measles virus can synergize with fludarabine. Cancer Res 67:10939–10947. https://doi.org/10.1158/0008-5472.CAN-07-1252

Ungerechts G, Springfeld C, Frenzke ME, Lampe J, Parker WB, Sorscher EJ, Cattaneo R (2007b) An immunocompetent murine model for oncolysis with an armed and targeted measles virus. Mol Ther 15:1991–1997. https://doi.org/10.1038/sj.mt.6300291

Ungerechts G, Frenzke ME, Yaiw K-C, Miest T, Johnston PB, Cattaneo R (2010) Mantle cell lymphoma salvage regimen: synergy between a reprogrammed oncolytic virus and two chemotherapeutics. Gene Ther 17:1506–1516. https://doi.org/10.1038/gt.2010.103

Ungerechts G, Bossow S, Leuchs B, Holm PS, Rommelaere J, Coffey M, Coffin R, Bell J, Nettelbeck DM (2016) Moving oncolytic viruses into the clinic: clinical-grade production, purification, and characterization of diverse oncolytic viruses. Mol Ther Methods Clin Dev 3: 16018. https://doi.org/10.1038/mtm.2016.18

Veinalde R, Grossardt C, Hartmann L, Bourgeois-Daigneault M-C, Bell JC, Jäger D, von Kalle C, Ungerechts G, Engeland CE (2017) Oncolytic measles virus encoding interleukin-12 mediates potent antitumor effects through T cell activation. Onco Targets Ther 6:e1285992. https://doi.org/10.1080/2162402X.2017.1285992

Weiss K, Salzig D, Röder Y, Gerstenberger J (2013) Influences of process conditions on measles virus stability. Am J Biochem Biotechnol 9:243–254. https://doi.org/10.3844/ajbbsp.2013.243.254

WHO. Measles vaccines: WHO position paper – April 2017 (Note de synthèse de l'OMS sur les vaccins contre la rougeole – avril 2017). Wkly Epidemiol Rec 2017, 92, 205–227

Woller N, Gürlevik E, Ureche C-I, Schumacher A, Kühnel F (2014) Oncolytic viruses as anticancer vaccines. Front Oncol 4:188. https://doi.org/10.3389/fonc.2014.00188

Zaoui K, Bossow S, Grossardt C, Leber MF, Springfeld C, Plinkert PK, Kalle CV, Ungerechts G (2012) Chemovirotherapy for head and neck squamous cell carcinoma with EGFR-targeted and CD/UPRT-armed oncolytic measles virus. Cancer Gene Ther 19:181–191. https://doi.org/10.1038/cgt.2011.75

Chapter 17
Recent Advancements and Challenges in Recombinant Expression for Commercial Production of Virus-Like Particles (VLPs)

Abhilasha K. Rani, Wajihul Hasan Khan, Manidipa Banerjee, and Anurag S. Rathore

Abbreviations

BHK	Baby Hamster Kidney
BMMY	Buffered Methanol-Complex Medium
CFDA	Chinese Food and Drug Administration
CHO	Chinese Hamster Ovary
CPV	Canine parvovirus virus
DNA	Deoxyribonucleic acid
ELISA	Enzyme-linked immunosorbent assay
EMA	European medicine agency
eVLPs	Enveloped Virus-like particles
FDA	Food and Drug Administration

Abhilasha K. Rani and Wajihul Hasan Khan contributed equally with all other contributors.

A. K. Rani
School of Interdisciplinary Research, Indian Institute of Technology, New Delhi, India

W. H. Khan
Department of Chemical Engineering, Indian Institute of Technology, New Delhi, India

M. Banerjee
School of Interdisciplinary Research, Indian Institute of Technology, New Delhi, India

Kusuma School of Biological Sciences, Indian Institute of Technology, New Delhi, India
e-mail: manidipa.banerjee@bioschool.iitd.ac.in

A. S. Rathore (✉)
School of Interdisciplinary Research, Indian Institute of Technology, New Delhi, India

Department of Chemical Engineering, Indian Institute of Technology, New Delhi, India
e-mail: asrathore@biotechcmz.com

S. Gautam et al. (eds.), *Bioprocess and Analytics Development for Virus-based Advanced Therapeutics and Medicinal Products (ATMPs)*,
https://doi.org/10.1007/978-3-031-28489-2_17

FMDV	Foot-and-mouth disease virus
GSK	Glaxosmithkline
HBsAg	Hepatitis B
HBV	Hepatitis B virus
HCV	Hepatitis C virus
HEK293T	Human embryonic kidney 293 cells
HEPES	4-(2-hydroxyethyl)-1-piperazineethanesulfonic acid
HEV	Hepatitis E virus
HIV	Human immunodeficiency virus
HPV	Human papillomavirus
HuNov	Human Norovirus
IMDM	Iscove's Modified Dulbecco's Medium
JEV	Japanese encephalitis virus
LB	Luria–Bertani
NMPA	National Medical Products Administration
NoV	Noroviruses
PTM	Post-translational modification
RNA	Ribonucleic acid
RSV	Respiratory syncytial virus
Sf9	Spodoptera frugiperda cell line
SUMO	Small Ubiquitin-like Modifier
VLPs	Virus-like particles
WHO	World Health Organization
YPD	Yeast Peptone Dextrose

17.1 Introduction

Virus-like particles (VLPs) are structural mimics of virus capsids that do not contain the viral genome, and therefore, cannot cause infection. The multivalent array of viral proteins displayed on the surface of VLPs makes them excellent immunogens, similar to native viruses (Urakami et al. 2017). The lack of genome in VLPs makes them safer vaccine candidates compared to infectious viruses (Saraswat et al. 2016). Several VLP-based recombinant vaccines have been approved for the prevention of viral diseases including Hepatitis B (Recombivax®, Engerix®), cervical cancer caused by Human Papilloma Virus (Cervarix® Gardasil®), and Hepatitis E (Hecolin®) (Mittal et al. 2021). Several other VLP-based vaccines against viruses such as Norwalk Virus, Parvovirus, and Influenza virus are currently being investigated in preclinical trials (Bachmann and Jennings 2010; WHO 2014; Effio et al. 2015; Roldão et al. 2010). VLP vaccinations have been observed in clinical trials to have a long half-life (~9 years) in human blood circulation, thereby offering stable antibody production (Slifka and Amanna 2019; Yang and Zhang 2018). Long-term immunity is achieved by combining a VLP-based vaccination with an appropriate adjuvant and repeated VLP antigen exposure as a booster dose (Slifka and Amanna 2019). Despite the many advantages associated with VLPs and their proven effectiveness in preclinical studies, their use is mostly limited to lab-scale research studies

(Roldão et al. 2010). To be commercially successful, vaccines must be cost-effective with highly scalable manufacturing processes, a major deterrent for most VLP-based vaccines that are under development (Vicente et al. 2011). It is imperative that low-cost production platforms are created to overcome the high costs of plasmids and reagents that are otherwise associated with large-scale transient transfection (Alvim et al. 2019).

A variety of VLPs are presently under development, with a few available commercially (Effio et al. 2016). Mammalian, plant, bacteria, yeast, and insect cells have all been explored as expression hosts for VLPs production, with each expression system offering its own benefits and drawbacks (Roldão et al. 2010). Figure 17.1 illustrates the various critical areas of VLP development. Upstream processing (USP) offers one of the most significant bottlenecks for VLPs production with low titers and complex processes. In mammalian expression hosts, VLP outputs may vary from 0.2 mg/L to 100s mg/L. Utilization of bacterial and yeast expression hosts could enhance the yield; however, these expression hosts require additional downstream processing (Roldão et al. 2010). The low titer in batch cultures and temperature sensitivity presently restrict the commercial production of VLPs (Grein et al. 2017). Manufacturing of VLPs requires knowledge of structural integrity such as structural epitope stability and multimeric state maintenance during large-scale production, purification, formulation, and storage (Peacey et al. 2007). Commercially available VLP-based vaccines for some virus families are summarized in Table 17.1.

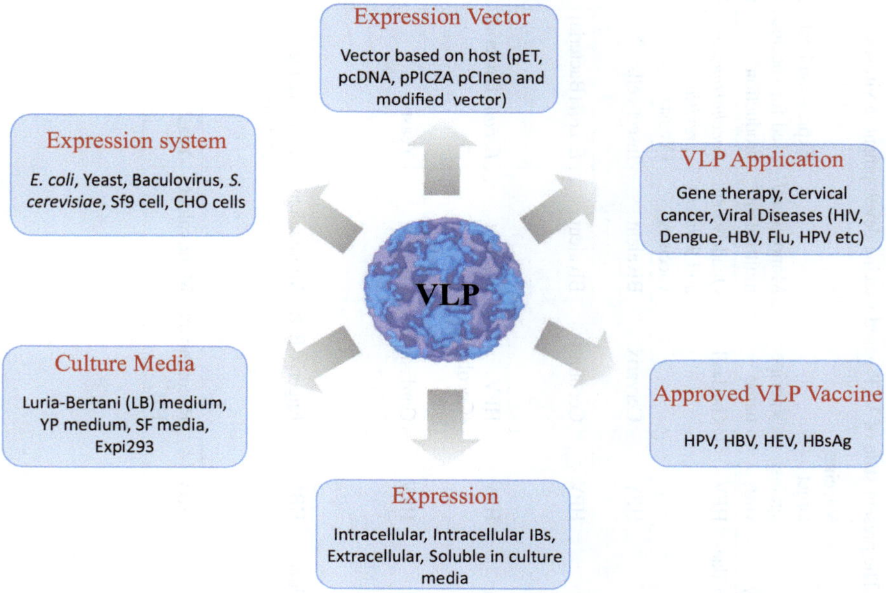

Fig. 17.1 Various aspects of VLP development. The picture depicted the VLPs process from conception to application. Many VLP-based vaccines are commercially available

Table 17.1 The present state of VLP-based vaccine development in viruses

Virus family	Vaccine target against virus	Vaccine name	Monovalent/ multivalent	Expression host used for vaccine production	Antigen used for vaccine production	Company	Regulatory authority/ approval	Current status	References
Papillomaviridae	HPV	Gardasil	Quadrivalent and 9-valent vaccines	*Saccharomyces cerevisiae* (Yeast)	HPV6/11/ 16/18 L1	Merck Sharp and Dohme	FDA/2006	Commercially available	Thrane et al. (2016), Kis et al. (2019)
	HPV	Cervarix	Bivalent	Insect cells	HPV16 and 18 L1	GlaxoSmithKline Biologicals	EMA/2007	Commercially available	GSK (2021)
	HPV	Cecolin®	Bivalent	*E. coli* (Bacteria)	HPV16 and 18 L	Innovax	NMPA/ 2016	Available in China	Gu et al. (2017), Qiao et al. (2020)
	HPV	HPV (Cecolin-9®)	9-valent	*E. coli* (Bacteria)	HPV6/11/ 16/18/31/33/ 45/52/58 L1	Innovax	NA	Phase 2 NCT03935204	Wei et al. (2018)
	HPV	Gardasil-9	9-valent	Yeast	HPV6/11/ 16/18/31/33/ 45/52/58 L1 VLP	Merck Sharp and Dohme	FDA/1983	Commercially available	Le and Müller (2021)
Hepadnaviridae	HBV	Engerix-B ®	Monovalent	*S. cerevisiae* (Yeast)	SHBs, HBsAg	GlaxoSmithKline Biologicals, Belgium	FDA/1989	Commercially available	Kis et al. (2019), Thrane et al. (2016)
	HBV	Recombivax HB	Monovalent	Yeast	HBV HBsAg	Merck Sharp and Dohme	FDA/1983	Commercially available	Hilleman (1987)

	HEV	Hecolin ®	Univalent	*E. coli*	HEV truncated E2, 239 aa.	Xiamen Innovax	CFDA/ 2011	Available in China	Thrane et al. (2016), Kis et al. (2019), Zhang et al. (2014), Li et al. (2005), Zhu et al. (2010)
	HEV	HEV	Univalent	*E. coli*	HEV E2, 495 aa.	NA	NA	Preclinical evaluation	Zheng et al. (2018)
	HBV	Hepavax-Gene®	Bivalent	*H. polymorpha* (Yeast)	HBsAg SHBs,	Crucell	Chinese Authorities	Licensed	Qian et al. (2020)
	HEV	rHEV	NA	Insect cells	HEV (aa 112–607	NA	NA	Phase 2	Qian et al. (2020)
	HBV	Sci-B-Vac®	Trivalent	Mammalian (CHO cells)	SHBs, MHBs, LHB	(Bio-Hep-B®) (SciGen, Israel)	Approved by Israeli/ 2020	Licensed	Qian et al. (2020)
	HEV	p179	NA	Bacteria (*E. coli*)	HEV (aa 439–617)	Changchun Institute of Biological Products Co. Ltd.	NA	Phase 1	Qian et al. (2020)
	HBV	Hepatitis B	Multivalent	Cell-free synthesis system	HBV truncated HBcAg	NA	NA	Preclinical evaluation	Bundy et al. (2008)
	HBV	Heplisav-B	Monovalent	Yeast (*H. polymorpha*)	SHBs	(Dynavax)	FDA/2017	Licensed	Qian et al. (2020)
Coronaviridae	Corona virus	SARS-CoV-2	S-trimer nanoparticle	Insect cells (Sf 9 cell)	SARS-CoV-2 Spike	NVX-CoV2373, Novavax	Regulatory approve year and authority	Phase 3 NCT04611802	Tian et al. (2020), Keech et al. (2020)
	Corona virus	MERS-CoV and SARS-CoV	S nanoparticle	Insect cells (Sf9 cell)	MERS-CoV/ SARS-CoV Spike	Novavax	NA	Preclinical evaluation	Coleman et al. (2014)

(continued)

Table 17.1 (continued)

Virus family	Vaccine target against virus	Vaccine name	Monovalent/ multivalent	Expression host used for vaccine production	Antigen used for vaccine production	Company	Regulatory authority/ approval	Current status	References
Orthomyxoviridae	Influenza virus	Influenza	HA nanoparticle	Insect cells (Sf 9 cell)	Influenza virus HA	NanoFluTM, Novavax	NA	Phase 1/2 NCT04120194	Smith et al. (2017), Portnoff et al. (2020)
	Influenza virus	Influenza A	23 Valent	Bacteria	Virus matrix protein	ACAM-FLU-A	NA	Phase 1	Trial (2021)
	Influenza A	Inflexal®V	Trivalent	Cell-free	Virus envelope proteins		FDA approved	Licensed	Herzog et al. (2009)
Paramyxoviridae	RSV	RSV/F	F nanoparticle	Insect cells (Sf 9 cell)	RSV F	Novavax	NA	Phase 1 NCT03026348	Glenn et al. (2013), Fries et al. (2017)
Parvoviridae	CPV	CPV vaccine	Multivalent	*E. coli*	CPV VP2	NA	NA	Preclinical evaluation	Xu et al. (2014)

NA not available

Stable expression provides a broad range of options for optimizing the process for better production, as well as the development of continuous processes such as perfusion, which can operate at high cell density at a steady state, providing a constant and ideal environment for the cells. This can lead to considerable cost efficiencies as well as consistency in product quality. Use of a stable cell line in a stirred-tank perfusion bioreactor controlled by an ATF-2 cell retention device resulted in a constitutive generation of Zika virus VLPs and high titers (300 mg/mL) (Alvim et al. 2019). Several commercial VLP-based human vaccines are now in a developmental phase.

17.2 Expression Platforms for Large-Scale Production

A platform that may lead to the creation of a low-cost vaccination must be developed to facilitate the cost-effective production of VLPs. Long-term stable expression of VLPs in perfusion bioreactors could allow for steady-state production, thereby lowering manufacturing costs, increasing process efficiency, and ensuring consistency in product quality (Alvim et al. 2019). Selecting an optimal expression system is a key step toward creating a commercially successful VLP. Bacterial (Xiao et al. 2016), yeast (Bredell et al. 2018), insect (Sequeira et al. 2018), and mammalian cell lines (Nishida et al. 2017) have been explored for large-scale VLP synthesis; however, plant-based expression systems are also being investigated. This article focuses on major developments in recombinant expression of VLPs with bacterial, yeast (*Pichia pastoris, Hansenula polymorpha, Saccharomyces cerevisiae*), and mammalian cells (HEK293 and CHO cells) as expression hosts (Zeltins 2013; Roldão et al. 2010).

17.2.1 Bacterial Expression System

Bacterial systems are the most widely used expression platforms for generating recombinant proteins. However, they are not appropriate for the production of eVLPs (enveloped VLP), due to issues related to product solubility and a lack of post-translational modifications (PTM) (Rosano and Ceccarelli 2014). Regulated in vitro assembly of viral proteins into capsids is a promising VLP production technique (Effio et al. 2016). The *E. coli* expression system is commonly employed for recombinant protein production because of its fast growth, low maintenance cost, high productivity, and well-established genetics and biochemical properties of the bacterial host (Gupta and Shukla 2016; Baeshen et al. 2015). There are many disadvantages with bacterial expression system including the formation of inclusion bodies, presence of endotoxins, lack of post-translational modification (Liu et al. 2019), lack of appropriate disulfide bond formation (Zeltins 2013), and codon bias (Liu et al. 2019). VLP expression in the *E. coli* system is possible when soluble capsid proteins self-assemble, as in the case of the Hepatitis B Virus (HBV) core antigen generated in *E. coli* with a yield of 372 μg/mL after purification (Zeltins 2013). *E. coli* was used to express and purify Ebola protein (VP40 antigen) and HPV-16 L1 protein. After purification from *E. coli*, the yield of HPV-16 L1 was 0.75 μg/mL (Roldão et al. 2010). While *E. coli* offered higher yields, extra denaturation, and refolding steps were required for the proper formation of the VLPs (Yazdani et al. 2019).

Researchers have investigated the co-expression of FMDV full-length capsid proteins (VP0, VP1, and VP3) fused with an SUMO tag (Small Ubiquitin-like Modifier) clone in pET28b vector transformed in an *E. coli* system, with the expression of each protein controlled by a different plasmid and selected by use of a single antibiotic (Xiao et al. 2016). Without any optimization, up to 50 mg/L of FMDV (Foot-and-mouth disease virus) particles were produced in a 10-L bioreactor. The use of high throughput bacterial colonies screening method for upstream process development of VLPs manufacturing has been described recently (Effio et al. 2016) and production level of 1.63 ± 0.13 mg/mL murine polyomavirus capsid (VP1) protein has been reported at a high temperature (37 °C), acidic pH (Slifka and Amanna 2019), and low oxygen supply at a shaker speed of 600 rpm (Effio et al. 2016).

17.2.2 Yeast Expression System

Yeast is an ideal expression system for recombinant protein production because of low-cost culture medium, rapid growth, existing knowledge of genome modification, and availability of post-translational modifications (Fletcher et al. 2016; Vieira Gomes et al. 2018; Baghban et al. 2019; Huertas Romera and Michán Doña 2019). For biopharmaceutical development, *S. cerevisiae* and *P. pastoris* are the most prominent yeast expression systems under use. The well-established genome sequence and pathways of *S. cerevisiae* offer a significant opportunity to enhance recombinant protein production from mg/L to gm/L concentrations (Huang et al. 2018). Nuclear envelopes, endosomes, Golgi complex, and ER membranes have been suggested as potential platforms for virus exocytosis. As a result, secretory VLP synthesis in yeasts could be mediated by intracellular membrane secretions with an alternative pathway (Welsch et al. 2007). *P. pastoris,* a methylotrophic yeast has been recently demonstrated to be an impressive host for large-scale recombinant protein production. The tightly regulated alcohol oxidase promoter in *P. pastoris* is highly expressive in nature (Farnós et al. 2005). *P. pastoris* has the capability to achieve high cell density and produce properly folded glycosylated proteins (Juturu and Wu 2018; Looser et al. 2015; Yang and Zhang 2018; Werten et al. 2019).

Researchers have utilized the *P. pastoris* (Bg11) system to generate secreted Norovirus VLPs (NoV VLPs) by expressing the capsid protein VP1. The expression of VP1 from a PJ012 expression vector was induced by methanol, which resulted in a yield of 0.6 g/L of VLPs in a 1.5L bioreactor. In the fermentation medium, these

VLPs are completely assembled into virions (Tomé-Amat et al. 2014). Generation of HPV-16 L1 VLPs at a concentration of 11 mg/L in *P. pastoris* (KM71H), using 0.5% methanol for 96 h has been reported (Alireza et al. 2018). Other researchers have expressed unique HPV16 L1 and L2 chimeric VLPs in different stains of *Pichia, H. polymorpha* NCYC 495, GS115 (Mut^s) and KM71 (Mut^+) for bioreactor compatibility (Bredell et al. 2018). The impact of DO (Dissolved oxygen) set point for control feeding strategies was evaluated. Of all the strains that were examined, *H. polymorpha* was found to be the best-performing strain with respect to the yield of VLPs. Researchers have replaced the pPICZαB, AOX 1 promoter with the PEX8 promotor of GS115 strain of *P. pastoris* for the production of EV71 VLPs. After promoter replacement, the generation of ~270 mg/L EV71 VLPs in an F22 bioreactor could be achieved (Yang et al. 2020).

The secretory VLP, mostly produced by *P. pastoris*, typically suffers from low titer, often due to the retention of some part of VLPs in cell lysate as an insoluble fraction because of its larger size and complicated structure (Potvin et al. 2012). This could be minimized by optimizing both the expression level (e.g., promoter strength, folding helper co-expression signal sequence) and the cultivation process (physical and chemical parameters) (Eto et al. 2021).

17.2.3 Mammalian Expression System

Mammalian cells are well-established expression systems for the production of biopharmaceuticals (Owczarek et al. 2019). Complex recombinant proteins with relatively large sizes can be expressed in mammalian cells in the correctly folded form. As host systems like bacteria and yeast do not contain the essential factors for the production of enveloped VLPs, mammalian systems are needed for eVLP production (Zeltins 2013). Gene delivery and clone selection for mammalian cell expression is tedious and time-consuming when compared to bacterial and yeast systems (Gupta and Shukla 2016). One of the major advantages of a mammalian expression system is the availability of authentic post-translational modification, which assists in attaining correct tertiary and quaternary protein structure (Nishida et al. 2017). Many different mammalian cell lines, such as Vero, AGE1.CR, EB6, PER.C6, and HEK293 are utilized for VLP expression (Genzel 2015). HEK293 cells are simple to culture and easy to grow in serum-free media (Stepanenko and Dmitrenko 2015). These cells are typically used for transient gene expression because they contain a SV40 large T antigen that makes them highly transfectable (Büssow 2015). While HEK293F cells, with transient or stable transfection, have been utilized to generate VLPs of adeno-associated virus, Marburg, HIV, norovirus, and Ebola VLPs derived from HEK293T cells, CHO cells have been used to manufacture HBV VLP (Roldão et al. 2010) (Cervera Gracia et al. 2018). Figure 17.2 depicts VLP production in a mammalian cell line.

Researchers have created four types of VLPs: Zika, JEV, YEF, and CHIKV, using bicistronic expression and lentiviral transfection of HEK293T cells. They also

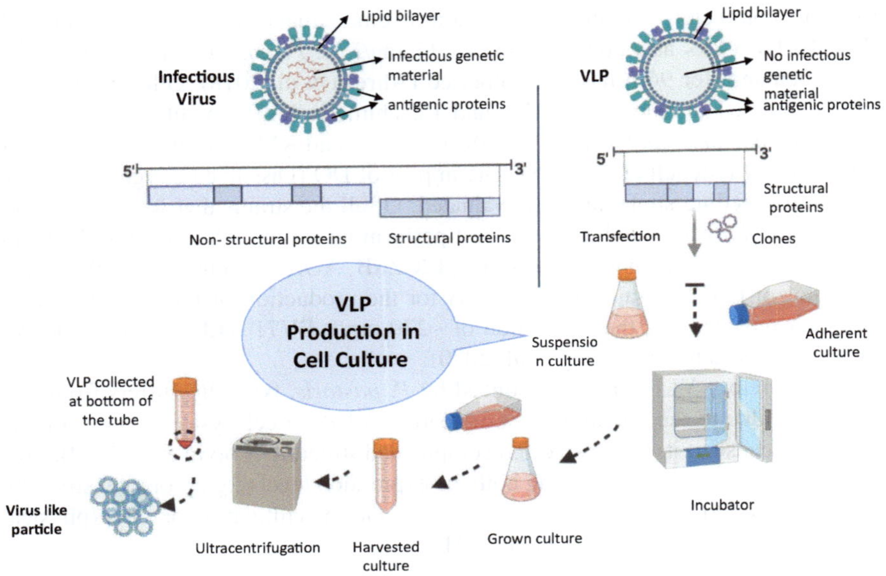

Fig. 17.2 Production of virus-like particle in cell culture. The clone gene encoding structural protein was transfected in suspension or adherent cell line for the generation of VLPs. The culture was collected and purified using an ultracentrifugation method

created stable cell lines by adapting HEK293T adherent cells to suspension in EX-CELL 293T (Sigma) serum-free media with 25 mM HEPES buffer, antibiotic, and 6 mM L-glutamine (Garg et al. 2017). The pCIneo (Promega, USA) vector has been also used to create stable HEK293SF-3F6 (suspension-adapted) cells that constantly produce Zika Virus VLPs (Alvim et al. 2019). The researchers used batch and stepwise feeding processes for large-scale VLP synthesis and found that stepwise feeding resulted in four times better expression (~3.58×10^{12} VLPs/mL) than batch feeding (~3.5×10^{9} VLPs/mL), indicating that perfusion enhances VLP productivity (Boix-Besora et al. 2022). A schematic representation of the process development of VLPs is presented in Fig. 17.3.

A stable HEK293T cell line for Zika virus VLP has also been developed. The genes for prM-E were transfected in HEK293T cells using a pLenti6/V5 vector for transfection, and 10 μg/mL blasticidin was used for the selection of VLP-producing cells. After 8–10 passages, the highly expressing cells were isolated by serial dilution and VLP expression was checked at regular intervals with western blotting and found ~1.7–2.3 mg/mL VLPs. Cells were sorted by flow cytometry for checking E antigen expression (Garg et al. 2017).

Suspension CHO cell line, grown in serum-free medium, has been widely used in large-scale biotherapeutic production. S and PreS2 proteins of HBV VLPs have been produced and assembled in CHO cells (Legastelois et al. 2017). VLPs produced through CHO cell line show structural heterogeneity after post-translational

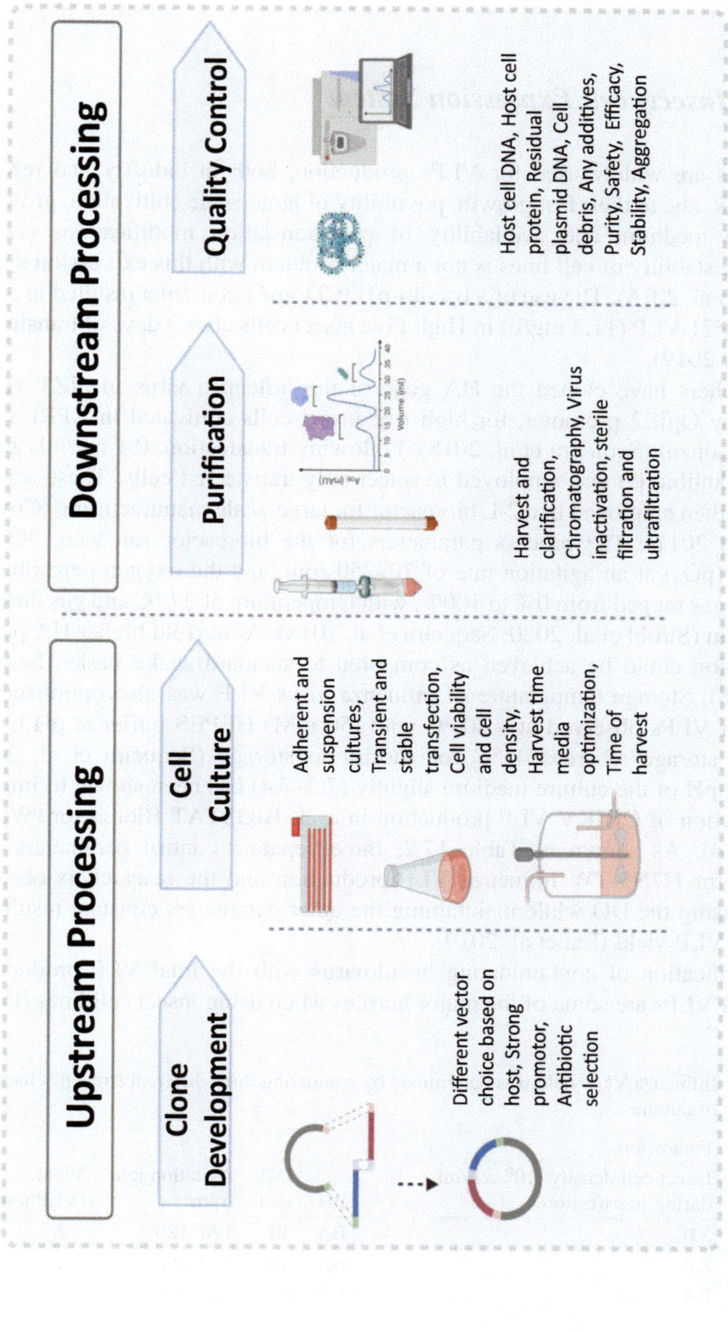

Fig. 17.3 The main steps in the manufacturing of VLPs. Clone development, cell culture, purification, and quality control are the most essential steps in the manufacturing of VLPs. The aim of the procedure is to reduce the process time and cost while maintaining the VLPs purity and quality at a higher scale

modifications (PTM), due to inherent variability, structural complexity, and environmental fluctuation during protein expression (Shrivastava et al. 2022).

17.2.4 Insects cell Expression System

Insect cells are widely used for VLPs production, both in industry and research laboratories, due to their fast growth, possibility of large-scale cultivation, growth in serum-free medium, and availability of post-translation modifications (Zeltins 2013). The stability of cell lines is not a major problem with this expression system (Besnard et al. 2016). The use of a baculo-p1-3CD-gp41 construct resulted in a high yield of EV71 VLP (11.3 mg/L) in High Five insect cells after 3 days of transfection (Kim et al. 2019).

Researchers have cloned the HA gene of the influenza virus in pIZT vector, operated by OpIE2 promoter, for high five insect cells cultivated in XPRESSTM medium (Lonza) (Sequeira et al. 2018). Following transfection, 0.1 mg/mL zeocin (selection antibiotic) was employed to select only transfected cells. These selected cells were then employed in a 2-L bioreactor for large-scale manufacturing (Cox and Hashimoto 2011). The process parameters for the bioreactor run were 30% air saturation (pO_2) at an agitation rate of 70–250 rpm, and the oxygen percentage in the mixed gas ranged from 0% to 100%, with temperature of 27 °C and gas flow rate of 0.01 vvm (Strobl et al. 2020; Sequeira et al. 2018). A fourfold higher HA protein concentration could be achieved as compared to standard shake flasks (Sequeira et al. 2018). Storage temperature of influenza virus VLP was also optimized and stability of VLPs observed at −80 °C with (50 mM) HEPES buffer at pH 6.6 for long-term storage whereas 4 °C for short-term storage (Sequeira et al. 2018). Increasing pH of the culture medium slightly (7.2–7.4) has been shown to improve the production of CHIKV VLP production in a 2L BIOSTAT Bioreactor (Wagner et al. 2014). As shown in Table 17.2, three separate control parameters were identified for H7N9-TW Influenza VLP production and the researchers observed that increasing the DO while maintaining the other parameters constant resulted in improved VLP yield (Lai et al. 2019).

Co-purification of contaminating baculovirus with the final VLP product and stability of VLPs are some of the major hurdles when using insect cell lines (Possee

Table 17.2 Influenza VLP yield was determined by comparing three different strategies harvested on third day of culture

	Parameters				
Strategies	Insect cell density (10^6 cell/mL) during transfection	pH	DO (%)	Agitation rate (rpm)	Yield (VLP unit/μL)
1	2.0	6.4	30	120–180	8
2	8.0	6.4	30	120–180	>2
3	2.4	6.4	80	120–180	512

et al. 2019). Further, insect cells are typically lysed after 3–4 days of baculovirus infection, and as a result, the proteases released from the lysed cells have been known to cause considerable product degradation (Besnard et al. 2016).

17.2.5 Cell-Free VLP Production

Cell-free biosynthesis is an alternate and flexible method of VLP production. The protein synthesis machinery from cells is extracted and exploited for in vitro biosynthesis of proteins. This technique does not require cell culture because it is based on the introduction of mRNA into the cell-free extract for the production of the desired protein (Sheng et al. 2017). A cell extract with ten additional translation factors such as tRNA synthetases (20 aminoacyl), T7 RNA polymerase, ribosome, myokinase, creatine kinase, nucleoside diphosphatases kinase, and pyrophosphatase are required for cell-free protein synthesis (Gregorio et al. 2019).

Researchers have used the *P. pastoris* (*Komagataella phaffi*) system to demonstrate the assembly of Hepatitis B Virus VLPs in a cell-free environment (Spice et al. 2020). Cells were collected and homogenized from a primary culture of *P. pastoris* (FHL1) grown in YPD medium. The cell lysate was utilized to synthesize Hepatitis B virus-like proteins in vitro (Spice et al. 2020; Aw et al. 2020). Similarly, HuNoV (Human Norovirus) capsid protein (s) were synthesized and VLPs assembled in a cell-free *E. coli* system (Sheng et al. 2017). For this cell-free VLP generation, S30 and CK-T7 cell extracts were used. After 2 h of IPTG induction, cells were collected and the cell lysate was utilized to generate HuNoV-VLP, which resulted in 1.0 g/L VLP production (Sheng et al. 2017). Table 17.3 offers a summary of the production elements for the various VLP-based biotherapeutics.

17.3 Other Factors Affecting Large-Scale VLP Production

There are many other factors that influence the commercial manufacturing of VLPs (Dai et al. 2018). These include expression plasmid (Guo et al. 2021) and composition of growth medium (Zeltins 2013). The production and stability of VLPs is affected by temperature, pH, ionic strength, chemical treatment, and shear force (Pattenden et al. 2005). Despite the fact that wild-type viruses are inherently stable, VLPs lack the viral genome and can therefore be relatively unstable (Dai et al. 2018). Appropriate monitoring during fermentation, product loss during centrifugation and cell lysis, and storage temperature are some of the key challenges that are faced during VLP manufacturing (Josefsberg and Buckland 2012). Researchers have extensively published on the optimization of the physical parameters and the media components as means to enhance the production of VLPs (Zeltins 2013). Further, the generation of multiple proteins with diverse post-translational modifications that

Table 17.3 Summary of the production elements for the various VLP-based biotherapeutics

Virus-like particle biotherapeutics particle type	Host expression system	Vector used/ promoter used	Protein components	Culture medium used in VLP production	VLP size (nm) Diameter	Harvest fraction	Highest expression level	Vaccine used for	References
NVLP (Norwalk Virus-Like Particles)	*Escherichia coli* (GI 724) strain	pTrxFus vector	Capsid protein VP1, VP2	M9 minimal medium	38 nm	Soluble (cytoplasm)	56 mg/L	Viral gastroenteritis	Merlin et al. (2014)
	Pichia pastoris	pPICZalphaA/ AOX1 promoter	Capsid protein VP1, VP2	NA	38 nm	Soluble	10 mg/L	Viral gastroenteritis	Merlin et al. (2014)
	Sf9 cells	Baculoviral vector	Capsid protein VP1, VP2	NA	38 nm	Soluble	125 mg/L	Viral gastroenteritis	Merlin et al. (2014)
EV71-VLP	Pichia pastoris	pPICZαB (Invitrogen) plasmid/ PEX8 promoter	P1 and 3C	YPD medium and BMMY	~35 nm	Soluble	270 mg/L	Hand, foot, and mouth disease	Yang et al. (2020)
FMDV-VLP	*Escherichia coli* BL21(DE3)	pET28b vector with SUMO protein tag	VP0, VP1 and VP3 genes	Luria–Bertani (LB) medium	~30 nm	Intracellular expression	~50 mg/L	Hand, foot, and mouth disease	Xiao et al. (2016)
HIV-1	HEK293SF-3F6	pGag:eGFP plasmid/	Gag polyprotein	CD293	100 nm	Soluble	3617 ± 17 monomers per VLP	AIDS	Lavado-García et al. (2020), Fuenmayor et al. (2018)
AAV virus-like particles	*Escherichia coli* BL21 (DE3)	pET24b/T7 promoter	Protein VP3	Luria–Bertani (LB) medium	34.4 nm or 38.2 nm	Intracellular (IB's)	0.23 mg or 5 × 109	Used in Gene therapy	Le et al. (2019)

Chimeric L1/L2 HPV-16 VLP	*Pichia pastoris* KM71H strain	pPICZA plasmid/ AOX1 promoter	L1/L2 protein	BMMY media	60 nm in diameter	Intracellular	NA	Cervical cancers	Alireza et al. (2018)
HPV-16 VLP L1	*Pichia pastoris* X-33 strain	pPGKΔ3 expression vector/*PGK1* promoter (PPGK1)	L1 capsid protein	YPD	55 nm	Intracellular expression	NA	Cervical cancers	Mariz et al. (2015)
RHD vaccine	Pichia pastoris PVP12-11 strain	–	Capsid protein	BMMY	30–40 nm	Intracellular	860 mg/L	Rabbit hemorrhagic disease virus	Fernández et al. (2013)
	Pichia pastoris MP36 strain	Pps7 expression vector/ AOX promoter	VP60 capsid protein	YP medium with 2% glycerol	35 nm	Intracellular	1.5 g/L	(RHDV)	Schellenbacher et al. (2017)
Influenza vaccine	HEK293 suspension, serum free	BacMam PR8/CMV promoter	HA, NA, and M1 genes	SFM4Transfx 293™ (HyQ) media (HyClone)	100 ± 20 nm	Soluble	1.5×10^8 VLPs/mL	Influenza Virus	Thompson et al. (2015)
	Insect spodoptera frugiperda (Sf9)	pVL1393 plasmid/polyhedron promoter (polh)	HA, NA, and M1 genes	Sf900 II medium (GIBCO)	~100 nm	Soluble	5.85×10^9 VLPs/mL	Influenza Virus	Thompson et al. (2015)
Dengue-2 virus-like particle	Expi293TM cells	pcDNA3.4/ CMV promoter	Structural proteins CprME with NS2B/NS3 protease,	Expi293 medium (Gibco A1435101)	50 nm	In suspension	3 mg/L	Dengue	Boigard et al. (2018)

(continued)

Table 17.3 (continued)

Virus-like particle biotherapeutics particle type	Host expression system	Vector used/ promoter used	Protein components	Culture medium used in VLP production	VLP size (nm) Diameter	Harvest fraction	Highest expression level	Vaccine used for	References
ZIKV	HEK293SF-3F6 cell line	pCIneo vector	prM-E gene	HEK TF medium (Xell AG, Germany)	50 nm	In suspension	80 lg/L on day 11 of perfusion bioreactor	Zika virus	Alvim et al. (2019)
Hantavirus	CHO cell	Modified vector	Gn, Gc, NP	IMDM medium	100 nm	In suspension	1 mg/L	Hantavirus	Li et al. (2010)
Rotavirus VLPs	Spodoptera frugiperda Sf-9 cell line	Tricistronic baculovirus	Rotaviral vp2, vp6 and vp7 genes	SF900 II medium (Gibco, UK)	75 nm	In suspension	1.6 mg VLPs per liter in bioreactor	Rotavirus	Peixoto et al. (2007)
NoV VLPs	*P. pastoris* strain, Bg11	Methanol inducible pJ912 expression vector	VP1 protein	BMMY	40 nm	Secreted form	0.6 g/L. NoV VLPs	Norovirus	Tomé-Amat et al. (2014)

NA not available

assemble into complex, enveloped, or non-enveloped virions is a considerable challenge in large-scale production (Brune and Howarth 2018; Rey and Lok 2018).

17.4 Conclusions and Future Prospects

In this review, we present key developments that have occurred with respect to the recombinant expression of VLPs for their commercial production. VLPs were created as a replacement for conventional virus-based vaccines due to their safety, flexibility, and immunogenic properties. However, if VLPs are to be utilized as vaccine candidates for large-scale immunization, more needs to be done to facilitate their commercial production. Key factors include choice of vector and expression host, creation of a stable cell line, reactor-based production, downstream process development, and characterization of the final product. It is likely that these will continue to be areas of active research for academia and industry in the coming decades.

Acknowledgments Authors would like to acknowledge funding support from the *Council of Scientific and Industrial Research* (CSIR) and the University Grants Commission (UGC), Human Resource Development Group (UGC-Ref. No.: 598/CSIR-UGC NET DEC.2017).

References

Alireza S, Khosrow A, Bahman AK, Alireza S-Z, Ataollah G, Manoocher M (2018) VLP production from recombinant L1/L2 HPV-16 protein expressed in Pichia pastoris. Protein Pept Lett 25: 783–790

Alvim RGF, Itabaiana I Jr, Castilho LR (2019) Zika virus-like particles (VLPs): Stable cell lines and continuous perfusion processes as a new potential vaccine manufacturing platform. Vaccine 37: 6970–6977

Aw R, Spice AJ, Polizzi KM (2020) Methods for expression of recombinant proteins using a Pichia pastoris cell-free system. JCPiPS 102:e115

Bachmann MF, Jennings GT (2010) Vaccine delivery: a matter of size, geometry, kinetics and molecular patterns. Nat Rev Immunol 10:787–796

Baeshen MN, Al-Hejin AM, Bora RS, Ahmed MMM, Ramadan HAI, Saini KS, Baeshen NA, Redwan EM (2015) Production of biopharmaceuticals in E. coli: current scenario and future perspectives. J Microbiol Biotechnol 25:953–962

Baghban R, Farajnia S, Rajabibazl M, Ghasemi Y, Mafi A, Hoseinpoor R, Rahbarnia L, Aria M (2019) Yeast expression systems: overview and recent advances. Mol Biotechnol 61:365–384

Besnard L, Fabre V, Fettig M, Gousseinov E, Kawakami Y, Laroudie N, Scanlan C, Pattnaik P (2016) Clarification of vaccines: an overview of filter based technology trends and best practices. Biotechnol Adv 34:1–13

Boigard H, Cimica V, Galarza JM (2018) Dengue-2 virus-like particle (VLP) based vaccine elicits the highest titers of neutralizing antibodies when produced at reduced temperature. Vaccine 36: 7728–7736

Boix-Besora A, Lorenzo E, Lavado-García J, Gòdia F, Cervera L (2022) Optimization, production, purification and characterization of HIV-1 GAG-based virus-like particles functionalized with SARS-CoV-2. JV 10:250
Bredell H, Smith JJ, Görgens JF, van Zyl WH (2018) Expression of unique chimeric human papilloma virus type 16 (HPV-16) L1-L2 proteins in Pichia pastoris and Hansenula polymorpha. Yeast 35:519–529
Brune KD, Howarth M (2018) New routes and opportunities for modular construction of particulate vaccines: stick, click, and glue. JFii 9:1432
Bundy BC, Franciszkowicz MJ, Swartz JR (2008) Escherichia coli-based cell-free synthesis of virus-like particles. Biotechnol Bioeng 100:28–37
Büssow K (2015) Stable mammalian producer cell lines for structural biology. Curr Opin Struct Biol 32:81–90
Cervera Gracia L, Fuenmayor J, Gòdia F, Gonzalez-Domínguez I (2018) Extended gene expression for HIV-1 VLPs scale-up and production enhancement using shRNA and chemical additives. Vaccine Technology VII
Coleman CM, Liu YV, Mu H, Taylor JK, Massare M, Flyer DC, Glenn GM, Smith GE, Frieman MB (2014) Purified coronavirus spike protein nanoparticles induce coronavirus neutralizing antibodies in mice. Vaccine 32:3169–3174
Cox MM, Hashimoto YJ (2011) A fast track influenza virus vaccine produced in insect cells. Joip 107:S31–S41
Dai S, Wang H, Deng FJ (2018) Advances and challenges in enveloped virus-like particle (VLP)-based vaccines. JoIS 2
Effio CL, Wenger L, Ötes O, Oelmeier SA, Kneusel R, Hubbuch J (2015) Downstream processing of virus-like particles: single-stage and multi-stage aqueous two-phase extraction. J Chromatogr A 1383:35–46
Effio CL, Baumann P, Weigel C, Vormittag P, Middelberg A, Hubbuch J (2016) High-throughput process development of an alternative platform for the production of virus-like particles in Escherichia coli. J Biotechnol 219:7–19
Eto Y, Saubi N, Ferrer P, Joseph-Munné J (2021) Expression of chimeric HPV-HIV protein L1P18 in Pichia pastoris; purification and characterization of the virus-like particles. JP 13:1967
Farnós O, Boué O, Parra F, Martín-Alonso JM, Valdés O, Joglar M, Navea L, Naranjo P, Lleonart R (2005) High-level expression and immunogenic properties of the recombinant rabbit hemorrhagic disease virus VP60 capsid protein obtained in Pichia pastoris. J Biotechnol 117:215–224
Fernández E, Toledo JR, Méndez L, González N, Parra F, Martín-Alonso JM, Limonta M, Sánchez K, Cabrales A, Estrada MP (2013) Conformational and thermal stability improvements for the large-scale production of yeast-derived rabbit hemorrhagic disease virus-like particles as multipurpose vaccine. PLoS One 8:e56417
Fletcher E, Krivoruchko A, Nielsen J (2016) Industrial systems biology and its impact on synthetic biology of yeast cell factories. Biotechnol Bioeng 113:1164–1170
Fries L, Shinde V, Stoddard JJ, Thomas DN, Kpamegan E, Lu H, Smith G, Hickman SP, Piedra P, Glenn GM (2017) Immunogenicity and safety of a respiratory syncytial virus fusion protein (RSV F) nanoparticle vaccine in older adults. Immun Ageing 14:1–14
Fuenmayor J, Cervera L, Rigau C, Gòdia F (2018) Enhancement of HIV-1 VLP production using gene inhibition strategies. JAm, biotechnology 102:4477–4487
Garg H, Sedano M, Plata G, Punke EB, Joshi A (2017) Development of virus-like-particle vaccine and reporter assay for Zika virus. J Virol 91:e00834–e00817
Genzel Y (2015) Designing cell lines for viral vaccine production: where do we stand? Biotechnol J 10:728–740
Glenn GM, Smith G, Fries L, Raghunandan R, Lu H, Zhou B, Thomas DN, Hickman SP, Kpamegan E, Boddapati S (2013) Safety and immunogenicity of a Sf9 insect cell-derived respiratory syncytial virus fusion protein nanoparticle vaccine. Vaccine 31:524–532
Gregorio NE, Oza JP, Levine MZ (2019) Protocols: a user's guide to cell-free protein synthesis. JM 2:24

Grein TA, Weidner T, Czermak P (2017) Concepts for the production of viruses and viral vectors in cell culture. New insights into cell culture technology 10:62590

GSK (2021) Vaccines. https://au.gsk.com/en-au/about-us/what-we-do/vaccines/

Gu Y, Wei M, Wang D, Li Z, Xie M, Pan H, Wu T, Zhang J, Li S, Xia N (2017) Characterization of an Escherichia coli-derived human papillomavirus type 16 and 18 bivalent vaccine. Vaccine 35: 4637–4645

Guo C, Fordjour FK, Tsai SJ, Morrell JC, Gould SJ (2021) Choice of selectable marker affects recombinant protein expression in cells and exosomes. JJoBC 297

Gupta SK, Shukla P (2016) Advanced technologies for improved expression of recombinant proteins in bacteria: perspectives and applications. Crit Rev Biotechnol 36:1089–1098

Herzog C, Hartmann K, Künzi V, Kürsteiner O, Mischler R, Lazar H, Glück RJV (2009) Eleven years of Inflexal® V—a virosomal adjuvanted influenza vaccine 27:4381–4387

Hilleman MR (1987) Yeast recombinant hepatitis B vaccine. Infection 15:3–7

Huang M, Wang G, Qin J, Petranovic D, Nielsen J (2018) Engineering the protein secretory pathway of Saccharomyces cerevisiae enables improved protein production. Proc Natl Acad Sci 115:E11025–E11032

Huertas Romera MJ, Michán Doña CM (2019) Paving the way for the production of secretory proteins by yeast cell factories. Microb Biotechnol 1–2

Josefsberg JO, Buckland B (2012) Vaccine process technology. Biotechnol Bioeng 109:1443–1460

Juturu V, Wu JC (2018) Heterologous protein expression in Pichia pastoris: latest research progress and applications. ChemBioChem 19:7–21

Keech C, Albert G, Cho I, Robertson A, Reed P, Neal S, Plested JS, Zhu M, Cloney-Clark S, Zhou H (2020) Phase 1–2 trial of a SARS-CoV-2 recombinant spike protein nanoparticle vaccine. N Engl J Med 383:2320–2332

Kim H-J, Son HS, Lee SW, Yoon Y, Hyeon J-Y, Chung GT, Lee J-W, Yoo JS (2019) Efficient expression of enterovirus 71 based on virus-like particles vaccine. JPO 14:e0210477

Kis Z, Shattock R, Shah N, Kontoravdi C (2019) Emerging technologies for low-cost, rapid vaccine manufacture. Biotechnol J 14:1800376

Lai C-C, Cheng Y-C, Chen P-W, Lin T-H, Tzeng T-T, Lu C-C, Lee M-S, Hu AY-C (2019) Process development for pandemic influenza VLP vaccine production using a baculovirus expression system. JJobe 13:1–9

Lavado-García J, González-Domínguez I, Cervera L, Jorge I, Vázquez JS, Gòdia F (2020) Molecular characterization of the coproduced extracellular vesicles in HEK293 during virus-like particle production. J Proteome Res 19:4516–4532

Le DT, Müller KM (2021) In vitro assembly of virus-like particles and their applications. Life 11: 334

Le DT, Radukic MT, Müller KM (2019) Adeno-associated virus capsid protein expression in Escherichia coli and chemically defined capsid assembly. Sci Rep 9:1–10

Legastelois I, Buffin S, Peubez I, Mignon C, Sodoyer R, Werle B (2017) Non-conventional expression systems for the production of vaccine proteins and immunotherapeutic molecules. Human Vaccines Immunotherap 13:947–961

Li SW, Zhang J, Li YM, Ou SH, Huang GY, He ZQ, Sheng XG, Xian YL, Pang SQ, Ng MH (2005) A bacterially expressed particulate hepatitis E vaccine: antigenicity, immunogenicity and protectivity on primates. Vaccine 23:2893–2901

Li C, Liu F, Liang M, Zhang Q, Wang X, Wang T, Li J, Li D (2010) Hantavirus-like particles generated in CHO cells induce specific immune responses in C57BL/6 mice. Vaccine 28 (26):4294–4300

Liu M, Wang B, Wang F, Yang Z, Gao D, Zhang C, Ma L, Yu X (2019) Soluble expression of single-chain variable fragment (scFv) in Escherichia coli using superfolder green fluorescent protein as fusion partner. Appl Microbiol Biotechnol 103:6071–6079

Looser V, Bruhlmann B, Bumbak F, Stenger C, Costa M, Camattari A, Fotiadis D, Kovar K (2015) Cultivation strategies to enhance productivity of Pichia pastoris: a review. Biotechnol Adv 33: 1177–1193

Mariz FC, Coimbra EC, Jesus ALS, Nascimento LM, Torres FAG, Freitas AC (2015) Development of an IP-Free biotechnology platform for constitutive production of HPV16 L1 capsid protein using the Pichia pastoris PGK1 promoter. Biomed Res Int 2015

Merlin M, Gecchele E, Capaldi S, Pezzotti M, Avesani L (2014) Comparative evaluation of recombinant protein production in different biofactories: the green perspective. Biomed Res Int 2014

Mittal M, Banerjee M, Lua LH, Rathore AS (2021) Current status and future challenges in transitioning to continuous bioprocessing of virus-like particles. JJoCT, Biotechnology

Nishida T, Kubota S, Takigawa M (2017) Production of recombinant CCN2 protein by mammalian cells. In: CCN Proteins. Springer, pp 95–105

Owczarek B, Gerszberg A, Hnatuszko-Konka K (2019) A brief reminder of systems of production and chromatography-based recovery of recombinant protein biopharmaceuticals. Biomed Res Int

Pattenden LK, Middelberg APJ, Niebert M, Lipin DI (2005) Towards the preparative and large-scale precision manufacture of virus-like particles. Trends Biotechnol 23:523–529

Peacey M, Wilson S, Baird MA, Ward VK (2007) Versatile RHDV virus-like particles: Incorporation of antigens by genetic modification and chemical conjugation. Biotechnol Bioeng 98: 968–977

Peixoto C, Sousa MFQ, Silva AC, Carrondo MJT, Alves PM (2007) Downstream processing of triple layered rotavirus like particles. J Biotechnol 127(3):452–461

Portnoff AD, Patel N, Massare MJ, Zhou H, Tian J-H, Zhou B, Shinde V, Glenn GM, Smith G (2020) Influenza hemagglutinin nanoparticle vaccine elicits broadly neutralizing antibodies against structurally distinct domains of H3N2 HA. Vaccine 8:99

Possee RD, Chambers AC, Graves LP, Aksular M, King LA (2019) Recent developments in the use of baculovirus expression vectors. JCIiMB 34:215–230

Potvin G, Ahmad A, Zhang Z (2012) Bioprocess engineering aspects of heterologous protein production in Pichia pastoris: a review. Biochem Eng J 64:91–105

Qian C, Liu X, Xu Q, Wang Z, Chen J, Li T, Zheng Q, Yu H, Gu Y, Li S (2020) Recent progress on the versatility of virus-like particles. Vaccine 8:139

Qiao Y-L, Wu T, Li R-C, Hu Y-M, Wei L-H, Li C-G, Chen W, Huang S-J, Zhao F-H, Li M-Q (2020) Efficacy, safety, and immunogenicity of an Escherichia coli-produced bivalent human papillomavirus vaccine: an interim analysis of a randomized clinical trial. J Natl Cancer Inst 112:145–153

Rey FA, Lok S-M (2018) Common features of enveloped viruses and implications for immunogen design for next-generation vaccines. JC 172:1319–1334

Roldão A, Mellado MCM, Castilho LR, Carrondo MJT, Alves PM (2010) Virus-like particles in vaccine development. Expert Rev Vaccines 9:1149–1176

Rosano GL, Ceccarelli EA (2014) Recombinant protein expression in Escherichia coli: advances and challenges. Front Microbiol 5:172

Saraswat S, Athmaram TN, Parida M, Agarwal A, Saha A, Dash PK (2016) Expression and characterization of yeast derived chikungunya virus like particles (CHIK-VLPs) and its evaluation as a potential vaccine candidate. PLoS Negl Trop Dis 10:e0004782

Schellenbacher C, Roden RBS, Kirnbauer R (2017) Developments in L2-based human papillomavirus (HPV) vaccines. Virus Res 231:166–175

Sequeira DP, Correia R, Carrondo MJT, Roldão A, Teixeira AP, Alves PM (2018) Combining stable insect cell lines with baculovirus-mediated expression for multi-HA influenza VLP production. Vaccine 36:3112–3123

Sheng J, Lei S, Yuan L, Feng X (2017) Cell-free protein synthesis of norovirus virus-like particles. RSC Adv 7:28837–28840

Shrivastava A, Joshi S, Guttman A, Rathore AS (2022) N-Glycosylation of monoclonal antibody therapeutics: a comprehensive review on significance and characterization. JACA 339828

Slifka MK, Amanna IJ (2019) Role of multivalency and antigenic threshold in generating protective antibody responses. Front Immunol 10:956

Smith G, Liu Y, Flyer D, Massare MJ, Zhou B, Patel N, Ellingsworth L, Lewis M, Cummings JF, Glenn G (2017) Novel hemagglutinin nanoparticle influenza vaccine with Matrix-M™ adjuvant induces hemagglutination inhibition, neutralizing, and protective responses in ferrets against homologous and drifted A (H3N2) subtypes. Vaccine 35:5366–5372

Spice AJ, Aw R, Bracewell DG, Polizzi KM (2020) Biotechnology: synthesis and assembly of hepatitis B virus-like particles in a Pichia pastoris cell-free system. JFiB 8:72

Stepanenko AA, Dmitrenko VV (2015) HEK293 in cell biology and cancer research: phenotype, karyotype, tumorigenicity, and stress-induced genome-phenotype evolution. Gene 569:182–190

Strobl F, Ghorbanpour SM, Palmberger D, Striedner G (2020) Evaluation of screening platforms for virus-like particle production with the baculovirus expression vector system in insect cells. JSr 10:1–9

Thompson CM, Petiot E, Mullick A, Aucoin MG, Henry O, Kamen AA (2015) Critical assessment of influenza VLP production in Sf9 and HEK293 expression systems. BMC Biotechnol 15:1–12

Thrane S, Janitzek CM, Matondo S, Resende M, Gustavsson T, De Jongh WA, Clemmensen S, Roeffen W, van de Vegte-Bolmer M, Van Gemert GJ (2016) Bacterial superglue enables easy development of efficient virus-like particle based vaccines. J Nanobiotechnol 14:1–16

Tian J-H, Patel N, Haupt R, Zhou H, Weston S, Hammond H, Logue J, Pornoff A, Norton J, Guebre-Xabier M (2020) M SARSCoV-2 spike glycoprotein vaccine candidate NVX-CoV2373 elicits immunogenicity in baboons and protection in mice. Nat Commun

Tomé-Amat J, Fleischer L, Parker SA, Bardliving CL, Batt CA (2014) Secreted production of assembled Norovirus virus-like particles from Pichia pastoris. Microb Cell Factories 13:134

Trial UC (2021) U.S. Clinical Trials. Safety study of recombinant M2e influenza – a vaccine in healthy adults (FLU-A). http://www.clinicaltrials.gov/ct2/show/NCT00819013?term=sanofi+and+influenza+and+m2e&rank=1. Accessed 12 Apr 2012

Urakami A, Sakurai A, Ishikawa M, Yap ML, Flores-Garcia Y, Haseda Y, Aoshi T, Zavala FP, Rossmann MG, Kuno S (2017) Development of a novel virus-like particle vaccine platform that mimics the immature form of alphavirus. Clin Vaccine Immunol 24:e00090–e00017

Vicente T, Roldão A, Peixoto C, Carrondo MJT, Alves PM (2011) Large-scale production and purification of VLP-based vaccines. J Invertebr Pathol 107:S42–S48

Vieira Gomes AM, Souza Carmo T, Silva Carvalho L, Mendonça Bahia F, Parachin NS (2018) Comparison of yeasts as hosts for recombinant protein production. Microorganisms 6:38

Wagner JM, Pajerowski JD, Daniels CL, McHugh PM, Flynn JA, Balliet JW, Casimiro DR, Subramanian S (2014) Enhanced production of Chikungunya virus-like particles using a high-pH adapted spodoptera frugiperda insect cell line. PLoS One 9(4):e94401

Wei M, Wang D, Li Z, Song S, Kong X, Mo X, Yang Y, He M, Li Z, Huang B (2018) N-terminal truncations on L1 proteins of human papillomaviruses promote their soluble expression in Escherichia coli and self-assembly in vitro. Emerg Microb Infect 7:1–12

Welsch S, Müller B, Kräusslich H-G (2007) More than one door–budding of enveloped viruses through cellular membranes. FEBS Lett 581:2089–2097

Werten MWT, Eggink G, Stuart MAC, de Wolf FA (2019) Production of protein-based polymers in Pichia pastoris. Biotechnol Adv 37:642–666

WHO (2014) Guidelines on the quality, safety and efficacy ofrecombinant malaria vaccines targeting the pre-erythrocytic and blood stagesof Plasmodium falciparum. World Health Organization

Xiao Y, Chen H-Y, Wang Y, Yin B, Lv C, Mo X, Yan H, Xuan Y, Huang Y, Pang W (2016) Large-scale production of foot-and-mouth disease virus (serotype Asia1) VLP vaccine in Escherichia coli and protection potency evaluation in cattle. BMC Biotechnol 16:1–9

Xu J, Guo H-C, Wei Y-Q, Dong H, Han S-C, Ao D, Sun D-H, Wang H-M, Cao S-Z, Sun S-Q (2014) Self-assembly of virus-like particles of canine parvovirus capsid protein expressed from Escherichia coli and application as virus-like particle vaccine. Appl Microbiol Biotechnol 98: 3529–3538

Yang Z, Zhang Z (2018) Engineering strategies for enhanced production of protein and bio-products in Pichia pastoris: a review. Biotechnol Adv 36:182–195

Yang Z, Gao F, Wang X, Shi L, Zhou Z, Jiang Y, Ma X, Zhang C, Zhou C, Zeng X (2020) Development and characterization of an enterovirus 71 (EV71) virus-like particles (VLPs) vaccine produced in Pichia pastoris. Human Vaccines Immunotherap 16:1602–1610

Yazdani R, Shams-Bakhsh M, Hassani-Mehraban A, Arab SS, Thelen N, Thiry M, Crommen J, Fillet M, Jacobs N, Brans A (2019) Production and characterization of virus-like particles of grapevine fanleaf virus presenting L2 epitope of human papillomavirus minor capsid protein. JBb 19:1–12

Zeltins A (2013) Construction and characterization of virus-like particles: a review. Mol Biotechnol 53:92–107

Zhang X, Wei M, Pan H, Lin Z, Wang K, Weng Z, Zhu Y, Xin L, Zhang J, Li S (2014) Robust manufacturing and comprehensive characterization of recombinant hepatitis E virus-like particles in Hecolin®. Vaccine 32:4039–4050

Zheng M, Jiang J, Zhang X, Wang N, Wang K, Li Q, Li T, Lin Q, Wang Y, Yu H (2018) Characterization of capsid protein (p495) of hepatitis E virus expressed in Escherichia coli and assembling into particles in vitro. Vaccine 36:2104–2111

Zhu F-C, Zhang J, Zhang X-F, Zhou C, Wang Z-Z, Huang S-J, Wang H, Yang C-L, Jiang H-M, Cai J-P (2010) Efficacy and safety of a recombinant hepatitis E vaccine in healthy adults: a large-scale, randomised, double-blind placebo-controlled, phase 3 trial. Lancet 376:895–902

Chapter 18
Oncolytic Viruses and Viral Gene Therapy Vectors: Principles of Safety

Bart Spiesschaert and Saurabh Gautam

18.1 Introduction

Oncolytic viruses (OVs) and some of the viral gene therapy vectors (GTVs) originate from virus field isolates that have the inherent capability to induce disease in the species they are endemic in, or which is reflecting the "natural habitat." To make these viruses suitable for therapeutic use, multiple approaches have been employed. Initially, improved safety of viruses, with a human pathogen background, was obtained by using vaccine strains. These isolates were previously attenuated through serial passage in cell culture. A key example of this is the Edmonston measles virus strain (Katz et al. 1962) and the poliomyelitis virus strain derived oncolytic virus. Another approach is to base the therapeutics on animal viruses that are apathogenic in humans, such as orthoreoviruses (Van Den Wollenberg et al. 2009).

In more recent history, there has been an effort to adapt viruses with a higher level of accuracy and control, facilitated by genetic engineering. This enabled the introduction of numerous modifications which rendered potentially pathogenic wildtype viruses suitable for safe use in the clinic.

For the different OVs and GTVs, varying mechanisms of targeted attenuation were employed. The mechanisms for OVs were focused on making the viruses more onco-selective and mitigating any potential shedding into the environment (Buijs et al. 2015). While for viral vectors, used purely as gene delivery systems, replication deficiency or tissue-restricted transgene expression, as well as retargeted tropism, predominantly achieved the required levels of safety (Chira et al. 2015). These

B. Spiesschaert (✉) · S. Gautam (✉)
Boehringer Ingelheim International GmbH, Ingelheim, Germany

ViraTherapeutics GmbH, Rum, Austria
e-mail: bart.spiesschaert@boehringer-ingelheim.com;
saurabh.gautam@boehringer-ingelheim.com

S. Gautam et al. (eds.), *Bioprocess and Analytics Development for Virus-based Advanced Therapeutics and Medicinal Products (ATMPs)*,
https://doi.org/10.1007/978-3-031-28489-2_18

modifications also needed to be stable without any risk of reverting or recombining back to wildtype.

In this chapter, we will first provide an overview of the different criteria for intrinsic safety by which certain viruses were selected to serve as the platform for OVs and GTVs. Secondly, targeted methods of modification, by which viruses obtain an improved safety profile, will be discussed. However, the large number of viral vectors, evaluated for numerous indications, only allows the presentation of the different safety principles through selected examples.

18.2 Intrinsic Principles of Safety

Certain types of viruses have evolved to target normal cell types that have characteristics in common with cancerous cells. Several of these viruses have been chosen as the basis for OVs or GTVs in order to exploit these evolutionary features.

One feature, that intrinsically provides a higher level of safety, is replication capabilities limited to dividing cells, a property characteristic for cancer. For instance, parvovirus, when delivered locally to the tumor, conveys high safety since it requires active cell division for a productive infection (Hartley et al. 2020).

Also, aberrant cell signaling in the cytosolic compartment can create a favorable environment for the replication of certain viruses, otherwise incapable of productive infection. For instance, mammalian orthoreoviruses (mORVs) can only replicate in human cells with an activated Kras pathway (Maitra et al. 2014).

Apart from signaling pathways that drive the malignant transformation, many cancers also develop defects in their innate immune signaling. This can keep cancer cells from achieving an antiviral state, a feature that predominantly determines the host range of OVs, such as MYXV, in normal cells (Spiesschaert et al. 2011).

The requirement of specific cellular receptors for virus cell entry can also be exploited as a safety feature. Moreover, some viruses were selected as OVs because they utilize cellular receptors that are upregulated on cancer cells. For instance, Poliovirus requires CD155 for cell entry, which is upregulated on neuroectodermal cancers (Nandi et al. 2021). Measles virus, on the other hand, requires CD46 which is overexpressed in multiple myeloma (Ong et al. 2006). Laminin, important for tumor development and invasion, is the receptor for the oncolytic Sindbis virus (Wang et al. 1992; Scherwitzl et al. 2020). Another positive stranded RNA virus, limited by its cell entry receptor, is the Seneca valley virus. Its dependency on the anthrax toxin receptor, tumor endothelial marker-8 (TEM-8), for cell entry limits its tropism to neuroendocrine cancers (Miles et al. 2017; Cook and Chauhan 2020). Contrary to B3, Coxsackievirus (CVA) 21 also has a similar safety feature. Its binding and uptake occur through the intercellular adhesion molecule-1 (ICAM-1) (Geisler et al. 2021). This immunoglobulin-like molecule is commonly expressed at high levels in a wide range of cancers (Bradley et al. 2014).

When it comes to GTVs, a different type of intrinsic differences in safety exists. The mechanisms through which GTVs achieve stable expression of their transgenes can have implications on their safety. In the case of adeno-associated viruses (AAVs), stable expression of the transgene is predominantly achieved as an episome, avoiding the need for genome integration (McCarty et al. 2004). Lentiviral vectors seem to be biased toward transcriptional units (Ronen et al. 2011), rather than 5′ regulatory sequences, as is the case for retroviral vectors (McCormack and Rabbitts 2004). The latter can be problematic as cases of leukemia have been linked to treatment with these types of vectors (Ronen et al. 2011).

18.3 Targeted Principles of Safety

The rise of genetic engineering allowed for viruses, that are inherently pathogenic, to be modified into safe backbones for OVs and GTVs. Numerous ways of targeted attenuation are employed for the improvement of these therapeutic viruses.

Below, we describe the targeting of a selection of tumor–virus specific interactions, in relation to the different cellular compartments (membrane, cytosol, and nucleus) where they take place. Through these modifications, the toxic effects on the natural target tissues in the host are mitigated, while still preserving the ability to kill tumor cells.

18.3.1 Targeting Nuclear Mechanisms

18.3.1.1 Tissue-Specific Promotors

One way to restrict DNA OVs is to put essential viral genes under the control of tissue-specific promotors. For instance, the oncolytic AdV genes responsible for replication, like E1, were put under the control of the human telomerase reverse transcriptase (hTERT) promoter element. The hTERT gene is usually silenced in normal cells, but it is significantly expressed in ~90% of human cancers (Fujiwara 2019). Furthermore, additional approaches are pursued to further control transgene expression in order to limit off-target effects. For instance, mechanisms for promotor-based temporal control, and safety switches, are under development for GTVs (Page et al. 2020).

18.3.1.2 Transgene Positioning

Transgene positioning within the genome can also lead to improved safety through general attenuation. For instance, positioning the transgene at the 3′ end of the vesicular stomatitis virus (VSV) genome led to attenuated replication while still inducing a potent immune response (Wollmann et al. 2010; van den Pol and Davis 2013).

18.3.1.3 Virulence Factor Deletion

A method to restrict larger oncolytic viruses, such as herpes- and pox-based OVs, is to remove the virulence factors responsible for their ability to replicate in non-dividing cells. These types of OVs, deficient for thymidine kinase (TK) and ribonucleotide reductase (RR), are limited to replicate in cells with active DNA synthesis, such as cancer cells (Kirn and Thorne 2009; Potts et al. 2017).

In the case of AdV, deletion of E1A (pRb-binding) and E1B (inhibitor of the cellular tumor suppressor protein p53) genes restricts these vectors to replication in pRb-pathway disrupted and p53-deficient cancers, respectively (Fukuda et al. 2009).

18.3.1.4 Mitigating Genotoxicity of GTV Integration

In the case of GTVs, prolonged expression of the therapeutic transgene is desired. The presence of the GTVs as an episome is sufficient for stable expression in non-dividing cells. However, in the case of dividing cells, repeated applications or integration of the transgene into the cellular genome is required for long-term expression (Chira et al. 2015). This process comes with potential safety risks. When GTVs integrate randomly in the host genome, they can disrupt genes, regulatory sequences or even proto-oncogene sites (Rosas et al. 2012). For instance, the integration of a retrovirus-delivered therapeutic gene into the LMO2 proto-oncogene region triggered development of leukemia in some individuals (Ronen et al. 2011). To increase the safety of this integration process, multiple mechanisms are utilized. An important reason for the activation of oncogenes after integration is the strong promoter activity of the LTR U3 region. The removal of the U3 region from the 3′ LTR, resulting in loss of the LTR promoter activity, but does not affect vector titers or transgene expression (Zufferey et al. 1998). Site-specific homologous sequences flanking the transgene have been proposed to reduce the risk of insertional mutagenesis (Chira et al. 2015). In order to completely avoid the risk of insertional mutagenesis, non-integrating lentiviral vectors were also developed (Luis 2020).

18.3.2 Targeting Cytosolic Mechanisms

Several innate antiviral effector molecules and corresponding signaling pathways reside in the cytosol. Numerous viruses have evolved mechanisms to inhibit these antiviral mechanisms, leading to increased virulence. However, many cancer types have defects in these mechanisms, such as dysfunctional interferon (IFN) signaling or upregulation of the Ras pathway. The removal of these virulence factors is therefore a powerful tool to improve oncoselectivity.

18.3.2.1 Virulence Factor Deletion and Immune Modulation

In the case of oncolytic VSV, a loss of function mutation has been introduced into the matrix gene M (e.g., ΔM51, M51R). This mutation makes VSV more sensitive to the effects of IFN signaling, limiting its replication potential in normal tissues. Theoretically, the occurrence of compensatory mutations or a reversion to wildtype could occur negating these benefits. Although compensatory mutations have been reported for modifications subject to selection pressure in vitro (Stanifer et al. 2011; Harouaka and Wertz 2012), this has not yet been reported for ΔM51 or M51R. This attenuation also reduces the neuropathogenic characteristic for VSV (Wollmann et al. 2012).

Alternatively, genes can be added to OVs that actively stimulate antiviral immunity in cells still sensitive. VSV has been armed with IFNβ for that purpose. However, loss of function mutations in the OV can negate the safety benefit provided by these transgenes (Quiñones-Kochs et al. 2001; Dinh et al. 2012).

18.3.2.2 Virulence Factor Substitution

For oncolytic polio virus, neuro-attenuation is achieved through a different mechanism. Here, its internal ribosome entry site (IRES) element is responsible for its neuropathogenicity. Substitution with the IRES of a rhinovirus can mitigate this safety risk (Brown et al. 2014). Although reversion to wildtype is a point of concern (Buijs et al. 2015).

18.3.2.3 Posttranscriptional Gene Regulation

Another approach is to destabilize the viral genomes by the insertion of microRNA response elements (MREs). MicroRNA plays a role in posttranscriptional gene regulation (Ankasha et al. 2018). This approach has been comprehensively reviewed by Autumn Ruiz and Steven Russell (Ruiz and Russell 2015). Briefly, transgene expression by GTVs can be successfully limited to their target tissues by the use of MREs (Suzuki et al. 2008; Geisler et al. 2010; Xie et al. 2011). Also, OV replication has been controlled in this way, improving their safety.

First, a more general approach was employed. Targeting ubiquitous microRNA, like Let-7 and miR7, that are downregulated in numerous cancers, was used to limit the tropism of several viruses, such as VSV, measles, VACV, AdV, and HSV (Edge et al. 2008; Sugio et al. 2011; Leber et al. 2011; Hikichi et al. 2011; Mazzacurati et al. 2015). However, the expression levels of the targeted microRNAs are variable across different tissue types. Focusing on one ubiquitous microRNA can run the risk of saturating this microRNA in certain cell types.

A preferred method would therefore be to limit expression in normal cells by incorporating multiple MREs. For instance, the use of muscle-specific MREs mitigated the risk of CVA21 induced myositis (Kelly et al. 2008). The incorporation of microRNA-Let-7, microRNA-122, and microRNA-124 specific MREs, controlling the glycoprotein H gene expression of HSV, resulted in an OV that specifically targets hepatocellular carcinoma cells (Fu et al. 2012).

One concern regarding this approach is the risk of escape mutations (Barnes et al. 2008). However, incorporating multiple MREs throughout the genome can minimize this. For negative-stranded RNA viruses, an additional concern is that this mechanism is likely limited to the positive-stranded mRNA molecules (Kelly et al. 2010), allowing continued production by the negative-stranded genome copies potentially leading to saturation of the miRNAs. This contributes to why the majority of miRNA-targeted viruses are positive-sense, single-stranded RNA viruses (Ruiz and Russell 2015).

18.3.3 Tissue Tropism Restriction

Limiting the ability of the virus to enter cell types by modifying its surface receptor has also been a frequently used method to improve the safety of viral vectors. One approach has been to completely exchange the wildtype surface protein with that of a virus with a more restricted cell tropism. For instance, the glycoprotein of VSV has successfully been exchanged with that of the lymphocytic choriomeningitis virus in order to minimize the inherent neurotropism of VSV (Muik et al. 2011).

More complex modifications of the viral cell entry proteins have been tested to improve oncoselectivity. In oncolytic HSV, the cell entry glycoproteins C and D were modified by disrupting the native binding regions and introducing a single-chain variable fragment (scFv), specific for tumor cell surface targets (Menotti et al. 2006; Grandi et al. 2010).

For the RNA viruses, measles virus, and VSV, a similar approach was tested (Nakamura et al. 2005; Ayala-Breton et al. 2012). This approach was also successfully used on AdV (Belousova et al. 2008).

Also for AdV, exchanging hexon proteins between serotypes helped to reduce liver toxicity (Short et al. 2010). In addition, modifying the fiber protein of AdV or PEGylation/chemical shielding of the virus particles led to reduction in liver tropism (Uusi-Kerttula et al. 2015).

These types of mutations, when used in viral vectors more prone to mutate, do run the risk of rapidly accumulating mutations adapting the new surface protein to its target cell (Gao et al. 2006). This initial instability needs to be controlled for during development.

18.3.4 Targeting Extracellular Mechanisms

Some viruses have also been engineered to be activated by factors in the tumor micro-environment. Matrix metalloprotease (MMP) are overexpressed in a wide variety of human tumors and degrade the extracellular matrix surrounding the tumor, thereby facilitating invasive growth (Stamenkovic 2000). Variants of OVs have been generated which contain substrate motifs, linked to a blocking domain in the viral surface proteins. Only by cleavage of this motif can the virus bind with its cellular receptors and enter the surrounding cells. Limiting the activation of OVs to the TME greatly increases its tumor specificity and safety (Mühlebach et al. 2010; José et al. 2014).

Another approach is to remove virulence factors from the OVs that exert their effect in the extracellular environment and are not needed in the TME. This approach has been employed on multiple occasions for oncolytic VACV. For instance, Vaccinia growth factor, which normally activates the epidermal growth factor receptor pathway was deleted from the OV's genome, since this pathway is constitutively activated in a wide range of cancers. Also, immunosuppressive proteins, such as B18R, which sequester type I IFN, can be deleted when treating cancers with defective IFN signaling (Kirn and Thorne 2009).

18.4 Environmental Safety

Safety of OVs and GTVs primarily relates to the absence of toxicity of the administration. But also the absence of environmental shedding is an important parameter of safety (Schenk-Braat et al. 2007).

Most of the above discussed safety modifications are generally attenuating and also result in the absence of shedding (e.g., oncolytic HSV, MeV, etc.). Although, HSV leakage from a treated lesion has been reported (Senzer et al. 2009).

However, for pathogens where their safety does not come from reduced potency, but rather their animal origin, shedding is more of a concern. When these viruses, such as orthoreoviruses, do not cause disease in humans or negatively impact animal populations, the risk is generally low. However, shedding of mesogenic or velogenic Newcastle Disease virus strains needs to be closely monitored since this could adversely affect avian livestock and wildlife (Buijs et al. 2015).

Shedding from DNA viruses, such as AdV, could pose safety issues through another mechanism (Griscelli et al. 2003). Homologous recombination between closely related AdV has been proposed in humans (Walsh et al. 2009). Were this to happen with modified AdVs, this could lead to rescue of replication competence or even tropism changes (Buijs et al. 2015).

18.5 Manufacturing Safety for Viral Vectors

There are multiple aspects and parameters which should be ensured to ascertain the safety of the active pharmaceutical ingredient (API), including but not limited to cell lines and cell banks, plasmids or vectors used, virus seed banks, raw material, manufacturing facility and equipment, process-related impurities, product-related impurities, adventitious agents, etc. However, there is an excellent amount of literature available when it comes to the generic safety aspects of APIs during the manufacturing process (Sahoo et al. 2009; Cox Gad 2016; European Medicines Agency 2018; Gröner 2018; Fda and Cber 2020). Therefore, and in alignment with this book, this section will focus on the safety aspects during manufacturing of virus-based therapeutics which are unique to viruses as APIs.

When it comes to virus-based advanced therapy medicinal products (ATMPs) (i.e., OVs and GTVs), viral sequences are extremely important to ensure safety and are chosen and modified carefully to avoid activation of any oncogenes by the methods mentioned earlier in this chapter. Another aspect for ensuring safety of these virus-based therapeutics during manufacturing is the generation/engineering of cell lines to be used for upstream production. In case of lentiviruses, there has been an evolution of lentiviral vectors used for safe production of virus particles, from first to fourth generation, fourth being the latest and safest variation (Ferreira et al. 2021). The most used third-generation vector system consists of a four-plasmid system, wherein first being the transfer plasmid, second carries the *gag-pol* retroviral genes, third carries the *rev* gene, and a fourth helper plasmid with envelope gene (e.g., VSV-G). To further enhance the safety aspects, a split genome approach is utilized to significantly reduce the incidents of recombination events leading to replication competent viruses (Warnock et al. 2006; Schambach et al. 2013). For example, having viral structural and functional genes (i.e., *gag-pol* and *env*) in the cell line and deleted from the viral vector, which rather contains the therapeutic gene of interest along with the sequences essential for efficient packaging of viral particles (Ferreira et al. 2021). These production cell lines are often referred as packaging cell lines. Moreover, during cell line generation, it is also important to reduce the risk of recombination and avoid the transfer of oncogenes from the producer cell line to the therapeutic virus particles (Chadeuf and Salvetti 2010; Chen et al. 2020).

Further attributes for ensuring safety of these virus-based ATMPs during manufacturing are avoiding product-related impurities. These are introduced during the manufacturing process and have a high safety implication for the API (Segura et al. 2013; Wright 2014; Perry and Rayat 2021). If we look at lentivirus (Fig. 18.1) or AAV vectors (Table 18.1), there are various types of product-related impurities reported so far with giving rise to immunotoxicity and genotoxicity. Moreover, the presence of helper viruses, used during the upstream process, in the final API formulation is clearly a safety issue. In fact, it is always recommended to test for the presence of nucleic acid or expressed proteins intended to be used only in the initial virus transfection or rescue process. Therefore, minimizing the amount of product-related impurities in final drug product is always an aim as well as a challenge. Such challenges can only be overcome by having a well-characterized and robust manufacturing process along with a suite of analytics to support the process.

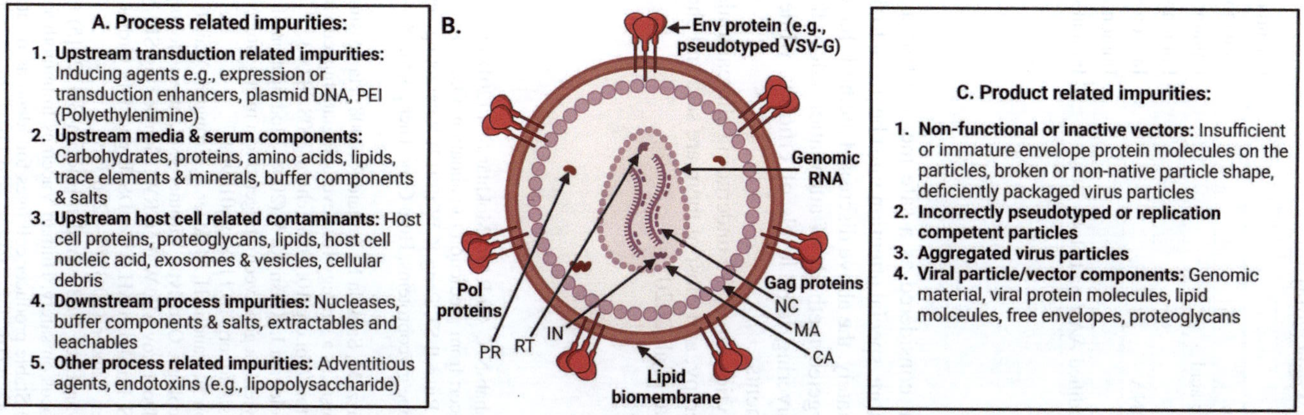

Fig. 18.1 Lentivirus structure and associated process and product-related impurities. (**a**). Process-related impurities during manufacturing of lentiviruses, (**b**) typical structure of lentivirus particles, and (**c**) product-related impurities for lentiviruses. Figure created with BioRender.com

Table 18.1 Potential toxicity due to product-related impurities with API during the manufacturing of AAVs. Table adapted from Wright (2014)

Product-related impurities	Potential toxicity
Non-infectious AAV particles	Immunotoxicity
Encapsidated host cell nucleic acid	Immunotoxicity, genotoxicity
Replication competent AAV	Immunotoxicity
Encapsidated helper vector DNA	Immunotoxicity, genotoxicity
Empty AAV capsids	Immunotoxicity
Aggregated, deformed, or modified AAV particles	Immunotoxicity, genotoxicity

18.6 Conclusion

OVs and GTVs are often considered as a safe means of local delivery of potent therapeutic cargos. Therefore, special attention must be given to their oncoselectivity and general safety. Fortunately, the above discussed methods have proven successful in many cases. These targeted methods of attenuation allow the use of otherwise more potent viruses, delivering high levels of efficacy, while also fulfilling the necessary safety requirements.

Currently, many oncolytic viruses are undergoing clinical trials for applications in single or combination therapy, and most of them are safe and show almost no dose-limiting toxicities (Mondal et al. 2020).

References

Ankasha SJ, Shafiee MN, Wahab NA, Ali RAR, Mokhtar NM (2018) Post-transcriptional regulation of microRNAs in cancer: from prediction to validation. Oncol Rev 12:39–44

Ayala-Breton C, Barber GN, Russell SJ, Peng K-W (2012) Retargeting vesicular stomatitis virus using measles virus envelope glycoproteins. Hum Gene Ther 23(5):484–491. https://doi.org/10.1089/hum.2011.146

Barnes D, Kunitomi M, Vignuzzi M, Saksela K, Andino R (2008) Harnessing endogenous miRNAs to control virus tissue tropism as a strategy for developing attenuated virus vaccines. Cell Host Microbe 4(3):239–248. https://doi.org/10.1016/j.chom.2008.08.003

Belousova N, Mikheeva G, Gelovani J, Krasnykh V (2008) Modification of adenovirus capsid with a designed protein ligand yields a gene vector targeted to a major molecular marker of cancer. J Virol 82(2):630–637. https://doi.org/10.1128/jvi.01896-07

Bradley S, Jakes A, Harrington K, Pandha H, Melcher A, Errington-Mais F (2014) Applications of coxsackievirus A21 in oncology. Oncolytic Virother 3:47. https://doi.org/10.2147/ov.s56322

Brown MC, Dobrikova EY, Dobrikov MI, Walton RW, Gemberling SL, Nair SK, Desjardins A, Sampson JH, Friedman HS, Friedman AH, Tyler DS, Bigner DD, Gromeier M (2014) Oncolytic polio virotherapy of cancer. Cancer 120:3277–3286

Buijs PRA, Verhagen JHE, van Eijck CHJ, van den Hoogen BG (2015) Oncolytic viruses: from bench to bedside with a focus on safety. Human Vaccines Immunotherapeutics 11:1573–1584

Chadeuf G, Salvetti A (2010) Stable producer cell lines for adeno-associated virus (AAV) assembly. Cold Spring Harb Protoc 5(10). https://doi.org/10.1101/pdb.prot5496

Chen YH, Pallant C, Sampson CJ, Boiti A, Johnson S, Brazauskas P, Hardwicke P, Marongiu M, Marinova VM, Carmo M, Sweeney NP, Richard A, Shillings A, Archibald P, Puschmann E, Mouzon B, Grose D, Mendez-Tavio M, Chen MX, Warr SRC, Senussi T, Carter PS, Baker S, Jung C, Brugman MH, Howe SJ, Vink CA (2020) Rapid lentiviral vector producer cell line generation using a single DNA construct. Mol Ther Methods Clin Dev 19:47–57. https://doi.org/10.1016/J.OMTM.2020.08.011

Chira S, Jackson CS, Oprea I, Ozturk F, Pepper MS, Diaconu I, Braicu C, Raduly LZ, Calin GA, Berindan-Neagoe I (2015) Progresses towards safe and efficient gene therapy vectors. Oncotarget 6(31):30675–30703. https://doi.org/10.18632/oncotarget.5169

Cook M, Chauhan A (2020) Clinical application of oncolytic viruses: a systematic review. Int J Mol Sci 21:1–36

Cox Gad S (2016) Drug safety evaluation

Dinh PX, Panda D, Das PB, Das SC, Das A, Pattnaik AK (2012) A single amino acid change resulting in loss of fluorescence of eGFP in a viral fusion protein confers fitness and growth advantage to the recombinant vesicular stomatitis virus. Virology 432(2):460–469. https://doi.org/10.1016/j.virol.2012.07.004

Edge RE, Falls TJ, Brown CW, Lichty BD, Atkins H, Bell JC (2008) A let-7 microRNA-sensitive vesicular stomatitis virus demonstrates tumor-specific replication. Mol Ther 16(8):1437–1443. https://doi.org/10.1038/MT.2008.130

European Medicines Agency (2018) Clinical efficacy and safety guidelines | European Medicines Agency. https://www.ema.europa.eu/en/human-regulatory/research-development/scientific-guidelines/clinical-efficacy-safety/clinical-efficacy-safety-general. Accessed 26 Sep 2022

Fda, Cber (2020) Chemistry, manufacturing, and control (CMC) information for human gene therapy investigational new drug applications (INDs) guidance for industry

Ferreira MV, Cabral ET, Coroadinha AS (2021) Progress and perspectives in the development of lentiviral vector producer cells. Biotechnol J 16:2000017

Fu X, Rivera A, Tao L, De Geest B, Zhang X (2012) Construction of an oncolytic herpes simplex virus that precisely targets hepatocellular carcinoma cells. Mol Ther 20(2):339–346. https://doi.org/10.1038/mt.2011.265

Fujiwara T (2019) Multidisciplinary oncolytic virotherapy for gastrointestinal cancer. Annals of Gastroenterological Surgery 3:396–404

Fukuda K, Abei M, Ugai H, Kawashima R, Seo E, Wakayama M, Murata T, Endo S, Hamada H, Hyodo I, Yokoyama KK (2009) E1A, E1B double-restricted replicative adenovirus at low dose greatly augments tumor-specific suicide gene therapy for gallbladder cancer. Cancer Gene Ther 16(2):126–136. https://doi.org/10.1038/cgt.2008.67

Gao Y, Whitaker-Dowling P, Watkins SC, Griffin JA, Bergman I (2006) Rapid adaptation of a recombinant vesicular stomatitis virus to a targeted cell line. J Virol 80(17):8603–8612. https://doi.org/10.1128/jvi.00142-06

Geisler A, Jungmann A, Kurreck J, Poller W, Katus HA, Vetter R, Fechner H, Müller OJ (2010) microRNA122-regulated transgene expression increases specificity of cardiac gene transfer upon intravenous delivery of AAV9 vectors. Gene Ther 18(2):199–209. https://doi.org/10.1038/gt.2010.141

Geisler A, Hazini A, Heimann L, Kurreck J, Fechner H (2021) Coxsackievirus b3—its potential as an oncolytic virus. Viruses 13:718

Grandi P, Fernandez J, Szentirmai O, Carter R, Gianni D, Sena-Esteves M, Breakefield XO (2010) Targeting HSV-1 virions for specific binding to epidermal growth factor receptor-vIII-bearing tumor cells. Cancer Gene Ther 17(9):655–663. https://doi.org/10.1038/cgt.2010.22

Griscelli F, Opolon P, Saulnier P, Mami-Chouaib F, Gautier E, Echchakir H, Angevin E, Le Chevalier T, Bataille V, Squiban P, Tursz T, Escudier B (2003) Recombinant adenovirus shedding after intratumoral gene transfer in lung cancer patients. Gene Ther 10:386–395

Gröner A (2018) Pathogen safety. In: Biopharmaceutical processing: development, design, and implementation of manufacturing processes. Elsevier, pp 1075–1104

Harouaka D, Wertz GW (2012) Second-site mutations selected in transcriptional regulatory sequences compensate for engineered mutations in the vesicular stomatitis virus nucleocapsid protein. J Virol 86(20):11266–11275. https://doi.org/10.1128/jvi.01238-12
Hartley A, Kavishwar G, Salvato I, Marchini A (2020) A roadmap for the success of oncolytic parvovirus-based anticancer therapies. Annu Rev Virol 7:537–557
Hikichi M, Kidokoro M, Haraguchi T, Iba H, Shida H, Tahara H, Nakamura T (2011) MicroRNA regulation of glycoprotein B5R in oncolytic vaccinia virus reduces viral pathogenicity without impairing its antitumor efficacy. Mol Ther 19(6):1107–1115. https://doi.org/10.1038/MT.2011.36
José A, Rovira-Rigau M, Luna J, Giménez-Alejandre M, Vaquero E, García De La Torre B, Andreu D, Alemany R, Fillat C (2014) A genetic fiber modification to achieve matrix-metalloprotease-activated infectivity of oncolytic adenovirus. J Control Release 192:148–156. https://doi.org/10.1016/j.jconrel.2014.07.008
Katz SL, Enders JF, Holloway A (1962) The development and evaluation of an attenuated measles virus vaccine. Am J Public Health 52. https://doi.org/10.2105/ajph.52.suppl_2.5
Kelly EJ, Hadac EM, Greiner S, Russell SJ (2008) Engineering microRNA responsiveness to decrease virus pathogenicity. Nat Med 14(11):1278–1283. https://doi.org/10.1038/nm.1776
Kelly EJ, Nace R, Barber GN, Russell SJ (2010) Attenuation of vesicular stomatitis virus encephalitis through microRNA targeting. J Virol 84(3):1550–1562. https://doi.org/10.1128/JVI.01788-09
Kirn DH, Thorne SH (2009) Targeted and armed oncolytic poxviruses: a novel multi-mechanistic therapeutic class for cancer. Nat Rev Cancer 9:64–71
Leber MF, Bossow S, Leonard VHJ, Zaoui K, Grossardt C, Frenzke M, Miest T, Sawall S, Cattaneo R, von Kalle C, Ungerechts G (2011) MicroRNA-sensitive oncolytic measles viruses for cancer-specific vector tropism. Mol Ther 19(6):1097–1106. https://doi.org/10.1038/mt.2011.55
Luis A (2020) The old and the new: prospects for non-integrating lentiviral vector technology. Viruses 12:1103
Maitra R, Seetharam R, Tesfa L, Augustine TA, Klampfer L, Coffey MC, Mariadason JM, Goel S (2014) Oncolytic reovirus preferentially induces apoptosis in KRAS mutant colorectal cancer cells, and synergizes with irinotecan. Oncotarget 5(9):2807–2819. https://doi.org/10.18632/oncotarget.1921
Mazzacurati L, Marzulli M, Reinhart B, Miyagawa Y, Uchida H, Goins WF, Li A, Kaur B, Caligiuri M, Cripe T, Chiocca N, Amankulor N, Cohen JB, Glorioso JC, Grandi P (2015) Use of miRNA response sequences to block off-target replication and increase the safety of an unattenuated, glioblastoma-targeted oncolytic HSV. Mol Ther 23(1):99–107. https://doi.org/10.1038/MT.2014.177
McCarty DM, Young SM, Samulski RJ (2004) Integration of adeno-associated virus (AAV) and recombinant AAV vectors. Annu Rev Genet 38:819–845
McCormack MP, Rabbitts TH (2004) Activation of the T-cell oncogene LMO2 after gene therapy for X-linked severe combined immunodeficiency. N Engl J Med 350(9):913–922. https://doi.org/10.1056/nejmra032207
Menotti L, Cerretani A, Campadelli-Fiume G (2006) A herpes simplex virus recombinant that exhibits a single-chain antibody to HER2/neu enters cells through the mammary tumor receptor, independently of the gD receptors. J Virol 80(11):5531–5539. https://doi.org/10.1128/jvi.02725-05
Miles LA, Burga LN, Gardner EE, Bostina M, Poirier JT, Rudin CM (2017) Anthrax toxin receptor 1 is the cellular receptor for Seneca Valley virus. J Clin Investig 127(8):2957–2967. https://doi.org/10.1172/JCI93472
Mondal M, Guo J, He P, Zhou D (2020) Recent advances of oncolytic virus in cancer therapy. Human Vaccines Immunotherapeutics 16:2389–2402
Mühlebach MD, Schaser T, Zimmermann M, Armeanu S, Hanschmann KMO, Cattaneo R, Bitzer M, Lauer UM, Cichutek K, Buchholz CJ (2010) Liver cancer protease activity profiles

support therapeutic options with matrix metalloproteinase-activatable oncolytic measles virus. Cancer Res 70(19):7620–7629. https://doi.org/10.1158/0008-5472.CAN-09-4650

Muik A, Kneiske I, Werbizki M, Wilflingseder D, Giroglou T, Ebert O, Kraft A, Dietrich U, Zimmer G, Momma S, von Laer D (2011) Pseudotyping vesicular stomatitis virus with lymphocytic choriomeningitis virus glycoproteins enhances infectivity for glioma cells and minimizes neurotropism. J Virol 85(11):5679–5684. https://doi.org/10.1128/JVI.02511-10

Nakamura T, Peng KW, Harvey M, Greiner S, Lorimer IAJ, James CD, Russell SJ (2005) Rescue and propagation of fully retargeted oncolytic measles viruses. Nat Biotechnol 23(2):209–214. https://doi.org/10.1038/nbt1060

Nandi SS, Gohil T, Sawant SA, Lambe UP, Ghosh S, Jana S (2021) CD155: a key receptor playing diversified roles. Curr Mol Med 22(7):594–607. https://doi.org/10.2174/1566524021666210910112906

Ong HT, Timm MM, Greipp PR, Witzig TE, Dispenzieri A, Russell SJ, Peng KW (2006) Oncolytic measles virus targets high CD46 expression on multiple myeloma cells. Exp Hematol 34(6): 713–720. https://doi.org/10.1016/j.exphem.2006.03.002

Page A, Fusil F, Cosset FL (2020) Toward tightly tuned gene expression following lentiviral vector transduction. Viruses 12:1427

Perry C, Rayat ACME (2021) Lentiviral vector bioprocessing. Viruses 13(2):268. https://doi.org/10.3390/v13020268

Potts KG, Irwin CR, Favis NA, Pink DB, Vincent KM, Lewis JD, Moore RB, Hitt MM, Evans DH (2017) Deletion of F4L (ribonucleotide reductase) in vaccinia virus produces a selective oncolytic virus and promotes anti-tumor immunity with superior safety in bladder cancer models. EMBO Mol Med 9(5):638–654. https://doi.org/10.15252/emmm.201607296

Quiñones-Kochs MI, Schnell MJ, Buonocore L, Rose JK (2001) Mechanisms of loss of foreign gene expression in recombinant vesicular stomatitis viruses. Virology 287(2):427–435. https://doi.org/10.1006/viro.2001.1058

Ronen K, Negre O, Roth S, Colomb C, Malani N, Denaro M, Brady T, Fusil F, Gillet-Legrand B, Hehir K, Beuzard Y, Leboulch P, Down JD, Payen E, Bushman FD (2011) Distribution of lentiviral vector integration sites in mice following therapeutic gene transfer to treat β-thalassemia. Mol Ther 19(7):1273–1286. https://doi.org/10.1038/mt.2011.20

Rosas LE, Grieves JL, Zaraspe K, La Perle KMD, Fu H, McCarty DM (2012) Patterns of scAAV vector insertion associated with oncogenic events in a mouse model for genotoxicity. Mol Ther 20(11):2098–2110. https://doi.org/10.1038/mt.2012.197

Ruiz AJ, Russell SJ (2015) MicroRNAs and oncolytic viruses. Current opinion in virology 13:40–48. https://doi.org/10.1016/j.coviro.2015.03.007

Sahoo N, Choudhury K, Manchikanti P (2009) Manufacturing of biodrugs: need for harmonization in regulatory standards. BioDrugs 23:217–229

Schambach A, Zychlinski D, Ehrnstroem B, Baum C (2013) Biosafety features of lentiviral vectors. Hum Gene Ther 24:132–142

Schenk-Braat EAM, van Mierlo MKB, Wagemaker G, Bangma CH, Kaptein LCM (2007) An inventory of shedding data from clinical gene therapy trials. J Gene Med:910–921

Scherwitzl I, Opp S, Hurtado AM, Pampeno C, Loomis C, Kannan K, Yu M, Meruelo D (2020) Sindbis virus with Anti-OX40 overcomes the immunosuppressive tumor microenvironment of low-immunogenic tumors. Mol Ther Oncolytics 17:431–447. https://doi.org/10.1016/J.OMTO.2020.04.012

Segura MM, Mangion M, Gaillet B, Garnier A (2013) New developments in lentiviral vector design, production and purification. Expert Opin Biol Ther 13:987–1011

Senzer NN, Kaufman HL, Amatruda T, Nemunaitis M, Reid T, Daniels G, Gonzalez R, Glaspy J, Whitman E, Harrington K, Goldsweig H, Marshall T, Love C, Coffin R, Nemunaitis JJ (2009) Phase II clinical trial of a granulocyte-macrophage colony-stimulating factor-encoding, second-generation oncolytic herpesvirus in patients with unresectable metastatic melanoma. J Clin Oncol 27(34):5763–5771. https://doi.org/10.1200/JCO.2009.24.3675

Short JJ, Rivera AA, Wu H, Walter MR, Yamamoto M, Mathis JM, Curiel DT (2010) Substitution of adenovirus serotype 3 hexon onto a serotype 5 oncolytic adenovirus reduces factor X binding, decreases liver tropism, and improves antitumor efficacy. Mol Cancer Ther 9(9):2536–2544. https://doi.org/10.1158/1535-7163.MCT-10-0332

Spiesschaert B, McFadden G, Hermans K, Nauwynck H, Van de Walle GR (2011) The current status and future directions of myxoma virus, a master in immune evasion. Vet Res 42(1):76. https://doi.org/10.1186/1297-9716-42-76

Stamenkovic I (2000) Matrix metalloproteinases in tumor invasion and metastasis. Semin Cancer Biol 10(6):415–433. https://doi.org/10.1006/scbi.2000.0379

Stanifer ML, Cureton DK, Whelan SPJ (2011) A recombinant vesicular stomatitis virus bearing a lethal mutation in the glycoprotein gene uncovers a second site suppressor that restores fusion. J Virol 85(16):8105–8115. https://doi.org/10.1128/jvi.00735-11

Sugio K, Sakurai F, Katayama K, Tashiro K, Matsui H, Kawabata K, Kawase A, Iwaki M, Hayakawa T, Fujiwara T, Mizuguchi H (2011) Enhanced safety profiles of the telomerase-specific replication-competent adenovirus by incorporation of normal cell-specific microRNA-targeted sequences. Clin Cancer Res 17(9):2807–2818. https://doi.org/10.1158/1078-0432.CCR-10-2008/83570/AM/ENHANCED-SAFETY-PROFILES-OF-THE-TELOMERASE

Suzuki T, Sakurai F, Nakamura SI, Kouyama E, Kawabata K, Kondoh M, Yagi K, Mizuguchi H (2008) miR-122a-regulated expression of a suicide gene prevents hepatotoxicity without altering antitumor effects in suicide gene therapy. Mol Ther 16(10):1719–1726. https://doi.org/10.1038/MT.2008.159

Uusi-Kerttula H, Hulin-Curtis S, Davies J, Parker AL (2015) Oncolytic adenovirus: strategies and insights for vector design and immuno-oncolytic applications. Viruses 7:5987–6020

van den Pol AN, Davis JN (2013) Highly attenuated recombinant vesicular stomatitis virus VSV-12′GFP displays immunogenic and oncolytic activity. J Virol 87(2):1019–1034. https://doi.org/10.1128/jvi.01106-12

Van Den Wollenberg DJM, Van Den Hengel SK, Dautzenberg IJC, Kranenburg O, Hoeben RC (2009) Modification of mammalian reoviruses for use as oncolytic agents Modification of mammalian reoviruses for use as oncolytic agents. Expert Opin Biol Ther 9:1509–1520

Walsh MP, Chintakuntlawar A, Robinson CM, Madisch I, Harrach B, Hudson NR, Schnurr D, Heim A, Chodosh J, Seto D, Jones MS (2009) Evidence of molecular evolution driven by recombination events influencing tropism in a novel human adenovirus that causes epidemic keratoconjunctivitis. PLoS One 4(6):e5635. https://doi.org/10.1371/journal.pone.0005635

Wang KS, Kuhn RJ, Strauss EG, Ou S, Strauss JH (1992) High-affinity laminin receptor is a receptor for Sindbis virus in mammalian cells. J Virol 66(8):4992–5001

Warnock JN, Merten OW, Al-Rubeai M (2006) Cell culture processes for the production of viral vectors for gene therapy purposes. Cytotechnology 50:141–162

Wollmann G, Rogulin V, Simon I, Rose JK, van den Pol AN (2010) Some attenuated variants of vesicular stomatitis virus show enhanced oncolytic activity against human glioblastoma cells relative to normal brain cells. J Virol 84(3):1563–1573. https://doi.org/10.1128/JVI.02040-09

Wollmann G, Ozduman K, van den Pol AN (2012) Oncolytic virus therapy for glioblastoma multiforme: concepts and candidates. Cancer J (Sudbury, MA) 18(1):69–81. https://doi.org/10.1097/PPO.0b013e31824671c9

Wright JF (2014) Product-related impurities in clinical-grade recombinant AAV vectors: characterization and risk assessment. Biomedicine 2:80–97

Xie J, Xie Q, Zhang H, Ameres SL, Hung JH, Su Q, He R, Mu X, Seher Ahmed S, Park S, Kato H, Li C, Mueller C, Mello CC, Weng Z, Flotte TR, Zamore PD, Gao G (2011) MicroRNA-regulated, systemically delivered rAAV9: a step closer to CNS-restricted transgene expression. Mol Ther 19(3):526–535. https://doi.org/10.1038/MT.2010.279

Zufferey R, Dull T, Mandel RJ, Bukovsky A, Quiroz D, Naldini L, Trono D (1998) Self-inactivating lentivirus vector for safe and efficient in vivo gene delivery. J Virol 72(12):9873–9880. https://doi.org/10.1128/jvi.72.12.9873-9880.1998

MIX
Papier aus verantwortungsvollen Quellen
Paper from responsible sources
FSC® C105338

If you have any concerns about our products,
you can contact us on
ProductSafety@springernature.com

In case Publisher is established outside the EU,
the EU authorized representative is:
Springer Nature Customer Service Center GmbH
Europaplatz 3, 69115 Heidelberg, Germany

Printed by Libri Plureos GmbH
in Hamburg, Germany